Mycotoxins In Grain

Compounds Other Than Aflatoxin

Edited by

J. D. Miller
Plant Research Centre
Agriculture Canada
Ottawa, Ontario, Canada

H. L. Trenholm
Centre for Food and Animal Research
Agriculture Canada
Ottawa, Ontario, Canada

eagan press
St. Paul, Minnesota, USA

Library of Congress Catalog Card Number: 93-73623
International Standard Book Number: 0-9624407-5-2

Printed in the United States of America on acid-free paper

Eagan Press
3340 Pilot Knob Road
St. Paul, MN 55121-2097, USA

Contributors

D. Abramson, Research Station, Agriculture Canada, Winnipeg, Manitoba R3T 2M9, Canada

J. W. ApSimon, Faculty of Graduate Studies and Research, Carleton University, Ottawa, Ontario K1S 5B6, Canada

J. B. Beardall, Faculty of Graduate Studies and Research, Carleton University, Ottawa, Ontario K1S 5B6, Canada

B. A. Blackwell, Plant Research Centre, Agriculture Canada, Ottawa, Ontario K1A 0C6, Canada

G. S. Bondy, Toxicology Research Division, Health Protection Branch, Health Canada, Ottawa, Ontario K1A 0L2, Canada

L. Charmley, Centre for Food and Animal Research, Agriculture Canada, Ottawa, Ontario K1A 0C6, Canada

T. Kuiper-Goodman, Toxicological Evaluation Division, Health Protection Branch, Health Canada, Ottawa, Ontario K1A 0L2, Canada

R. R. Marquardt, Department of Animal Science, University of Manitoba, Winnipeg, Manitoba R3T 2N2, Canada

J. D. Miller, Plant Research Centre, Agriculture Canada, Ottawa, Ontario K1A 0C6, Canada

K. H. Ominski, Department of Animal Science, University of Manitoba, Winnipeg, Manitoba R3T 2N2, Canada

J. J. Pestka, Department of Food Science and Human Nutrition, Michigan State University, East Lansing, Michigan 48824, U.S.A.

D. B. Prelusky, Centre for Food and Animal Research, Agriculture Canada, Ottawa, Ontario K1A 0C6, Canada

A. Rosenberg, Economic Analysis Unit, Ridgetown College, Ridgetown, Ontario N0P 2C0, Canada

B. A. Rotter, Centre for Food and Animal Research, Agriculture Canada, Ottawa, Ontario K1A 0C6, Canada

R. G. Rotter, Toxicological Evaluation Division, Health Protection Branch, Health Canada, Ottawa, Ontario K1A 0C6, Canada

M. E. Savard, Plant Research Centre, Agriculture Canada, Ottawa, Ontario K1A 0C6, Canada

P. M. Scott, Food Research Division, Health Canada, Ottawa, Ontario K1A 0L2, Canada

A. Sibilia, Istituto Tossine e Micotossine da Parassiti Vegetali, CNR, 70125 Bari, Italy

R. N. Sinha, Research Station, Agriculture Canada, Winnipeg, Manitoba R3T 2M9, Canada

C. H. A. Snijders, Centre for Plant Breeding and Reproduction Research (DLO), NL-6700 AA Wageningen, The Netherlands

H. L. Trenholm, Centre for Food and Animal Research, Agriculture Canada, Ottawa, Ontario K1A 0C6, Canada

A. Visconti, Istituto Tossine e Micotossine da Parassiti Vegetali, CNR, 70125 Bari, Italy

Preface

This may be the first book on mycotoxins that does not deal with the most notorious of mycotoxins, aflatoxins. We recognize the continued economic importance and health significance of aflatoxins. However, after four decades of extensive research, our knowledge of aflatoxin has become so immense that many good treatments on various aspects of aflatoxins are already available.

At the Eighth International Union of Pure and Applied Chemistry (IUPAC) Symposium on Mycotoxins and Phycotoxins in Mexico, November 1992, Pieter Stein, president of the Applied Chemistry Division of IUPAC used the term "fumonisin generation" of mycotoxin researchers. His statement reflects the substantial shift of research effort from aflatoxins to other agriculturally important mycotoxins produced by several genera of fungi. These toxic metabolites include the trichothecenes deoxynivalenol and nivalenol, zearalenone, and fumonisins from various *Fusarium* species that occur worldwide. Ochratoxin from *Aspergillus ochraceus* and *Penicillium verrucosum* is a problem in western Europe, but it can occur elsewhere including North America. Several toxins from *Alternaria* species appear to be important in certain geographical areas.

Our intention is to publish a reference text on the chemistry and biology of mycotoxins other than aflatoxin, with emphasis on those that affect food safety and animal health and productivity. A broad range of topics is included from chemistry to economic and human health problems associated with mycotoxin-contaminated grain. The chapter on the chemistry of mycotoxins outlines the molecular structures and spectroscopic data where appropriate. Mass spectroscopy and nuclear magnetic resonance spectroscopy data acquired on modern instruments are included for those *Fusarium* metabolites discovered during the past decade. Since *Fusarium* species are primarily plant pathogens, the text highlights plant pathology and plant breeding for fungal resistance in affected crops. The species of *Aspergillus* (other than *A. flavus* or *A. parasiticus*) and *Penicillium* that produce important storage mycotoxins are discussed in terms of their ecology and physiology. In contrast, toxigenic *Alternaria* species are largely saprophytic on senescent cereal tissue.

Mycotoxin-contaminated grain is associated with food safety issues and the health and productivity of farm animals. In this book, several perspectives are included: overall toxicology; immunotoxicology; and the important new area of study, the interactive effects of mixtures of mycotoxins. An appraisal of decontamination technology is also provided.

Finally, the consequences of mycotoxin-contaminated grain for humans are considered: strategies used in risk assessments for mycotoxins in food, mycotoxin residues in animal products used for human food, and the economic impact of mycotoxins on agriculture and food production. After almost 30 years, the human carcinogenicity of aflatoxin has been established (International Agency for Research on Cancer Monographs on the Evaluation of Carcinogenic Risk of Chemicals to Humans no. 56. 1993. IARC, Geneva.). The role of the other mycotoxins in human disease has not been firmly elucidated. A surprising number of human diseases have been blamed on mycotoxins in food. These are reviewed. As was the case for aflatoxin, the importance of fungal metabolites, such as fumonisin, deoxynivalenol and/or nivalenol, and ochratoxin will require much more study before their impact on human health is fully known.

We appreciate the willing authors who have contributed many excellent chapters. Jan Beardall deserves special credit for her dedication to the project. Lorraine Cusson is thanked for her secretarial support. We thank Ed Weber and Mary Beaussart of the International Development and Research Centre, Ottawa, for providing the stimulus to undertake this project and partial funding. The Research Branch of Agriculture Canada provided us the opportunity to work in this exciting area of science.

We hope that this text will serve as a reference point for the study of mycotoxins other than aflatoxin for the "fumonisin generation" of mycotoxin researchers. The participation of scientists from many disciplines to prepare this text demonstrates the need to work together on this important agricultural and human health problem.

J. David Miller
H. Locksley Trenholm

Contents

Part One
Fusarium

Chapter 1

The Biosynthetic Diversity
of Secondary Metabolites

John W. ApSimon

The pathogenic nature of certain species of fungi to plants has been observed virtually since the beginning of agriculture. These plant pathogens can produce metabolites that show toxic effects when they are ingested. Several examples in recent history exemplify this property. In 1960, Turkey X disease killed 100,000 turkeys, 14,000 ducklings, and thousands of partridge and pheasant poults in England. In the mid-1930s and late 1970s there were outbreaks of a sickness in horses called equine leukoencephalomalacia in the United States (Hasseltine and Mehlhman 1978, Roddricks 1978, Richard and Thurston 1986), and alimentary toxic aleukia has been responsible for the distress and death of thousands of people since it was first recorded in the 19th century (Moss and Smith 1985). Although these syndromes are all very different, they have one thing in common; they are all caused by mycotoxins.

Mycotoxins are metabolites that are produced by fungi growing on cereals, nuts, soybeans, and several other crops including fruit. The Turkey X disease outbreak in England was traced to contaminated peanuts from Brazil and led to the discovery of aflatoxins produced primarily by *Aspergillus flavus* Link:Fr. and *A. parasiticus* Speare. Equine leukoencephalomalacia is doubtless caused by toxins from *Fusarium moniliforme* J. Sheld., and alimentary toxic aleukia is thought to be caused by the trichothecene mycotoxins. Since mycotoxin-producing fungi grow on some of the staple foods of both humans and animals, both populations are affected by them. Also, products such as eggs, milk, dairy products, and meat can be contaminated through the ingestion of feed-containing mycotoxins (Moss and Smith 1985).

Toxigenic *Fusarium* species are now recognized to be a major agricultural problem. The extensive research carried out over the past two decades has revealed a large number of toxic *Fusarium* secondary metabolites. There

3

are at least 24 species of *Fusarium* that have been associated with human
and/or animal health problems. These are associated with a large number
of secondary metabolites of varied concentrations (Marasas et al 1984, Joffe
1986, Miller 1988). The majority of *Fusarium* toxins have been discovered
in the last decade and represent an amazing variety of biosynthetic origins.
A compilation of these compounds is provided in chapter 4, and it is the
intention of this contribution to bring these together by demonstrating the
biosynthetic commonality of origins for these chemical species. The genus
Fusarium is a versatile biosynthetic apparatus capable of producing secondary metabolites by all of the major known routes for secondary metabolite
formation (ApSimon et al 1991). This compilation organizes the secondary
metabolites by biosynthetic routes.

Isoprenoids

The most widespread biosynthetic pathway to secondary (and some
primary) metabolites is the isoprenoid pathway (Herbert 1989). Metabolites
from this route are based on patterns of the repeating isoprene (isopentene)
units **1** joined in a head-to-tail arrangement. Compounds of isoprenoid
origins are known generically as terpenes, and there are thousands of these
species reported to date. The relationship of terpenes to isoprenes is seen
in compounds such as geraniol **2** and farnesol **3** (dotted lines indicate the
attachments between isoprene units).

isoprene	geraniol	farnesol
1	**2**	**3**

Terpenes arising from farnesol **3** (3 × isoprene) are known as sesquiterpenes and lead to the greatest variety of *Fusarium* secondary metabolites.
All of these appear to derive from an obligatory precursor trichodiene **4**,
which arises as shown in Scheme I from farnesol pyrophosphate (the biologically activated form of farnesol implicated in the biosynthetic pathway).
An added point to emphasize in this route is the subsequent migration
of methyl groups after the initial cyclization of the open chain farnesol.
This is a common occurrence in terpene biosynthesis that leads to compounds

whose skeletons do not appear to be simply derived by isoprene condensation and cyclization.

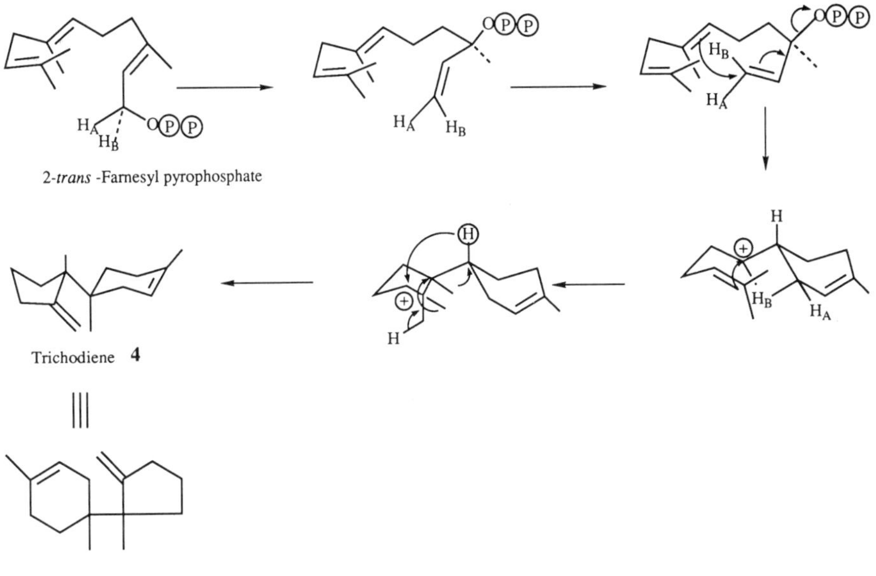

2-*trans*-Farnesyl pyrophosphate

Trichodiene **4**

4

Scheme I. Biosynthetic route to trichodiene.

Trichodienoids

The term trichodienoid is used to describe the variety of compounds that derive from trichodiene as an obligatory precursor and present a plethora of structural types (ApSimon et al 1990). These are summarized

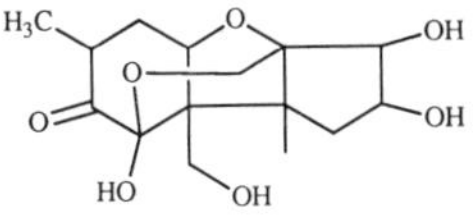

5 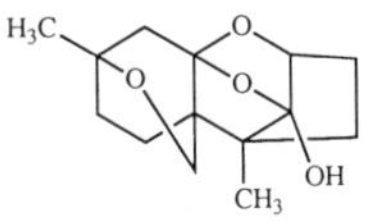   6 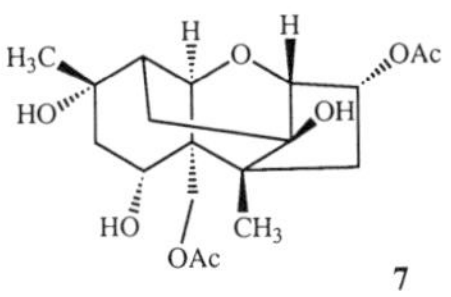   7

in Scheme II, and the commonly used class name is illustrated in this figure. All of the trichodienoids described to date derive from one of these skeletons, which are often subsequently oxygenated and esterified or acylated (see chapter 4). More intricate rearrangements of the observed molecular frameworks can lead to compounds such as gramilaurone **5** (Bekker et al 1991) from *F. graminearum* Schwabe, sporol **6** (Metcalf et al 1992), and the cyclo-species **7** (Grove 1986; Ament et al 1988; ApSimon, *unpublished*).

As a greater variety of closely related trichodienoids are described, many of which are present in very small concentrations, the question as to the artifactual nature of some of these must be raised. The processes of extraction and separation do occasion acidic or basic conditions, and the juxtaposition of reactive functional groups raise the question in many cases as to the natural origin of a compound such as **7** (Grove 1986; Ament et al 1988; ApSimon, *unpublished*). Workers in this area must demonstrate the natural source of their materials.

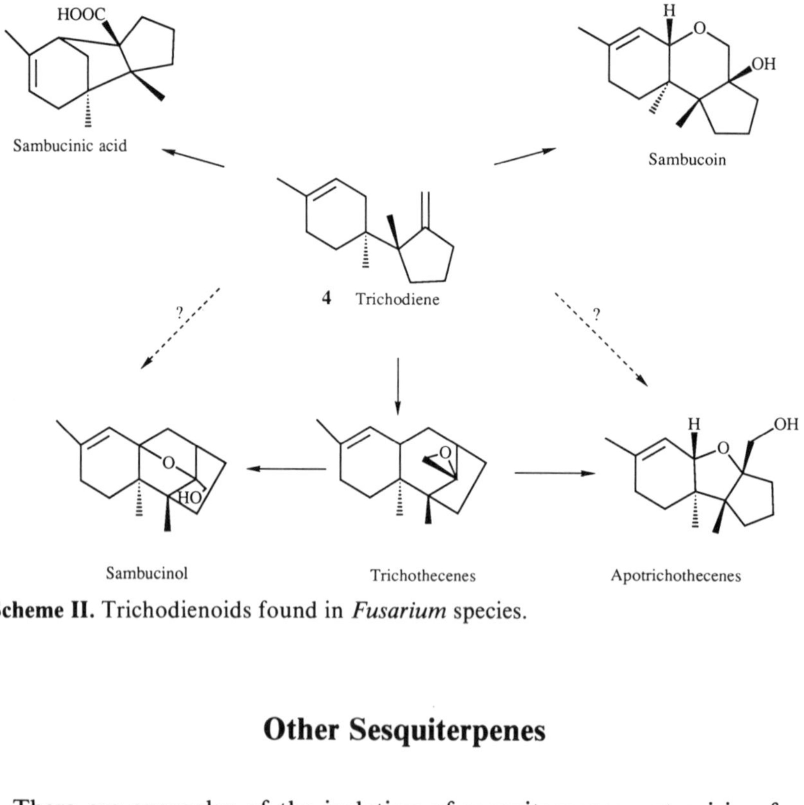

Scheme II. Trichodienoids found in *Fusarium* species.

Other Sesquiterpenes

There are examples of the isolation of sesquiterpenes not arising from the trichodiene pathway in *Fusarium*. A different folding arrangement for farnesol **3** leads in *F. culmorum* (Wm.G.Sm.) Sacc. to the tricyclic species culmorin **8**, which has recently been shown to co-occur with a number of oxygenated analogs (ApSimon et al 1992b).

The same species also reveals a number of metabolites based on a single cyclization of nerolidol **9** to yield a series of compounds identified as cyclonerols **10** (Goi et al 1970, 1971; J. W. ApSimon, B. Blackwell, G. Kasitu, J. D. Miller, and M. Savard, *unpublished*). The structural varieties

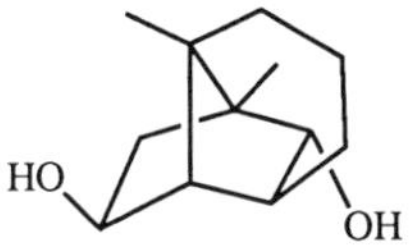

8 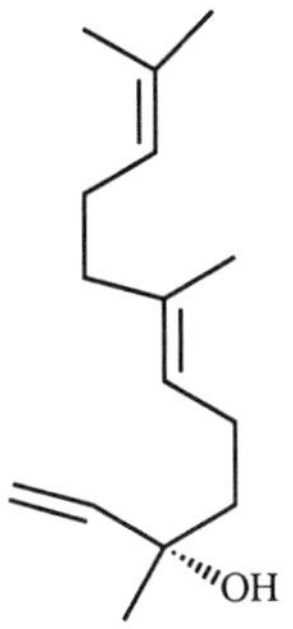   9 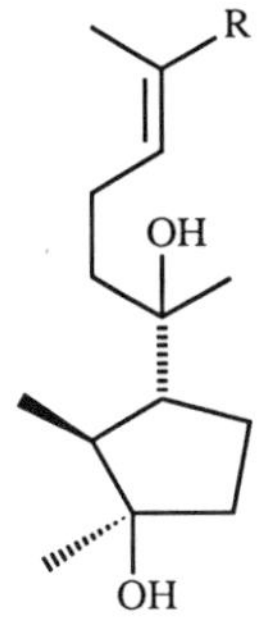   10: R = CH$_3$ or CH$_2$OH

demonstrated above represent over 80 compounds reported to date, and it is undoubtedly certain that many more are yet to be reported.

Polyketides

Coenzyme A thioester **11** plays a pivotal role in the biosynthesis of long-chain fatty acids and their derivatives and of a large group of natural products whose structures are based, like fatty acids, on the combination of acetate units into chains of polyketides, which can then suffer a variety of reductive

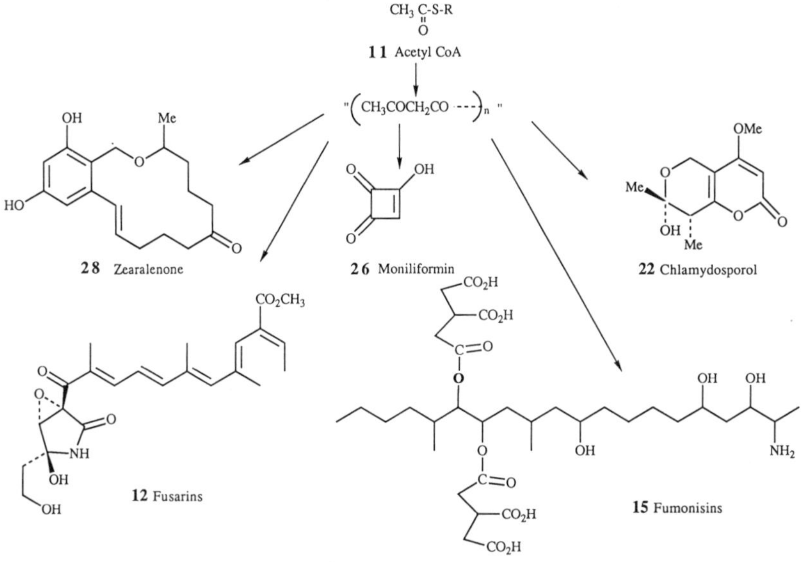

Scheme III. Polyketide-derived secondary metabolites.

processes, cyclization, etc. (Herbert 1989). The genus *Fusarium* uses this pathway to elaborate a number of secondary metabolites of diverse structures, although their occurrence is not as widespread as the isoprenoid-derived trichodienoids. Typical structures are shown in Scheme III, exemplifying the breadth of structures produced by fusaria. Attention is also drawn to the incorporation of extra molecular fragments leading to further structural intricacies.

Examples of the complex secondary biosynthetic operations that occur after original polyketide assembly are provided by these compounds and are considered separately here.

Fusarin Biosynthesis

The fusarins were originally isolated by Bjeldanes and Wiebe (1991) from a culture of *Fusarium moniliforme*.

Fusarin C **12** has been found to be produced on corn cultures by a number of other *Fusarium* species (Farber and Sander 1986), such as *F. sporotrichioides, F. graminearum, F. avenaceum, F. culmorum, F. sambucinum, F. poae,* and *F. crookwellense* (Golinski et al 1988).

Although fusarin C is a mutagen, it is not a cancer initiator despite the fact that crude extracts of *F. moniliforme* cultures are carcinogenic. It has also been shown that metabolic activation was required for maximum mutagenic activity of *F. moniliforme* extracts as purification of fusarin C proceeded. This loss of mutagenicity and the lack of carcinogenicity of fusarin C led eventually to the discovery of the fumonisins (discussed later). Fusarin C **12** is the only fusarin reported to be mutagenic. Other fusarins are known: fusarin A **13** and fusarin D **14** and more recently isomers E and F have been described. The biosynthesis of the fusarins by *F. moniliforme* was studied by Steyn and Vleggaar (1985), who concluded that the fusarins were derived from the condensation of a C_{14}-polyketide and a C_4 unit, possibly oxaloacetate, obtained from the Krebs tricarboxylic acid cycle. The difference in the enrichment levels obtained after labeling with $[1,2-^{13}C_2]$acetate also suggested that the polyketide was formed from an acetyl-coenzyme A (CoA) starter unit and six malonyl-CoA units.

FUMONISINS

The fumonisins form the most recently discovered family of *Fusarium* toxins (Gelderblom et al 1984, 1986; Steyn and Vleggaar 1985). Their isolation was prompted by the death of many horses from equine leukoencephalomalacia, mainly in New Caledonia, and elevated esophageal cancer in South Africa. The feed ingested by these horses was found to be contaminated by the fungus *Fusarium moniliforme*. The failure to find the toxic components produced by *F. moniliforme* cultures in the usual organic extracts caused researchers to look more closely at aqueous fractions. Two groups in South Africa and New Caledonia independently isolated the most abundant fumonisin, fumonisin B_1 **15**, from such feed. The South Africans were the first to publish their results in 1988, also describing three other fumonisins, B_2, A_1, and A_2—structures **16, 17,** and **18** (Bezuidenhout et al 1988). Since then, two more fumonisins have been described, fumonisins B_3 **19** and B_4 **20**. The amounts of fumonisins B_2 and B_3 usually found are about 15–25% that of fumonisin B_1, with other fumonisins produced

	R1	R2	R3
15	OH	OH	H
16	H	OH	H
17	OH	OH	Ac
18	H	OH	Ac
19	OH	H	H
20	H	H	H

in even smaller amounts. However, some *F. moniliforme* isolates have been found to produce more B_2 than B_1.

A systematic search for fumonisins followed their discovery. *F. moniliforme* and *F. proliferatum* are the only species reported to date to produce large amounts of fumonisins. One other species also from the section Liseola, *F. nygamai,* may produce smaller amounts (Nelson et al 1990, Gelderblom et al 1991). However, the common occurrence of *F. moniliforme* on corn means that there is potentially a worldwide problem.

Initial studies on the biosynthesis of fumonisins may have helped shed some light on the mechanism for the toxicity of fumonisins. Comparison of the structures of fumonisins and sphingosine **21** suggested that fumonisins may originate from the combination of a fatty acid and an amino acid, with its methyl groups originating from methionine, a well-known methyl group donor. It has since been shown that deuterium from CD_3-methionine is readily incorporated into fumonisins (Richard and Riley 1992). Despite the structural similarity of fumonisins and sphingolipids, it seems that the biosynthesis appears to involve acetate forming a polyketide and the addition of glutamic acid, methionine, and serine (Blackwell et al 1993). Other experiments (Bacon et al 1991) have shown that fumonisin B_1 inhibited the incorporation of serine into sphingosine and reduced the concentration of sphingosine, consequently causing the accumulation of its precursor, sphinganine, up to toxic levels and reducing the transformation of sphingosine to ceramide. The involvement of sphingolipids as cell membrane components, especially in brain and nerve tissue, could explain the effects of fumonisins on horses. Equine leukoencephalomalacia results in extensive degeneration of brain tissue. Sphingolipids are also involved in cell-cell communication, cell growth, differentiation, and transformation, and they inhibit protein kinase C. The influence of fumonisins on some of these steps may play a role in their apparent carcinogenicity and other pharmacological effects.

Chlamydosporol and Acuminatopyrone

Chlamydosporol actually refers to a pair of compounds, α- and β-chlamydosporols. These compounds were first reported in 1990 (Miller et al 1990). Soon thereafter, Grove and Hitchcock (1991) also isolated these compounds and were able to obtain crystals. Crystallographic analysis showed that the previously published structure, based on nuclear magnetic resonance (NMR) and mass spectroscopic information, had been incorrect. The accepted structure of the major compound is **22**.

The toxicity of these compounds was found to be relatively low. When tested in brine shrimp and HeLa cell assays, the LC_{50} for a mixture of the isomers was approximately 400 g/L in both systems (Miller et al 1990).

Along with chlamydosporol, Grove and Hitchcock (1991) reported the isolation of a basic compound acuminatopyrone from *Fusarium acuminatum* to which structure **23** was assigned from its NMR and ultraviolet spectra and by analysis of its degradation products.

22

23

24

25

Biosynthetically, chlamydosporol is viewed as an acetogenin as indicated by the heavy lines in **24**. It might well be argued that the congener acuminatopyrone would be similarly derived, thus leading to structure **25** instead of **23**, and recent work in the author's laboratory confirms this.

Moniliformin

The unusual cyclobutane derivative moniliformin **26** isolated from *F. subglutinans* appears to arise from the acetate pathway by the usual coupling of two acetate units as demonstrated by both radioactive carbon and C-13 labeling studies (Breiphol and Frank 1984; G. Kasitu, J. W. ApSimon, B. Blackwell, and J. D. Miller, *unpublished*).

There are a variety of biosynthetic origins that may be envisioned for this unusual structure, including derivation from carbohydrates, amino acids, and pyruvate. The labeling studies described show unequivocally that two acetate units are involved in the original formation to cyclobutadione, which on subsequent oxidation and dehydration leads to moniliformin 26. Labeling experiments indicate uniform activity at all carbon atoms, a result that is consistent with an equilibrating intermediate 26a.

26a   26

In these laboratories, a feeding experiment with ^{13}C-labeled acetate showed the same behavior.

Zearalenone and the Zearalenols

The chemistry of zearalenone 27 and the zearalenols 28a and 28b has been thoroughly investigated since the characterization of zearalenone in 1966 (Hidy et al). Two review articles discussed these compounds in the mid-1970s (Shipchandler 1975, Mirocha and Pathre 1976). The most interesting recent synthetic development has been the synthesis of chiral zearalenone (Hitchcock and Pattenden 1990, Carreno et al 1991, Hegedus et al 1991).

In the recent NMR study of the biosynthesis of zearalenone, Blackwell et al (1985) studied the distribution of [1-^{13}C] and [2-^{13}C]acetate precursors in the compound when it was produced by Fusarium graminearum grown in liquid culture. [1-^{13}C] and [2-^{13}C]acetate were incorporated in alternate carbon atoms in zearalenone, consistent with the head-to-tail condensation of nine acetate units. This series of molecules represents a classic case of an acetate-derived natural product.

27

	R_1	R_2
28 a	H	OH
28 b	OH	H

Metabolites Derived from Amino Acids

It is usually accepted that the amino acids are not directly involved in secondary metabolism, but there exist a large number of natural products that are obviously derived from these compounds. In *Fusarium* species, there are two classes of compounds that are directly derived from amino acids yet are accumulated as secondary metabolites.

BUTENOLIDE

The structure of butenolide **29** suggests that it may be derived from an amino acid. A recent study using ^{13}C and ^{15}N labels showed that **29** was derived from glutamic acid (Kendall 1988). The pathway from glutamic acid to butenolide is not yet known; however, the suggestion was made in 1977 that 4-acetamido-2-butenoic acid could be an intermediate (Ciegler et al 1977).

29

THE ENNIATINS

The enniatins are cyclic hexadepsipeptides, meaning that they are cyclic combinations of three amino and three hydroxy acids, joined by amide or ester bonds. The latest review on these compounds was published in 1985 (Steinrauf 1985). In summary, enniatins were first noticed because of their antimicrobial properties and more recently because they have shown insecticidal properties, more specifically against adult blowflies and mosquito larvae (Grove and Pople 1980) and the spruce budworm (Miller et al 1985, Calhoun et al 1988). They have also shown some phytotoxicity by being involved with Fusarium wilt in tomatoes and inhibiting wheat seed germination (Burmeister and Plattner 1987).

All enniatins contain three residues of the same hydroxy acid, α-hydroxy-isovaleric acid. The four amino acid residues can be either N-methyl-isoleucine, N-methyl-valine, N-methyl-leucine, or valine. The distribution of these amino acids between the known enniatins is shown in structures **30–37**.

The presence of N-methyl groups is apparently important for the biological activity of the enniatins, because they greatly increase the lipophilicity of the enniatins. The presence of N-methyl groups, along with the absence of free carboxylic acid or amino groups because of their cyclic nature, makes the enniatins very lipophilic. They are essentially insoluble in water and most soluble in hexane. The biological activity of the enniatins also rests in their ionophoric character. They behave as passive ion carriers, disrupting ionic balances and inducing swelling of cells and rupture of cell walls. The biological activity of enniatin enantiomorphs, being equal to that of the enniatins, also illustrates their role as passive ion-carriers (Ivanov et al 1969). The first enniatins, A **30** and B **31** were isolated in the late 1940s (Ettlinger et al 1947); however, the correct structure of these compounds was not established until 1963 (Keller-Schierlein et al; Quitt et al 1963; Ivanov et al 1963a,b). Until 1991, these enniatins were the only ones to have been isolated and fully characterized. Enniatins A1 **32** and B1 **33** were also known but had never been isolated. These have now also been isolated and characterized, as well as four new enniatins, A2 **34** the first natural N-methyl-leucine-containing enniatin, and B2 **35**, B3 **36**, and B4 **37** (ApSimon et al 1992a, 1992c). *F. acuminatum* was the only species to produce all of the above enniatins, whereas only enniatins A, A1, B, B1, and B2 were found in cultures of *F. avenaceum* and *F. compactum*. Enniatins B2 and B3 both appear to be precursors of enniatin B, lacking one or two N-methyl groups; enniatin B4 contains an N-methyl-leucine group like enniatin A2.

Although the biosynthesis of enniatins appears rather obvious at first glance, certain interesting points have emerged. For instance, they are not obtained by chain expansion followed by cyclization but rather by condensation of three identical peptides (in the case of symmetrical enniatins) (Keller et al 1983). It has also been shown that valine is the precursor for both the N-methyl-valine and the hydroxy-isovaleric acid residues. N-methyl-valine is not incorporated in enniatin B, suggesting that valine is

	R_1	R_2	R_3	R_4	R_5
30 (A)	Me	s-Bu	Me	s-Bu	s-Bu
31 (B)	Me	i-Pr	Me	i-Pr	i-Pr
32 (A1)	Me	s-Bu	Me	s-Bu	i-Pr
33 (B1)	Me	i-Pr	Me	i-Pr	s-Bu
34 (A2)	Me	s-Bu	Me	i-Bu	s-Bu
35 (B2)	H	i-Pr	Me	i-Pr	i-Pr
36 (B3)	H	i-Pr	H	i-Pr	i-Pr
37 (B4)	Me	i-Pr	Me	i-Bu	i-Pr

not methylated to *N*-methyl-valine until after being joined to the hydroxy acid (Kleinkauf et al 1983).

The hydroxyvaleric acid fragment is reminiscent of a single isoprene unit or a methylated C_4 acetogenin arising by true secondary metabolic pathways which then combine with the amino acid portion (leucine, isoleucine, or valine) to give these compounds.

Conclusion

The seemingly broad range of secondary metabolites reported in the *Fusarium* species treated in this book are structurally derived from common biosynthetic origins. Doubtless many more secondary metabolites will emerge as these economically important species are examined, but it is likely that these compounds will fall into the biosynthetic classification already noted.

Acknowledgments

Support for our contribution to *Fusarium* chemistry has been generously provided by Agriculture Canada, The Natural Sciences and Engineering Research Council of Canada, and Carleton University.

Literature Cited

Ament, J. E., Grove, J. F., and Henson, J. R. 1988. ^{13}C NMR spectra of some trichothecene mycotoxins and derivatives. Magn. Reson. Chem. 26:475.

ApSimon, J. W., Blackwell, B. A., Blais, L., Fielder, D. A., Greenhalgh, R., Kasitu, G., Miller, J. D., and Savard, M. 1990. Mycotoxin from *Fusarium* species: Detection, determination, and variety. Pure Appl. Chem. 62(7):1339-1346.

ApSimon, J. W., Blackwell, B., Blais, L., Fielder, D. A., Greenhalgh, R., Kasitu, G., Kendal, J. E., Miller, J. D., and Savard, M. 1991. Pages 201-208 in: Studies in Natural Product Chemistry. Vol 9. Atta-Ur-Rzahman, ed. Elselvier Science Publishers, Amsterdam.

ApSimon, J. W., Blackwell, B. A., Blais, L. A., Greenhalgh, R., and Miller J. D. 1992a. Isolation and characterization of enniatins from *Fusarium avenaceum* DAOM 196490. Can. J. Chem. 70:1281-1287.

ApSimon, J. W., Blackwell B. A., Fielder D. A., Greehalgh R., Kasitu G., and Miller J. D. 1992b. Isolation and characterization of culmorin derivatives produced by *Fusarium culmorum* CMI 14764. Can. J. Chem. 70:1308-1316.

ApSimon, J. W., Blais, L., Greenhalgh, R., Miller, J. D., and Visconti, A. 1992c. Production of enniatins by *Fusarium acuminatum* and *Fusarium compactum* in liquid cultures of three new enniatins B2, B3, B4. J. Agric. Chem. 40:1076-1082.

Bacon, C. W., Merril, A. H., Jr., Norred, W. P., Riley, R. T., and Wang, E. 1991. Inhibition of sphingolipids biosynthesis by fumonisins: Implication for diseases associated with *Fusarium moniliforme*. J. Biol. Chem. 266:14486-14490.

Bekker, A. R., Kononenko G. P., Leonov N. A., and Soboleva N. A. 1991. Gramilaurone, A novel natural sesuiterpenoid from *Fusarium gramineanum* Schw. Tetrahedron Lett. 32(16):1893.

Bezuidenhout, S. C., Gelderblom, W. C. A., Gorst-Allman, C. P., Horak, R. M., Marasas W. F. O., Spiteller G., and Vleggaar R. 1988. Structure elucidation of the fumonisins, a mycotoxin from *Fusarium moniliforme*. J. Chem. Soc. Chem. Comm. 743-745.

Bjeldanes, L. F., and Wiebe, L. A. 1991. Fusarin C, a mutagen from *Fusarium moniliforme* growth on corn. J. Food Sci. 46:1424-1426.

Blackwell, B. A., Greenhalgh, R., and Miller, J. D. 1985. ^{13}C NMR study of the biosynthesis of toxins by *Fusarium graminearum*. J. Biol. Chem. 260:4243-4247.

Blackwell, B. A., Miller, J. D., and Savard, M. E. 1993. Production of carbon-14 labelled fumonisin in liquid culture. J. Off. Anal. Chem. Int. (In press.)

Breiphol, G., and Frank B. 1984. Biosynthesis of moniliformin, a fungal toxin with cyclobutanedione structure. Angew. Chem. Int. Ed. Engl. 23:996-998.

Burmeister, H. R., and Plattner R. D. 1987. Enniatin production by *Fusarium tricinctum* and its effects on germinating wheat seeds. Phytopathology 77:1483-1487.

Calhoun, L., Giguere, P., Strongman, D. B., Strunz, G. M., and Yu, C.-M. 1988. Enniatins from *Fusarium avenaceum*, isolated from balsam fir foliage and their toxicity to spruce budworm larvae *Choristoneura fumiferana* (CLEM) (Lepidoptera: Tortrocodae). J. Chem. Ecol. 14:753-764.

Carreno, M. C., Maestro, M. C., Pedregal, C., Ruano, J. L. G., Rubio A., and Sollande, G. 1991. Asymmetric synthesis of of (S)-zearalenone dimethyl ether, an orsellinic acid type macrolide. J. Org. Chem. 56:2317-2322.

Ciegler, A. , Spencer, G. F., Tjarks, L. W., Vesonder, R. F., and Wallen, L. L. 1977. 4-Acetomido-2-butenoic acid from *Fusarium graminearum*. Phytochemistry 16:1296-1297.

Ettlinger, L., Gaumann, E., Nager U., Plattner, P. A., and Roth S. 1947. Enniatins, ein neues, gegen Mykobakterien Wirsames Antibiotikum. Experientia 3:202-203.

Farber, L. F., and Sander G. W. 1986. Production of fusarin C by *Fusarium* spp. J. Agric. Food Chem. 34:963-966.

Gelderblom, W. C. A., Marasas, W. F. O., Steyn, P. S., Thiel, P. G., van der Merwe, K. J., van Rooyen, P. H., Vleggaar, R., and Wessels, P. L. 1984. Structure of fusarin C, a mutagen produced by *Fusarium moniliforme*. J. Chem. Soc. Chem. Commun. 122-124.

Gelderblom, W. C. A., Jaskiewicz, K., Marasas, W. F. O., and Thiel, P. G. 1986. Investigation of the carcinogenecity of fusarin C—A mutagenic metabolite of *Fusarium moniliforme*. Carcinogenesis 7:1899-1901.

Gelderblom, W. C. A., Marasas, W. F. O., Nieuwenhuis, J. J., Shephard, G. S., Sydenham, E. W., and Thiel, P. G. 1991. Survey of fumonisin production by *Fusarium* species. Appl. Environ. Microbiol. 57:1089-1093.

Goi, M., Morisaki N., and Nozoe S. 1970. Structure of cyclonerodiol. Tetrahedron Lett. 1293.

Goi, M., Morisaki N., and Nozoe S. 1971. Synthesis and stereochemistry of cyclonerodiol. Tetrahedron Lett. 3701.

Golinski, P., Latus-Zietkiewicz, D., Perkowski, J., and Versonder, R. F. 1988. Formation of fusarenone X, nivalenol, zearalenone, α-*trans* zearalenol, β-*trans* zearalenol and fusarin C by *Fusarium crookwellense*. Appl. Environ. Microbiol. 54:2147-2148.

Grove, J. F. 1986. Phytotoxic compounds produced by *Fusarium equiseti* Part 8. Acid catalysed reaarangement of 12,13 epichlorothe-9-enes. J. Chem. Soc. Perkin Trans. I, 647.

Grove, J. F., and Pople, M. 1980. The insecticidal activity of beauvericin and enniatin complex. Mycopathologia 70:103-105.

Grove, J. F., and Hitchcock, P. B. 1991. Metabolic products of *Fusarium acuminatum*: acuminatopyrone and clamydosporol. J. Chem. Soc. Perkin Trans. I, 997-999.

Hasseltine, C. W., and Mehlhman, eds. 1978. Mycotoxins: In Human and Animal Health. Pathotox, Parck Forest South, IL.

Hegedus, L. S., Kalivretenos, A., and Stille, J. K. 1991. Synthesis of β-resorcyclic macrolide via organopalladium chemistry. Application to the synthesis of (S)-zearalenone. J. Org. Chem. 56:2883-2894.

Herbert, R. D. 1989. The biosynthesis of secondary metabolites. 2nd ed. Chapman and Hall, London.

Hidy, P. H., Hodge, E. B., Urry, W. H., and Wehrmeister, H. L. 1966. The structure of zearalenone. Tetrahedron Lett. 27:3109-3114.

Hitchcock, S. A., and Pattenden, G. 1990. Synthesis of macrocylic via allylic radical intermediates. A total synthesis of (−)-zearalenone. Tetrahedron Lett. 31:3641-3644.

Ivanov, V. T., Kiryushkin, A. A., Ovichinikov, Y. A., and Shemyakin, M. M. 1963a. The structure and total synthesis of enniatin B. Tetrahedron Lett. 885-890

Ivanov, V. T., Kiryushkin, A. A., Ovichinikov, Y. A., and Shemyakin, M. M. 1963b. Synthesis of enniatin A. Bull. Acad. Sci. USSR. Div. Chem. Sci. (Engl. Transl.) (Izv. Akad. Nauk SSSR Ser. Khim.) 1148.

Ivanov, V. T., Ovichinnikov, Y. A., and Shemyakin M. M. 1969. Topochemical investigation of peptide systems. Angew. Chem. Int. Ed. Engl. 8:492-499.

Joffe, A. Z. 1986. *Fusarium* Species, Their Biology and Toxicology. Wiley Interscience, New York.

Keller, U., Kleinkauf, H., and Zocher, R. 1983. Mechanism of depsipeptide formation catalysed by enniatin synthetase. Biochem. Biophys. Res. Comm. 110:292-299.

Keller-Schierlein, W., Plattner, P. A., Quitt, P., Studer, R. O., and Vogler, K. 1963. Synthesen in der Depsipeptid-Reihe Mitteilung 2. Synthese von enniatin B. Helv. Chim. Acta 46:927-935.

Kendall, J. E. 1988. Biosynthesis of butenolide by *Fusarium culmorum*, HLX 1503. M.Sc. thesis, Carleton University, Ottawa, Canada.

Kleinkauf, H., Salnikow J., and Zocher R. 1983. Biosynthesis of enniatin B. FEBS Lett. 71:13-17.

Marasas, W. F. O., Nelson, P. E., and Toussoun 1984. Toxigenic *Fusarium* species: Identity and Mycotoxicology. Pennsylvania State University Press, University Park.

Metcalf, C. A., Shulter G., and Ziegler F. E. 1992. Configuration by total synthesis of the revised structure of sporol: An application of cyclic thionocarbonate initiated radical cyclization. Tetrahedron Lett. 33(22):3117-3120 and references therein.

Miller, J. D. 1988. Pages 65-67 in: Issues in Food Safety, Toxicology Forum, Washington, DC.

Miller, J. D., Strongman, D. B., and Whitney, N. J. 1985. Observation of fungi associated with spruce budworm in infested balsam fir needles. Can J. For. Res. 15:896-901.

Miller, J. D., Salleh, B., Savard, M. E., and Strange, R. N. 1990. Chlamydosporol, a new metabolite from *Fusarium chlamydosporum*. Mycopathologia 110:177-181.

Mirocha, C. J., and Pathre, S. V. 1976. Pages 178-227 in: Mycotoxin and Other Fungal Related Food Problems. J. V. Rodicks, ed. Am. Chem. Soc. Adv. Chem. Ser. 149.

Moss, M. O., and Smith, J. E. 1985. Mycotoxins. Wiley, Toronto.

Nelson, P. E., Osweiler, G. D., Plattner, R. D., Richard, J. L., Rise, L. G., Ross P. F., and Wilson T. M. 1990. Production of *Fumonisins moniliforme* and *Fusarium proliferatum*. Isolates associated with equine leukoencephalomalacia and pulmonary edema syndrome in swine. Appl. Environ. Microbiol. 56:3225-3226.

Quitt, P., Studer, R. O., and Vogler, K. 1963. Synthesen in der Depsipeptid-Reihe 2. Mitteilung Synthesen von Enniatin A. Helv. Chim. Acta. 46:1715-1720.

Richard, J. L., and Thurston, J. R., eds. 1986. Diagnosis of Mycotoxicoses. Martin Nijhoff, Boston.

Richard, J. T., and Riley, R. T., eds. 1992. Fumonisins: A current perspective and a view to the future. Mycopathologia 117(1 and 2).

Roddricks, J. V., ed. 1978. Diagnosis of Mycotoxicoses. Pathotox, Park Forest South, IL.

Shipchandler, M. T. 1975. Chemistry of zearalenone and some of its derivatives. Heterocycle 3:471-520.

Steinrauf, L. K. 1985. Pages 139-171 in: Antibiotics and Their Complexes. Vol. 19, Metal Ions in Biological Systems. H. Sigel, ed. Marcel Dekker, New York.

Steyn, P. S., and Vleggaar R. 1985. Biosynthetic studies on fusarins, metabolites of *Fusarium moniliforme*. J. Chem. Soc. Chem. Commun. 1:189-1191.

Chapter 2

Epidemiology of *Fusarium* Ear Diseases of Cereals

J. David Miller

The epidemiology of *Fusarium graminearum* ear diseases of wheat and corn is the subject of an excellent review by Sutton (1982). Since that time, few critical experiments have been reported on the epidemiology of the *Fusarium* diseases of small grains and corn. Wheat, corn, and barley appear to be most affected by *Fusarium* toxins, and these three crops constitute two-thirds of the world production of cereals. Oats, rye, and triticale have all been reported to contain *Fusarium* mycotoxins (Chelkowski 1989a, Miller et al 1985, Scott 1989); however, it is generally regarded that these crops, with the exception of some triticale varieties, are resistant or escape significant contamination (Chelkowski 1989a, Miller et al 1985).

The end point of an epidemic of head blight or Gibberella ear rot caused by *F. graminearum* or *F. culmorum* is deoxynivalenol in the kernels. Hence a rather modest epidemic can produce a crop that contains too much mycotoxin for human food or animal feed. Canadian guidelines call for no more than 2 ppm of deoxynivalenol in uncleaned soft white winter wheat for adult food and 1.2 ppm in uncleaned soft white winter wheat intended for baby food (Van Egmond 1989). Animal feed should contain no more than 1 ppm of deoxynivalenol in the diet. An end point of Fusarium kernel rot caused by *F. moniliforme* is the presence of fumonisin (Thiel et al 1991). Safe concentrations of fumonisin for human or animal feed have not been determined.

This review attempts to build on the information covered by Sutton (1982), concentrating on three general areas: 1) factors affecting the fusaria that cause disease in wheat and corn, 2) insights gained from studies of agronomic practice in relation to Fusarium head blight and corn ear and kernel rots, and 3) the processes of initial infection in wheat and corn and factors that promote epidemics. The species discussed in this chapter follow the system of Nelson et al (1983).

Wheat

SPECIES, TEMPERATURE, AND MOISTURE

The term Fusarium head blight is a description of a disease of wheat and other small grains that is caused by several species of the genus *Fusarium*. Hence any discussion of the epidemiology of such diseases must begin with an examination of the fusaria isolated from cereals. Although *F. graminearum* diseases of wheat and corn have been studied the most, this species is not always the prime cause of disease in all regions where cereals are grown. Additionally, within regions, there is year-to-year variability in the species isolated. The species involved are not equally pathogenic nor are the toxins they produce equally toxic (Table 2.1).

A general problem arises when discussing the epidemiology of *Fusarium* diseases of cereals—the association of different species with different climatological conditions. For example, *F. graminearum* is more associated with cereals grown in warmer areas than *F. culmorum*. Temperature and moisture interact to influence the growth of all fungi including fusaria (Magan and Lacey 1984). In all probability, the influence of temperature relates to conditions that allow a sustained period of warm conditions (daytime temperatures greater than 30°C) regardless of daily means. For example, although the Maritime Provinces of Canada are generally considered cool, *F. graminearum* is the dominant head blight species. The daytime maxima can be higher than 30°C in wheat-growing areas. The use of terms such as "warm" and "cool" are generalizations. Unfortunately, there is little basis to allow more precision.

TABLE 2.1
Fusaria Isolated from Wheat Kernels

Species	Pathogenicity[a]	Mycotoxins[b]
F. graminearum	High	Deoxynivalenol or nivalenol, zearalenone
F. culmorum	High to moderate	Deoxynivalenol, zearalenone
F. crookwellense	Low to moderate	Nivalenol, zearalenone
F. avenaceum	Low	Moniliformin
F. acuminatum	Low	Acuminatin[c]
F. sporotrichioides	Low	T-2, HT-2
F. poae	Very low	Diacetoxyscirpenol
F. equiseti	Very low	Diacetoxyscirpenol

[a] Data from Chelkowski (1989a), Lamprecht et al (1990), Stack and McMullen (1985), and Scott et al (1988).
[b] Data from Marasas et al (1984).
[c] Data from Visconti et al (1989).

A number of studies have reported the fusaria isolated from wheat seeds. Five or six species are consistently isolated from wheat affected by Fusarium head blight. The most pathogenic species, *F. graminearum* and *F. culmorum*, are generally the most common species found. *F. graminearum* is common in wheat from North America and China (Table 2.2; Wang and Miller 1988). *F. culmorum* is the dominant species in cooler wheat-growing areas such as Finland, France, Poland, and The Netherlands (Table 2.2; Saur 1991, Snijders and Perkowski 1990). *F. graminearum* has two biotypes; one forms the ascomycete *Gibberella zeae* and is primarily associated with ear diseases. The second type does not form perithecia and is associated with crown rots in dryland wheat production (Cook 1981). A sexual stage has not been reported for *F. culmorum*.

F. avenaceum is also common in wheat from all regions studied. This species has a sexual stage, *G. avenacea*. The distribution of the newly described species *F. crookwellense* has not been properly examined. It is rare in wheat from Canada and Poland but is the primary cause of head blight in irrigated land in the Orange Free State, South Africa (Abramson et al 1987, Chelkowski 1989a, Scott et al 1988). No sexual state of this fungus has been identified. *F. poae* and *F. sporotrichioides* are also isolated from wheat kernels in low to moderate frequencies more often under cooler conditions (Table 2.2 and references cited therein). In one study, the fusaria were isolated from wheat kernels collected weekly from anthesis to harvest over four years in the same location. The frequency of occurrence of *F. avenaceum* and *F. sporotrichioides* is negatively correlated to temperature (calculated from data in R. A. Martin, *unpublished*).

The distribution of the head blight species is broadly affected by pathogenicity. The principal species, ranked from most pathogenic to least, are *F. graminearum, F. culmorum, F. crookwellense,* and *F. avenaceum* (Table 2.1). The regional and annual variation of the pathogenic species are most affected by temperature; from coolest to warmest conditions, the order of species found is *F. culmorum, F. crookwellense, F. avenaceum,* and *F. graminearum.* As noted by Cook (1981), the dominant head blight species is "determined by temperature more than any other factor." The less pathogenic species, including *F. sporotrichioides, F. poae,* and *F. nivale,* occur under cooler conditions. In general terms, only *F. graminearum* and *F. culmorum* produce significant visual symptoms.

Although increased rainfall promotes Fusarium head blight, incidence is said to be most affected by moisture at anthesis (Cook 1981, Duthie et al 1986, Sutton 1982). In one study, however, the frequencies of *F. graminearum* and *F. avenaceum* were not correlated to precipitation (calculated from data in R. A. Martin, *unpublished*). This implies that disease is a function of the timing of moisture rather than amount of rain. The effect of water potential on growth and sporulation is similar for *G. zeae, F. culmorum,* and *F. avenaceum* (Sung and Cook 1981).

TABLE 2.2
Rank Order of *Fusarium* Species Isolated from Wheat

Place / Type and Year	First	Second	Third	Fourth	Fifth	Sixth
Manitoba						
Durum, 1986[a]	*graminearum*	*avenaceum*	*sporotrichioides*	*poae*	*acuminatum*	*equiseti*
Durum, 1987[a]	*avenaceum*	*graminearum*	*sporotrichioides*	*acuminatum*	*poae*	*equiseti*
Spring wheat, 1986[a]	*graminearum*	*avenaceum*	*sporotrichioides*	*poae*	*equiseti*	*culmorum*
Red spring, 1986[a]	*graminearum*	*avenaceum*	*sporotrichioides*	*poae*	*acuminatum*	*culmorum*
Red spring, 1986[b]	*graminearum*	*sporotrichioides*	*acuminatum*	*equiseti*	*poae*	*avenaceum*
Ontario						
White winter, 1983[c]	*poae*	*equiseti*	*graminearum*	*sporotrichioides*	*culmorum*	*avenaceum*
White winter, 1984[c]	*equiseti*	*sporotrichioides*	*poae*	*graminearum*	*acuminatum*	*avenaceum*
White winter, 1986[a]	*graminearum*	*poae*	*avenaceum*	*sporotrichioides*	*equiseti*	*culmorum*
White winter, 1987[a]	*graminearum*	*sporotrichioides*	*acuminatum*	*poae*	*avenaceum*	*equiseti*
Red winter, 1987[a]	*graminearum*	*avenaceum*	*acuminatum*	*sporotrichioides*	*acuminatum*	*equiseti*
Red spring, 1987[a]	*graminearum*	*avenaceum*	*poae*	*sporotrichioides*	*equiseti*	
Quebec[c]						
White winter, 1984	*sporotrichioides*	*avenaceum*	*poae*	*equiseti*	*acuminatum*	*graminearum*
Maritime Provinces[c]						
White winter, 1984	*avenaceum*	*sporotrichioides*	*poae*	*equiseti*	*graminearum*	*acuminatum*
Prince Edward Island[d]						
Spring wheat, 1985	*avenaceum*	*sporotrichioides*	*graminearum*	*poae*		
Spring wheat, 1986	*graminearum*	*avenaceum*	*poae*	*sporotrichioides*		
Spring wheat, 1987	*avenaceum*	*poae*	*sporotrichioides*	*graminearum*		
Spring wheat, 1988	*avenaceum*	*sporotrichioides*	*graminearum*	*poae*		
North Dakota[e]						
Spring wheat, 1981	*graminearum*	*culmorum*	*poae*	*equiseti*	*acuminatum*	*tricinctum*
Minnesota[f]						
Spring wheat, 1984–1986	*graminearum*	*poae*	*equiseti*	*sporotrichioides*	*acuminatum*	
Finland[g]						
Wheat, 1966–1971	*culmorum*	*poae*	*avenaceum*	*acuminatum*	*graminearum*	

Poland[h]				
Wheat, 1984	*culmorum*	*graminearum*	*avenaceum*	*nivale*	*crookwellense*
Wheat, 1985	*avenaceum*	*culmorum*	*nivale*	*graminearum*	*crookwellense*
Wheat, 1986	*culmorum*	*avenaceum*	*nivale*	*graminearum*
Wheat, 1987	*nivale*	*avenaceum*	*culmorum*	*graminearum*

[a] Data from Clear and Patrick (1990).
[b] Data from Abramson et al (1987).
[c] Data from Duthie et al (1986).
[d] Data from Martin (*unpublished*).
[e] Data from Stack and McMullen (1985).
[f] Data from Wilcoxson et al (1988).
[g] Data from Ylimaki (1981).
[h] Data from Chelkowski (1989a).

BIOTIC FACTORS

In the case of *F. graminearum,* both ascospores and macroconidia infect the plant. Although Fusarium head blight has been considered monocylic, it is more likely polycyclic. Ascospores can be released from debris to initiate infection, which is followed by the production of macroconidia under favorable conditions (Anderson 1948, Khonga and Sutton 1986). Dispersal of ascospores has been found to be related to high relative humidities (Ayers et al 1975). High concentrations of ascospores and macroconidia of *G. zeae* have been found in the air under epidemic conditions (Ayers et al 1975; Martin 1988; Tshanz et al 1975; Miller et al, *unpublished data*). *F. culmorum* is dispersed by macroconidia. This is apparently also the mode of dispersal for *F. avenaceum* (Cook 1981).

Wheat cultivars having resistance to Fusarium head blight are most susceptible at anthesis (Cook 1981, Snijders 1993, Sutton 1982). However, susceptible cultivars can be infected through to head maturity (Anderson 1948; Hart et al 1984; R. A. Martin, *unpublished data*). Compounds such as betaine and choline in anthers and other head organs have been demonstrated to affect Fusarium head blight. Their concentrations vary according to cultivar (reviewed in Sutton 1982). Susceptibility to glume blotch, powdery mildew, and leaf rust have been shown to promote *Fusarium* infection (Martin and Johnston 1982, Mesterhazy 1984, Mesterhazy and Rowaished 1977). Early colonization of the lemmae by *F. poae* and *F. sporotrichioides* may promote subsequent colonization by *F. graminearum,* Sturz and Johnston 1983; R. A. Martin, *unpublished data*). Damage caused by insects may also be a point of entry for the *Fusarium* species that cause head blight (Sutton 1982; R. A. Martin, *unpublished data*). Fusaria including *F. graminearum* are common on insects and birds (Warner and French 1970, Windels et al 1976).

Corn

SPECIES, TEMPERATURE, AND MOISTURE

On the basis of symptoms, there are two general kinds of damage to corn ears caused by *Fusarium* species—Gibberella or pink ear rot and Fusarium ear rot or kernel rot. The former is prevalent in northern temperate climates, especially in wet years, and is caused by *G. zeae* and *F. culmorum.* The latter is associated with warm, dry years and insect damage and is caused by *F. subglutinans* (*G. subglutinans*), *F. moniliforme* (syn. *G. fujikuroi*), and *F. proliferatum* (Table 2.3; Shurtleff 1980).

Unlike the situation for wheat, extensive studies of the occurrence of fusaria on kernels have not been reported. *F. subglutinans* and *F. moniliforme* are reported to be very common in corn kernels in Canada, the United States, and Europe. *F. graminearum* is also important in the northern United States and Canada and parts of Europe but is displaced by *F. culmorum* in eastern Europe (Abbas et al 1988, Chelkowski 1989b, Neish et al 1983). In warmer parts of the United States and lowland tropics, *F. moniliforme* causes one of the most important ear diseases (De Leon and Pandey 1989, King and Scott 1981, Ochor et al 1987). *F. sporotrichiodes* is found in colder areas (Chelkowski 1989b, Neish et al 1983).

As is the case for wheat, prevalence of the different fusaria in corn is governed by temperature. One important difference is the apparent ubiquity of *F. moniliforme* and related species (Bacon and Williamson 1992). The significance of this fungus in healthy tissue remains poorly understood (King and Scott 1981). For many years, *F. moniliforme* has been known to occur systemically in leaves, stems, roots, and kernels (Foley 1962). This led to the suggestion that some strains of *F. moniliforme* produce disease symptoms and others do not (Bacon and Williamson 1992).

F. graminearum disease incidence is affected by moisture at silk emergence, and prevalence is increased with wet weather later in the season (Al-Heeti 1987, Sutton et al 1980, Tuite et al 1974). Continuous rain in northern temperate areas is usually associated with cold temperatures not conducive to the growth of *F. graminearum* (Sutton 1982). Monitoring the growth of *F. graminearum* in experimentally infected ears showed the growth rate of the fungus was sensitive to temperature. A period in which only about 10 days had average growing degree-days greater than 5° C virtually halted growth. An average value of approximately 15 growing degree-days greater than 5° C resulted in rapid growth (Miller et al 1983).

In summary, moderate rain at silk emergence coupled with periods of warm temperature appears to be required for initial ear colonization. As noted, incidence of *F. moniliforme* disease symptoms appears to be related to dry conditions and may be related to insect herbivory (see next section).

TABLE 2.3
Fusaria Isolated from Corn

Species	Pathogenicity[a]	Mycotoxins[b]
F. subglutinans	High	Moniliformin
F. moniliforme	Moderate	Fumonisin[c]
F. graminearum	High	Deoxynivalenol or nivalenol, zearalenone
F. culmorum	High to moderate	Deoxynivalenol, zearalenone
F. sporotrichioides	Low	T-2, HT-2
F. crookwellense	Moderate	Nivalenol, zearalenone
F. proliferatum	Moderate	Fumonisin, moniliformin[c]
F. avenaceum	Low	Moniliformin
F. poae	Very low	Diacetoxyscirpenol
F. equiseti	Very low	Diacetoxyscirpenol

[a] Data from Al-Heeti (1987), Chelkowski (1989a), Gendloff et al (1986), and Mesterhazy (1982).
[b] Data from Marasas et al (1984).
[c] Data from Thiel et al (1991).

BIOTIC FACTORS

The fungi overwinter in debris. In *F. moniliforme* infections, most kernels contain the fungus (Bacon and Williamson 1992, Sutton 1982). Airborne ascospores and conidia of *F. graminearum* and *F. moniliforme* (conidia in the other species) are able to infect cobs by growing down the silk. Susceptibility increases with silk emergence (Sutton 1982). This is not a result of changes in silk moisture or phenolic content (Reid et al 1992a,b). The phenomenon appears to be related to a decline in nutrients including those in pollen (Naik and Busch 1978; Reid et al 1992c; Sutton 1982; Schaafsma et al, *in press*).

As with Fusarium head blight of wheat, there is some evidence that the ear fusaria interact. *F. moniliforme* has been reported to suppress the growth of other ear fungi (Rheeder et al 1990). Co-isolation of *F. poae* and *F. sporotrichioides* with *F. graminearum* resulted in the suggestion that simultaneous infection might be conducive to invasion by *F. graminearum* (Al-Heeti 1987).

In one study, inoculations of ears of corn grown in growth chambers with either ascospores or conidia of *G. zeae* failed to result in infection. No mechanical damage (insect or bird) occurred in the chambers. The authors suggested that this meant that kernels must first be damaged before *G. zeae* can infect the plant (Wicklow and Caldwell 1990). In field experiments, spraying conidial suspensions on silks has been reported to induce disease, although infection was highly dependent on rain or irrigation after inoculation (Koehler 1959, Ullstrup 1970). Experimental infections by this route are also highly dependent on the spores being present in the silk channel very near silk emergence (Schaafsma et al, *in press*).

Although some authors have assumed that placing the spores on the silks simulates natural infection (Reid et al 1992a,b; Ullstrup 1970), this remains unclear. One problem with such experimental inoculations is that high numbers of conidia are used (approximately 10^5 per dose) as well as that nutrients are inadvertently added from the media used to make the spores. Neither could occur in nature (Reid et al 1992a,c; Ullstrup 1970).

Bird and insect damage of the ear appear to be important in the infection of corn by *F. graminearum*. Bird damage predisposes corn to infections by *F. graminearum*, and *Fusarium* spores were found to be common on the bird species involved (Warner and French 1970, Whitney 1954). The accumulation of zearalenone was shown to be related to bird damage (Sutton et al 1980).

Mechanical damage to kernels by the European corn borer (*Ostrinia nubilalis*) appears to increase incidence of *F. graminearum* by predisposing the ear to infection. Other insects, including sap beetles (*Glischrochilus quadrisignatus*), carry *Fusarium* spores and are attracted to the wounds made by the European corn borer (Attwater and Busch 1983, Christensen and Schneider 1950, Windels et al 1976). Introduction of sap beetles to ears that were unwounded or wounded resulted in increased incidence of ear rot (Attwater and Busch 1983). Various other insect species that overwinter in debris are attracted to the developing cob and can likely vector *F. graminearum* (Sutton 1982).

In plants under irrigation and with high natural inoculum, simulated bird damage resulted in good infections of cobs by *F. graminearum* (Schaafsma et al, *in press*). In a plot where corn followed hay and thus had low natural inoculum, and without irrigation, wounding ears with sterile toothpicks did not result in *F. graminearum* infections (Miller et al 1983). A field survey of the relationship of European corn borer damage and *Fusarium* did not find a correlation between the insect and *F. graminearum* (Lew et al 1991). Hence, for this mechanism to apply, there must be both insects and inoculum in soil debris.

The relationship of insect damage to *F. moniliforme* appears more certain. A field survey demonstrated that the incidence of the European corn borer increased *F. moniliforme* disease and fumonisin concentrations (Lew et al 1991). Disease incidence was also shown to relate to populations of thrips (*Frankliniella occidentalis*). The application of insecticides reduced thrip populations and disease incidence. Husk looseness at the "brown silk stage" was also demonstrated to reduce incidence of *F. moniliforme* (Farrar and Davis 1991). Other kinds of kernel damage increase disease caused by *F. moniliforme*. Hybrids with an increased propensity of kernel splitting have more disease (Odvody et al 1990).

In studies of experimental inoculation techniques, ear rot severities were related to wound size (Drepper and Renfro 1990). Inoculations with *F. moniliforme* appeared not to result in increased disease over natural levels (Rheeder et al 1990). Wound inoculation with *F. graminearum* resulted in high fumonisin concentrations. This suggested that disease-stress also promotes *F. moniliforme* infection (Schaafsma et al, *in press*). Since *F. moniliforme* is endemic in kernels, the sum of these findings supports the view that kernel damage per se promotes the disease (and fumonisin formation; Thiel et al 1991).

Agronomic Practice

The use of *Fusarium*-susceptible wheat cultivars and corn hybrids is mostly responsible for incidence of *F. graminearum*. High soil nitrogen was suggested to promote Fusarium head blight, but the data are variable (Ivashenko and Nazarovskaya 1990, Levitin et al 1990, Teich 1989). In a controlled study, supplementary nitrogen as ammonia increased infection over control plots. A plant growth regulator was demonstrated to increase Fusarium head blight (Martin et al 1991). One report suggested that the increased use of dwarf varieties and growth retardants played a role by putting the heads closer to the soil (Levitin et al 1990). However, the use of growth regulators and added nitrogen have a variety of physiological effects. It seems most likely that increased disease incidence is due to stress on the plant (Martin et al 1991).

Soilborne *F. graminearum* inoculum has been found to increase in fields in continuous wheat crops, but this does not affect ear infections (Fernandez and Fernandes 1990, Sturz and Johnston 1983, Teich and Nelson 1984, Wilcoxson et al 1989). Treatment of seed with fungicides has little impact on Fusarium head blight, suggesting infected seed is not an important source of inoculum (Martin and Johnston 1982, Mihuta-Grimm and Forster 1989, Teich and Hamilton 1985, Wong et al 1992).

Foliar fungicides have been shown to have little effect on Fusarium head blight incidence (Martin et al 1991). Indeed, there is evidence that application of some fungicides could have an undesirable effect on deoxynivalenol production. In a trial involving the application of foliar fungicides to wheat, symptoms were reduced compared with controls but deoxynivalenol content was not (Martin and Johnston 1982). In laboratory experiments, certain concentrations of the fungicide tridemorph were demonstrated to result in greater production of T-2 toxin per unit dry weight by *F. sporotrichioides* (Moss and Frank 1985). Barley grown in field plots was inoculated with *F. sporotrichioides* and treated with tridemorph. Some treatments resulted in greater toxin concentrations in the crop than control values (Moss and Frank 1986).

The fact that *G. zeae* infects both wheat and corn was recognized in 1898, which led to the suggestion that control could be affected by not planting wheat after corn (Teich 1989). Wheat following a corn crop was reported to have significantly more Fusarium head blight (Ivanshenko and Nazarovskaya 1990, Teich 1989). This may be because the fungus is able to persist on the pieces of debris that come from infected corn, which are larger than those of wheat straw (Teich 1989, Windels and Kommedahl 1984).

A variety of nonhost plants are sources of *F. graminearum* inoculum. These include sorghum, soybeans, wild oats, and various common weeds (Fernandez 1991, Ivashenko and Nazarovskaya 1990, Martin and Johnston 1982). A corn-soybean rotation yielded apparently less severe disease than corn following corn (Lipps and Deep 1991). High weed density was found to increase Fusarium head blight (Teich and Nelson 1984).

Epidemic Conditions

WHEAT

In areas where corn and wheat are both grown, epidemics of *F. graminearum* disease seldom occur in both crops in the same year. This reinforces the crucial role of the timing of precipitation. If there is sufficient moisture after anthesis, head blight–susceptible wheat cultivars will become infected with a minimum of inoculum (provided the temperature is appropriate for the growth of *F. graminearum*). The concept that minimal inoculum is required, is evidenced by data from experimental inoculations and agronomic studies. Data from studies of crop rotation do not demonstrate an overwhelming impact of inoculum density.

Beyond the very susceptible period at anthesis, most cultivars of wheat continue to be receptive to infection as long as the correct moisture and temperature conditions are present (Andersen 1948, Wilcoxson et al 1992). Epidemics where an appreciable percentage of the heads in a field are infected appear to require moisture for some time after anthesis. This allows the fungus to sporulate on the heads and infect adjacent plants and late-flowering tillers, which corresponds with data showing large concentrations of airborne conidia over wheat plots. In most years, however, the most important factor in determining severity of infection remains moisture in the susceptible period at anthesis (Clear and Patrick 1990). As noted by Sutton (1982), knowledge of the temporal relationships of ascospore and conidia production on overwintered debris and weeds would be valuable (Ivashenko and Nazarovskaya 1990).

CORN

F. graminearum

At silk emergence, initial infections in corn can come from at least two routes. Spores can land on the exposed part of the silk at silk emergence plus perhaps one week. Initial infections resulting from this mechanism require the presence of spores in the air. These could come from crop residues or from conidia or ascospores released from growth on weeds or grasses near the corn field. Sap beetles can carry spores down the silk channel without kernel injury. Sufficient moisture and adequate temperature are needed to establish colonization. This route therefore requires the congruence of three factors: airborne or insect-borne spores, inoculum at the correct time, and appropriate moisture and temperature.

A second route involves damage to the kernels plus a vector for the fungal inoculum. The kernels can be infected directly by debris-feeding picnic beetles carrying spores attracted to wounds made by tunneling insect species such as the European corn borer. Windblown spores could also infect such wounds. The timing of this is related to the population dynamics of the insect species involved (which vary from year to year). Inoculum must be present in the debris or air. Broadly speaking, moisture would not be crucial for initial infection by this route. Adequate moisture for plant growth is adequate for fungal growth (Miller et al 1983). Infection of the plant from bird damage can arise only after kernels are formed (perhaps two weeks after silk emergence).

Perhaps more so than with wheat, there is a second set of factors that governs the occurrence of an epidemic in corn. Data collected on the response of 19 hybrids to silk inoculation and inoculation by placing spores in a mechanically damaged portion of the ear are shown in Figure 2.1. Silk inoculations were done at silk emergence, which, in principle, approximates the first general mechanism described above. The second method is similar to vectoring via bird or insect damage. These results indicated only the most general relationship between the two methods with respect to tolerance of *F. graminearum*. The ranking of the hybrids evaluated by the two methods was different (Hough 1991).

Data for the same hybrids collected during severe Gibberella ear rot epidemics that occurred in Ontario in 1986 and 1987 are also shown in Figure 2.1 (adapted from Hough 1991). Somewhat more disease was seen

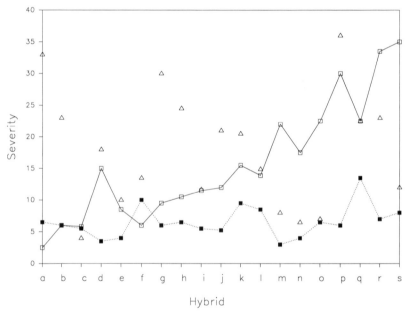

Fig. 2.1. Reaction of various corn hybrids to inoculation with *Fusarium graminearum* in the silk channel (percent severity, □) (Reid et al 1992a), or inoculation by placing spores in a wound made by a screwdriver (percent severity, ■), and average performance of the hybrids in the 1986 and 1987 Gibberella ear rot epidemics (severity index/4, △) (Hough 1991). **a,** Hyland LG2350; **b,** Pickseed 4555; **c,** Funks G-4010; **d,** Northrup King 9214; **e,** Funks 4106; **f,** Funks 4027; **g,** Cargill SX 123; **h,** Speares SO 109; **i,** Cargill 839; **j,** First Line 1656; **k,** Pioneer 3704; **l,** Hyland HL 2280; **m,** Dekalb DK524; **n,** Pioneer 3790; **o,** Pickseed 8898; **p,** Pioneer 3737; **q,** Pride K1184; **r,** Hyland HL2803; **s,** Pride K5574

in 1986. The correlation between the experimental inoculations and the survey are poor with the exception of a few hybrids. At the "tolerant" end of the scale, the best correlation existed for the fairly resistant hybrids Funks 4010 and Funks 4106, which share one inbred. The data from a number of more susceptible hybrids converge. These hybrids are quite resistant to wound inoculations as done over many years in many locations (Miles, *unpublished data*; Miller, *unpublished data*).

Another point of note was the response of Pioneer 3790, which along with the above two hybrids had the least infection in both epidemic years across the survey area (Hough 1991). For this hybrid, silk channel inoculation indicated greater susceptibility than evidenced from wound inoculation or the natural epidemics. For Pioneer 3790, analysis of natural and experimental infections suggests that *F. graminearum* can be present on the kernels without further growth (Schaafsma et al, *in press*).

Weather data for southern Ontario in 1986 and 1987 indicate that July temperatures were 7% warmer and rainfall was 25% higher than the 30-year averages. The most striking difference between the two years and the 30-year averages was rainfall in August and September. In 1986, there was 100% more rainfall than normal in August and September, and in 1987, rainfall was 38% higher than normal. Temperatures were average for those months and fell in the range for ascospore production (Clear and Patrick 1990, Sutton 1982).

Data from the 1986 and 1987 Gibberella ear mold epidemics suggest the importance of resistance to growth of *F. graminearum* after initial infection has taken place. Regardless of how the fungus finds its way into the plant (airborne spores or vectoring by insects or birds), weather conditions that promote the spread and reinfection of adjacent plants late in the summer are necessary for an epidemic (Sutton 1982, Tuite et al 1974).

Weather conditions in the early part of the season play a role in determining the inoculum pressure. That is, conditions suitable for ascospore production from debris must occur such that receptive plants can be infected. There are no data that allow a critical evaluation of the relative importance of airborne spores in relation to the initial infection process (i.e., at or near silk emergence) versus spores vectored by birds or insects then or later in the year. In years when August and September rainfall was average, zearalenone concentration was found to be correlated with the presence of damaged ears (Sutton et al 1980). This suggests that insect and bird damage is an important predisposing factor to ear rot. However, it is clear that severe epidemics require high rainfall and normal to above normal temperatures to allow reinfection from rain-splashed, airborne, and/or insect-vectored propagules (or propagules that land on the husks or silks and become dormant). These late-season events emphasize the prime importance of resistance to hyphal spread within the cob.

Sutton (1982) ended his review on *F. graminearum* disease as follows.

> From the point of view of disease management, it would be useful to know much more of the spatial relationships of inoculum sources and suscepts. Distances that the pathogen may spread within the crop and cross infection among fields of various susceptible crops are particularly pertinent.

This still remains true, 12 years later.

F. sporotrichioides is associated with extended periods of wet and cold weather and rarely causes significant kernel rot (Chelkowski 1989b, Marasas et al 1984). The principal toxin of this species, T-2, rarely occurs in corn in significant concentrations (Scott 1989). However, there have been a few occurrences of T-2 toxicosis in the United States (Al-Heeti 1987).

F. moniliforme

The weight of evidence is that *F. moniliforme* is endemic in corn kernels. Incidence of Fusarium kernel rot is higher in warmer climates under dry conditions. In such environments, insect damage is well recognized as a collateral factor. Regardless of moisture stress, insects appear to promote *F. moniliforme* occurrence. Even low concentrations of fumonisin are likely to be a problem for human and animal health (Bacon and Williamson 1992, Thiel et al 1991). More research on the factors that control this disease is urgently needed.

Literature Cited

Abbas, H. K., Mirocha, C. J., Pokorney, J. D., Gould, S. L., and Kommedahl, T. 1988. Mycotoxins of *Fusarium* spp. associated with infected ears of corn in Minnesota. Appl. Environ. Microbiol. 54:1039-1933.

Abramson, D., Clear, R. M., and Nowicki, T. W. 1987. *Fusarium* species and trichothecene mycotoxins in suspect samples of 1985 Manitoba wheat. Can. J. Plant Sci. 67:611-619.

Al-Heeti, A. A. 1987. Pathological, toxicological and biological evaluations of *Fusarium* species associated with ear rot of maize. Ph.D. thesis, University of Wisconsin-Madison. Univ. Microfilms Int. Diss. Inf. Serv. 8727220.

Anderson, A. L. 1948. The development of *Gibberella zeae* head blight of wheat. Phytopathology 38:599-611.

Attwater, W. A., and Busch, L. V. 1983. Role of the sap beetle *Glischrochilus quadrisignatus* in the epidemiology of Gibberella ear rot. Can. J. Plant Pathol. 5:158-163.

Ayers, J. E., Pennypacker, S. P., Nelson, P. E., and Pennypacker, B. W. 1975. Environmental factors associated with airborne ascospores of *Gibberella zeae* in corn and wheat fields. Phytopathology 65:835.

Bacon, C. W., and Williamson, J. W. 1992. Interactions of *Fusarium moniliforme*, its metabolites and bacteria with corn. Mycopathologia 117:65-71.

Chelkowski, J. 1989a. Formation of mycotoxins produced by *Fusarium* in heads of wheat, triticale and rye. Pages 63-84 in: *Fusarium* Mycotoxins, Taxonomy and Pathogenicity. J. Chelkowski, ed. Elsevier, Amsterdam.

Chelkowski, J. 1989b. Mycotoxins associated with corn cob fusariosis. Pages 53-62 in: *Fusarium* Mycotoxins, Taxonomy and Pathogenicity. J. Chelkowski, ed. Elsevier, Amsterdam.

Christensen, J. J., and Schneider, C. L. 1950. European corn borer (*Pyrausta nubilalis*) in relation to shank, stalk and ear rots of corn. Phytopathology 40:284-191.

Clear, R. M., and Patrick, S. K. 1990. *Fusarium* species isolated from wheat samples containing tombstone (scab) kernels from Ontario, Manitoba and Saskatchewan. Can. J. Plant Sci. 70:1057-1069.

Cook, R. J. 1981. *Fusarium* diseases of wheat and other small grains in North America. Pages 39-52 in: *Fusarium* Diseases, Biology and taxonomy. P. E. Nelson, T. A. Toussoun, and R. J. Cook, eds. Pennsylvania State University Press, University Park.

De Leon, C., and Pandey, S. 1989. Improvement of resistance to ear and stalk rots and agronomic traits in tropical maize gene pools. Crop Sci. 29:12-17.

Drepper, W. J., and Renfro, B. L. 1990. Comparison of methods for inoculation of ears and stalks of maize with *Fusarium moniliforme*. Plant Dis. 74:952-956.

Duthie, J. A., Hall, R., and Asselin, A. V. 1986. *Fusarium* species from seed of winter wheat in eastern Canada. Can. J. Plant Pathol. 8:282-288.

Farrar, J. J., and Davis, R. M. 1991. Relationships among ear morphology, western flower thrips, and Fusarium ear rot of corn. Phytopathology 81:661-666.

Fernandez, M. R. 1991. Recovery of *Cochliobolus sativus* and *Fusarium graminearium* from living and dead wheat and nongramineous winter crops in southern Brazil. Can. J. Bot. 69:1900-1906.

Fernandez, M. R., and Fernandes, J. M. C. 1990. Survival of wheat pathogens in wheat and soybean residues under conservation tillage systems in southern and central Brazil. Can. J. Plant Pathol. 12:289-294.

Foley, D. C. 1962. Systemic infection of corn by *Fusarium moniliforme*. Phytopathology 68:1331-1335.

Gendloff, E. H., Rossman, E. C., Casale, W. L., Islieb, T., and Hart, L. P. 1986. Components of resistance to Fusarium ear rot in field corn. Phytopathology 76:684-688.

Hart, L. P., Pestka, J. J., and Liu, M. T. 1984. Effect of kernel development and wet periods on production of deoxynivalenol in wheat infected with *Gibberella zeae*. Phytopathology 74:1415-1418.

Hough, K. D. 1991. Improved evaluation of *Fusarium graminearium* ear mold tolerance and assessment of differential mycotoxin accumulation in corn (*Zea mays* L.) hybrids. M.Sc. thesis, University of Guelph, Guelph, Ontario, Canada.

Ivashenko, V. G., and Nazarovskaya, L. A. 1990. Characteristics of the ascomycetous stage of the Fusarium head blight pathogen of different crops in Krasnodar Krai. Rep. All-Union Acad. Agric. Sci. 12:11-14.

Khonga, E. B., and Sutton, J. C. 1986. Survival and inoculum production by *Gibberella zeae* in wheat and corn residues. Can. J. Plant Pathol. 8:351.

King, S. B., and Scott, G. E. 1981. Genotypic differences in maize to kernel infection by *Fusarium moniliforme*. Phytopathology 71:1245-1247.

Koehler, B. 1959. Corn ear rots in Illinois. Ill. Agric. Exp. Stn. Bull. 639.

Lamprecht, S. C., Marasas, W. F. O., Knox-Davis, P. S., and Calitz, F. J. 1990. Cross-pathogenicity of *Fusarium avenaceum* and *F. graminearium* gr. 1 to *Medicago truncatula* and wheat. Phytophylactia 22:209-211.

Lew, H., Adler, A., and Edinger, W. 1991. Moniliformin and the European corn borer (*Ostrinia nubilalis*). Mycotoxin Res. 7:71-76.

Levitin, M. M., Ivashchenco, V. G., and Shipilova, N. P. 1990. Fusarium ear blight of wheat. Fuzarioz Kolosa Pshenitsy 24:446-453.

Lipps, P. E., and Deep, I. W. 1991. Influence of tillage and crop rotation on yield, stalk rot and recovery of *Fusarium* and *Trichoderma* spp. from corn. Plant Dis. 75:828-833.

Magan, N., and Lacey, J. 1984. Water relations of some *Fusarium* species from infected wheat ears and grain. Trans. Br. Mycol. Soc. 83:281-285.

Marasas, W. F. O., Nelson, P. E., and Toussoun, T. A. 1984. Toxigenic *Fusarium* species: Identity and mycotoxicology. Pennsylvania State University Press, University Park.

Martin, R. A. 1988. Use of a high-through-put jet sampler for monitoring viable airborne propagules of *Fusarium* in wheat. Can. J. Plant Pathol. 10:359-360.

Martin, R. A., and Johnston, H. W. 1982. Effects and control of Fusarium disease of cereal grains in the Atlantic provinces. Can. J. Plant Pathol. 4:210-216.

Martin, R. A., MacLeod, J. A., and Caldwell, C. 1991. Influences of production inputs on incidence of infection by *Fusarium* species on cereal seed. Plant Dis. 75:784-788.

Mesterhazy, A. 1982. Resistance of corn to Fusarium ear rot and its relation to seedling resistance. Phytopathol. Z. 103:218-231.

Mesterhazy, A. 1984. A laboratory method to predict pathogenicity of *Fusarium graminearum* in field and resistance of wheat to scab. Acta Phytopathol. Acad. Sci. Hung. 19:205-218.

Mesterhazy, A., and Rowaished, A. K. 1977. Analysis of symptoms caused by *Fusarium graminearum* and its relation to powdery mildew infection in wheat. Acta Phytopathol. Acad. Sci. Hung. 12:289-301.

Mihuta-Grimm, and Forster, R. L. 1989. Scab of wheat and barley in southern Idaho and evaluation of seed treatments for eradication of *Fusarium* spp. Plant Dis. 73:769-771.

Miller, J. D., Young, J. C., and Trenholm, H. L. 1983. *Fusarium* toxins in field corn I:. Time course of fungal growth and production of deoxynivalenol and other mycotoxins. Can. J. Bot. 61:3080-3087.

Miller, J. D., Young, J. C., and Sampson, D. R. 1985. Deoxynivalenol and Fusarium head blight resistance in spring cereals. Phytopathol. Z. 113:359-367.

Moss, M. O., and Frank, J. M. 1985. Influence of the fungicide tridemorph on T-2 toxin production by *Fusarium sporotrichioides*. Trans. Br. Mycol. Soc. 84:585-590.

Moss, M. O., and Frank, J. M. 1986. Production of trichothecenes in winter barley inoculated with *Fusarium sporotrichioides*. Proc. World Congr. Foodborne Infections and Intoxications 2nd, vol. 2. Berlin.

Naik, D. M., and Busch, L. V. 1978. Stimulation of *Fusarium graminearum* by maize pollen. Can. J. Bot. 56:1113-1117.

Neish, G. A., Farnworth, E. R., Greenhalgh, R., and Young, J. C. 1983. Observations on the occurrence of *Fusarium* species and their toxins in corn in eastern Ontario. Can. J. Plant Pathol. 5:11-16.

Nelson, P. E., Toussoun, T. A., and Marasas, W. F. O. 1983. *Fusarium* Species: An Illustrated Manual for Identification. Pennsylvania State University Press, University Park.

Ochor, T. E., Trevathan, L. E., and King, S. B. 1987. Relationship of harvest date and host genotype to infection of maize kernels by *Fusarium moniliforme*. Plant Dis. 71:311-313.

Odvody, G. N., Remmers, J. C., and Spencer, N. M. 1990. Association of kernel splitting with kernel and ear rots of corn in a commercial hybrid grown in the coastal bend of Texas. Phytopathology 80:1045.

Reid, L. M., Bolton, A. T., Hamilton, R. I., Woldemariam, T., and Mather, D. E. 1992a. Effect of silk age on resistance of maize to *Fusarium graminearium* Can. J. Plant Pathol. 14:293-298.

Reid, L. M., Mather, D. E., Arnason, J. T., Hamilton, R. I., and Bolton, A. T. 1992b. Changes of host phenolics of maize silks infected with *Fusarium graminearium*. Can. J. Bot. 70:1697-1702.

Reid, L. M., Mather, D. E., Hamilton, R. I., and Bolton, A. T. 1992c. Genotypic differences in the resistance of maize silk to *Fusarium graminearium*. Can. J. Plant Pathol. 14:211-214.

Rheeder, J. P., Marasas, W. F. O., van Wyk, P. S., and van Schalkwyk, D. J. 1990. Reaction of South African maize cultivars to ear inoculation with *Fusarium moniliforme, F. graminearum* and *Diplodia maydis* Phytophylactica 22:213-218.

Saur, L. 1991. Recherche de geniteurs de resistance a la fusariose de l'epi causee par *Fusarium culmorum* chez le ble et les especes voisines. Agronomie 11:535-541.

Schaafsma, A. W., Miller, J. D., Savard, M. E., and Ewing, R. 1993. Ear rot development and mycotoxin production in corn in relation to inoculation method and corn hybrid for three species of *Fusarium*. Can. J. Plant Pathol. (In press).

Scott, D. B., De Jager, E. J. H., and van Wyk, P. S. 1988. Head blight of irrigated wheat in South Africa. Phytophylactica 20:317-319.

Scott, P. M. 1989. The natural occurrence of trichothecenes. Pages 2-26 in Trichothecene Toxicosis: Pathophysiological Effects, vol. 1. V. R. Beasley, ed. CRC Press, Boca Raton, FL.

Shurtleff, M. C. 1980. Compendium of Corn Diseases. American Phytopathological Society, St. Paul, MN.

Snijders, C. H. A. 1993. Breeding for resistance to *Fusarium* in wheat and maize. Pages 37-58 in: Mycotoxins in Grain: Compounds Other than Aflatoxin. J. D. Miller and H. L. Trenholm, eds. Eagan Press, St. Paul, MN.

Snijders, C. H. A., and Perkowski, J. 1990. Effects of head blight caused by *Fusarium culmorum* on toxin production and weight of wheat kernels. Phytopathology 80:566-570.

Stack, R. W., and McMullen, M. 1985. Head blighting potential of *Fusarium* species associated with spring wheat head. Can. J. Plant Pathol. 7:79-82.

Sturz, A. V., and Johnston, H. W. 1983. Early colonization of the ears of wheat and barley by *Fusarium poae*. Can. J. Plant Pathol. 5:107-110.

Sung, J. M., and Cook, R. J. 1981. Effect of water potential on reproduction and spore germination by *Fusarium roseum* 'graminearum,' 'culmorum' and 'avenaceum'. Phytopathology 71:499-504.

Sutton, J. C. 1982. Epidemiology of wheat head blight and maize ear rot caused by *Fusarium graminearium*. Can. J. Plant Pathol. 4:195-209.

Sutton, J. C., Baliko, W., and Liu, H. J. 1980. Fungal colonization and zearalenone accumulation in maize ears injured by birds. Can. J. Plant Sci. 60:453-461.

Thiel, P. G., Marasas, W. F. O., Sydenham, E. W., Shephard, G. S., Gelderblom, W. C. A., and Nieuwenhuis, J. J. 1991. Survey of fumonisin production by *Fusarium* species. Appl. Environ. Microbiol. 57:1089-1093.

Teich, A. H. 1989. Epidemiology of wheat (*Triticum aestivum* L.) scab caused by *Fusarium* spp. Pages 269-282 in: *Fusarium* Mycotoxins, Taxonomy and Pathogenicity. J. Chelkowski, ed. Elsevier, Amsterdam.

Teich, A. H., and Nelson, K. 1984. Survey of Fusarium head blight and possible effects of cultural practices in wheat fields in Lambton County in 1983. Can. Plant Dis. Surv. 64:11-13.

Teich, A. H., and Hamilton, J. R. 1985. Effect of cultural practices, soil phosphorus, potassium and pH on the incidence of Fusarium head blight and deoxynivalenol levels in wheat. Appl. Environ. Microbiol. 49:1429-1431.

Tschanz, A. T., Horst, R. K., and Nelson, P. E. 1975. Ecological aspects of ascospore discharge in *Gibberella zeae*. Phytopathology 65:597-599.

Tuite, J. G., Shaner, G., Rambo, G., Foster, J., and Caldwell, R. W. 1974. The Gibberella ear rot epidemics in corn in Indiana in 1965 and 192. Cereal Sci. Today 19:238-241.

Ullstrup, A. J. 1970. Methods for inoculating ears with *Gibberella zeae* and *Diplodia maydis*. Plant Dis. Rep. 54:658-662.

Van Egmond, H. P. 1989. Current situation on regulations for mycotoxins. Overview of tolerances and status of standard methods of sampling and analysis. Food Addit. Contam. 6:139-188.

Visconti, A., Mirocha, C. J., Logrieco, A., Bottakico, A., and Solfrizzo, M. 1989. Mycotoxins produced by *Fusarium acuminatum:* Isolation and characterization of acuminatin, a new trichothecene. J. Agric. Food Chem. 37:1348-1351.

Warner, G. M., and French, D. W. 1970. Dissemination of fungi by migratory birds: Survival and recovery of fungi from birds. Can. J. Bot. 48:907-910.

Wang, Y. Z., and Miller, J. D. 1988. Screening techniques and sources of resistance to Fusarium head blight. Pages 239-250 in: Wheat production: Constraints in tropical environments. A. R. Khlatt, ed. Centro Internacional de Mejoramiento de Maiz y Trigo (CIMMYT), Mexico City.

Whitney, N. J. 1954. Ear rot in hybrid corn in Essex county, Ontario, in relation to damage by birds. Plant Dis. Rep. 38:384-387.

Wicklow, D. T., and Caldwell, R. W. 1990. Noninfection of maize ears inoculated with *Gibberella zeae* ascospores or macroconidia in a controlled environment. Trans. Mycol. Soc. Jpn. 31:29-34.

Wilcoxson, R. D., Kommedahl, T., Ozmon, E. A., and Windels, C. A. 1988. Occurrence of *Fusarium* species from Minnesota and their pathogenicity to wheat. Phytopathology 78:586-589.

Wilcoxson, R. D., Ozmon, E. A., and Pierce, A. R. 1989. Effect of necrotic spikelets and location of plots on infection of wheat cultivars by *Fusarium graminearium*. Int. J. Trop. Plant Dis. 7:55-60.

Wilcoxson, R. D., Busch, R. H., and Ozman, E. A. 1992. Fusarium head blight resistance in spring wheat cultivars. Plant Dis. 76:658-661.

Windels, C. E., and Kommedahl, T. 1984. Late-season colonization and survival of *Fusarium graminearium* group II in corn stalks in Minnesota. Plant Dis. 68:791-793.

Windels, C. E., Windels, M. B., and Kommedahl, T. 1976. Association of *Fusarium* species with picnic beetles on corn ears. Phytopathology 66:328-331.

Wong, L. S. L., Tekauz, A., Leisle, D., Abramson, D., and McKenzie, R. I. H. 1992. Prevalence, distribution and importance of Fusarium head blight in Manitoba. Can. J. Plant Pathol. 14:233-238.

Ylimaki, A. 1981. The mycoflora of cereal seeds and some feedstuffs. Ann. Agric. Fenn. 20:74-88.

Chapter 3

Breeding for Resistance to *Fusarium* in Wheat and Maize

C. H. A. Snijders

Introduction

EPIDEMIOLOGY

Fusarium head blight, a fungal disease of wheat (*Triticum aestivum* L.), is found in both temperate and semitropical regions. A number of species of *Fusarium* may be responsible, but generally *F. graminearum* Schwabe, with perfect state *Gibberella zeae* (Schwein), and *F. culmorum* (Wm.G.Sm.) Sacc., perfect state unknown, predominate as the causal factor for Fusarium head blight (Mesterhazy 1977, 1978; Stack and McMullen 1985; Wilcoxson et al 1988). Both *F. graminearum* and *F. culmorum* cause root rot, crown rot, foot rot, stem rot, and head blight in wheat. *Fusarium* in maize causes seedling blight and root, stalk, and ear rots. The most frequently isolated species are *F. graminearum* and *F. moniliforme* J. Sheld., followed by *F. culmorum, F. subglutinans* (Wollenweb. & Reinking) P.E. Nels., T.A. Toussoun, & Marasas, *F. equiseti* (Corda) Sacc. sensu Gordon and *F. proliferatum* (T. Matsushima) Nirenberg (Abbas et al 1988, Bottalico et al 1989, Cook 1978, Gilbertson 1985, Krüger 1985, Logrieco and Bottalico 1986, Mesterhazy 1982, Rheeder et al 1990a, Windels et al 1988). Species frequencies are influenced by geography, climate, and year. Bottalico et al (1989) observed that the presence of *Fusarium* species in infected maize ears (peduncles, cobs, and kernels) is almost the same as that found in the stalks, although the frequency of the species can differ in different plant growth stages.

Heavy *Fusarium* damage to wheat or maize often occurs in areas of continuous monoculture practice or where wheat is planted after maize or vice versa. The pathogen originates from seed, soil, and from inoculum on host residues, including those of small-grain cereals and certain grasses (Khonga and Sutton 1988). Different practices for minimizing initial inoculum, notably rotation with nonhost crops, plowing of infested residues, and low nitrogen inputs, have long been recommended for managing *Fusarium* in wheat and maize (Khonga and Sutton 1988, Martin et al 1991). There are no practical chemical treatments that are effective in preventing this infection. The only way to eliminate this problem is to introduce cultivars that are resistant to *Fusarium*.

PHYTOTOXINS

F. graminearum and *F. culmorum* are the *Fusarium* species with the highest pathogenicity (the ability to cause disease) and with the highest aggressiveness (the amount of disease caused). *F. graminearum* isolates are significantly more aggressive on corn stalks than *F. subglutinans* and *F. moniliforme* (Gilbertson 1985). Moreover, *F. graminearum* and *F. culmorum* are the most important producers of the phytotoxic nonmacro-cyclic trichothecenes deoxynivalenol (DON) and 3-acetyldeoxynivalenol (3-ADON) (Marasas et al 1984). These trichothecenes have been shown to inhibit eukaryotic protein synthesis by blocking the peptidyl transferase step (Carter et al 1980). Growth of wheat coleoptile tissue was strongly to fully inhibited at $10^{-3}M$ concentrations of DON and 3-ADON (Bruins et al, *in press*; Wang and Miller 1988), and wheat seedling growth was completely inhibited at a concentration of $10^{-4}M$ DON (Bruins et al, *in press*; Shimada and Otani 1990; Snijders 1988). A phytotoxic effect of DON during development of the seed is considered probable from comparisons of the data on phytotoxicity with the high concentrations of DON found in wheat kernels and maize tissue (Atlin et al 1983, Bennett et al 1988, Bottalico et al 1989, Snijders and Perkowski 1990, Snijders and Krechting 1992). DON produced with head blight in wheat or stalk and ear rot in maize can also have physiological effects in other parts of the plant. DON is water soluble, and translocation in the phloem as a bulk flow of solution is assumed to occur. DON was found in tissues of infected corn plants that were not invaded by *F. graminearum* (Miller et al 1983, Young and Miller 1985). Similar observations were made for *F. culmorum* in wheat (Snijders and Krechting 1992). *F. moniliforme* does not produce DON or related trichothecenes. It produces the water-soluble moniliformin and fumonisins (Gelderblom et al 1988, Marasas et al 1984). Only recently information on phytotoxicity of these toxins became available. Fumonisin B_1 (FB$_1$) reduced the growth of duckweed by 50% at a concentration of 0.9 μM. A 100-fold higher concentration of moniliformin suppressed the growth to only 16% (Vesonder et al 1992). Corn callus growth was inhibited by 50% at an FB$_1$ level aof 13 μM, while the same effect was obtained for moniliformin at 102 μM (Van Asch et al 1992). FB$_1$ concentrations of 4.1–1,388 μM applied to jimsonweed caused electrolyte leakage, autolysis, and photobleaching directly proportional to the concentration and duration of exposure. FB_1 levels of 0.1–100 μM caused leaf necrosis and dose-dependent reductions in shoot and root length of tomato and corn plants (Lamprecht et al 1993).

Variation in aggressiveness towards wheat heads and maize ears among isolates of *F. graminearum* and *F. culmorum* is large (Mesterhazy 1978, 1982, 1984). DON and 3-ADON are nonspecific toxins and are regarded as aggressiveness factors, i.e., they increase the extent of disease symptoms and colonization but are not involved in the primary interaction that determines the basic compatibility between host and pathogen (Al-Heeti 1987; Desjardins 1992; Manka et al 1985, 1989; Snijders and Krechting 1992). DON is produced in the cells immediately adjacent to the hyphal tip (Miller and Greenhalgh 1988). An active defense mechanism of the host in reaction to hyphal invasion after penetration requires protein synthesis (e.g., for the production of enzymes). Low concentrations of trichothecenes such as DON prevent protein synthesis, and as a result further colonization is not inhibited. Plants resistant to these toxins will not show complete resistance but an increased resistance to colonization.

So far, no data are available on the possible involvement of fumonisins in pathogenesis.

PHYSIOLOGICAL SPECIALIZATION

Significant interactions between strains of *F. graminearum* or *F. culmorum* and wheat genotypes have been reported (Mesterhazy 1984, 1988; Snijders and Van Eeuwijk 1991). However, interaction patterns were not stable over experiments, and genotype ranking was only slightly influenced by the strains. No evidence has been found for the occurrence of races of *F. culmorum* or *F. graminearum* adapted to different wheat genotypes. Snijders and Van Eeuwijk (1991) applied a statistical model for the description of genotype × strain interaction over years that provided a means to check whether the environmental "lability" of interaction in the aforementioned studies was really part of the pattern in the data, and hence merited agricultural interpretation. This study concluded that it was merely noise. Snijders and Van Eeuwijk (1991) described the Fusarium head blight resistance in wheat as horizontal resistance (in terms of Vanderplank). Small interactions between corn hybrids or inbreds with *F. graminearum* strains were observed for ear infection, but no important rank reversals resulted from these interactions (Al-Heeti 1987, Atlin et al 1983, Gendloff et al 1986, Mesterhazy and Kovacs 1986). The absence of such an interaction indicates that in maize, as in wheat, there is no evidence for strain-specific or race-specific resistance. Any reasonably aggressive strain should be satisfactory for screening purposes (Al-Heeti 1987, Snijders and Van Eeuwijk

1991). However, as aggressiveness of strains also depends on the environment, it is advisable to work with several pure strains to decrease the chance of an aggressiveness level too low or too high for differentiation and to neutralize a possible statistical interaction (Mesterhazy 1982, Mesterhazy and Kovacs 1986, Snijders and Van Eeuwijk 1991).

Resistance in Wheat

DISEASE AND DAMAGE

A spikelet penetrated by *F. culmorum* shows light brown, water-soaked spots on the glumes. Soon, infected spikelets lose the water-soaked appearance and take the color of the ripe heads. From the inoculated spikelet a necrosis spreads to adjacent spikelets. Several forms of damage can be distinguished. An infection of the head by *Fusarium* can reduce kernel set and kernel weight, causing a yield reduction (Snijders and Perkowski 1990). Invasion of the kernel by *Fusarium* destroys the starch granules and cell walls, resulting in a poor quality product (Bechtel et al 1985). Study of the effects of *F. culmorum* infection on endosperm storage proteins (high molecular weight glutenins) of a susceptible wheat line in the period three to eight weeks after flowering and inoculation showed that from the fifth week on, the proteins disappeared. This could occur as a result of consumption by the fungus and inhibition of new protein production by highly accumulated DON (Snijders and Krechting 1992; Snijders and Kolster, *unpublished*). The embryo is usually not infected except in heavily invaded kernels. However, slightly infected kernels with apparently uninfected embryos exhibit reduced germination and vigor (Bechtel et al 1985). Quality loss because of mycotoxin content will be discussed in other chapters.

EXPERIMENTAL INOCULATION

Since wheat is most susceptible to Fusarium head blight at anthesis (Schroeder and Christensen 1963), experimental inoculations are made at flowering time. A spore suspension is sprayed on the heads from 0.25 m above the crop. To ensure a high relative humidity during the nights after inoculation, the field may be sprinkled each day in the evening (Snijders and Perkowski 1990), or a bunch of heads may be covered by polyethylene bags for 24 h (Mesterhazy 1978). Head blight ratings are determined as the product of the percentage of heads infected and the proportion of infected (bleached) spikelets (Snijders and Perkowski 1990). Mesterhazy (1977, 1984) used yield reduction in resistance assessment. However, other studies showed that heritability for yield reduction is smaller than for visually assessed Fusarium head blight (Snijders 1990e). Costly ergosterol analyses, indicating the amount of fungal biomass, or DON analyses give a very accurate estimate of the resistance level (Snijders and Perkowski 1990; Snijders and Krechting 1992). Ergosterol analyses indicated that an estimate for infection level from a field plot gives a satisfactory estimate of the amount of fungal biomass and can therefore be used for selection (Snijders and Krechting 1992). In one Chinese study inoculum was dripped onto a spikelet, and the resistance to spreading was assessed by the number of spikelets infected at the wax stage (Zhang and Pan 1982). Gu (1983) inoculated by scattering diseased, sporulating wheat grains in the field 15 days and five days before heading, leading to head inoculation as a result of wind and water splash.

GENETIC VARIATION

Genetic variation with respect to resistance to head blight has been reported for winter wheat, and for spring wheat from both temperate and (sub-)tropical areas (Mesterhazy 1987; Schroeder and Christensen 1963; Snijders 1990e, 1992; Wang and Miller 1988). To what extent resistance to *F. culmorum* is related to resistance to *F. graminearum* is not clear. Mesterhazy (1983, 1987, 1988) found correlation coefficients of up to 0.90 between the reaction of wheat genotypes to *F. culmorum* isolates and the reaction to *F. graminearum* isolates. Spring wheat genotypes that had been reported to be resistant to head blight caused by *F. graminearum* were also resistant to *F. culmorum* (Snijders 1990e). Snijders (1990e) divided resistant germ plasm into three gene pools: winter wheats from Eastern Europe (Praag 8, Novokrumka 0102), spring wheats from China and Japan (Ning 8343, Wuhan #1, Nobeokabozu komugi), and spring wheats from Brazil (FT 83-326, Frontana).

COMPONENTS OF RESISTANCE

Fusarium can be isolated as endophyte from roots, culms, leaves, glumes, and kernels of wheat, and infection can remain latent during some plant growth stages (Burgess et al 1987, Cook 1968, Purss 1971, Sieber et al 1988). Systemic fungal growth of *F. culmorum* in stem tissue of wheat was found in both mechanically injured and soil-inoculated plants, resulting in a high level of infection (Snijders 1990a). Crown and foot rot can therefore lead to infection of the upper stem internodes under conditions not suitable for *Fusarium* dispersal. However, no evidence was found for systemic fungal growth leading to infected heads. Resistance to Fusarium head blight can be characterized as consisting of two components: resistance to initial penetration (component 1) and resistance to spreading of the pathogen in host tissue (component 2) (Schroeder and Christensen 1963, Snijders 1990b). Existence of genetic variation for resistance to spreading of *F. culmorum* in wheat stems was demonstrated by Snijders (1990a). Resistance to systemic fungal growth belongs to resistance component 2. The Zhang and Pan (1982) inoculation method provides a means for evaluation of resistance component 2. The percentage of infected heads in a field plot describes component 1; the proportion of infected spikelets is a resultant of components 1 and 2 (Snijders 1990b).

Morphological characters can also influence resistance. Mesterhazy (1987) ascribed the higher susceptibility of the awned types to the fact that humidity, caused by dew and rain, would last longer between awns, providing longer optimal conditions for the pathogen.

GENETICS

The resistance to *Fusarium* in wheat is quantitative; complete resistance has not been discovered. In various studies in spring wheat, the pattern of inheritance of resistance to *F. graminearum* was dominant to over-dominant and was controlled by two to five genes (Gu 1983, Lin et al 1992, Yu 1982, Zhang and Pan 1982, Zhou et al 1987). Diallel analysis of 10 homozygous winter wheat genotypes representing different levels of resistance to Fusarium head blight showed that resistance to *F. culmorum* was uniformly transmitted to all offspring (Snijders 1990b). Resistance was governed by several minor genes. The individual genes had large effects on resistance and were inherited in a mainly additive fashion (Snijders 1990c). With respect to the nonadditive effects, dominance of resistance predominated. Transgression for resistance was observed in progenies of crosses between resistant parents (Snijders 1990d), showing that accumulation of resistance genes is possible.

Resistance in Maize

DISEASE AND DAMAGE

Stalk rot originates in the lower part of the stalk, extending both up along the stem and down into the proximal regions of the brace roots. Severe rot causes premature senescence of stalks, which turn light brown externally while the internal tissues become shredded and often develop a pink or red color. Lodging is related to stalk rot susceptibility and is particularly serious when harvesting is delayed (Burgess et al 1981, Cook 1978). Christensen and Wilcoxson (1966) give an extensive review on stalk rot characteristics.

F. graminearum causes a generalized ear rot, which starts at the tip; occasionally the rot originates at the base of the ear. This fungus causes a pronounced reddish discoloration of the rotted grain and husk tissues and produces a growth of pinkish white to red mycelium on the surface of colonized grain (Burgess et al 1981). *F. moniliforme* and *F. subglutinans* may cause a generalized ear rot, which usually is initiated through natural wounds (i.e., from insects or birds) at the tip of the ear and moves down the ear as a contiguous mold front. More commonly these two species cause rotting of individual grains or localized areas of grain. These fungi form conspicuous, pale pink to white mycelium between the grains and under the husk. They commonly infect the grain internally at the embryo end, usually without causing visible symptoms (asymptomatic infection) (Burgess et al 1981, King and Scott 1981). The kernels are most susceptible at about the blister stage of development, the period when they are filling rapidly but are still tender (Davis et al 1989). In California, ears are often completely consumed by *F. moniliforme*, leaving a light-weight ear with inner husks cemented to the kernels by mycelia (Farrar and Davis 1991). Ear rot infection is more severe in plants infected with maize dwarf mosaic virus, and when ears are damaged by hail, birds, or insects such as the European corn borer, corn earworm, and picnic beetle (Fisher et al 1983, Gulya et al 1980, Tosić et al 1977).

PHYSIOLOGY

The interaction of host, pathogen, and environment greatly influences corn stalk rots. An early season water deficit accelerated root senescence of maize plants and permanently increased the likelihood of chronic water stress, resulting in an earlier senescence and increased infection by *F. moniliforme* and its systemic colonization (Schneider and Pendery 1983). Rot generally begins at the lower internodes. Whether a plant will lodge is determined by strength dynamics, including weight and height of the ear, amount of stalk deterioration, rind strength and thickness, and wind force (Dodd 1980, Mesterhazy and Kovacs 1986, Mostafa et al 1990). As long as cells remain vigorous, most host genotypes have genetic components for resistance to the potential pathogens. Apparently synthesis of cellular resistance substances decreases with senescence (Dodd 1980). Pappelis et al (1975) recommended that inbreds should be selected for stalk rot resistance by using pith condition ratings. Cellular senescence is caused by carbohydrate deficiency (Dodd 1980). The imposition of stress, like high population densities and late defoliation, results in a reduction of total sugars in pith tissue of the stalk at physiological maturity, and causes an increase in the amount of stalk rot (Mortimore and Ward 1964). Abiotic factors that influence translocation of photosynthate stored as sugars within the stem and from the stem to the grain are suggested to influence resistance. In experiments, lower disease incidence and severity were obtained by inoculating the internode below the ear than by inoculating the first elongated internode from the base (Drepper and Renfro 1990), perhaps because there is a lower carbohydrate concentration in the lower internodes. When kernels were prevented from developing or when population densities were reduced, sugars in the pith were maintained or increased and resistance to stalk rot increased (Mortimore and Ward 1964). Resistant hybrids are able to tolerate more stress than susceptible hybrids before becoming diseased and generally have a higher sugar content in the lower stems than susceptible hybrids (Mortimore and Ward 1964). The whole ear/whole plant ratio was higher in susceptible genotypes during the last period of growth (Barrière 1985). Stalk rot of corn appears also to be affected by the plant's mineral nutritional status. Differential accumulation of phosphorus in corn populations selected for this characteristic may have a significant effect on stalk rot parameters; the largest amount of stalk rot after inoculation with *F. graminearum* and *F. moniliforme* and subsequent lodging occurred in low-phosphorus-accumulating populations (Porter et al 1981).

As stalk rot resistance is related to the carbohydrate concentration in the stalk and plant vigor, large improvements in stalk quality may lead to competition between the stalk and ear for photosynthate, which may reduce yield. Though some studies demonstrated this yield penalty for stalk rot resistance, Mostafa et al (1990) and Dolstra et al (*in press*) showed that improved *F. graminearum* stalk rot resistance by simple recurrent selection had no consistent effect on grain yield. Selection for Fusarium stalk rot should therefore lead to superior genotypes without sacrifice of agronomic potential.

EXPERIMENTAL INOCULATION

Historically, corn breeders have had no selectable criterion for Fusarium ear rot resistance. To select for resistance, breeders plant corn hybrids in areas with a known high incidence of Fusarium ear rot. Hybrids then are rated for ear rot at maturity, and a determination of relative resistance is made at that time (Davis et al 1989, Farrar and Davis 1991). Because the occurrence of the ear mold disease is sporadic, a reliable technique of producing artificial epidemics is preferable. Inoculation methods can be divided into two types: inoculation with mechanical injury (type 1) and inoculation without mechanical injury (type 2). Type 1 refers to the toothpick method and its derivatives as first described by Young (1943). About 10 days after midsilk date, ears are inoculated by inserting *Fusarium*-encrusted toothpicks, pellets, or nail punches through the husk perpendicular to the ear axis and midway between the butt and ear tip, where they remain until harvest (Drepper and Renfro 1990, Gulya et al 1980, Ullstrup 1970). Ear and stalk rot incidences were positively correlated with the diameter of the inoculation tool (Drepper and Renfro 1990). In type 2 methods, a spore suspension is sprayed onto the maize silks with an atomizer until runoff (Gulya et al 1980, Warren 1978) or injected into the silk tuft (Reid et al 1992b, Styer and Cantliffe 1984), or toothpicks colonized by *Fusarium* are placed in the silk channel near the cob tip (Enerson and Hunter 1980a,

Sutton and Baliko 1981). The silks remain uncovered (Enerson and Hunter 1980a, Gulya et al 1980, Sutton and Baliko 1981) or are covered immediately after inoculation with moist papers and then bagged (Ullstrup 1970, Warren 1978). Al-Heeti (1987) observed that ear rot development was much reduced when the inoculated ears were bagged. The best differential infections were obtained when inoculations were made four to 17 days after pollination (Ullstrup 1970, Warren 1978). Reid et al (1992a) reported that the best differentiation between resistant and susceptible genotypes was obtained when inoculation was conducted within the first week after silking. Sufficient moisture in the silk and kernel is necessary for the germination of conidia of *F. moniliforme* and for subsequent invasion of the kernel. After silks become dry, little or no infection occurs (Warren 1978). Gulya et al (1980) distinguished resistant inbred lines by measuring ear rot development (rotted area) on ears incubated in moisture chambers for three days. Rheeder et al (1990b) concluded that the percentage of diseased kernels and percentage of *Fusarium* isolations were the best assessment methods to differentiate for *F. moniliforme* and *F. graminearum* resistance between hybrids. However, the observed low variation for resistance among the hybrids makes the conclusions of little value. Ratings of ear rot caused by *F. moniliforme* were correlated with the number of asymptomatically infected seeds (Styer and Cantliffe 1984). Al-Heeti (1987) used visible ear rot at harvest to assess severity of ear rot after inoculation with *F. graminearum, F. sporotrichioides,* and *F. poae.* Also, Reid et al (1992b) used visible ear rot incidence and severity to carry out a diallel analysis of resistance in maize to *F. graminearum* inoculations. As in wheat, an accurate estimate of the resistance level can be obtained by ergosterol analyses, or by DON analyses for *F. graminearum* or *F. culmorum* infections.

The most desirable method for stalk inoculation is with a toothpick or injection needle at the second internode aboveground during the growth stage between tasseling and pollination. The amount and extent of rot can be observed by splitting the stalk open (Drepper and Renfro 1990, Lal and Singh 1984). However, several European studies have reported that artificially induced stalk rots do not correspond with the natural pathological symptoms (Bausch et al 1982).

Similar to *Fusarium* resistance in wheat, resistance to *Fusarium* in maize can be seen as consisting of two components: 1) resistance to initial penetration, and 2) resistance to spreading of the pathogen in host tissue (Hunter et al 1986, Snijders and Perkowski 1990). Inoculation method type 1 screens generally for resistance component 2 only. Wound-type inoculations simulate insect attack to some degree as they bypass morphological barriers (Drepper and Renfro 1990). Inoculation type 2 closely simulates natural infection of a nonwounded host plant. If the percentage of infected plants is observed, resistance component 1 is measured. If ear rot or percentage of infected kernels is observed, a resultant of resistance components 1 and 2 is observed (Snijders 1990b). Several authors have reported that inoculation type 1 was ineffective for differentiating genotypes for resistance to stalk and ear rots caused by *F. graminearum* and *F. moniliforme* (Tosić et al 1977, Ullstrup 1970). Gulya et al (1980) concluded that severity of ear rot caused by *F. moniliforme* was greater, and the resistant genotypes were more easily distinguished when inoculated by method type 1 than by type 2, but this conclusion was based on a very narrow genetic variation for resistance. Wounds were necessary to ensure invasion of a susceptible cultivar by the relatively nonaggressive *F. sporotrichioides* and *F. poae* (Al-Heeti 1987). Farrar and Davis (1991) reported that inoculations in conjunction with kernel or ear wounding are successful although of limited value because important natural barriers to infection may be compromised. However, it is noteworthy that this inoculation method screens for resistance to colonization (component 2). The fact that most studies have not succeeded in differentiating for resistance level results from a lack of genetic variation for this component in the tested populations. Ullstrup (1970) distinguished maize resistant to ear rot from *F. graminearum* from susceptible maize with inoculation type 2. Enerson and Hunter (1980b), Hunter et al (1986), and Reid et al (1992b) distinguished resistant from susceptible populations from observations of the percentage of infected plants and spread of infection on infected ears. Hunter et al (1986) concluded that heritability for the percent infected plants was higher, thereby indicating that in their population the variation for resistance component 2 was relatively very small or absent. Using diallel analyses, Reid et al (1992b) concluded that the results for mean disease incidence and disease ratings of infected ears were similar. This indicates that in their population the variation for *Fusarium* resistance was caused by a variation for resistance to penetration located in the silk channel (component 1), and variation in resistance to colonization was merely absent. Enerson and Hunter (1980b), however, observed differences in resistance components 1 and 2 between two hybrids.

GENETIC VARIATION

Differences among corn hybrids with respect to resistance to *F. graminearum* have been reported by several authors (Al-Heeti 1987; Atlin et al 1983; Cullen et al 1983; Enerson and Hunter 1980c; Hart et al 1984; Hunter et al 1986; Reid et al 1992b). Several inbreds showed partial resistance to kernel infection by *F. moniliforme* but none were immune (Headrick and Pataky 1989, Holley et al 1989, Warren 1978). Varieties from tropical South and Central America may provide resistance to corn kernel infection (Headrick and Pataky 1989, Holley et al 1989), since that is where *Zea* originated and genetic diversity will be largest. Expression of genetic variation among tropical sweet corn germ plasm has enabled the selection of satisfactory levels of resistance to *F. moniliforme* in Hawaii. Al-Heeti (1987) observed highly significant correlations between corn cultivar responses to the invasion of *F. graminearum*, *F. sporotrichioides*, and *F. poae*. Resistance to *F. graminearum* stalk rot is known to confer partial resistance to *Diplodia maydis* (Berk.) Sacc. and probably to other stalk rot organisms (Hooker 1978, Mostafa et al 1990, White 1978). Rheeder et al (1990a) demonstrated that hybrids resistant to natural seed infection by *Fusarium* species also showed resistance to *D. maydis*. White (1978) explains this similarity of reactions to the two stalk rot fungi by the fact that breeders have selected for resistance to both pathogens. One could speculate on a non–host-specific resistance mechanism inhibiting colonization. However, no general relationship is found between seedling, stalk, and ear responses to *Fusarium* (Hooker 1978, Mesterhazy and Kovacs 1986, Ooka and Kommedahl 1977).

COMPONENTS OF RESISTANCE

In addition to local infections, corn plants can become systemically infected with *F. moniliforme*, *F. culmorum*, and *F. graminearum*, even without showing symptoms (Cook 1978, Foley 1962, Lawrence et al 1981, Windels et al 1988). Cook (1978) and Davis et al (1989) did not find any indications that *Fusarium* infection enters the kernel through the cob. However, scanning electron microscopy studies by Bacon et al (1992) demonstrated that *F. moniliforme* is associated with tip caps of asymptomatic kernels, appearing to have arrived there from growth occurring on or within the cob. It has been suggested that internal infections occur through the silks or from adjacent cob tissues (Burgess et al 1981, Davis et al 1989). Infection of the ear via the stalk probably occurs only infrequently compared with infection from airborne or insect-borne inoculum (Ooka and Kommedahl 1977). Several factors can prevent *Fusarium* from entering through the silk channel opening. Morphological characters can influence infection; pendant ears and ears with loose husk coverage at the brown silk stage had less ear mold than corresponding upright ears and ears with tight husks (Enerson and Hunter 1980c, Koehler 1959). Physiological factors may influence resistance; kernels that mature faster shorten the period of optimum susceptibility. In susceptible hybrids leakage of nutrients from young kernels may stimulate growth of the fungus (Davis et al 1989). Maize pollen stimulates germination of conidia and germ tube growth of *F. graminearum*. This suggests that pollen might enable the fungus to establish itself on silks and thereby enter maize ears, forming an important contributory factor to the infection of the host by *F. graminearum* (Naik and Busch 1978). Also, an indirect resistance as a result of resistance to insects can lead to a resistant phenotypic expression. Fusarium ear rot was positively correlated with the population of thrips in corn ears (Farrar and Davis 1991). Thrips may be acting as vectors of *F. moniliforme* or as wounding agents by feeding on plant tissue. The physical exclusion of thrips from the developing ear, as found with tight husks, is a major factor in reducing disease levels (Farrar and Davis 1991). This tight husk effect contradicts the results of Enerson and Hunter (1980c).

A resistance located in the silk channel belongs to resistance component 1. The silks are apparently an effective plug, blocking entry to spores or insects into the tip of the ear. Colonization of silks by *F. moniliforme* in sweet corn is related to the condition of the silks. Little colonization occurs until the exposed silks begin to senesce, and little growth of the fungus into the ear is observed until the silks covered by husks begin to senesce (Davis et al 1989, Scott and King 1984). Symptomatic and asymptomatic infection of kernels by *F. moniliforme* is less for inbreds with silks that are green and actively growing at inoculation than for inbreds with brown silks (Headrick et al 1990, Headrick and Pataky 1991). Delayed senescence of silks is one possible selection criterion that might lead to a reduced incidence of *F. moniliforme* kernel infection.

Resistance component 2 prevents the colonization of plant tissue by the fungus. It can show its effect in the seed coat or other areas of the kernels. Duvick et al (1992) isolated a small peptide from maize kernels inhibiting spore germination and hyphal elongation of *F. moniliforme* and *F. graminearum* in vitro that may contribute to the resistance to kernel infection. Several studies examined the influence of the protein and sugar-starch fraction of the endosperm on resistance. The incorporation of the *opaque-2* gene (*o2*) into elite inbred lines and synthetic populations to increase the lysine and tryptophan composition of the endosperm led to observations of higher susceptibility to *F. moniliforme*, *F. graminearum* and *Diplodia zeae* in *o2* material compared with their normal *N* endosperm counterparts (Loesch et al 1976, Warren 1978). Sivasankar et al (1975) showed that *F. moniliforme* seed rot increased with every additional dosage of *o2* allele. This intensified susceptibility in *o2* endosperm may be caused by a softer kernel (Warren 1978). Genetic studies demonstrating interactions between disease response and endosperm type, however, indicate that *o2* lines can be developed with relatively high levels of resistance to several ear-infecting organisms (Ooka and Kommedahl 1977, Loesch et al 1976, Warren 1978). Sweet corn hybrids with the recessive mutant gene *shrunken-2* (*sh2*) in place of the standard recessive sugary (*su*) gene (resulting in more sugar and less starch) became heavily infected with *F. moniliforme* in the field earlier than kernels of hybrid plants with the standard *su* gene (Styer and Cantliffe 1984). Styer and Cantliffe (1984) suggested that the higher sugar content of *sh2* kernels would provide an ideal substrate for pathogenic growth. However, Headrick and Pataky (1991) and Headrick et al (1990) showed that there was no distinct correlation between endosperm type (*sh2, su, se, Su*) in quantity and concentration of kernel carbohydrates, and infection by *F. moniliforme*. *F. moniliforme* enters the kernel through the tip cap. In *sh2* seeds, the pathogen appeared to enter also through small cracks in the pericarp or by appressoria. Due to the collapsed nature of *sh2* seeds, the pericarp pulled away from the aleurone layer forming small pockets into which hyphae grew and sporulated. These pockets were not seen in *su* seeds (Styer and Cantliffe 1984) and were the cause of the higher susceptibility. The heavier earworm infestation in *sh2* ears would increase pathogen access (Styer and Cantliffe 1984). These studies indicate that there is no intrinsic effect of endosperm type on resistance component 2; effects of endosperm on resistance are due to inherent morphological characters.

Ooka and Kommedahl (1977) concluded that Texas male-sterile cytoplasm did not confer greater susceptibility to *F. moniliforme* ear or kernel infection than the normal cytoplasm.

The brown midrib mutant of maize (*bm3/bm3*) is being developed as a potential feed source for improving nutrition in ruminant animals. The greater digestibility of the vegetative parts is due to significantly reduced lignification of the tissue. Ears of the *bm3/bm3* mutant were much more susceptible to *F. moniliforme* ear rot than the isogenic normal (*Bm3/Bm3*) type, which, however, does not necessitate a consistent genetic association with susceptibility to the fungus (Nicholson et al 1976).

Mesterhazy and Kovacs (1986) concluded that as stalk rot heavily influences ear rot response, genetically valid ear rot data can be obtained only on healthy stalks. However, as carbohydrate concentration did not influence ear rot infection, the stalk-ear carbohydrate ratio only had an effect on stalk rot resistance.

GENETICS

Resistance to Fusarium ear rot is quantitative; complete resistance has not been discovered. Genetic studies involving diallel crosses by artificial inoculation type 1, type 2, or natural infection have shown that resistance to *F. moniliforme* and *F. graminearum* is polygenic and inherited in a mainly additive fashion (Boling and Grogan 1965, Chiang et al 1987, Gendloff et al 1986, Hart et al 1984, Hooker 1978, Odimeh and Manninger 1984, King and Scott 1981, de Léon and Pandey 1989, Reid et al 1992b, Szél 1984). Among the nonadditive components, dominant gene action was noticeable (Reid et al 1992b). However, in general, selection based on general combining ability can be used to predict which hybrids will have the best resistance. The most effective way to create inbred lines with a high degree of resistance is through simple recurrent selection (Chiang et al 1987, de Léon and Pandey 1989).

Cullen et al (1983) and King and Scott (1981) observed that inbred lines were considerably more susceptible to *F. moniliforme* kernel infection (ear rot) than crosses of the lines, indicating that disease reaction may be influenced by plant vigor. Al-Heeti (1987) observed the same phenomenon for susceptibility of dent corn to *F. graminearum*. A maternal effect was apparent in one set of reciprocal crosses for *F. graminearum* (Gendloff et al 1986). Also the inheritance of resistance to kernel infection by *F. moniliforme* in starchy (*Su*) corn was under maternal control (Headrick and Pataky 1991). Scott and King (1984) determined that the factor(s) for resistance to kernel infection after natural inoculation by *F. moniliforme* were not conditioned by the genotype of the endosperm, embryo, or cytoplasm, but by the genotype of the pericarp (genetically identical to the female parent), of the silk or of other maternal tissues. Also the scanning electron microscopy results of Bacon et al (1992) suggest that the placental chalazian region and other maternal tissues might be important in the expression of resistance to this fungus.

Barrière (1979) observed a strong correlation between the stalk rot resistance of hybrids measured directly and that calculated from general combining ability estimates, indicating additivity. Hooker (1978) reported that two genes located on the short arm of chromosome 10 conferred resistance to *F. moniliforme* stalk rot. Rakha and El-deeb (1978) identified 17 genes for stalk rot resistance to *F. moniliforme*, which were located on both arms of chromosomes 1–4, 7, and 10; on the short arm of chromosome 8; on the long arms of chromosomes 5 and 6; and on the centromere regions of chromosomes 7 and 9.

Resistance to Toxin Accumulation

In wheat and corn, *Fusarium* infection (assessed visually or by ergosterol analyses) and DON content are correlated (Atlin et al 1983, Hart et al 1984, Snijders and Krechting 1992, Snijders and Perkowski 1990, Wang and Miller 1988). Besides the resistance mechanisms that determine the severity of head blight, there may be a second type of mechanism which influences kernel DON content. Snijders and Perkowski (1990) observed that for 10 wheat genotypes the ratio of Fusarium head blight rating to DON varied by a factor of 10, which was not correlated with Fusarium head blight resistance. Resistant wheat and rye cultivars inoculated with *F. graminearum* gave much higher fungal biomass to DON ratios than susceptible cultivars (Miller et al 1985). There seem to be at least two responses to the toxin; the first is degradation or conjugation of DON. The *Fusarium*-resistant wheat cultivar Frontana degraded 18% of ^{14}C deoxynivalenol added to fragmented embryo callus cultures after 72 h of incubation, whereas the susceptible wheat cultivar Casavant converted only 5% of the added DON (Miller and Arnison 1986). Declines of DON in vivo have also been observed (Miller and Young 1985, Scott et al 1984). The second type of response to DON is tolerance; germ plasm of various crops including wheat has been shown to be highly tolerant of trichothecenes. This has been found to be based on trichothecene degradation, increased membrane stability, and a modified peptidyl transferase (Jarvis et al 1991, Kuti et al 1989, Miller 1989, Wang and Miller 1988). Snijders and Krechting (1992) demonstrated in wheat that DON is transported from the chaff to the young kernel and that the pathogen colonizes the kernel later. A Fusarium head blight resistant wheat line that possessed resistance to hyphal invasion (component 2) appeared to inhibit DON translocation from chaff to kernel. Evidently this is a membrane-based trichothecene tolerance. As DON is regarded as an aggressiveness factor, DON tolerance will result in an increased level of resistance and inherent prevention of trichothecene accumulation.

Atlin et al (1983) demonstrated that in the tested maize hybrids, adjusted for mold level, hybrids did not differ significantly for toxin accumulation. Further information on DON tolerance in maize has not been published.

Conclusions

It is obvious that *Fusarium* resistance in wheat and maize show important similarities. Both crops can be infected with *Fusarium* species lacking external and internal symptoms. *F. graminearum* and *F. culmorum* are non–host specific, i.e., they are pathogenic to wheat and maize without showing specialization for crop or crop genotype. *F. moniliforme* is less pathogenic, but is also non–host specific because it is sporadically isolated from blighted wheat (Mesterhazy 1977, 1978; Wilcoxson et al 1988). *Fusarium* resistance in wheat and in maize does not show *Fusarium* species or strain specificity; even the opposite is reported, i.e., horizontal resistance to several stalk and ear rot pathogens in maize. Reviewing inoculation methods and genetics has shown that breeding for phenotypic resistance is difficult, time-consuming, and often without results. Although some good results were obtained for selection based on visual symptoms in wheat (Snijders 1990d), poor relations between fungal biomass and visual symptoms, especially for the extremes on the head blight assessment scale (Snijders and Krechting 1992; Wang and Miller 1988), make a well-defined selection tool necessary. *Fusarium* resistance is a complex that can be split up in two components. Study of the individual components may lead to further splitting until resistance mechanisms can be determined that are qualitative in expression, that can be studied on tissue or at the cell level, that are monogenic or oligogenic, and that can be labeled. The fact that trichothecenes play a role as aggressiveness factors is an important conclusion. It opens the door for study of the trichothecene tolerance mechanism (or mechanisms) as a part of *Fusarium* resistance and for study of its specificity and its potential use in other crops. Accumulation of genes encoding for different mechanisms might lead to cultivars with an acceptable level of resistance and absence of toxin accumulation.

Literature Cited

Abbas, H. K., Mirocha, C. J., Kommedahl, T., Burnes, P. M., Meronuck, R. A., and Gunther, R. 1988. Toxigenicity of *Fusarium proliferatum* and other *Fusarium* species isolated from corn ears in Minnesota. Phytopathology 78:1258-1260.

Abbas, H. K., Paul, R. N., Boyette, C. D., Duke, S. O., and Vesonder, R. F. 1992. Physiological and ultrastructural effects of fumonisin on jimsonweed leaves. Can. J. Bot. 70:1824-1833.

Al-Heeti, A. A. 1987. Pathological, toxicological and biological evaluations of *Fusarium* species associated with ear rot of maize. Ph.D. thesis, University of Wisconsin-Madison. Univ. Microfilms Int. Diss. Inf. Serv. 8727220.

Atlin, G. N., Enerson, P. M., McGirr, L. G., and Hunter, R. B. 1983. Gibberella ear rot development and zearalenone and vomitoxin production as affected by maize genotype and *Gibberella zeae* strain. Can. J. Plant Sci. 63:847-853.

Bacon, C. W., Bennett, R. M., Hinton, D. M., and Voss, K. A. 1992. Scanning electron microscopy of *Fusarium moniliforme* within asymptomatic corn kernels and kernels associated with equine leukoencephalomalacia. Plant Dis. 76:144-148.

Barrière, Y., 1985. Aspects physiologiques de la pourriture des tiges du maïs. Agronomie 5:361-368.

Bausch, P., Schuster, W., and Schlösser, E. 1982. Über die Anfälligkeit für Wurzel- und Stengelfäule von Mais. Angew. Bot. 56:9-24.

Bechtel, D. B., Kaleikau, L. A., Gaines, R. L., and Seitz, L. M. 1985. The effects of *Fusarium graminearum* infection on wheat kernels. Cereal Chemistry 62:191-197.

Bennett, G. A., Wicklow, D. T., Caldwell, R. W., and Smalley, E. B. 1988. Distribution of trichothecenes and zearalenone in *Fusarium graminearum*: Rotted corn ears grown in a controlled environment. J. Agric. Food Chem. 36:639-642.

Boling, M. B., and Grogan, C. O. 1965. Gene action affecting host resistance to Fusarium ear rot of maize. Crop Sci. 5:305-307.

Bottalico, A., Logrieco, A., and Visconti, A. 1989. *Fusarium* species and their mycotoxins in infected corn in Italy. Mycopathologia 107:85-92.

Bruins, M. B. M., Karsaï, I., Schepers, J., and Snijders, C. H. A. Phytotoxicity of deoxynivalenol to wheat tissue with regard to in vitro selection for Fusarium head blight resistance. Plant Sci. (In press.)

Burgess, L. W., Dodman, R. L., Pont, W., and Mayers, P. 1981. Fusarium diseases of wheat, maize and grain sorghum in eastern Australia. Pages 64-76 in: Fusarium Diseases, Biology and Taxonomy. P. E. Nelson, T. A. Toussoun, and R. J. Cook, eds. Pennsylvania State University Press, University Park.

Burgess, L. W., Klein, T. A., Bryden, W. L., and Tobin, N. F. 1987. Head blight of wheat caused by *Fusarium graminearum* group 1 in New South Wales in 1983. Australas. Plant Pathol. 16:72-78.

Carter, C. J., Cannon, M., and Jimanez, A. 1980. A trichodermin-resistant mutant of *Saccharomyces cerevisiae* with an abnormal distribution of native ribosomal subunits. Eur. J. Biochem. 107:173-183.

Chiang, M. S., Hudon, M., Devaux, A., and Ogilvie, I. 1987. Inheritance of resistance to Gibberella ear rot in maize. Phytoprotection 68:29-33.

Christensen, J. J., and Wilcoxson, R. D. 1966. Stalk Rot of Corn. Monogr. 3. American Phytopathological Society, St. Paul, MN.

Cook, R. J. 1968. Fusarium root and foot rot of cereals in the Pacific Northwest. Phytopathology 58:127-131.

Cook, R. J. 1978. The incidence of stalk rot (*Fusarium* spp.) on maize hybrids and its effect on yield of maize in Britain. Ann. Appl. Biol. 88:23-30.

Cullen, D., Caldwell, R. W., and Smalley, E. B. 1983. Susceptibility of maize to *Gibberella zeae* ear rot: Relationship to host genotype, pathogen virulence, and zearalenone contamination. Plant Dis. 67:89-91.

Davis, R. M., Kegel, F. R., Sills, W. M., and Farrar, J. J. 1989. Fusarium ear rot of corn. Calif. Agric. 43:4-5.

de Léon, C., and Pandey, S. 1989. Improvement of resistance to ear and stalk rots and agronomic traits in tropical maize gene pools. Crop Sci. 29:12-17.

Desjardins, A. E. 1992. Genetic approaches to the chemical ecology of phytopathogenic *Fusarium* species. Handb. Appl. Mycol. 5:333-357.

Dodd, J. L. 1980. The role of plant stresses in development of corn stalk rots. Plant Dis. 64:533-537.

Dolstra, O., Marton, C., Menzi, M., Mohr, I., Plienegger, D. I., and Pronczuk, M. Evaluation of recurrent selection for stalk rot resistance in a synthetic maize population. In: Breeding and Molecular Biology: Accomplishments and Future Promises. Proc. Conf. Eucarpia Maize Sorghum Sect. XVI, Bergamo, Italy, June 1993. (In press.)

Drepper, W. J., and Renfro, B. L. 1990. Comparison of methods for inoculation of ears and stalks of maize with *Fusarium moniliforme*. Plant Dis. 74:952-956.

Duvick, J. P., Rood, T., Rao, A. G., and Marshak, D. R. 1992. Purification and characterization of a novel antimicrobial peptide from maize (*Zea mays* L.) kernels. J. Biol. Chem. 267:18814-18820.

Enerson, P. M., and Hunter, R. B. 1980a. A technique for screening maize (*Zea mays* L.) for resistance to ear mold incited by *Gibberella zeae* (Schw.) Petch. Can. J. Plant Sci. 60:1123-1128.

Enerson, P. M., and Hunter, R. B. 1980b. Differential apparent infection rates of ear mold incited by *Gibberella zeae* between two artificially inoculated hybrids of maize. Can. J. Plant Sci. 60:1459-1461.

Enerson, P. M., and Hunter, R. B. 1980c. Response of maize hybrids to artificially inoculated ear mold incited by *Gibberella zeae*. Can. J. Plant Sci. 60:1463-1465.

Farrar, J. J., and Davis, R. M. 1991. Relationships among ear morphology, western flower thrips, and Fusarium ear rot of corn. Phytopathology 81:661-666.

Fisher, N. L., Gregory, L. V., and Ayers, J. E. 1983. Ear rot of sweet corn caused by *Fusarium* species. (Abstr.). Phytopathology 73:366.

Foley, D. C. 1962. Systemic infection of corn by *Fusarium moniliforme*. Phytopathology 52:870-872.

Gelderblom, W. C. A., Jaskiewicz, K., Marasas, W. F. O., Thiel, P. G., Horak, R. M., Vleggaar, R., and Kriek, N. P. J. 1988. Fumonisins—Novel mycotoxins with cancer-promoting activity produced by *Fusarium moniliforme*. Appl. Environ. Microbiol. 54:1806-1811.

Gendloff, E. H., Rossman, E. C., Casale, W. L., Isleib, T. G., and Hart, L. P. 1986. Components of resistance to Fusarium ear rot in field corn. Phytopathology 76:684-688.

Gu, J. 1983. A study on the genetics of resistance to wheat scab. Scientia Agric. Sinica 6:61-64.

Gilbertson, R. L., Brown, W. M., and Ruppel, E. G. 1985. Prevalence and virulence of *Fusarium* spp. associated with stalk rot of corn in Colorado. Plant Dis. 69:1065-1068.

Gulya, T. J., Jr., Martinson, C. A., Loesch, P. J., Jr. 1980. Evaluation of inoculation techniques and rating dates for Fusarium ear rot of opaque-2 maize. Phytopathology 70:1116-1118.

Hart, L. P., Gendloff, E., and Rossman, E. C. 1984. Effect of corn genotypes on ear rot infection by *Gibberella zeae*. Plant Dis. 68:296-298.

Headrick, J. M., and Pataky, J. K. 1989. Resistance to kernel infection by *Fusarium moniliforme* in inbred lines of sweet corn and the effect of infection on emergence. Plant Dis. 73:887-892.

Headrick, J. M., and Pataky, J. K. 1991. Maternal influence on the resistance of sweet corn lines to kernel infection by *Fusarium moniliforme*. Phytopathology 81:268-274.

Headrick, J. M., Pataky, J. K., and Juvik, J. A. 1990. Relationships among carbohydrate content of kernels, condition of silks after pollination, and the response of sweet corn inbred lines to infection of kernels by *Fusarium moniliforme*. Phytopathology 80:487-494.

Holley, R. N., Hamilton, P. B., and Goodman, M. M. 1989. Evaluation of tropical maize germ plasm for resistance to kernel colonization by *Fusarium moniliforme*. Plant Dis. 73:578-580.

Hooker, A. L. 1978. Genetics of disease resistance in maize. Pages 319-332 in: Maize Breeding and Genetics. D. B. Walden, ed. John Wiley and Sons, New York.

Hunter, R. B., G. N. Atlin, and J. F. Muldoon, 1986. Genotype x environment interactions for ear mold resistance and its subcomponents in maize hybrids. Can. J. Plant Sci. 66:291-297.

Jarvis, B. B., Mokhtari-Rejali, N., Schenkel, E. P., Barros, C. S., and Matzenbacher, N. I. 1991. Trichothecene mycotoxins from Brazilian *Baccharis* species. Phytochemistry 30:789-797.

Khonga, E. B., and Sutton, J. C. 1988. Inoculum production and survival of *Gibberella zeae* in maize and wheat residues. Can. J. Plant Pathol. 10:232-239.

King, S. B., and Scott, G. E. 1981. Genotypic differences in maize to kernel infection by *Fusarium moniliforme*. Phytopathology 71:1245-1247.

Koehler, B. 1959. Corn ear rots in Illinois. Ill. Exp. Stn. Bull. 639.

Krüger, W. 1985. Fusarium resistance in maize. Zea Inf. Bull. FAO Res. Netw. Maize 2:9-13.

Kuti, J. O., Ng, T. J., and Bean, G. A. 1989. Possible involvement of a pathogen-produced trichothecene metabolite in Myrothecium leaf spot of muskmelon. Physiol. Mol. Plant Pathol. 34:41-54.

Lal, S., and Singh, I. S. 1984. Breeding for resistance to downy mildews and stalk rots in maize. Theor. Appl. Genet. 69:111-119.

Lamprecht, S. C., Marasas, W. F. O., Alberts, J. F., Cawood, M. E., Gelderbloom, W. C. A., Shaphard, G. S., Thiel, P. G., and Calitz, F. J. 1993. Phytotoxicity of fumonisins and TA-toxin to corn and tomato. Int. Congr. Plant Pathol. (Abstr. 14.1.2) 6:247.

Lawrence, E. B., Nelson, P. E., and Ayers, J. E. 1981. Histopathology of sweet corn seed and plants infected with *Fusarium moniliforme* and *F. oxysporum*. Phytopathology 71:379-386.

Lin, Y., Yang, Z., and Wu, Z. 1992. Genetic analysis of resistance to scab (*Gibberella zeae*) in wheat varieties from different regions. Acta Agric. Shanghai 8(1):31-36.

Loesch, P. J., Jr., Foley, D. C., and Cox, D. F. 1976. Comparative resistance of opaque-2 and normal inbred lines of maize to ear-rotting pathogens. Crop Sci. 16:841-842.

Logrieco, A., and Bottalico, A. 1986. Specie di *Fusarium* e micotossine associate al marciume del culmo del mais, in Basilicata. Phytopathol. Mediterr. 25:26-32.

Manka, M., Chelkowski, J., Brayford, D., Visconti, A., Kwasna, H., and Perkowski, J. 1989. *Fusarium graminearum* Schwabe (teleomorph *Gibberella zeae* Schw. Petch)— Cultural characteristics, pathogenicity towards cereal seedlings and ability to produce mycotoxins. J. Phytopathol. 124:143-148.

Manka, M., Visconti, A., Chelkowski, J., and Bottalico, A. 1985. Pathogenicity of *Fusarium* isolates from wheat, rye and triticale towards seedlings and their ability to produce trichothecenes and zearalenone. J. Phytopathol. 113:24-29.

Marasas, W. F. O., Nelson, P. E., and Toussoun, T. A. 1984. Toxigenic *Fusarium* species, identity and mycotoxicology. Pennsylvania State University Press, University Park.

Martin, R. A., MacLeod, J. A., and Caldwell, C. 1991. Influences of production inputs on incidence of infection by *Fusarium* species on cereal seed. Plant Dis. 75:784-788.

Mesterhazy, A. 1983. Breeding wheat for resistance to *Fusarium graminearum* and *Fusarium culmorum*. Z. Pflanzenzuecht. 91:295-311.

Mesterhazy, A. 1977. Reaction of winter wheat varieties to four *Fusarium* species. J. Phytopathol. 90:104-112.

Mesterhazy, A. 1978. Comparative analysis of artificial inoculation methods with *Fusarium* spp. on winter wheat varieties. J. Phytopathol. 93:12-25.

Mesterhazy, A. 1982. Resistance of corn to Fusarium ear rot and its relation to seedling resistance. J. Phytopathol. 103:218-231.

Mesterhazy, A. 1983. Breeding wheat for resistance to *Fusarium graminearum* and *Fusarium culmorum*. Z. Pflanzenzuecht. 91:295-311.

Mesterhazy, A. 1984. A laboratory method to predict pathogenicity of *Fusarium graminearum* in field and resistance of wheat to scab. Acta Phytopathol. Acad. Sci. Hung. 19:205-218.

Mesterhazy, A. 1987. Selection of head blight resistant wheats through improved seedling resistance. Plant Breed. 98:25-36.

Mesterhazy, A. 1988. Expression of resistance of wheat to *Fusarium graminearum* and *F. culmorum* under various experimental conditions. J. Phytopathol. 123:304-310.

Mesterhazy, A., and Kovacs, K. 1986. Breeding corn against fusarial stalk rot, ear rot and seedling blight. Acta Phytopathol. Entomol. Hung. 21:231-249.

Miller, J. D. 1989. Effect of *Fusarium graminearum* metabolites on wheat cells. Pages 449-452 in: Phytotoxins and Plant Pathogenesis A. Graniti, R. D. Durbin, and A. Ballio. eds. NATO ASI Ser. H27.

Miller, J. D., and Arnison, P. G. 1986. Degradation of deoxynivalenol by suspension cultures of the Fusarium head blight resistant wheat cultivar Frontana. Can. J. Plant Pathol. 8:147-150.

Miller, J. D., and Greenhalgh, R. 1988. Metabolites of fungal pathogens and plant resistance. Pages 117-129 in: Biotechnology for Crop Protection. P. A. Hedin, J. J. Menn, and R. M. Hollingworth, eds. ACS Symp. Ser. 379.

Miller, J. D., and Young, J. C. 1985. Deoxynivalenol in an experimental *Fusarium graminearum* infection of wheat. Can. J. Plant Pathol. 7:132-134.

Miller, J. D., Young, J. C., and Trenholm, H. L. 1983. Fusarium toxins in field corn. I. Time course of fungal growth and production of deoxynivalenol and other mycotoxins. Can. J. Bot. 61:3080-3087.

Miller, J. D., Young, J. C., and Sampson, D. R. 1985. Deoxynivalenol and Fusarium head blight resistance in spring cereals. J. Phytopathol. 113:359-367.

Mortimore, C. G., and Ward, G. M. 1964. Root and stalk rot of corn in south-western

Ontario. III. Sugar levels as a measure of plant vigor and resistance. Can. J. Plant Sci. 44:451-457.

Mostafa, M. A. N., Coors, J. G., and Drolsom, P. N. 1990. Correlated changes in grain yield and agronomic traits from selection for resistance to stalk rot in maize caused by *Gibberella zeae* (Schw.) Petch. Maydica 35:253-258.

Naik, D. M., and Busch, L. V. 1978. Stimulation of *Fusarium graminearum* by maize pollen. Can. J. Bot. 56:1113-1117.

Nicholson, R. L., Bauman, L. F., and Warren, H. L. 1976. Association of *Fusarium moniliforme* with brown midrib maize. Plant Dis. Rep. 60:908-910.

Odimeh, M., and Manninger, I. 1984. Diallel analysis of resistance of maize to ear-rotting pathogens. Acta Agron. Acad. Sci. Hung. 33:441-444.

Ooka, J. J., and Kommedahl, T. 1977. Kernels infected with *Fusarium moniliforme* in corn cultivars with opaque-2 endosperm or male-sterile cytoplasm. Plant Dis. Rep. 61:162-165.

Pappelis, A. J., Mumford, P. M., Abney, T. S., and Pappelis, G. A. 1975. Classification of corn inbreds using pith cell death patterns and the prediction of stalk rot response. Cereal Res. Commun. 3:227-232.

Porter, R. M., Ayers, J. E., Johnson, M. W., Jr., and Nelson, P. E. 1981. Influence of differential phosphorus accumulation on corn stalk rot. Agron. J. 73:283-287.

Purss, G. S. 1971. Pathogenic specialization in *Fusarium graminearum*. Aust. J. Agric. Res. 22:553-561.

Rakha, F. A., and El-deeb, M. H. 1978. Locating genes for resistance to stalk-rot of maize using chromosomal translocation. Alexandria J. Agric. Res. 26:375-386.

Reid, L. M., Bolton, A. T., Hamilton, R. I., Woldemariam, T., and Mather, D. E. 1992a. Effect of silage on resistance of maize to *Fusarium graminearum*. Can. J. Plant Pathol. 14:293-298.

Reid, L. M., Mather, D. E., Hamilton, R. I., and Bolton, A. T. 1992b. Diallel analysis of resistance in maize to *Fusarium graminearum* infection via the silk. Can. J. Plant Sci. 72:915-923.

Rheeder, J. P., Marasas, W. F. O., Van Wyk, P. S., du Toit, W., Pretorius, A. J. D., and Van Schalkwyk, J. 1990a. Incidence of *Fusarium* and *Diplodia* species and other fungi in naturally infected grain of South African maize cultivars. Phytophylactica 22:97-102.

Rheeder, J. P., Marasas, W. F. O., Van Wyk, P. S., and Van Schalkwyk, D. J. 1990b. Reaction of South African maize cultivars to ear inoculation with *Fusarium moniliforme, F. graminearum* and *Diplodia maydis*. Phytophylactica 22:213-218.

Schneider, R. W., and Pendery, W. E. 1983. Stalk rot of corn: Mechanism of predisposition by an early season stress. Phytopathology 73:863-871.

Schroeder, H. W., and Christensen, J. J. 1963. Factors affecting resistance of wheat to scab caused by *Gibberella zeae*. Phytopathology 53:831-838.

Scott, G. E., and King, S. B. 1984. Site of action of factors for resistance to *Fusarium moniliforme* in maize. Plant Dis. 68:804-806.

Scott, P. M., Nelson, K., Kanhere, S. R., Karpinski, K. F., Hayward, S., Neish, G. A., Teich, A. H. 1984. Decline in deoxynivalenol (vomitoxin) concentrations in 1983 Ontario winter wheat before harvest. Appl. Environ. Microbiol. 48:884-886.

Shimada, T., and Otani, M. 1990. Effects of *Fusarium* mycotoxins on the growth of shoots and roots at germination in some Japanese wheat cultivars. Cereal Res. Commun. 18:229-233.

Sieber, T., Riesen, T. K., Miller, E., and Fried, P. M. 1988. Endophytic fungi in four winter wheat cultivars (*Triticum aestivum* L.) differing in resistance against

Stagonospora nodorum (Berk.) Cast. and Germ. = *Septoria nodorum* (Berk.) Berk. J. Phytopathol. 122:289-306.

Sivasankar, D., Asnani, V. L., Lal, S., and Agarwal, B. D. 1975. Dosage effects of opaque-2 gene on the susceptibility in maize to seed rots by *Cephalosporium acremonium* and *Fusarium moniliforme*. Indian Phytopathol. 28:235-237.

Snijders, C. H. A. 1988. The phytotoxic action of deoxynivalenol and zearalenone on wheat seedlings. Pages 103-104 in: Mycotoxins and Phycotoxins. IUPAC '88 and ICPP '88. K. Aibara, S. Kumagai, K. Ohtsubo, and T. Yoshizawa, eds. Proc. Jpn. Assoc. Mycotoxicol. Suppl. 1.

Snijders, C. H. A. 1990a. Systemic fungal growth of *Fusarium culmorum* in stems of winter wheat. J. Phytopathol. 129:133-140.

Snijders, C. H. A. 1990b. Diallel analysis of resistance to head blight caused by *Fusarium culmorum* in winter wheat. Euphytica 50:1-9.

Snijders, C. H. A. 1990c. The inheritance of resistance to head blight caused by *Fusarium culmorum* in winter wheat. Euphytica 50:11-18.

Snijders, C. H. A. 1990d. Response to selection in F_2 generations of winter wheat for resistance to head blight caused by *Fusarium culmorum*. Euphytica 50:163-169.

Snijders, C. H. A. 1990e. Genetic variation for resistance to Fusarium head blight in bread wheat. Euphytica 50:171-179.

Snijders, C. H. A. 1992. Breeding for resistance to Fusarium head blight in wheat. Pages 142-147 in: Proc. Regional Wheat Workshop for Eastern, Central and Southern Africa, 7th. Centro Internacional Mejoramiento Maiz y Trigo (CIMMYT), Mexico City.

Snijders, C. H. A., and Perkowski, J. 1990. Effects of head blight caused by *Fusarium culmorum* on toxin content and weight of wheat kernels. Phytopathology 80:566-570.

Snijders, C. H. A., and Van Eeuwijk, F. A. 1991. Genotype × strain interactions for resistance to Fusarium head blight caused by *Fusarium culmorum*. Theor. Appl. Genet. 81:239-244.

Snijders, C. H. A., and Krechting, C. F. 1992. Inhibition of deoxynivalenol translocation and fungal colonization in Fusarium head blight resistant wheat. Can. J. Bot. 70:1570-1576.

Stack, R. W., and McMullen, M. P. 1985. Head blighting potential of *Fusarium* species associated with spring wheat heads. Can. J. Plant Pathol. 7:79-82.

Styer, R. C., and Cantliffe, D. J. 1984. Infection of two endosperm mutants of sweet corn by *Fusarium moniliforme* and its effect on seedling vigor. Phytopathology 74:189-194.

Sutton, J. C., and Baliko, W. 1981. Methods for quantifying partial resistance to *Gibberella zeae* in maize ears. Can. J. Plant Pathol. 3:26-32.

Szél, S. 1984. Study of resistance to *Fusarium graminearum* in opaque maize lines and in hybrids produced from them by diallel crossing. Acta Agron. Acad. Sci. Hung. 33:444-448.

Tošić, M., Panić, M., and Ivanović, M. 1977. Effect of previous mosaic virus infection on corn susceptibility to the stalk and ear rot caused by *Gibberella zeae*. Ann. Phytopathol. 9:395-401.

Ullstrup, A. J. 1970. Methods for inoculating corn ears with *Gibberella zeae* and *Diplodia maydis*. Plant Dis. Rep. 54:658-662.

Van Asch, M. A. J., Rijkenberg, F. H. J., and Coutinho, T. A. 1992. Phytotoxicity of fumonisin B_1, moniliformin, and T-2 toxin to corn callus cultures. Phytopathology 82:1330-1332.

Vesonder, R. F., Labeda, D. P., and Peterson, R. E. 1992. Phytotoxic activity of selected water-soluble metabolites of *Fusarium* against *Lemna minor* L. (duckweed). Mycopathologia 118:185-189.

Wang, Y. Z., and Miller, J. D. 1988. Effects of *Fusarium graminearum* metabolites on wheat tissue in relation to Fusarium head blight resistance. J. Phytopathol. 122:118-125.

Warren, H. L. 1978. Comparison of normal and high-lysine maize inbreds for resistance to kernel rot caused by *Fusarium moniliforme*. Phytopathology 68:1331-1335.

White, D. G. 1978. Correlation of corn stalk rot reactions caused by *Diplodia maydis* and *Gibberella zeae*. Plant Dis. Rep. 62:1016-1018.

Wilcoxson, R. D., Kommedahl, T., Ozmon, E. A., and Windels, C. E. 1988. Occurrence of *Fusarium* species in scabby wheat from Minnesota and their pathogenicity to wheat. Phytopathology 78:586-589.

Windels, C. E., Kommedahl, T., Stienstra, W. C., and Burnes, P. M. 1988. Occurrence of *Fusarium* species in symptom-free and overwintered cornstalks in northwestern Minnesota. Plant Dis. 72:990-993.

Young, H. C. 1943. The toothpick method of inoculating corn for ear and stalk rot. Phytopathology 33:16.

Young, J. C., and Miller, J. D. 1985. Appearance of fungus, ergosterol and *Fusarium* mycotoxins in the husk, axial stem and stalk after ear inoculation of field corn. Can. J. Plant Sci. 65:47-53.

Yu, Y. J. 1982. Monosomic analysis for scab resistance and yield components in the wheat cultivar Soo-moo-3. Cereal Res. Commun. 10:185-190.

Zhang, L.-Q., and Pan, X.-P., 1982. A study on resistance to colonization of *Gibberella zeae* in wheat varieties. J. S. China Agric. Coll. 3:21-29.

Zhou, C. F., Xua, S. S., Qian, C. M., Yao, G. C., and Shen, J. X. 1987. On the problem of breeding wheat for scab resistance. Scientia Agric. Sinica 20:19-25.

Chapter 4

Spectral Characteristics of Secondary Metabolites from *Fusarium* Fungi

M. E. Savard and B. A. Blackwell

This chapter is a compilation of spectral data for a series of secondary metabolites from fungi of the genus *Fusarium*. During the 10 years our group has spent on mycotoxin research, many more new secondary metabolites have been isolated, structurally characterized, and reported in the literature than were previously tabulated in the original *Handbook of Toxic Fungal Metabolites* published by Cole and Cox in 1981. The present compilation is not intended to replace this handbook, which covers toxigenic compounds from several genera of fungi, but to provide an update on only those compounds produced by *Fusarium* fungi. Cole and Cox (1981) provides researchers with an extensive tabulation of chemical, physical, spectral, and biological data on a wide variety of compounds, whereas here we have tabulated only spectral data. Other aspects of these metabolites, namely biosynthesis, chemistry, occurrence, epidemiology, and toxicity are discussed elsewhere. In the process of investigating the toxicity, biosynthesis, and production of several of the known and important mycotoxins, many novel compounds have been isolated, leading to a greater understanding of the role these compounds play in the physiology of fungal contamination. Unambiguous structure determination with the use of nuclear magnetic resonance (NMR) spectroscopy and mass spectrometry (MS) is the primary evidence required in order to conduct biosynthetic studies, to identify these compounds in extracts, and to identify new and related compounds.

The metabolites presented here are not a complete listing but represent most of the classes of secondary metabolites that may be found in cultures of *Fusarium* fungi and illustrate the biosynthetic structural diversity that these fungi possess. Only those compounds that have been shown to be natural products are included; synthetic derivatives reported by this group or others have been purposely omitted. The toxicity of many of these metabolites has not yet been investigated. The following classes are included.

Trichothecenes

The trichothecenes are sesquiterpenes characterized by a double bond at position C-9, an epoxide ring at C-12, and various patterns of hydroxy and acetoxy substitutions at positions C-3, C-4, C-15, C-7, and C-8. As a result of the large number of trichothecenes isolated to date, an extensive NMR data base has been developed that finely correlates both ^{1}H NMR data (Savard et al 1987) and ^{13}C NMR data (Greenhalgh et al 1989a) to structural features and molecular conformation, both for the purposes of easy identification of new compounds and for the study of structure-toxicity relationships. Some attempt has also been made to relate the substitution pattern around the trichothecene ring structure to the species of *Fusarium* (Greenhalgh et al 1986c, 1990).

Modified Trichothecenes

The "trichodienoid" compounds are also sesquiterpenes and include the apotrichothecenes, sambucinol, and sambucoin derivatives. The most important characteristic of the apotrichothecene series lies in the configuration at C-11, where H-11 possesses a β-configuration, as opposed to the α-H of the trichothecenes. This rules out the possibility that these compounds could be artifacts due to rearrangements of the trichothecene ring during the isolation process. Limited patterns of hydroxyl substitution have been observed at the C-8 and the C-2 positions, but the metabolites are common to all species of *Fusarium* studied to date. These compounds are thought to arise from different cyclization patterns from those observed for the trichothecenes of the common precursor trichodiene (Greenhalgh et al 1990, Savard et al 1989, ApSimon et al 1991).

Culmorins

The culmorins are oxidized derivatives of the metabolite culmorin, first characterized by Barton and Werstiuk (1967), which occurs as a major secondary metabolite in *F. culmorum* as well as a more minor component in other species. Biosynthetically, these metabolites compete for the pool of farnesyl pyrophosphate with the other sesquiterpenes of much greater toxicity. The role of culmorin as a possible synergistic compound is suggested in other chapters. Several hydroxylated and reduced derivatives have been recently characterized (Kasitu et al 1992).

Enniatins

The enniatins are a series of cyclohexadepsipeptides derived from hydroxy and amino acids joined by amide and ester linkages, produced by strains of *F. avenaceum* and *F. acuminatum.* They consist of three *d*-2 hydroxy-isovaleric acid (HyIv) residues linked alternatively to *l*-amino acids or *N*-methyl amino acids to give an 18-membered cyclic skeleton. The A- and B-type enniatins consist of combinations of *N*-methylisoleucine (NMeIle) and/or *N*-methylvaline (NMeVal) residues and are the only known natural products of this type. The antibiotic, phytotoxic, and insecticidal properties of enniatins are related to their ionophoric structure. Several new A- and B-type enniatins have been isolated and recently characterized (Blais et al 1992, Visconti et al 1992).

Fusarins

The fusarins are C-24 mixed-function compounds, which are produced by several species of *Fusarium* and have been shown to be products of combined amino acid and polyketide biosynthesis. Of the five fusarins that have been characterized (fusarin A, C, D, E and most recently F; Savard and Miller 1992), only fusarin C has been reported to be mutagenic. The other fusarins have either shown no activity or have not yet been investigated.

Fumonisins

The fumonisins are a series of C-20 long-chain amino-polyols esterified with two tricarboxylic acid groups. They have been implicated in a wide variety of toxicological effects associated with the ingestion of contaminated corn and corn products. Fumonisins have a structure similar to that of sphingosine, which forms the base backbone of sphingolipids, and have been shown to inhibit sphingolipid biosynthesis. Six fumonisin structures have been identified: FB_1, FB_2, FB_3, FB_4, FA_1, and FA_2. Of these, only the B-series has been confirmed as natural products. The A-series consists of *N*-acyl derivatives, which may be produced during the isolation process. Fumonisin B_1 is usually the most abundant, and FB_2 and FB_3 are produced in much smaller quantities.

Miscellaneous Compounds

Other compounds including moniliformin, butenolide, chlamydosporol, zearalenone and its derivatives, cyclonerodiol and derivatives, and equisetin arise from independent biosynthetic pathways or are plausible precursors to the major end-product metabolites. A biosynthetic relationship between butenolide and moniliformin as products from glutamic acid metabolism was suggested by ApSimon et al (1991).

A detailed study of the relationships between all of these metabolites supports the concept of a metabolic grid present in *Fusarium* fungi rather than unique oxygenation or cyclization pathways, and enzyme systems that are very flexible in substrate selection.

Spectral Format

The detailed discussion of the NMR and MS experiments required for unambiguous assignment and structural determination of many of these compounds is presented elsewhere (Blackwell et al 1984; Greenhalgh et al 1986c, 1989a,b; Savard et al 1987). Where a sufficient quantity of compound was obtained, the ^{1}H spectra were assigned using the techniques of ^{1}H/^{1}H correlation spectroscopy (COSY) adjusted for both vicinal and long-range couplings, relayed coherence spectroscopy, total correlation spectroscopy, and nuclear Overhauser enhancement difference spectroscopy. Carbon resonance assignments were aided by the use of polarization transfer (DEPT) and ^{1}H/^{13}C heteronuclear correlation spectra. On occasion, quaternary carbon assignments were determined by long-range ^{1}H/^{13}C correlation acquired in normal mode (COLOC) or inverse mode (HMBC). When only trace quantities were available, assignments were made by comparison with similar compounds using known trends.

With only a few exceptions, if the spectra are presented, the data have been recorded as outlined below. Otherwise the data have been reproduced from the supplied references and may not be complete. Coupling constants that are reported in the listings are not specifically defined; greater detail may be obtained in the associated references. Small coupling constants that are reported in the spectral listings are often not visible in the spectra, especially those at 500 MHz. However, these values have been measured on an expansion with extra resolution enhancement and data point resolution, and confirmed by COSY correlations.

The molecular weight quoted for each metabolite is the calculated molecular weight that would be expected in a high-resolution mass spectrum (i.e., based on the most abundant isotopes). The mass spectra are presented in electron impact mode unless otherwise noted for the individual metabolite as being acquired by chemical ionization or fast atom bombardment techniques. Mass spectrometry was performed on a Finnigan MAT model 4500 gas chromatograph mass spectrometer equipped with a DS 400 (INCOS) data system. High-resolution electron impact, fast atom bombardment, and chemical ionization experiments were performed on a Finnigan MAT 312 mass spectrometer.

Nuclear magnetic resonance spectra were acquired on Bruker AM 250 or AM 500 NMR spectrometers operating at 250 (500) MHz and 62.8 (125.7) MHz, respectively, for ^{1}H and ^{13}C, using 5-mm dual probes. While most spectra were recorded in CDCl$_3$, alternative solvents are noted in the tables of spectral assignments. Chemical shifts are all reported relative to tetramethylsilane. Shifts are referenced to CHCl$_3$ at 7.24 ppm and CDCl$_3$ at 77.0 ppm for ^{1}H and ^{13}C, respectively, and similarly to 3.30 ppm and 49.0 ppm for CD$_3$OD. Typical ^{1}H spectra used a 32K data point set, 60° pulse, and 6–8 s total recycle time. ^{1}H spectra are often resolution enhanced. COSY-45 spectra were generally accumulated using 512K FIDs (free induction decay units) of 2K data points each, using the appropriate 90° pulse (12 μs for 250 MHz and 14 μs for 500 MHz). The data matrix was processed using sine bell windows, shifted $\pi/2$ in F_1, and zerofilled to give a final symmetrizable matrix of 1K by 2K. Data point resolution varied from 1 to 3 Hz per point as sweep widths were optimized for the individual samples. Similar parameters were used for the acquisition and processing of other ^{1}H two-dimensional experiments.

^{13}C NMR spectra were acquired using a 32K data point set, 60–70° pulse, 30 KHz sweep width, and total recycle time of 1.6–2.0 s. Heteronuclear correlation experiments were acquired using 256 FIDs (free induction decay units) of 4K data points, 96 or 128 transients, and 3.4 s total aquisition time. The data set was zerofilled once in F_1, and processed using a sine bell squared window shifted by $\pi/4$ in F_1 and a cosine bell squared in F_2, for a final data point resolution of <9 Hz/point in ^{13}C and <6 Hz/point in ^{1}H.

For each metabolite, the ^{1}H, ^{13}C, and mass spectra are presented in descending order from the top of the page. There is no standard display for the NMR spectra because they are plotted appropriately for the spectral window. When a compound has been identified by gas chromatography-mass spectrometry only and was either not further purified or of insufficient quantity for NMR characterization, only the mass spectrum appears. While there are a few commercial samples, most of the compounds presented have been isolated in our laboratories from large-scale fungal cultures originally extracted to produce either deoxynivalenol or T-2 toxin for toxicity studies, and which formed the basis of several theses and postdoctoral projects.

Acknowledgments

We thank the many students, technicians, and postdoctoral fellows who contributed their efforts in producing these metabolites. We also gratefully acknowledge the efforts of J. Nikiforuk for the acquisition and presentation of the NMR spectra, P. Lafontaine for a similar contribution of mass spectra, and L. Blais for assistance in assembling the data. We also thank J. D. Miller and J. W. ApSimon for their discussions, and the artists of Agriculture Canada's Research Program Services for their assistance in preparing the many spectra and molecular structures. We would also like to thank R. Greenhalgh, who, several years ago, proposed the preparation of an updated spectral compilation of fungal secondary metabolites. This represents publication no. 1479 from the Plant Research Centre.

Literature Cited

Abbas, H. K., Mirocha, C. J., Meronuck, R. A., Pokorny, J. D., Gould, S. L., and Kommedahl, T. 1988. Mycotoxins and *Fusarium* spp. associated with infected ears of corn in Minnesota. Appl. Environ. Microbiol. 54:1930-1933.

Abbas, H. K., Mirocha, C. J., Vesonder, R. F. and Gunther, R. 1990. Acute toxic effects of an isolate of moniliformin-producing *Fusarium oxysporum* and purified moniliformin on rats. Arch. Environ. Contam. Toxicol. 19:433-436.

ApSimon, J. W., Blackwell, B., Blais, L., Fielder, D. A., Greenhalgh, R., Kasitu, G., Kendall, J. E., Miller, J. D., and Savard, M. 1991. The *Fusarium* genus: A versatile biosynthetic engine. Pages 201-218 in: Studies in Natural Product Chemistry, vol. 9. Atta-ur-Rahman, ed. Elsevier Science Publishers, Amsterdam.

Ashley, J. N., Hobbs, B. C., and Raistrick, H. 1937. LV. Studies in the biochemistry of micro-organisms LIII. The crystalline colouring matters of *Fusarium culmorum* (W.G. Smith) Sacc. and related forms. Biochem. J. 31:385-397.

Audhya, T. K., and Russell, D. W. 1973. Production of enniatin A. Can. J. Microbiol. 19:1051-1054.

Bamburg, J. R., and Strong, F. M. 1969. Mycotoxins of the trichothecane family produced by *Fusarium tricinctum* and *Trichoderma lignorum*. Phytochemistry 8:2405-2410.

Bamburg, J. R., Riggs, N. V., and Strong, F. M. 1968. The structures of toxins from two strains of *Fusarium tricinctum*. Tetrahedron 24:3329-3336.

Barton, D. H. R., and Werstiuk, N. H. 1967. The constitution and stereochemistry of culmorin. Chem. Commun. pp. 30-31.

Bennett, G. A., Wicklow, D. T., Caldwell, R. W., and Smalley, E. B. 1988. Distribution of trichothecenes and zearalenone in *Fusarium graminearum*: Rotted corn ears grown in a controlled environment. J. Agric. Food Chem. 36:639-642.

Bezuidenhout, S. C., Gelderblom, W. C. A., Gorst-Allman, C. P., Horak, R. M., Marasas, W. F. O., Spiteller, G., and Vleggaar, R. 1988. Structure elucidation of the fumonisins, mycotoxins from *Fusarium moniliforme*. J. Chem. Soc. Chem. Comm. pp. 743-745.

Blackwell, B. A., Greenhalgh, R., and Bain, A. D. 1984. Carbon-13 and proton nuclear magnetic resonance spectral assignments of deoxynivalenol and other mycotoxins from *Fusarium graminearum*. J. Agric. Food Chem. 32, 1078-1083.

Blais, L. A., ApSimon, J. W., Blackwell, B. A., Greenhalgh, R., and Miller, J. D. 1992. Isolation and characterization of enniatins from *Fusarium avenaceum* DAOM 196490. Can. J. Chem. 70(5):1281-1287.

Blight, M. M., and Grove, J. F. 1974. New metabolic products of *Fusarium culmorum*: Toxic trichothec-9-en-8-ones and 2-acetylquinazolin-4(3H)-one. J. Chem. Soc. Perkin Trans. 1, pp. 1691-1693.

Bolliger, G., and Tamm, C. 1972. Vier neue metabolite von *Gibberella zeae*: 5-formyl-zearalenon, 7'-dehydrozearalenon, 8'-hydroxy- und 8'-epi-hydroxy-zearalenon. Helv. Chim. Acta 55:3030-3048.

Burmeister, H., and Plattner, R. D. 1987. Enniatin production by *Fusarium tricinctum* and its effect on germinating wheat seeds. Phytopathology 77:1483-1487.

Burmeister, H. R., Bennett, G. A., Vesonder, R. F., and Hesseltine, C. W. 1974. Antibiotic produced by *Fusarium equiseti* NRRL 5537. Antimicrob. Agents Chemother. 5:634-639.

Burmeister, H. R., Grove, M. D., Peterson, R. E., Weisleder, D., and Plattner, R. D. 1985. Isolation and characterization of two new fusaric acid analogs from *Fusarium moniliforme* NRRL 13,163. Appl. Environ. Microbiol. 50:311-314.

Cane, D. E., Iyengar, R., and Shiao, M.-S. 1981. Cyclonerodiol biosynthesis and the enzymatic conversion of farnesyl to nerolydil pyrophosphate. J. Am. Chem. Soc. 103:914-931.

Christensen, C. M., Nelson, G. H., and Mirocha, C. J. 1965. Effect on the white rat uterus of a toxic substance isolated from *Fusarium*. Appl. Microbiol. 13:653-659.

Cole, R. J., and Cox, R. H. 1981. Handbook of Toxic Fungal Metabolites. Academic Press, New York.

Cole, R. J., Kirksey, J. W., Cutler, H. G., Doubnik, B. L., and Peckham, J. C. 1973. Toxin from *Fusarium moniliforme*: Effects on plants and animals. Science 179:1324-1326.

Cole, R. J., Dorner, J. W., Gibert, J., Mortimer, D. N., Crews, C., Mitchell, J. C., Windingstad, R. M., Nelson, P. E., and Cutler, H. C. 1988. Isolation and identification of trichothecenes from *Fusarium compactum* suspected in the aetiology of a major intoxication of sandhill cranes. J. Agric. Food Chem. 36:1163-1167.

Cordier, C., Gruselle, M., Jaouen, G., Hughes, D. W., and McGlinchey, M. J. 1990. Structures of zearalenone and zeralanone in solution: A high-field NMR and molecular modelling study. Magn. Reson. Chem. 28:835-845.

Corley, D. G., Rottinghaus, G. E., Tracy, J. K., and Tempesta, M. S. 1986. New trichothecene mycotoxins of *Fusarium sporotrichioides* (MC-72083). Tetrahedron Lett. 27:4133-4136.

Corley, D. G., Rottinghaus, G. E., and Tempesta, M. S. 1987a. Toxic trichothecenes from *Fusarium sporotrichioides* (MC-72083). J. Org. Chem. 52:4405-4408.

Corley, D. G., Rottinghaus, G. E., and Tempesta, M. S. 1987b. Secondary metabolites from *Fusarium*. Two new modified trichothecenes from *Fusarium sporotrichioides* MC-72083. J. Nat. Prod. 50:897-902.

Farber, J. M., and Sanders, G. W. 1986. Production of fusarin C by *Fusarium* spp. J. Agric. Food Chem. 34:963-966.

Farber, J. M., Sanders, G. W., Lawrence, G. A., and Scott, P. M. 1988. Production of moniliformin by Canadian isolates of *Fusarium*. Mycopathologia 101:187-190.

Fielder, D. A. 1988. Isolation and characterization of secondary metabolites from *Fusarium sporotrichioides* DAOM 165006. M.Sc. thesis, Carleton University, Ottawa.

Freeman, G. G., Gill, J. E., and Waring, W. S. 1959. The structure of trichothecin and its hydrolysis products. J. Chem. Soc. pp. 1105-1132.

Fujimoto, Y., Morita, Y., and Tatsuno, T. 1972. Recherches toxicologiques sur les substances toxiques de *Fusarium nivale*: Étude chimique des toxines principales, nivalénol, fusarenon-X et nivalénol-4,15-di-O-acétate. Chem. Pharm. Bull. 20:1194-1203.

Gaumann, E., Roth, S., Ettlinger, L., Plattner, P. A., and Nager, U. 1947. Enniatin, ein neues, gegen mykobakterien wirksames antibiotikum. Experientia 3:202-203.

Gelderblom, W. C. A., Marasas, W. F. O., Steyn, P. S., Thiel, P. G., van der Merwe, K. J., van Rooyen, P. H., Vleggaar, R., and Wessels, P. L. 1984. Structure elucidation of fusarin C, a mutagen produced by *Fusarium moniliforme*. J. Chem. Soc. Chem. Comm. pp. 122-124.

Golinski, P., Vesonder, R. F., Latus-Zietkiewicz, D., and Perkowski, J. 1988. Formation of fusarenone X, nivalenol, zearalenone, α-trans-zearalenol, β-trans-zearalenol and fusarin C by *F. crookwellense*. Appl. Environ. Microbiol. 54:2147-2148.

Greenhalgh, R., Hanson, A. W., Miller, J. D., and Taylor, A. 1984a. Production and X-ray crystal structure of 3α-acetoxy-7α,15-dihydroxy-12,13-epoxytrichothec-9-en-8-one. J. Agric. Food Chem. 32:945-948.

Greenhalgh, R., Meier, R.-M., Blackwell, B. A., Miller, J. D., Taylor, A., and ApSimon, J. W. 1984b. Minor metabolites of *Fusarium roseum* (ATCC 28114). J. Agric. Food Chem. 32:1261-1264.

Greenhalgh, R., Meier, R.-M., Blackwell, B. A., Miller, J. D., Taylor, A., and ApSimon, J. W. 1986a. Minor metabolites of *Fusarium roseum* (ATCC 28114), part 2. J. Agric. Food Chem. 34:115-118.

Greenhalgh, R., Levandier, D., Adams, W., Miller, J. D., Blackwell, B. A., McAlees, A. J., and Taylor, A. 1986b. Production and characterization of deoxynivalenoland other secondary metabolites of *Fusarium culmorum* (CMI 14764, HLX 1503). J. Agric. Food Chem. 34:98-102.

Greenhalgh, R., Blackwell, B. A., Pare, J. R. J., Miller, J. D., Levandier, D., Meier, R. M., Taylor, A., and ApSimon, J. W. 1986c. Isolation and characterization by mass spectrometry and NMR spectroscopy of secondary metabolites of some *Fusarium* species. Pages 137-152 in: Mycotoxins and Phycotoxins. P. S. Steyn and R. Vleggaar, eds. Elsevier Press, Amsterdam.

Greenhalgh, R., Blackwell, B. A., Savard, M., Miller, J. D., and Taylor, A. 1988. Secondary metabolites produced by *Fusarium sporotrichioides* DAOM 165006 in liquid culture. J. Agric. Food Chem. 36:216-219.

Greenhalgh, R., Blackwell, B. A., and Savard, M. E. 1989a. The NMR spectra of trichothecenes and related fungal metabolites. Tetrahedron 45:2273-2283.

Greenhalgh, R., Fielder, D. A., Morrison, L. A., Charland, J.-P., Blackwell, B. A., Savard, M. E., and ApSimon, J. W. 1989b. Secondary metabolites of Fusarium species: Apotrichothecene derivatives. J. Agric. Food Chem. 37:699-705.

Greenhalgh, R., Fielder, D. A., Blackwell, B. A., Miller, J. D., Charland, J.-P., and ApSimon, J. W. 1990. Some minor secondary metabolites of *Fusarium sporotrichioides* DAOM 165006. J. Agric. Food Chem. 38:1978-1984.

Grove, J. F., and Hitchcock, P. B. 1991. Metabolic products of *Fusarium acuminatum*: Acuminatopyrone and chlamydosporol. J. Chem. Soc. Perkin Trans. 1, pp. 997-999.

Gyimesi, J., and Melera, A. 1967. On the structure of crotocin an antifungal antibiotic. Tetrahedron Lett. pp. 1665-1673.

Hacking, A., Rosser, W. R., and Dervish, M. T. 1976. Zearalenone-producing species of *Fusarium* on barley seed. Ann. Appl. Biol. 84:7-11.

Hagler, W. M., Mirocha, C. J., Pathre, S. V., and Behrens, J. C. 1979. Identification of the naturally occurring isomer of zearalenol produced by *Fusarium roseum* 'Gibbosum' in rice culture. Appl. Environ. Microbiol. 37:849-853.

Hanson, J. R., Marten, T., and Siverns, M. 1974. Studies in terpenoid biosynthesis. Part XII. Carbon-13 nuclear magnetic resonance spectra of the trichothecanes and the biosynthesis of trichothecolone from [2-^{13}C]mevalonic acid. J. Chem. Soc. Perkin

Trans., 1 pp. 1033-1036.

Hanson, J. R., Hitchcock, P. B., and Nyfeler, R. 1975. Cyclonerotriol [6-(3-hydroxy-2,3-dimethylcyclopentyl)-2-methylhept-2-ene-1,6-diol], a new sesquiterpenoid metabolite of Fusarium culmorum. J. Chem. Soc. Perkin Trans. 1, pp. 1586-1590.

Hesketh, A. R., Bycroft, B. W., Dewick, P. M., and Gilbert, J. 1993. Revision of the stereochemistry in trichodiol, trichotriol, and related compounds, and concerning their role in the biosynthesis of trichothecene mycotoxins. Phytochemistry 32:105-116.

Ishii, K., and Ueno, Y. 1981. Isolation and characterization of two new trichothecenes from Fusarium sporotrichioides strain M-1-1. Appl. Environ. Microbiol. 42:541-543.

Ishii, K., Pathre, S. V., and Mirocha, C. J. 1978. Two new trichothecenes produced by Fusarium roseum. J. Agric. Food Chem. 26:649-653.

Kasitu, G. C., ApSimon, J. W., Blackwell, B. A., Fielder, D. A., Greenhalgh, R., and Miller, J. D. 1992. Isolation and characterization of culmorin derivatives produced by Fusarium culmorum CM14764. Can. J. Chem. 70:1308-1316.

Kriek, N. P. J., Marasas, W. F. O., Steyn, P. S., van Rensburg, S. F., and Steyn, M. 1977. Toxicity of a moniliforming-producing strain of Fusarium moniliforme var. subglutinans isolated from maize. Food Cosmet. Toxicol. 15:579-587.

Lansden, J. A., Cole, R. J., Dorner, J. W., Cox, R. H., Cutler, H. C., and Clark, J. D. 1978. A new trichothecene mycotoxin isolated from Fusarium tricinctum. J. Agric. Food Chem. 26:246-249.

Lauren, D. R., Ashley, A., Blackwell, B. A., Greenhalgh, R., Miller, J. D., and Neish, G. A. 1987. Trichothecenes produced by Fusarium crookwellense DAOM 193611. J. Agric. Food Chem. 35:884-889.

Laurent, D., Lanson, M., Goasdoué, N., Kohler, F., Pellegrin, F., and Platzer, N. 1990. Étude en RMN ^{1}H et ^{13}C de la macrofusine, toxine isolée de mais infesté par Fusarium moniliforme Sheld. Analusis 18:172-179.

Marasas, W. F. O., Nelson, P. E., and Toussoun, T. A. 1984. Toxigenic Fusarium Species. Pennsylvania State University Press, University Park.

Mesilaakso, M., Moilanen, M., and Rahkamaa, E. 1989. ^{1}H and ^{13}C NMR analysis of some trichothecenes. Arch. Environ. Contam. Toxicol. 18:365-373.

Mirocha, C. J., Christensen, C. M., and Nelson, C. H. 1971. F-2 (zearalenone) estrogenic mycotoxin from Fusarium. Pages 107-138 in: Microbial Toxins, Vol.7. S. Kadis, A. Ciegler, and S. J. Ajl, eds. Academic Press, New York.

Mirocha, C. J., Pathre, S. V., Schauerhamer, B., and Christensen, C. M. 1976. Natural occurrence of Fusarium toxins in feedstuff. Appl. Environ. Microbiol. 32:553-556.

Mirocha, C. J., Abbas, H. K., Kommedahl, T., and Jarvis, B. B. 1989. Mycotoxin production by Fusarium oxysporum and Fusarium sporotrichioides isolated from Baccharis spp. from Brazil. Appl. Environ. Microbiol. 55:254-255.

Mohr, P., Tamm, C., Zurcher, W., and Zehnder, M. 1984. 48. Sambicinol and sambucoin, two new metabolites of Fusarium sambucinum possessing modified trichothecene structures. Helv. Chim. Acta 67:406-412.

Munoz, L., Castro, J. L., Cardelle, M., Castedo, L., and Riguera, R. 1989. Acetylated mycotoxins from Fusarium graminearum. Phytochemistry 28:83-85.

Mutert, W.-U., Lutfring, H., and Barz, W. 1981. Formation of fusaric acid by fungi of the genus Fusarium. Z. Naturforsch. C Biosci. 36C:338-339.

Nager, U. 1948. Über die Chemie der Enniatine. Diss. ETH (Basel, Birkhauser), 42S.

Nozoe, S., Goi, M., and Morisaki, N. 1970. Structure of cyclonerodiol. Tetrahedron Lett. pp. 1293-1296.

Ovchinnikov, Y. A., Ivanov, V. T., Mikhaleva, I. I., and Shemyakin, M. M. 1964. Izvest. Akad. Nauk. SSSR Khim. Ser. 1912. (English edition: Synthesis of eniatin

C. Bull. Acad. Sci. USSR Div. Chem. Sci. 1823.)

Ovchinnikov, Y. A., Ivanov, V. T., Evstratov, A. V., Mikhaleva, I. I., Bystrov, V. F., Portnova, S. L., Balashova, T. A., Meshcheryakova, E. N., and Tulchinsky, V. M. 1974. The enniatin ionophores. Conformation and ion biding properties. J. Pept. Protein Res. 6:465-498.

Pathre, S. V., and Mirocha, C. J. 1979. Zearalenone and related compounds. Adv. Chem. Ser. 149:178-227.

Pathre, S. V., Mirocha, C. J., Christensen, C. M., and Behrens, J. 1976. Monoacetoxyscirpenol. A new mycotoxin produced by *Fusarium roseum* Gibbosum. J. Agric. Food Chem. 24:97-103.

Phillips, N. J., Goodwin, J. T., Fraiman, A., Cole, R. J., and Lynn, D. G. 1989. Characterization of the *Fusarium* toxin equisetin: The use of phenylboronates in structure assignment. J. Am. Chem. Soc. 111:8223-8231.

Plattner, R. D., Weisler, D., Shackelford, D. D., Peterson, R., and Powell, R. G. 1992. A new fumonisin from solid cultures of *Fusarium moniliforme*. Mycopathologia 117:23-28.

• Richardson, K. E., Toney, G. E., Haney, C. A., and Hamilton, P. B. 1989. Occurrence of scirpentriol and its seven acetylated derivatives in culture extracts of *Fusarium sambucinum* NRRL 13495. J. Food Protection 52:871-876.

Ross, P. F., Nelson, P. E., Richard, J. L., Osweiler, G. D., Rice, L. G., Plattner, R. D., and Wilson, T. M. 1990. Production of fumonisins by *Fusarium moniliforme* and *Fusarium proliferatum* isolates associated with equine leukoencephalomalacia and a pulmonary edema syndrome in swine. Appl. Environ. Microbiol. 56:3225-3226.

Rosslein, L., Tamm., C., Zurcher, W., Riesen, A., and Zehnder, M. 1988. 66. Sambucinic acid, a new metabolite of *Fusarium sambucinum*. Helv. Chim. Acta 71:588-595.

Sanson, D. R. 1989. Chemistry of *Fusarium sambucinum* & *F. moniliforme*. Ph.D. thesis, University of Missouri, Columbia.

Sanson, D. R., Corley, D. G., Barnes, C. L., Searles, S., Schlemper, E. O., and Tempesta, M. S. 1989. New mycotoxins from *Fusarium sambucinum*. J. Org. Chem. 54:4313-4318.

Savard, M. E., and Greenhalgh, R. 1987. Synthesis and NMR analysis of new natural trichothecenes. J. Nat. Prod. 50:953-957.

Savard, M. E., and Miller, J. D. 1992. Characterization of fusarin F, a new fusarin from *Fusarium moniliforme*. J. Nat. Prod. 55:64-70.

Savard, M. E., Blackwell, B. A., and Greenhalgh, R. 1987. A ^{1}H nuclear magnetic resonance study of derivatives of 3-hydroxy-12,13-epoxytrichothec-9-enes. Can. J. Chem. 65:2254-2261.

Savard, M. E., Blackwell, B. A., and Greenhalgh, R. 1989. The role of ^{13}C-labelled trichodiene and bazzanene in the secondary metabolism of *Fusarium culmorum*. J. Nat. Prod. 52(6):1267-1278.

Savard, M. E., Miller, J. D., Salleh, B., and Strange, R. N. 1990. Chlamydosporol, a new metabolite from *Fusarium chlamydosporum*. Mycopathologia 110:177-181.

Scott, P. M., Abbas, H. K., Mirocha, C. J., Lawrence, G. A., and Weber, D. 1987. Formation of moniliformin by *Fusarium sporotrichioides* and *Fusarium culmorum*. Appl. Environ. Microbiol. 53:196-197.

Steyn, P. S., and Vleggaar, R. 1985. Biosynthetic studies of the fusarins, metabolites of *Fusarium moniliforme*. J. Chem. Soc. Chem. Comm. 1189-1191.

Stipanovic, R. D., and Schroeder, H. W. 1975. Zearalenone and 8'-hydroxy-zearalenone from *Fusarium roseum*. Mycopathologia 57:77-78.

Sugiura, Y., Watanabe, Y., Tanaka, T., Yamamoto, S., and Ueno, Y. 1990. Occurrence of *Gibberella zeae* strains that produce both nivalenol and deoxynivalenol. Appl. Environ. Microbiol. 56:3047-3051.

Sydenham, E. W., Thiel, P. G., Marasas, W. F. O., and Nieuwenhuis, J. J. 1989. Occurrence of deoxynivalenol and nivalenol in *Fusarium graminearum* infected undergrade wheat in South Africa. J. Agric. Food Chem. 37:921-926.

Tatsuno, T., Morita, Y., Tsunoda, H., and Uneda, M. 1970. Recherches toxicologiques des substances métaboliques du *Fusarium nivale*. VII. Latroisième substance métabolique de *F. nivale*, le diacétate de nivalénol. Chem. Pharm. Bull. 18:1485-1487.

Thiel, P. G., Meyer, C. J., and Marasas, W. F. O. 1982. Natural occurrence of moniliformin together with deoxynivalenol and zearalenone in Transkeian corn. J. Agric. Food Chem. 30:308-312.

Thiel, P. G., Marasas, W. F. O., Sydenham, E. W., Shephard, G. S., Gelderblom, W. C. A., and Nieuwenhuis, J. J. 1991. Survey of fumonisin production by *Fusarium* species. Appl. Environ. Microbiol. 57:1089-1093.

Tidd, B. K. 1967. Phytotoxic compounds produced by *Fusarium equiseti*. Part III. Nuclear magnetic resonance spectra. J. Chem. Soc. C, 218-220.

Ueno, Y., Ueno, I., Tatsuno, T., Okubo, K., and Tsunoda, H. 1969. Fusarenon-X, a toxic principle of *Fusarium nivale*-culture filtrate. Experientia 25:1062.

Urry, W. H., Wehrmeister, H. L., Lodge, E. B., Hidy, P. H. 1966. The structure of zearalenone. Tetrahedron Lett. pp. 3109-3114.

Vesonder, R. F., Ciegler, A., and Jensen, A. H. 1973. Isolation of the emetic principle from *Fusarium*-infected corn. Appl. Microbiol. 25:1008-1010.

Vesonder, R. F., Tjarks, L. W., Ciegler, A., Spencer, G. F., and Wallen, L. L. 1977. 4-Acetamido-2-butenoic acid from *Fusarium gràminearum*. Phytochemistry 16:1296-1297.

Visconti, A., Mirocha, C. J., Bottalico, A., and Chelkowski, J. 1985. Trichothecene mycotoxins produced by *Fusarium sporotrichioides* strain P-11. Mycotox. Res. 1:3-10.

Visconti, A., Mirocha, C. J., Logrieco, A., Bottalico, A., and Solfrizzo, M. 1989. Mycotoxins produced by *Fusarium acuminatum*. Isolation and characterization of acuminatin: A new trichothecene. J. Agric. Food Chem. 37:1348-1351.

Visconti, A., Blais, L., ApSimon, J. W., Greenhalgh, R., Miller, J. D. 1992. Production of enniatins by *Fusarium acuminatum* and *Fusarium compactum* in liquid culture: Isolation and characterization of three new enniatins, B2, B3 and B4. J. Agric. Food Chem. 40: 1076-1082.

Wei, C.-M., and McLaughlin, C. S. 1974. Structure-function relationship in the 12,13-eposytrichothecenes. Biochem. Biophys. Res. Commun. 57:838-844.

Wiebe, L. A., and Bjeldanes, L. F. 1981. Fusarin C, a mutagen from *Fusarium moniliforme* grown on corn. J. Food Sci. 46:1424-1426.

Yabuta, T., Kambe, K., and Hayaski, T. 1934. Biochemistry of the bakanae-fungus. I. Fusarinic acid, a new product of the bakanae-fungus. J. Agric. Chem. Soc. Jpn. 10:1059-1068. (Chem. Abstr. 29:1132, 1935).

Yates, S. G. 1971. Toxin-producing fungi from fescue pasture. Pages 191-206 in: Microbial Toxins, vol. 7. S. Kadis, A. Ciegler, and S. J. Ajl, eds. Academic Press, New York.

Yates, S. G., Tookey, H. L., Ellis, J. J., and Burkhardt, H. J. 1967. Toxic butenolide produced by *Fusarium nivale* (Fries) Cesati isolated from tall fescue (*Festuca arundinacea* Schreb.). Tetrahedron Lett. pp. 621-625.

Yoshizawa, T., Shirota, T., Morooka, N. J. 1978. Toxic substances in infected cereals. VI. Deoxynivalenol and its acetate as feed refusal principles in rice cultures with *Fusarium roseum* no. 117 (ATCC 28114). J. Food Hyg. Soc. Jpn. 19:178-184.

Alphabetical Listing of Compounds

COMPOUND   PAGES

ISOTRICHODERMIN

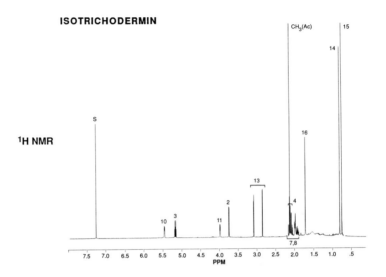

¹H NMR

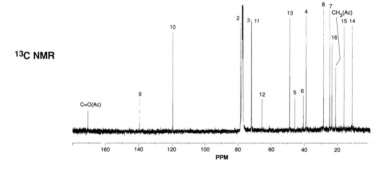

¹³C NMR

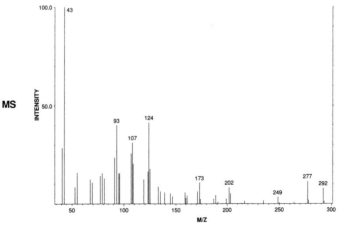

MS

ISOTRICHODERMIN

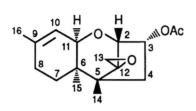

$C_{17}H_{24}O_4$

M.W. 292.1674

Species:

F. *culmorum*
F. *roseum*
F. *graminearum*
F. *crookwellense*

References: Greenhalgh et al. (1984b); Lauren et al. (1987).

Position	^{13}C	^1H
2	78.2	3.73 (d, 4.6Hz)
3	71.7	5.15 (ddd, 10.0, 4.6, 4.7)
4	38.6	2.1 (m)
5	45.3	
6	40.2	
7	24.4	1.90-2.00 (m)
8	28.2	1.90-2.00 (m)
9	139.5	
10	119.4	5.44 (dq, 5.4, 1.3)
11	71.5	3.96 (d, 5.5)
12	65.2	
13	48.4	2.84 (d, 4.0)
		3.07 (d, 4.0)
14	11.0	0.80 (s)
15	15.9	0.74 (s)
16	23.1	1.71 (br s)
CH_3(Ac)	20.9	2.13 (s)
C=O(Ac)	170.6	

7-HYDROXY-ISOTRICHODERMIN

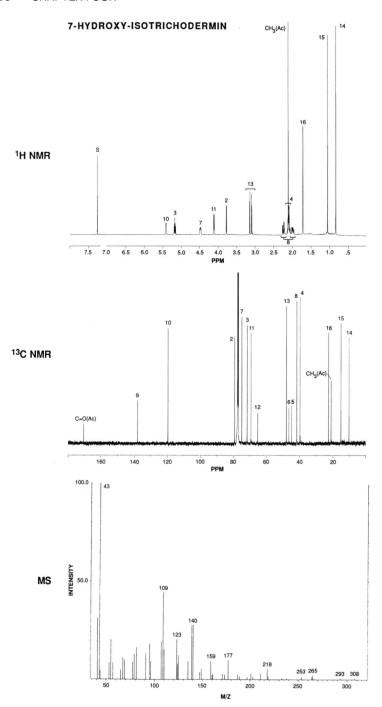

7-HYDROXY-ISOTRICHODERMIN

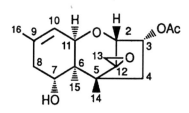

$C_{17}H_{24}O_5$

M.W. 308.1624

Species:

F. roseum
F. crookwellense

References: Greenhalgh et al. (1986a); Lauren et al. (1987).

Position	^{13}C	^1H
2	79.0	3.76 (d, 4.5)
3	71.4	5.16 (ddd, 9.8, 5.2, 4.5)
4	39.5	2.07 (m)
		2.10 (m)
5	44.6	
6	46.2	
7	74.7	4.47 (dd, 10.0, 6.0)
8	41.4	2.00 (dd, 17.2, 10.0)
		2.23 (dd, 17.2, 6.0)
9	138.0	
10	119.6	5.40 (br d, 5.4)
11	69.0	4.10 (d, 5.4)
12	64.9	
13	47.7	3.08 (d, 4.4)
		3.10 (d, 4.4)
14	10.2	0.82 (s)
15	15.0	1.04 (s)
16	22.4	1.70 (br s)
CH_3(Ac)	20.9	2.10 (s)
C=O(Ac)	170.6	

8-HYDROXY-ISOTRICHODERMIN

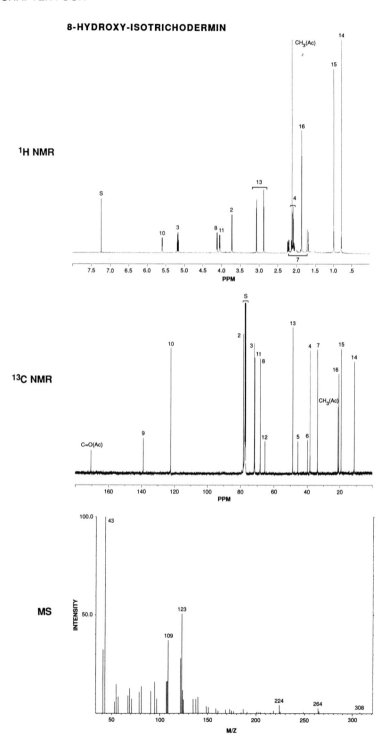

8-HYDROXY-ISOTRICHODERMIN

$C_{17}H_{24}O_5$

M.W. 308.1624

Species:

F. culmorum
F. roseum
F. crookwellense

References: Greenhalgh et al. (1986a); Lauren et al. (1987).

Position	^{13}C	^1H
2	78.1	3.72 (d, 4.6)
3	71.5	5.16 (ddd, 9.8, 5.5, 4.6)
4	38.1	2.09 (m, 9.8, 5.5)
5	45.4	
6	39.6	
7	33.6	1.70 (dd, 14.3, 1.4)
		2.21 (dd, 14.3, 5.7)
8	68.0	4.10 (br d, 5.7)
9	138.8	
10	122.1	5.59 (d, 5.7)
11	71.3	4.04 (d, 5.7)
12	65.2	
13	48.5	2.87 (d, 4.1)
		3.06 (d, 4.1)
14	11.0	0.77 (s)
15	19.0	0.98 (s)
16	20.4	1.86 (br s)
CH_3(Ac)	20.9	2.11 (s)
C=O(Ac)	170.6	

7,8-DIHYDROXY-ISOTRICHODERMIN

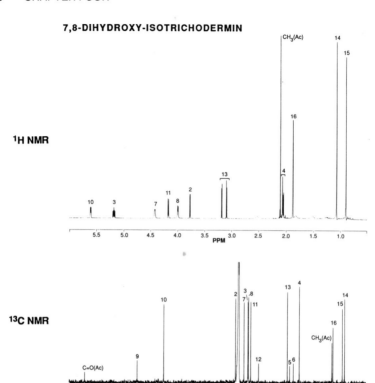

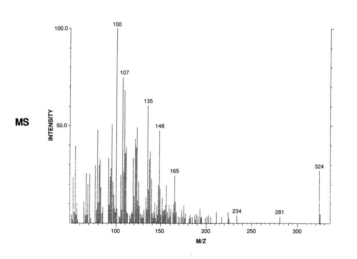

7,8-DIHYDROXY-ISOTRICHODERMIN

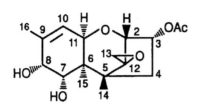

$C_{17}H_{24}O_6$

M.W. 324.1573

Species:

F. crookwellense

References: Lauren *et al.* (1987).

Position	^{13}C	^1H
2	78.9	3.76 (d, 4.5)
3	71.5	5.17 (m, 8.8, 6.9, 4.5)
4	40.6	α: 2.05 (m, 6.9, 7.8)
		β: 2.05 (m, 8.8, 7.8)
5	46.4	
6	44.2	
7	73.8	4.41 (dd, 9.7, 5.5)
8	71.1	3.98 (bt, 6.0)
9	138.8	
10	122.6	5.59 (dq, 5.9, 1.6)
11	69.8	4.16 (d, 5.9)
12	65.0	
13	47.8	3.10 (d, 4.3)
		3.17 (d, 4.3)
14	13.2	1.06 (s)
15	14.7	0.89 (s)
16	20.3	1.87 (br s)
CH_3(Ac)	20.9	2.10 (s)
C=O(Ac)	170.6	

Spectra not provided.

8β-HYDROXYTRICHOTHECENE

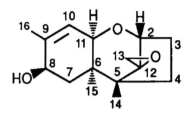

$C_{15}H_{22}O_3$

M.W. 250.1569

Species:

F. sporotrichioides

References: Corley et al. (1986).

Position	^{13}C	^1H
2	81.6	3.65 (d, 3.9)
3	30.2	1.91 (m)
		1.57 (dd, 13.0, 7.0)
4	36.8	2.78 (dd, 13.0, 7.0)
		1.68 (dd, 13.6, 7.0)
5	46.8	
6	44.8	
7	39.9	2.03 (dd, 12.8, 5.5)
		1.15 (dd, 12.5, 9.3)
8	69.2	4.12 (m)
9	136.2	
10	128.9	5.27 (m)
11	71.6	4.66 (m)
12	69.1	
13	49.3	3.37 (d, 4.1)
		3.06 (d, 4.1)
14	20.4	0.96 (s)
15	13.5	0.96 (s)
16	18.2	1.77 (s)

CALONECTRIN

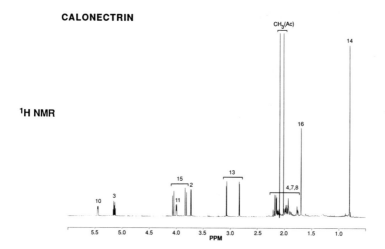

1H NMR

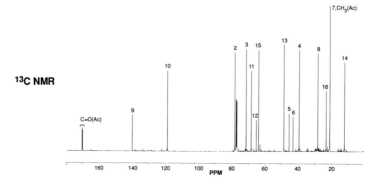

13C NMR

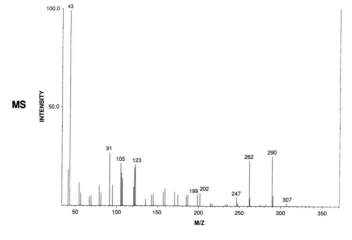

MS

CALONECTRIN

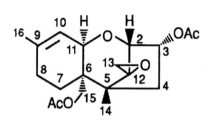

$C_{19}H_{26}O_6$

M.W. 350.1729

Species:

F. culmorum
F. roseum
F. graminearum

References: Greenhalgh *et al.* (1986a, 1986b).

Position	^{13}C	^1H
2	78.0	3.74 (d, 4.6)
3	71.2	5.15 (ddd, 9.6, 7.4, 4.6)
4	39.3	2.1-2.19 (m)
5	45.3	
6	42.9	
7	20.9	1.78-2.19 (m)
8	28.1	1.78-2.19 (m)
9	140.3	
10	118.9	5.45 (dq, 5.4, 1.3)
11	68.0	4.00 (d, 5.6)
12	64.9	
13	48.4	α: 2.84 (d, 4.0)
		β: 3.08 (d, 4.0)
14	12.1	0.81 (s)
15	63.5	α: 3.82 (d, 12.2)
		β: 4.06 (d, 12.2)
16	23.1	1.70 (br s)
CH_3(Ac)	20.9	2.01 (s)
	21.0	2.08 (s)
C=O (Ac)	170.4	
	170.7	

15-DEACETYLCALONECTRIN

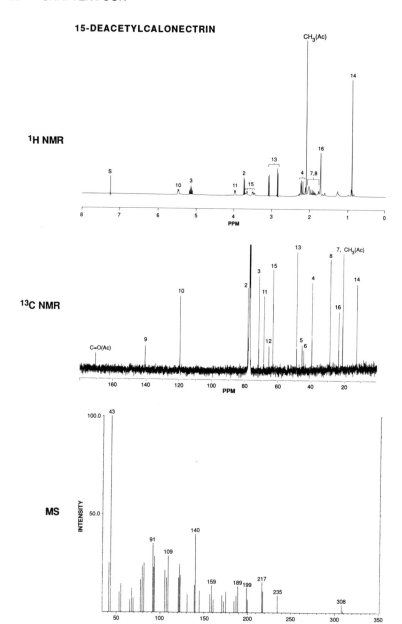

15-DEACETYLCALONECTRIN

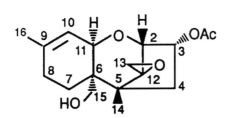

$C_{17}H_{24}O_5$

M.W. 308.1624

Species:

F. roseum
F. culmorum
F. graminearum

References: Greenhalgh et al. (1984b, 1986b).

Position	^{13}C	^1H
2	78.2	3.72 (d, 4.7)
3	71.5	5.13 (ddd, 10.0, 5.0, 4.6)
4	39.4	1.73-2.30 (m)
5	45.3	
6	44.2	
7	20.8	1.73-2.30 (m)
8	28.5	1.73-2.30 (m)
9	140.5	
10	119.5	5.46 (dd, 5.3, 1.5)
11	68.3	3.96 (d, 5.3)
12	65.2	
13	48.5	2.84 (d, 4.0)
		3.07 (d, 4.0)
14	12.3	0.90 (s)
15	62.8	3.47 (dd, 11.6, 4.5)
		3.66 (dd, 11.6, 3.2)
16	23.1	1.71 (d, 1.5)
CH_3(Ac)	20.8	2.09 (s)
C=O(Ac)	170.6	

3-DEACETYLCALONECTRIN

MS

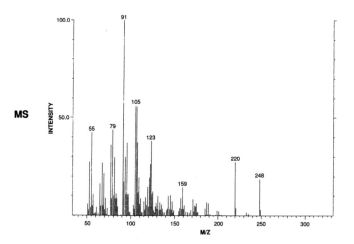

3-DEACETYLCALONECTRIN

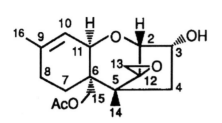

$C_{17}H_{24}O_5$

M.W. 308.1624

Species:

F. graminearum

References: Greenhalgh *et al.* (1986b).

Position	^{13}C	^1H
2	68.4	3.48 (d, 4.6)
3	69.1	4.43 (ddd, 4.6, 6, 9.5)
4	42.4	1.9-2.3 (m)
5	45.9	
6	45.9	
7	21.2	1.9-2.3 (m)
8	28.4	1.9-2.3 (m)
9	140.1	
10	119.3	5.48 (dd, 5.4, 1.4)
11	80.0	4.13 (d, 5.4)
12	65.6	
13	48.4	2.82 (d, 4.0)
		3.06 (d, 4.0)
14	12.3	0.79 (s)
15	62.8	3.82 (d, 12.2)
		4.06 (d, 12.2)
16	23.2	1.70 (s)
CH_3(Ac)	21.0	2.02 (s, 15-Ac)
C=O(Ac)	170.6	

7-HYDROXYCALONECTRIN

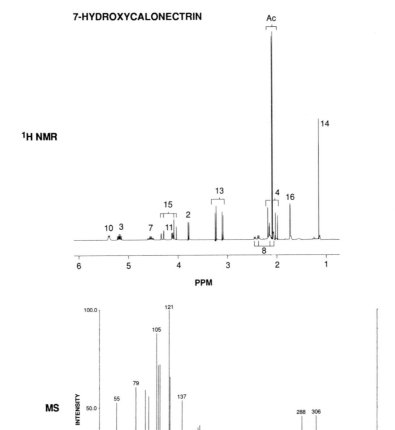

¹H NMR

MS

7-HYDROXYCALONECTRIN

$C_{19}H_{26}O_7$

M.W. 366.1678

Species:

F. roseum

References: Greenhalgh *et al.* (1986a).

Position	^{13}C	^1H
2	78.9	3.78 (d, 4.4)
3	71.1	5.17 (ddd, 10.0, 8.0, 4.5)
4	39.6	1.97-2.18 (m, 9.8)
5	46.1	
6	46.4	
7	71.0	4.55 (dd, 5.9, 11.2)
		1.41 (d, 7-OH, 11.2)
8	41.7	1.9-2.0 (m)
		2.41 (dd, 17.6, 5.9)
9	139.4	
10	119.0	5.38 (dd, 5.4, 1.3)
11	70.0	4.11 (5.4)
12	64.3	
13	47.8	3.09 (d, 4.4)
		3.23 (d, 4.4)
14	15.9	1.14 (s)
15	64.4	4.05 (d, 12.2)
		4.31 (m)
16	22.4	1.72 (br s)
CH_3(Ac)	21.0	2.08 (s)
	20.8	2.09 (s)
C=O(Ac)	170.0	
	170.1	

8-HYDROXYCALONECTRIN

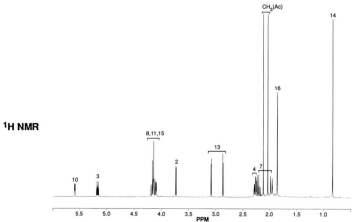

¹H NMR

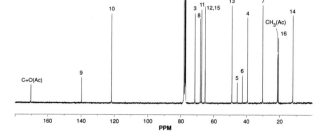

¹³C NMR

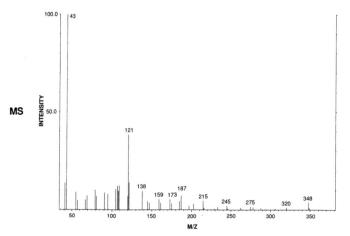

MS

8-HYDROXYCALONECTRIN

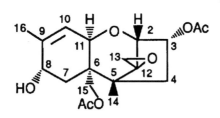

$C_{19}H_{26}O_7$

M.W. 366.1678

Species:

F. culmorum
F. roseum
F. graminearum

References: Greenhalgh et al. (1986a, 1986b).

Position	^{13}C	^1H
2	77.8	3.73 (d, 4.6)
3	71.1	5.15 (ddd, 10.0, 5.1, 5.0)
4	39.1	2.23-2.28 (m)
5	45.4	
6	42.3	
7	30.0	1.95 (d, 14.6)
		2.19 (dd, 14.6, 10.8)
8	67.8	4.08-4.18 (m)
9	139.5	
10	121.5	5.58 (dd, 5.7, 1.1)
11	67.0	4.08-4.18 (m)
12	64.9	
13	48.6	2.86 (d, 4.0)
		3.08 (d, 4.0)
14	11.9	0.83 (s)
15	65.0	4.08-4.18 (m)
16	20.5	1.85 (d, 0.6)
CH_3(Ac)	20.9	2.03 (s)
	21.0	2.11 (s)
C=O(Ac)	170.3	
	170.4	

7,8-DIHYDROXYCALONECTRIN

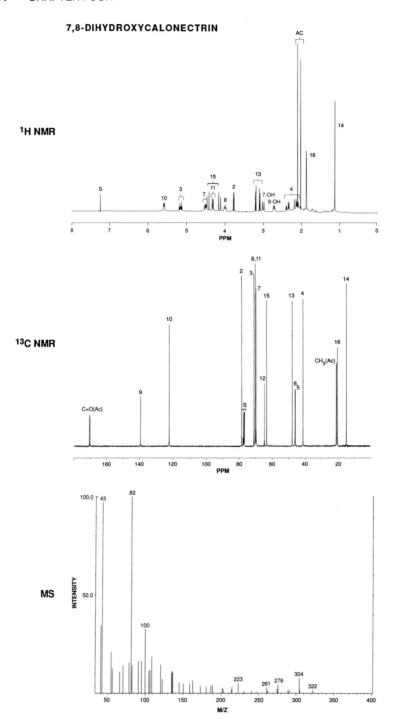

7,8-DIHYDROXYCALONECTRIN
(3,15-DIACETYL-7,8-DIHYDROXY-12,13-EPOXYTRICHOTHEC-9-ENE)

$C_{19}H_{26}O_8$

M.W. 382.1627

Species:

F. culmorum
F. roseum
F. graminearum

References: Greenhalgh et al. (1986a, 1986b).

Position	^{13}C	^1H
2	78.5	3.77 (d, 4.4)
3	71.0	5.15 (ddd, 11.1, 4.5, 4.4)
4	41.3	2.13 (m, 11.2)
		2.34 (dd, 15.0, 4.6)
5	46.1	
6	46.2	
7	69.6	4.50 (dd, 5.8, 9.6)
		3.00 (d, 7-OH, 9.7)
8	70.5	3.99 (dd, 5.8, 7.5)
		2.76 (d, 8-OH, 7.5)
9	139.2	
10	122.5	5.58 (dd, 5.8, 0.9)
11	70.1	4.32 (d, 5.8)
12	64.6	
13	47.8	3.09 (d, 4.3)
		3.20 (d, 4.3)
14	15.0	1.12 (s)
15	63.1	4.14 (d, 12.3)
		4.44 (d, 12.3)
16	20.4	1.86 (d, 1.0)
CH_3(Ac)	21.1	2.02 (s)
	21.0	2.10 (s)
C=O(Ac)	170.4	
	170.4	

Spectra not provided.

4-MONOACETOXYSCIRPENOL

$C_{17}H_{24}O_6$

M.W. 324.1573

Species:

F. roseum
F. sambucinum

References: Ishii *et al.* (1978); Richardson *et al.* (1989); Sanson *et al.* (1989).

Position	^{13}C	^1H
2		3.68 (d, 4.8)
3		4.26 (dd, 3.1, 4.8)
4		5.53 (d, 3.1)
5		
6		
7		1.7-2.2 (m)
8		1.7-2.2 (m)
9		
10		5.58 (br d, 5.0)
11		4.20 (br d, 5.0)
12		
13		2.78 (d, 4.0)
		3.06 (d, 4.0)
14		0.86 (s)
15		3.64 (d, 12)
		3.69 (d, 12)
16		1.74 (br s)
CH_3(Ac)		2.17 (s)
C=O(Ac)		

15-MONOACETOXYSCIRPENOL

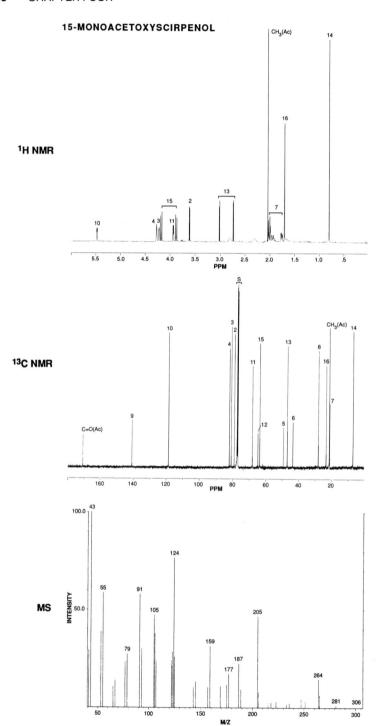

15-MONOACETOXYSCIRPENOL

$C_{17}H_{24}O_6$

M.W. 324.1573

Species:

F. sambucinum
F. roseum

References: Pathre et al. (1976); Richardson et al. (1989).

Position	^{13}C	^1H
2	78.9	3.60 (d, 4.6)
3	80.8	4.21 (dd, 2.7, 4.6)
4	81.9	4.26 (br d, 2.7)
5	49.2	
6	43.5	
7	21.1	1.76 (dd, 11.3, 4.8)
		2.01 (m)
8	28.0	1.98-2.00 (m)
9	140.7	
10	118.4	5.46 (d, 5.4)
11	68.0	3.92 (br d, 5.4)
12	64.7	
13	46.9	2.73 (d, 3.9)
		3.01 (d, 3.9)
14	6.9	0.79 (s)
15	63.6	3.87 (d, 12.3)
		4.18 (d, 12.3)
16	23.2	1.70 (br s)
CH_3(Ac)	21.3	2.04 (s)
C=O(Ac)	170.8	

4,15-DIACETOXYSCIRPENOL

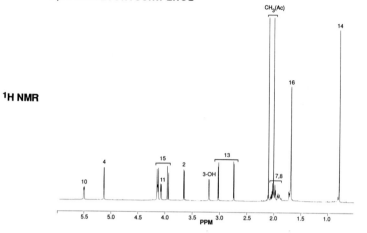

¹H NMR

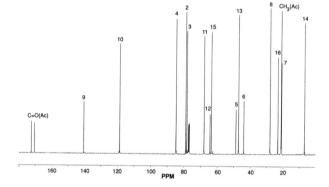

¹³C NMR

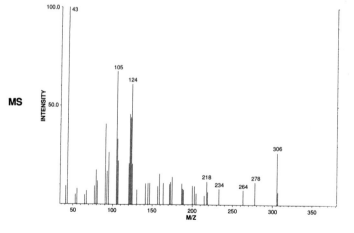

MS

4,15-DIACETOXYSCIRPENOL
(ANGUIDINE)

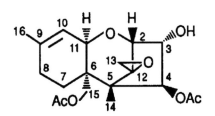

$C_{19}H_{26}O_7$

M.W. 366.1678

Species:

F. sambucinum

References:	Richardson *et al.* (1989).	
Position	**^{13}C**	**^1H**
2	78.9	3.65 (d, 4.9)
3	78.2	4.10 (dd, 4.9, 2.9)
4	84.6	5.11 (d, 2.9)
5	48.7	
6	43.8	
7	21.1	2.00 (d, 11.0)
		1.71 (d, 11.0)
8	27.8	1.93 (m)
9	140.4	
10	118.6	5.50 (d, 5.3)
11	67.9	4.02 (d, 5.3)
12	64.3	
13	47.0	3.00 (d, 4.0)
		2.74 (d, 4.0)
14	6.8	0.79 (s)
15	63.5	4.10 (d, 12.3)
		3.90 (d, 12.3)
16	23.0	1.68 (s)
$CH_3(Ac)$	20.8	2.00 (s)
		2.09 (s)
C=O(Ac)	170.5	
	172.5	

T-2 TETRAOL

MS

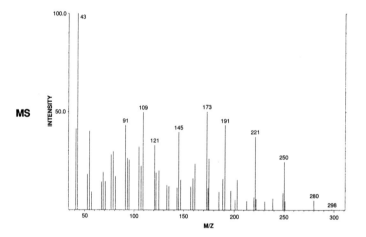

T-2 TETRAOL

$C_{15}H_{22}O_6$

M.W. 298.1416

Species:

F. sporotrichioides
F. poae

References: Cole and Cox (1981).

Position	$^{13}C^a$	$^1H^a$
2	80.6	3.60 (d, 4.8)
3	80.8	4.07 (dd, 4.8, 3.3)
4	82.3	4.12 (d, 3.3)
5	50.4	
6	45.9	
7	29.9	α: 1.97 (dt, 14.4, 1.6)
		β: 2.16 (dd, 14.4, 5.3)
8	67.0	4.00 (d, 4.2)
9	141.0	
10	122.6	5.52 (dq, 5.7, 1.5)
11	70.0	3.80 (d, 5.7)
12	65.8	
13	47.2	2.89 (d, 4.1)
		2.76 (d, 4.1)
14	7.3	0.84 (s)
15	63.1	3.72 (d, 12.4)
		3.40 (d, 12.4)
16	20.7	1.82 (br s)

[a]: spectra obtained in CD_3OD.

15-DEACETYLNEOSOLANIOL

¹H NMR

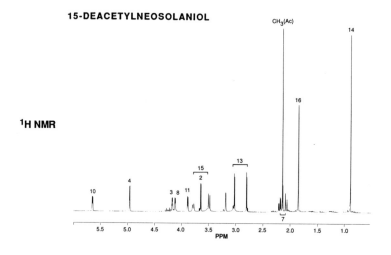

¹³C NMR

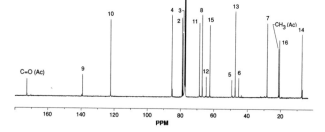

MS

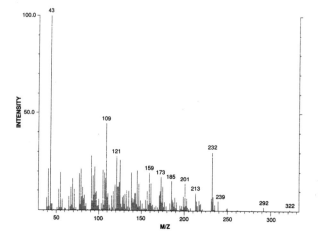

15-DEACETYLNEOSOLANIOL
(NT-2 TOXIN, 4-ACETYL-T-2 TETRAOL)

$C_{17}H_{24}O_7$

M.W. 340.1522

Species:

F. sporotrichioides
F. tricinctum
F. compactum

References: Visconti et al. (1989), Ishii and Ueno (1981); Savard et al. (1987).

Position	^{13}C	^1H
2	78.7	3.64 (d, 4.9)
3	78.3	4.17 (dd, 4.9, 3)
4	84.9	4.94 (d, 3)
5	49.1	
6	44.9	
7	27.9	α: 2.06 (dt, 14.3, 1.6)
		β: 2.19 (dd, 14.3, 5.2)
8	66.8	4.12 (m)
9	139.0	
10	122.0	5.65 (dq, 5.8, 1.5)
11	68.5	3.88 (d, 5.8)
12	64.4	
13	47.0	2.79 (d, 4)
		3.02 (d, 4)
14	7.0	0.89 (s)
15	62.2	3.48 (d, 12.6)
		3.76 (d, 12.6)
16	20.5	1.85 (br s)
CH_3(Ac)	21.0	2.13 (s)
C=O (Ac)	172.8	

NEOSOLANIOL

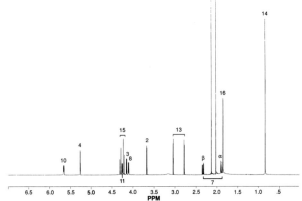

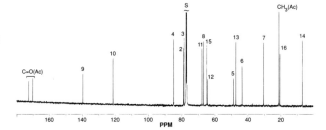

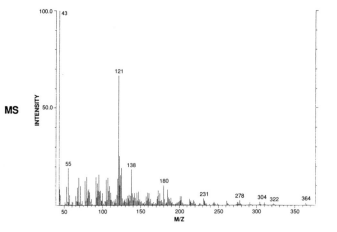

NEOSOLANIOL

$C_{19}H_{26}O_8$

M.W. 382.1627

Species:

F. sporotrichioides
F. culmorum
F. solani
F. poae
F. equiseti
F. avenaceum

References: Cole and Cox (1981); Mesilaakso et al. (1989).

Position	^{13}C	^1H
2	78.7	3.66 (d, 4.9)
3	78.4	4.14 (dd, 4.9, 2.9)
4	84.7	5.25 (d, 2.9)
5	48.7	
6	43.4	
7	30.4	α: 1.89 (dd, 14.5, 1.4)
		β: 2.33 (dd, 14.5, 5.7)
8	66.7	4.10 (d, 5.7)
9	139.6	
10	121.1	5.65 (d, 5.9)
11	67.7	4.25 (d, 5.9)
12	64.4	
13	47.2	2.78 (d, 4.0)
		3.04 (d, 4.0)
14	6.7	0.83 (s)
15	64.8	4.29 (d, 12.6)
		4.21 (d, 12.6)
16	20.4	1.85
CH_3(Ac)	21.0	2.02 (s)
	21.0	2.13 (s)
C=O(Ac)	169.8	
	172.0	

ISONEOSOLANIOL

¹H NMR

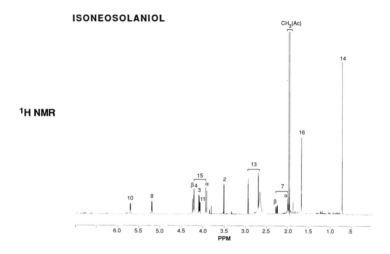

¹³C NMR

MS

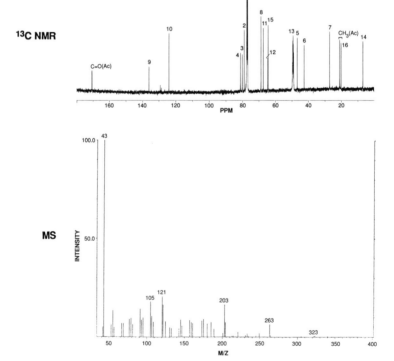

ISONEOSOLANIOL
(3,4-DIHYDROXY-8,15-DIACETOXY-12,13-EPOXYTRICHOTHEC-9-ENE;
ACUMINATIN; 8,15-DIACETYL T-2 TETRAOL)

$C_{19}H_{26}O_8$

M.W. 382.1627

Species:

F. tricinctum
F. equiseti
F. compactum

References: Cole *et al*. (1988); Visconti *et al*. (1989); Savard
and Greenhalgh (1987).

Position	^{13}C	^1H
2	78.9	3.50 (d, 5.0)
3	80.0	4.08 (dd, 5.0, 3.0)
4	81.1	4.20 (d, 3.0)
5	46.8	
6	42.4	
7	27.0	α: 1.98 (dd, 15.0, 1.4)
		β: 2.26 (dd, 15.0, 5.4)
8	68.7	5.18 (d, 5.4)
9	136.0	
10	124.0	5.69 (d, 5.9)
11	67.2	4.06 (d, 5.9)
12	64.6	
13	49.0	2.60 (d, 4.0)
		2.90 (d, 4.0)
14	6.8	0.79 (s)
15	64.3	α: 3.92 (d, 12.4)
		β: 4.20 (d, 12.4)
16	20.1	1.67 (s)
CH_3(Ac)	21.0	1.95 (s)
	20.9	1.97 (s)
C=O(Ac)	170.8	
	170.6	

8-ACETYLNEOSOLANIOL

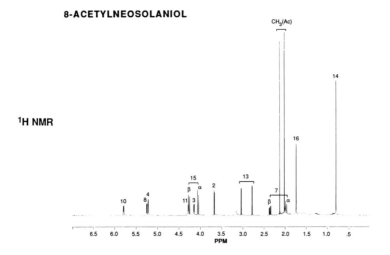

¹H NMR

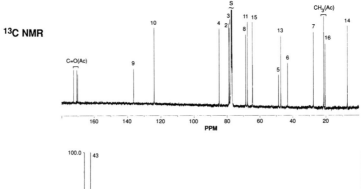

¹³C NMR

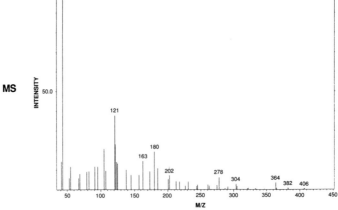

MS

8-ACETYLNEOSOLANIOL

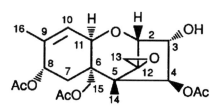

$C_{21}H_{28}O_9$

M.W. 424.1733

Species:

F. roseum
F. tricinctum
F. compactum

References: Lansden *et al.* (1978); Ishii *et al.* (1978); Visconti *et al.* (1989).

Position	^{13}C	^1H
2	78.7	3.67 (d, 4.9)
3	78.4	4.14 (dd, 4.9, 2.9)
4	84.6	5.21 (d, 2.9)
5	48.6	
6	43.0	
7	27.4	α: 1.97 (d, 15.1)
		β: 2.35 (dd, 15.1, 5.7)
8	68.5	5.24 (d, 5.7)
9	136.2	
10	123.8	5.79 (d, 5.9)
11	67.4	4.28 (d, 5.9)
12	64.3	
13	47.2	2.77 (d, 4.0)
		3.03 (d, 4.0)
14	7.0	0.80 (s)
15	64.4	α: 4.05 (d, 12.5)
		β: 4.26 (d, 12.5)
16	20.3	1.73 (s)
CH_3(Ac)	21.0	2.00 (s)
	21.1	2.01 (s)
	21.1	2.12 (s)
C=O(Ac)	172.6 (C-4)	
	170.5 (C-8)	
	170.0 (C-15)	

Spectra not provided.

SPOROTRICHIOL

$C_{20}H_{30}O_6$

M.W. 366.2042

Species:

F. sporotrichioides

References: Corley et al. (1986).

Position	^{13}C	^1H
2	79.6	3.49 (d)
3	69.1	4.49 (m)
4	42.1	2.05-2.25 (m)
5	45.8	
6	43.3	
7	26.8	1.96 (d)
		2.30 (dd)
8	68.4	5.50 (d)
9	135.7	
10	125.4	5.79 (d)
11	68.1	4.17 (d)
12	65.1	
13	48.5	2.57 (d, 4.1)
		3.06 (d, 4.1)
14	12.4	0.87 (s)
15	63.1	3.53 (d, 12.7)
		3.70 (d, 12.7)
16	20.4	1.75 (s)
17	171.9	
18	43.7	2.21 (m)
19	25.7	2.05-2.30 (m)
20,21	22.3	0.95 (d, 7)

4-DEACETOXY-T-2 TOXIN

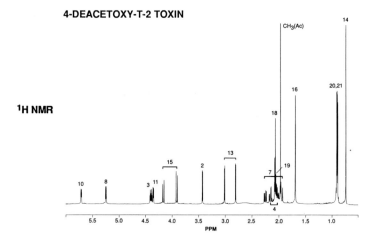

¹H NMR

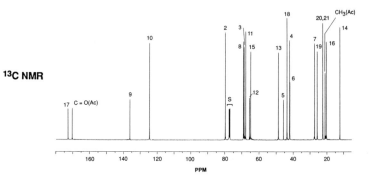

¹³C NMR

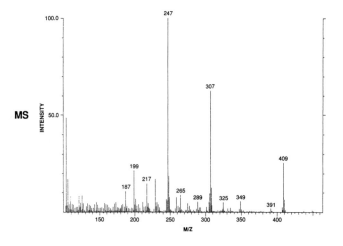

MS

4-DEACETOXY-T-2 TOXIN
(15-Ac-sporotrichiol)

$C_{22}H_{32}O_7$

M.W. 408.2148

Species:

F. sporotrichioides

References: Fielder (1988)

Position	^{13}C	^{1}H
2	79.6	3.42 (d, 4.6)
3	68.5	4.39 (ddd, 10.9, 4.6, 4.0)
4	42.0	2.16 (dd, 14.6, 4.0)
5	45.5	2.04 (m)
6	41.7	
7	27.1	2.25 (dd, 15.1, 5.7)
		1.94 (d, 15.1)
8	68.9	5.24 (d, 5.7)
9	136.0	
10	124.3	5.71 (d, 5.8)
11	67.6	4.35 (d, 5.8)
12	65.3	
13	48.4	2.80 (d, 3.9)
		3.00 (d, 3.9)
14	12.0	0.74 (s)
15	64.6	3.91 (d, 12.3)
		4.15 (d, 12.3)
16	20.2	1.69 (br s)
17	172.8	
18	43.4	2.06 (m)
19	25.6	2.03 (m)
20,21	22.3	0.90 (d, 6.3)
		0.89 (d, 6.3)
CH_3(Ac)	20.8	1.97 (s)
C=O(Ac)	170.3	

T-2 TRIOL

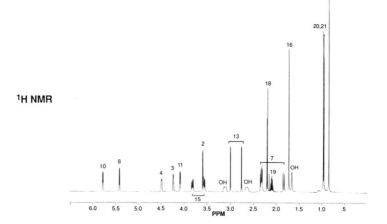

¹H NMR

¹³C NMR

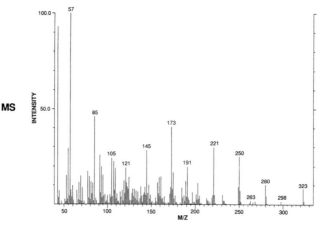

MS

T-2 TRIOL

$C_{20}H_{30}O_7$

M.W. 382.1991

Species:

F. sporotrichioides

References: Mesilaakso et al. (1989); Mirocha et al. (1989).

Position	^{13}C	^{1}H
2	78.8	3.59 (d, 4.9)
3	80.5	4.23 (dd, 4.9, 5.8)
4	81.6	4.48 (d, 6.8)
5	49.0	
6	44.0	
7	27.6	α: 1.84 (d, 15.1)
		β: 2.32 (dd, 15.1, 5.6)
8	68.2	β: 5.41 (d, 5.6)
9	136.0	
10	124.8	5.77 (d, 5.9)
11	67.6	4.08 (d, 5.9)
12	64.8	
13	46.9	2.75 (d, 3.9)
		2.99 (d, 3.9)
14	6.8	0.84 (s)
15	63.2	3.56 (dd, 12.7, 5.6)
		3.81 (dd, 12.7, 7.5)
16	20.3	1.72 (s)
17	172.0	
18	43.7	2.19 (app d, 7.2)
19	25.7	2.09 (m, 6.6, 7.2)
20, 21	22.4	0.96 (6.6)

T-2 TOXIN

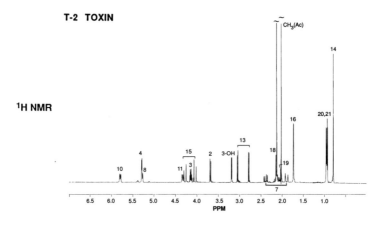

¹H NMR

¹³C NMR

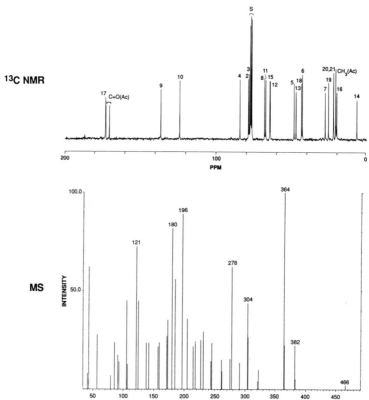

MS

T-2 TOXIN

$C_{24}H_{34}O_9$

M.W. 466.2202

Species:

F. sporotrichioides
F. poae
F. tricinctum

References: Bamburg et al. (1968)

Position	^{13}C	^1H
2	78.7	3.68 (d, 4.9)
3	78.2	4.13 (ddd, 4.9, 2.8, 2.8)
		3.18 (d, 2.8, 3-OH)
4	84.3	5.28 (d, 2.8)
5	48.3	
6	42.9	
7	27.7	2.38 (dd, 15.1, 5.8)
		1.87 (dd, 15.3, 1.5)
8	68.0	5.27 (d, 5.8)
9	136.3	
10	123.8	5.72 (dt, 5.8, 1.2)
11	67.3	4.33 (d, 5.8)
12	64.2	
13	47.0	3.04 (d, 3.9)
		2.78 (d, 3.9)
14	6.6	0.79 (s)
15	64.5	4.04 (d, 12.6)
		4.27 (d, 12.6)
16	20.1	1.72 (br s)
17	172.7	
18	43.4	2.10 (m)
19	25.6	2.01 (m)
20,21	22.2	0.94 (d, 6.5)
		0.93 (d, 6.5)
CH_3(Ac)	20.8	2.01, 2.12 (s)
C=O(Ac)	170.2, 172.5	

HT-2 TOXIN

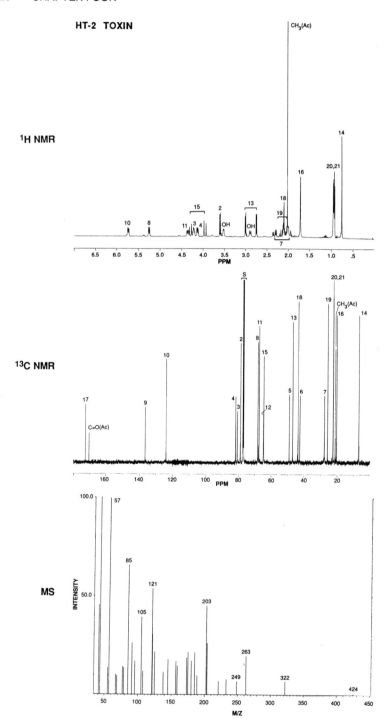

HT-2 TOXIN

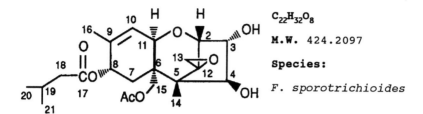

$C_{22}H_{32}O_8$

M.W. 424.2097

Species:

F. sporotrichioides

References: Bamburg and Strong (1969)

Position	^{13}C	^1H
2	78.7	3.60 (d, 4.9)
3	80.6	4.22 (m, 4.9, 3.0)
4	81.6	4.13 (br d, 3.0)
5	48.9	
6	42.5	
7	27.6	α: 2.00 (dt, 15.0, 1.5)
		β: 2.32 (dd, 15.0, 5.7)
8	68.1	5.25 (br d, 5.1)
9	136.4	
10	123.7	5.73 (br d, 6.0)
11	67.3	4.36 (br d, 6.0)
12	64.7	
13	46.9	2.74 (d, 4.0)
		3.00 (d, 4.0)
14	6.9	0.75 (s)
15	64.5	α: 3.97 (d, 12.5)
		β: 4.29 (d, 12.5)
16	20.3	1.71 (br s)
17	172.7	
18	43.6	2.09 (m)
19	25.7	2.05-2.10 (m)
20, 21	22.4	0.93 (d, 6.4)
		0.92 (d, 6.4)
$CH_3(Ac)$	21.1	2.01 (s)
$C=O(Ac)$	170.4	

ACETYL-T-2   TOXIN

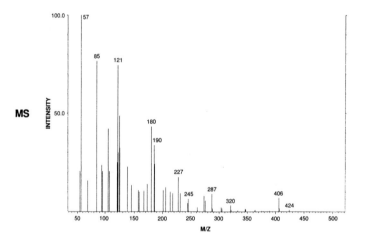

ACETYL-T-2 TOXIN

$C_{26}H_{36}O_{10}$

M.W. 508.2308

Species:

F. poae

References: Cole and Cox (1981).

Position	^{13}C	^1H
2	76.2	3.78
3	78.7	5.07
4	78.2	5.85
5	47.7	
6	42.0	
7	26.7	1.85-2.1
8	67.2	5.20
9	135.0	
10	124.0	5.65
11	66.3	4.14
12	63.6	
13	41.4	2.75
		2.95
14	6.1	0.65
15	63.6	4.00
		4.27
16	19.4	1.66
17	172.2	
18	42.4	1.93
19	25.3	2.0
20	21.7	0.96
21	21.7	0.96
CH_3(Ac)	20.3	2.03 (s)
C=O(Ac)	170.6	
	170.1	
	170.1	

Spectra not provided.

8-PROPIONYL-NEOSOLANIOL
8-ISOBUTYRYL-NEOSOLANIOL
8-BUTYRYL-NEOSOLANIOL

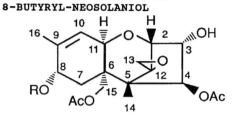

$C_{22}H_{30}O_9$ (Pr-NEO)
$C_{23}H_{32}O_9$ (i-Bu-NEO)
(Bu-NEO)

M.W. 438.1889 **(Pr-NEO)**
452.2046 **(i-Bu-NEO)**
(Bu-NEO)

Species:

F. sporotrichioides

References: Greenhalgh et al. (1988).

Position	$^{13}C^a$			^1H		
R=	**Pr-Neo**	**i-Bu-Neo**	**Bu-Neo**	**Pr-Neo**	**i-Bu-Neo**	**Bu-Neo**
17	174.1	176.6	173.4			
18	27.6	34.0	36.3	2.28	2.45	2.33
				(q,7.5)	(dd,7.0)	(t, 7.4)
19	8.9	18.9	18.4	1.13	1.16	1.64
				(t,7.5)	(d, 7.0)	(m, 7.4)
20		18.5	13.6		1.16	0.93
					(d, 7.0)	(t, 7.4)

[a]Chemical shifts for positions 2-16 and the acetate groups are the same as those reported for T-2 toxin.

4-PROPANOYL-HT-2

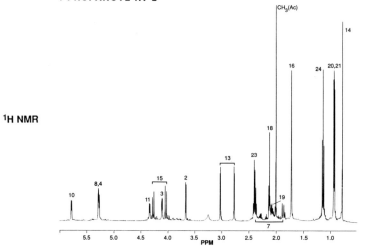

¹H NMR

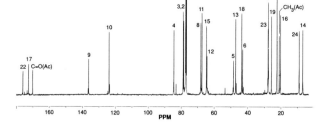

¹³C NMR

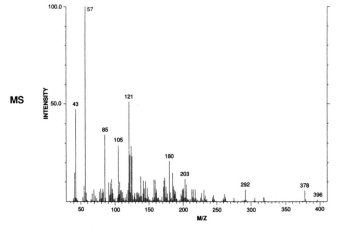

MS

4-PROPANOYL-HT-2

$C_{25}H_{36}O_9$

M.W. 480.2359

Species:

F. sporotrichioides

References: Corley et al. (1986).

Position	^{13}C	^1H
2	78.7	3.66 (d, 4.9)
3	78.5	4.10 (dd, 4.9, 2.9)
4	84.3	5.28 (d, 2.9)
5	48.4	
6	42.9	
7	27.6	2.38 (dd, 15.1, 6.0)
		1.88 (d, 15.1)
8	68.0	5.27 (d, 6.0)
9	136.3	
10	123.7	5.78 (d, 5.9)
11	67.3	4.33 (d, 5.9)
12	64.5	
13	47.2	3.03 (d, 3.9)
		2.76 (d, 3.9)
14	6.8	0.81 (s)
15	64.6	4.27 (d, 12.6)
		4.04 (d, 12.6)
16	20.2	1.72 (br s)
17	172.6	
18	43.6	2.12 (m)
19	25.8	2.02-2.13 (m)
20,21	22.4	0.94 (d,6.4)
	22.3	0.93 (d, 6.4)
22	176.0	
23	27.9	2.39 (q, 7.6)
24	9.0	1.14 (t, 7.6)
CH_3(Ac)	20.9	2.03 (s, 15-Ac)
C=O(Ac)	170.0	

3'-HYDROXY-T-2 TOXIN

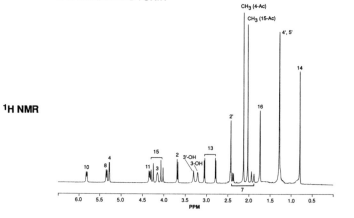

¹H NMR

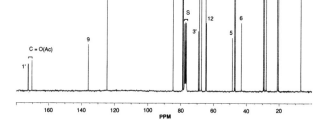

¹³C NMR

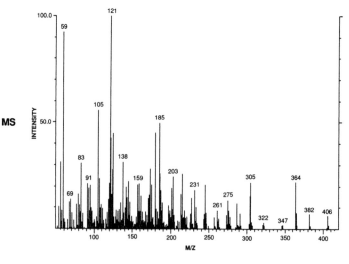

MS

3'-HYDROXY-T-2 TOXIN

$C_{24}H_{34}O_{10}$

M.W. 482.2151

Species:

F. sporotrichioides

References: Visconti *et al.* (1985), Greenhalgh *et al.* (1988), Fielder (1988).

Position	^{13}C	^1H
2	78.7	3.68 (d, 4.9)
3	78.3	4.15 (dd, 4.9,2.8)
4	84.5	5.27 (d, 2.8)
5	48.4	
6	42.9	
7	27.8	2.37 (dd, 15.1, 5.5)
		1.90 (br d, 15.1)
8	68.5	5.34 (d, 5.5)
9	135.7	
10	124.2	5.81 (br d, 5.7)
11	67.2	4.33 (br d, 5.7)
12	64.2	
13	47.1	3.05 (d, 3.9)
		2.78 (d, 3.9)
14	6.8	0.79 (s)
15	64.6	4.26 (d, 12.6)
		4.04 (d, 12.6)
16	20.3	1.73 (br s)
1'	172.5	
2'	46.6	2.42 (s)
3'	69.1	
4',5'	29.0, 29.5	1.27 (br s)
CH_3(Ac)	21.0, 20.9	2.02 (s), 2.12 (s)
C=O(Ac)	170.2, 172.3	

8-KETOISOTRICHODERMIN

MS

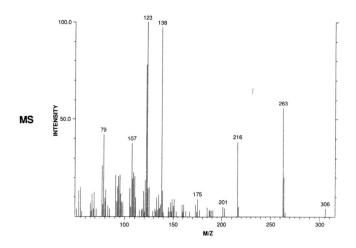

8-KETOISOTRICHODERMIN

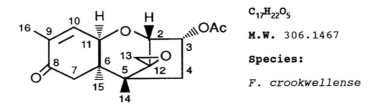

$C_{17}H_{22}O_5$

M.W. 306.1467

Species:

F. crookwellense

References: Lauren et al. (1987).

Position	^{13}C	^{1}H
2	78.5	3.83 (d, 4.6)
3	71.4	5.22 (m, 4.6, 10.9, 4.6)
4	42.1	2.03 (dd, 14.8, 10.9)
		2.17 (dd, 14.8, 4.6)
5	45.4	
6	43.4	
7	37.8	2.26 (dd, 15.3, 1.1)
		2.87 (d, 15.3)
8	189.0	
9	137.7	
10	137.6	6.51 (dd, 5.7, 1.5)
11	71.6	4.32 (d, 5.7)
12	57.5	
13	48.1	2.86 (d, 3.9)
		3.10 (d, 3.9)
14	10.6	0.75 (s)
15	18.3	0.92 (d, 1.1)
16	15.3	1.81 (d, 1.5)
CH_3(Ac)	20.8	2.14 (s)
C=O(Ac)	172.5	

Spectra not provided.

TRICHOTHECOLONE

$C_{15}H_{20}O_4$

M.W. 264.1361

Species:

F. roseum Link

References: Freeman *et al.* (1959); Gyimesi and Melera (1967);
Hanson *et al.*   (1974).

Position	^{13}C	^1H
2	70.2	3.90
3	40.2	1.93
		2.43
4	79.6	4.35
5	49.5	
6	43.6	
7	42.3	
8	199.4	
9	138.5	
10	137.7	6.49
11	73.1	3.94
12	66.0	
13	47.3	2.81
		3.12
14	6.2	
15	15.6	1.00
16	18.5	1.80

4,15-DIACETYL-7-DEOXYNIVALENOL

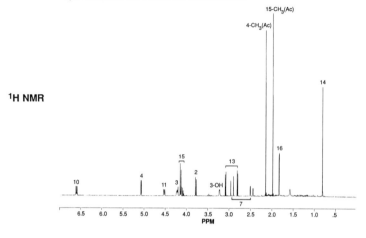

¹H NMR

¹³C NMR

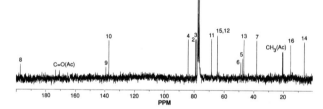

MS

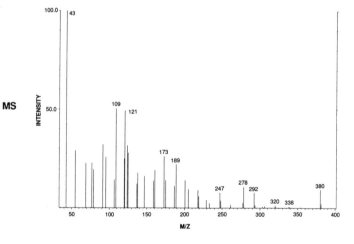

4,15-DIACETYL-7-DEOXYNIVALENOL

$C_{19}H_{24}O_8$

M.W. 380.1471

Species:

F. crookwellense
F. sporotrichioides

References: Lauren et al. (1987).

Position	^{13}C	^1H
2	79.0	3.77 (d, 4.9)
3	78.5	4.21 (ddd, 4.9, 2.9, 2.6)
		3.21 (d, 3-OH, 2.6)
4	83.8	5.06 (d, 2.9)
5	47.5	
6	48.8	
7	38.2	2.46 (dd, 15.9, 1.6)
		2.92 (dd, 15.9, 1.6)
8	196.7	
9	138.9	
10	136.9	6.59 (dd, 5.9, 1.5)
11	68.4	4.51 (d, 5.9)
12	64.4	
13	46.7	2.79 (d, 3.9)
		3.07 (d, 3.9)
14	6.1	0.80 (s)
15	64.5	4.10 (d, 12.4)
		4.17 (d, 12.4)
16	15.3	1.82 (d, 1.5)
CH_3(Ac)	20.8	1.97 (s, 15-Ac)
	20.5	2.14 (s, 4-Ac)
C=O(Ac)	170.2	
	172.7	

3-ACETYL-4,7-DIDEOXYNIVALENOL

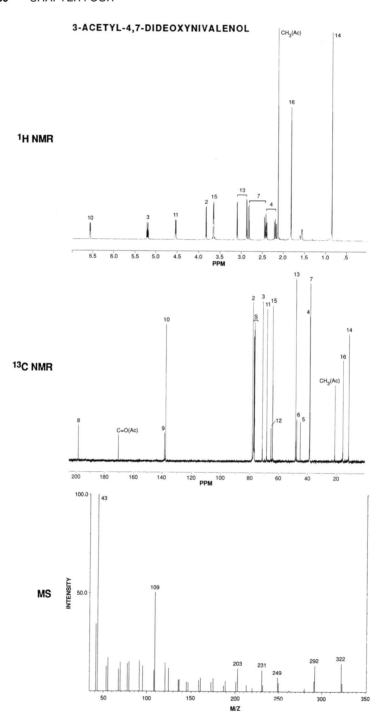

3-ACETYL-4,7-DIDEOXYNIVALENOL
(8-ketocalonectrin)

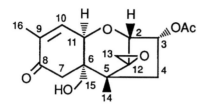

$C_{17}H_{22}O_6$

M.W. 322.1416

Species:

F. culmorum

References: Greenhalgh *et al.* (1986b)

Position	^{13}C	^1H
2	78.1	3.81 (d, 4.6)
3	71.2	5.18 (ddd, 10.2, 4.4, 4.6)
4	38.5	2.17 (dd, 14.7, 11.2)
		2.40 (dd, 14.7, 4.1)
5	45.0	
6	47.5	
7	38.4	2.42 (d, 16.8)
		2.86 (d, 16.8)
8	198.7	
9	138.3	
10	137.5	6.55 (dd, 5.7, 1.4)
11	68.2	4.52 (d, 5.7)
12	65.3	
13	48.3	2.86 (d, 5.9)
		3.09 (d, 5.9)
14	11.3	0.83 (s)
15	64.2	3.63 (d)
		3.65 (d)
16	15.4	1.80 (d, 1.4)
CH_3(Ac)	20.9	2.12 (s)
C=O(Ac)	170.5	

4-DEOXYNIVALENOL

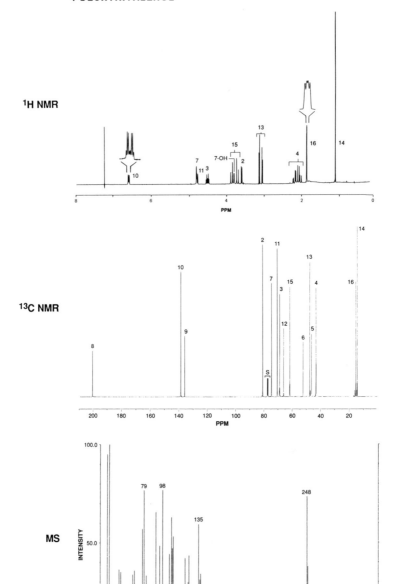

4-DEOXYNIVALENOL
(DON, vomitoxin)

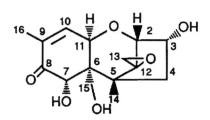

$C_{15}H_{20}O_6$

M.W. 296.1260

Species:

F. graminearum
F. culmorum

References: Vesonder et al. (1973), Mirocha et al. (1976),
Sugiura et al. (1990), Sydenham et al. (1989).

Position	^{13}C	^1H
2	80.6	3.62 (d, 4.5)
3	68.5	4.53 (dt, 10.7, 4.5)
4	43.0	α: 2.21 (dd, 14.8, 4.5)
		β: 2.07 (dd, 14.8, 10.7)
5	46.0	
6	52.1	
7	74.3	4.83 (d, 2)
8	200.3	
9	135.7	
10	138.3	6.61 (dq, 5.9, 1.5)
11	70.2	4.80 (d, 5.9)
12	65.7	
13	47.3	3.07 (d, 4.3)
		3.15 (d, 4.3)
14	14.1	1.13 (s)
15	61.4	3.73 (d, 11.7)
		3.89 (d, 11.7)
16	15.0	1.86 (br s)

3-ACETYL-4-DEOXYNIVALENOL

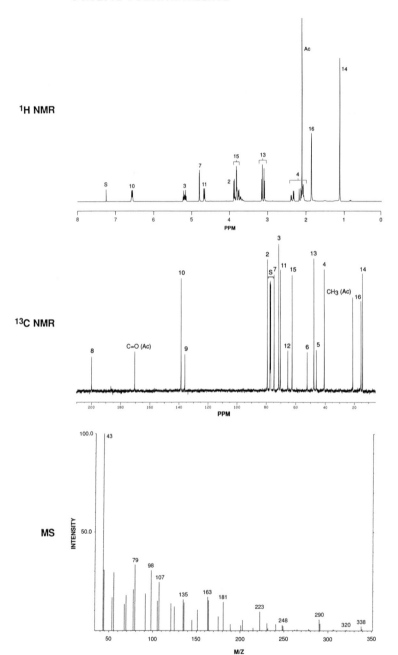

3-ACETYL-4-DEOXYNIVALENOL
(3α-acetoxy-7α,15-dihydroxy-8-keto-12,13-epoxytrichothec-9-ene)

$C_{17}H_{22}O_7$

M.W. 338.1365

Species:

F. culmorum
F. roseum
F. graminearum

References: Blight and Grove (1974), Wei and McLaughlin (1974), Greenhalgh et al. (1984a).

Position	^{13}C		^1H
2	79.1		3.87 (d, 4.5)
3	71.3		5.18 (dt, 11.2, 4.5)
4	40.5	α:	2.35 (dd, 15.1, 4.5)
		β:	2.12 (dd, 15.1, 11,2)
5	45.9		
6	52.0		
7	74.5		4.79 (d, 2.1)
8	199.8		
9	135.8		
10	138.4		6.56 (dq, 5.9, 1.5)
11	70.0		4.66 (d, 5.9)
12	65.1		
13	47.5	α:	3.08 (d, 4.3)
		β:	3.14 (d, 4.3)
14	14.1		1.12 (s)
15	62.2	α:	3.75 (d, 11.7)
		β:	3.81 (d, 11.7)
16	15.2		1.85 (br s)
CH_3(Ac)	20.9		2.10 (s)
C=O(Ac)	170.3		

15-ACETYL-4-DEOXYNIVALENOL

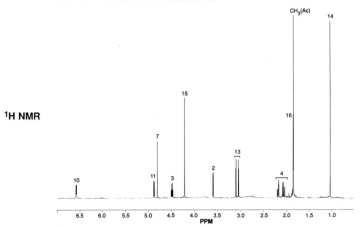

¹H NMR

¹³C NMR

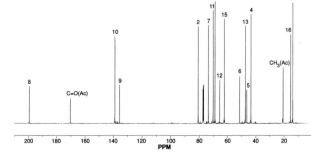

MS

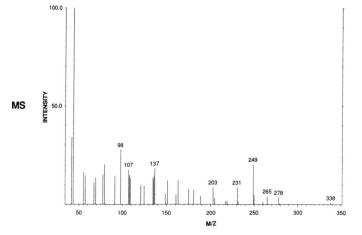

15-ACETYL-4-DEOXYNIVALENOL

$C_{17}H_{22}O_7$

M.W. 338.1365

Species:

F. graminearum

References: Bennett _et al._ (1988), Abbas _et al._ (1988).

Position	^{13}C	^1H
2	80.7	3.60 (d, 4.4)
3	68.9	4.50 (ddd, 10.6, 4.5, 4.4)
4	43.3	2.20 (dd, 14.8, 4.5)
		2.07 (dd, 14.8, 10.6)
5	46.4	
6	51.4	
7	73.5	4.81 (d, 1.9)
8	199.4	
9	135.5	
10	138.6	6.58 (dq, 1.3, 5.8)
11	70.1	4.87 (d, 5.8)
12	65.4	
13	47.3	3.11 (d, 4.2)
		3.06 (d, 4.2)
14	13.8	1.04 (s)
15	62.2	4.21 (s)
16	15.3	1.86 (br s)
CH_3(Ac)	20.6	1.85 (s)
C=O(Ac)	170.1	

3,15-DIACETYL-4-DEOXYNIVALENOL

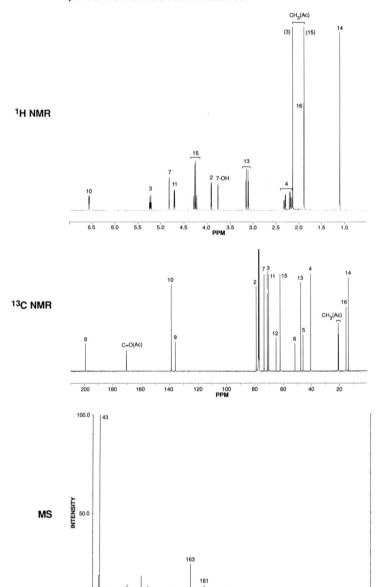

3,15-DIACETYL-4-DEOXYNIVALENOL

$C_{19}H_{24}O_8$

M.W. 380.1471

Species:

F. roseum
F. graminearum

References: Yoshizawa et al. (1978), Greenhalgh et al. (1986a),
Munoz et al. (1989).

Position	^{13}C	^{1}H
2	78.9	3.90 (d, 4.5)
3	71.1	5.22 (ddd, 11.2, 4.5, 4.4)
4	40.4	2.17 (dd, 11.2, 15.1)
		2.30 (dd, 15.1,4.4)
5	45.8	
6	51.5	
7	73.4	4.81 (d, 1.6)
		3.76 (d, 7-OH, 1.6)
8	199.2	
9	135.6	
10	138.4	6.57 (dd, 5.8, 1.6)
11	70.1	4.70 (d, 5.8)
12	64.9	
13	47.4	3.10 (d, 4.3)
		3.15 (d, 4.3)
14	13.6	1.09 (s)
15	62.1	4.22 (d, 12.0)
		4.28 (d, 12.0)
16	15.3	1.89 (d, 0.8)
CH_3(Ac)	20.9	1.88 (s)
	20.6	2.13 (s)
C=O(Ac)	170.2	
	170.0	

NIVALENOL

¹H NMR

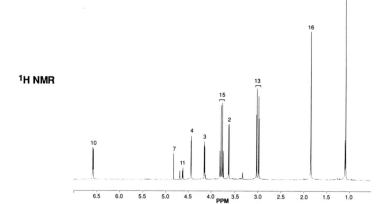

¹³C NMR

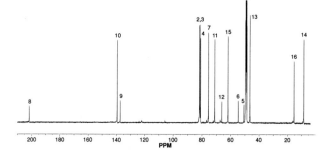

MS

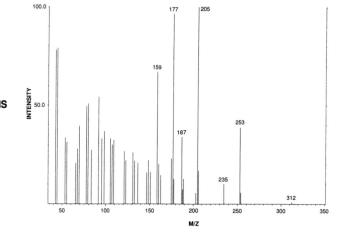

NIVALENOL
(3α,4β,7α,15-TETRAHYDROXY-12,13-EPOXYTRICHOTHEC-9-EN-8-ONE)

$C_{15}H_{20}O_7$

M.W. 312.1209

Species:

F. sporotrichioides
F. graminearum
F. crookwellense

References: Fujimoto *et al.* (1972), Sydenham *et al.* (1989),
Sugiura *et al.* (1990).

Position	$^{13}C^a$	$^1H^a$
2	81.6	3.61 (d, 4.8)
3	81.3	4.14 (dd, 4.8, 3.4)
4	80.8	4.43 (d, 3.4)
5	50.6	
6	54.5	
7	75.2	4.81 (s)
8	201.5	
9	137.1	
10	139.2	6.57 (dd, 6.0, 1.5)
11	71.0	4.72 (d, 6.0)
12	66.0	
13	46.3	3.01 (d, 4.4)
		2.97 (d, 4.4)
14	8.3	1.08 (s)
15	61.7	3.74 (d, 12.1)
		3.79 (d, 12.1)
16	15.2	1.84 (d, 1.5)

[a]: spectra obtained in CD_3OD

4-ACETYLNIVALENOL

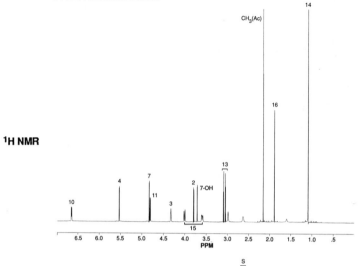

¹H NMR

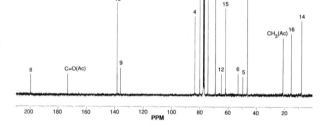

¹³C NMR

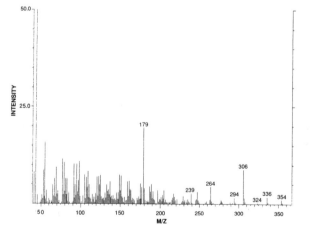

MS

4-ACETYLNIVALENOL
(FUSARENON-X)
(3α,7α,15-TRIHYDROXY-4β-ACETOXY-12,13-EPOXYTRICHOTHEC-9-EN-8-ONE)

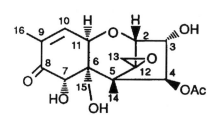

$C_{17}H_{22}O_8$

M.W. 354.1314

Species:

F. crookwellense
F. graminearum
F. equiseti
F. sporotrichioides

References: Ueno et al. (1969), Sugiura et al. (1990).

Position	^{13}C	^1H
2	77.5	3.79 (d, 4.7)
3	80.0	4.32 (m)
4	83.5	5.53 (d,3.4)
5	49.4	
6	52.9	
7	75.8	4.83 (d,2.0)
		3.71 (d, 2.0, 7-OH)
8	199.6	
9	135.9	
10	138.3	6.66 (d, 6.0)
11	69.1	4.81 (d, 6.0)
12	64.7	
13	46.3	3.04 (d, 4.2)
		3.09 (d, 4.2)
14	7.8	1.08 (s)
15	61.6	3.59 (dd, 12.5, 4.7)
		4.01 (br d, 12.5)
16	15.2	1.88 (br s)
CH_3(Ac)	21.0	2.14 (s)
C=O(Ac)	172.9	

4,15-DIACETYLNIVALENOL

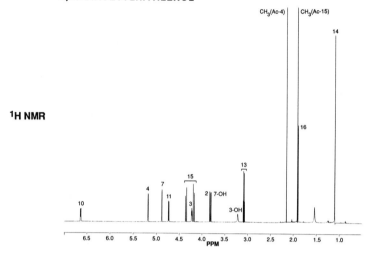

¹H NMR

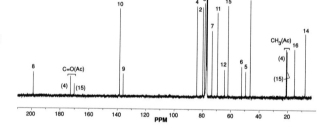

¹³C NMR

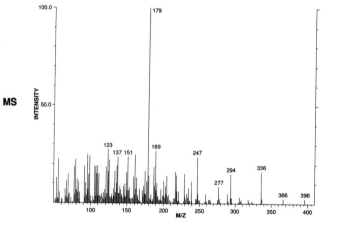

MS

4,15-DIACETYLNIVALENOL

$C_{19}H_{24}O_9$

M.W. 396.1420

Species:

F. sporotrichioides
F. crookwellense
F. equiseti

References: Tidd, B.K. (1967), Tatsuno et al. (1970); Fujimoto et al. (1972), Marasas et al. (1984); Lauren et al. (1987).

Position	^{13}C	^1H
2	79.7	3.82 (d, 4.7)
3	78.4	4.22 (ddd, 4.8, 3.1, 2.7)
		3.25 (d, 3-OH, 2.7)
4	83.9	5.16 (d, 3.1)
5	49.4	
6	52.2	
7	73.2	4.86 (d, 2.0)
		3.80 (d, 7-OH, 2.0)
8	198.9	
9	136.0	
10	138.3	6.63 (dd, 5.9, 1.6)
11	69.3	4.72 (d, 5.9)
12	64.3	
13	46.1	3.07 (d, 4.3)
		3.08 (d, 4.3)
14	7.6	1.09 (s)
15	61.8	4.19 (d, 12.3)
		4.32 (d, 12.3)
16	15.3	1.88 (d, 1.6)
CH_3(Ac)	20.6	1.90 (s, 15-Ac)
	21.0	2.14 (s, 4-Ac)
C=O(Ac)	170.0	
	172.7	

Spectra not provided.

TRICHODIOL
(TRICHO-10-ENE-2α,9α-DIOL)

$C_{15}H_{24}O_3$

M.W. 252.1725 (M^+-H_2O:234.1620)

Species:

F. sporotrichioides

References: Corley et al. (1987a); Hesketh et al. (1993).

Position	^{13}C	^1H
2	78.9	3.78 (dd, 5.1, 3.0)
3	30.0	1.97 (dddd, 13.2, 11.2, 7.2, 5.2)
		1.62 (dddd, 13.2, 7.2, 3.1, 3.0)
4	34.1	2.24 (ddd, 12.0, 12.0, 7.3)
5	45.6	1.48 (ddd, 12.9, 7.2, 2.9)
6	39.3	
7	27.1	1.32 (m)
		1.68-1.81 (m)
8	34.0	1.68-1.81 (m)
9	66.6	
10	131.1	5.55 (d, 10.2)
11	136.4	5.71 (br d, 10.2)
12	71.5	
13	50.1	3.57 (d, 4.0)
		2.92 (d, 4.0)
14	18.9	1.04 (s)
15	21.2	0.94 (s)
16	30.7	1.29 (s)

TRICHOTRIOL

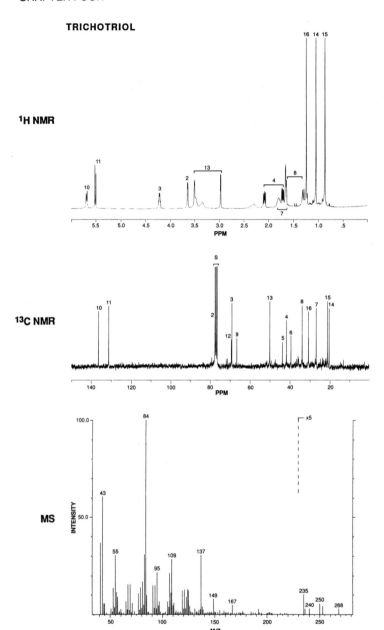

TRICHOTRIOL
(TRICHO-10-ENE-2α,3α,9α-TRIOL)

$C_{15}H_{24}O_4$

M.W. 268.1674 (M^+-H_2O:250.1569)

Species:

F. culmorum
F. sporotrichioides

References: Corley *et al.* (1987a); Kasitu *et al.* (1992); Hesketh *et al.* (1993).

Position	^{13}C		^1H
2	77.5		3.64 (d, 4.5)
3	69.0		4.21 (ddd, 4.5, 6.4, 8.0)
4	41.6	α:	2.08 (dd, 13.3, 8,0)
		β:	1.71 (dd, 13.3, 6.4)
5	43.7		
6	39.4		
7	26.9	α:	1.79 (m)
		β:	1.65 (m)
8	33.8	α:	1.65 (m)
		β:	1.30 (br d, 13.7)
9	66.5		
10	136.2		5.69 (br d, 10.1)
11	131.3		5.51 (d, 10.1)
12	69.3		
13	50.0	α:	3.50 (d, 4.0)
		β:	2.97 (d, 4.0)
14	20.4		1.05 (s)
15	21.2		0.87 (s)
16	30.7		1.24 (s)

FS-1

¹H NMR

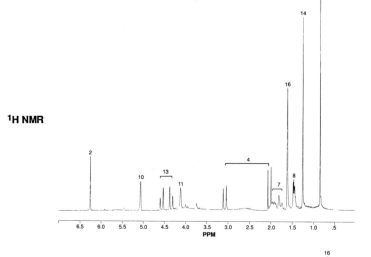

¹³C NMR

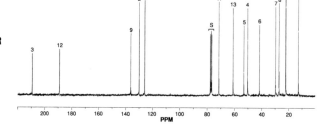

MS

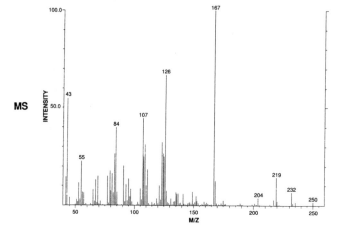

FS-1

$C_{15}H_{22}O_3$

M.W. 250.1569

Species:

F. sporotrichioides
F. sambucinum

References: Corley et al. (1986); Sanson et al. (1989).

Position	^{13}C	^1H
2	129.5	6.24 (t, 1.5)
3	208.4	
4	50.4	3.10 (d, 18.2)
		2.03 (d, 18.2)
5	53.1	
6	41.9	
7	29.9	1.95 (dd, 15, 4.6)
		1.79 (d, 15)
8	27.4	1.45 (m, 4.6)
9	135.7	
10	125.0	5.05 (br s)
11	71.3	4.11 (br s)
12	188.0	
13	61.0	4.55 (d, 17.3)
		4.34 (d, 17.3)
14	22.4	1.24 (s)
15	13.1	0.84 (s)
16	22.5	1.61 (br s)

Spectra not provided.

FS-2

$C_{15}H_{24}O_3$

M.W. 252.1725

Species:

F. sporotrichioides

References: Corley et al. (1987a).

Position	^{13}C	^{1}H
2	133.0	5.75 (dd, 1.5, 3.3)
3	73.4	4.79 (m)
4	47.4	2.10 (m)
5	54.6	
6	40.0	
7	27.8	1.75 (m)
		1.31 (m)
8	35.0	1.98 (m)
		1.73 (m)
9	65.7	
10	135.6	5.58 (d, 10.2)
11	133.2	5.67 (dd, 1.7, 10.2)
12	154.0	
13	60.7	4.35 (br d, 14.5)
		4.23 (br d, 14.5)
14	21.3	1.09 (s)
15	22.0	0.95 (s)
16	30.9	1.27 (s)

Spectra not provided.

FS-3

$C_{15}H_{20}O_3$

M.W. 248.1412

Species:

F. sambucinum

References: Sanson et al. (1989).

Position	^{13}C	^1H
2	129.6	6.29 (t, 1.5)
3	207.0	
4	50.6	2.32 (d, 18.8)
		2.60 (d, 18.8)
5	51.3	
6	47.1	
7	30.9	1.73 (dd, 12, 2.5)
		1.87 (dd, 12, 2.5)
8	28.3	2.20 (dd, 19, 2.5)
		1.63 (dd, 19, 2.5)
9	160.1	
10	126.4	5.75 (br s)
11	202.5	
12	186.3	
13	61.6	4.49 (dd, 17, 1.5)
		4.56 (dd, 17, 1.5)
14	23.8	1.50 (s)
15	18.2	1.17 (s)
16	29.0	1.93 (s)

Spectra not provided.

FS-4

$C_{15}H_{22}O_3$

M.W. 250.1569

Species:

F. sambucinum

References: Sanson et al. (1989).

Position	^{13}C	^1H
2	128.1	6.27 (t, 1.5)
3	205	
4	49	2.15 (d, 19.0)
		2.73 (d, 19.0)
5		
6		
7	28.2	1.40 (m)
		1.70 (m)
8	32.5	1.70 (m)
		1.85 (m)
9		
10	133.6	5.52 (d, 10)
11	130.9	5.61 (dd, 10, 1.5)
12	183	
13	62	4.69 (d, 17)
		4.35 (d, 17)
14	19.2	1.29 (s)
15	18.0	0.98 (s)
16	29.0	1.25 (s)

3α, 13-DIHYDROXY-EPIAPOTRICHOTHECENE

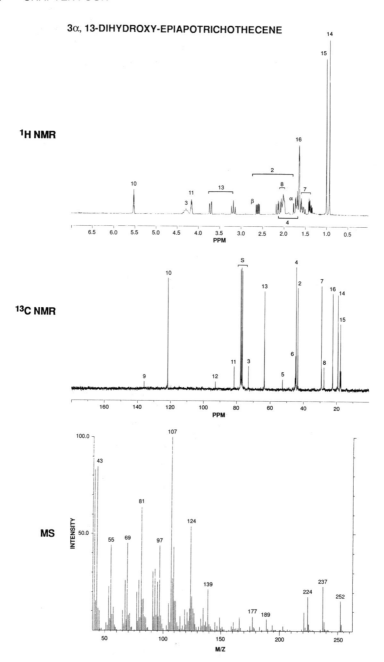

3α,13-DIHYDROXY-EPIAPOTRICHOTHECENE
(3α,13-DIHYDROXY-11-EPIAPOTRICHOTHEC-9-ENE)

$C_{15}H_{24}O_3$

M.W. 252.1725

Species:

F. crookwellense
F. graminearum
F. culmorum
F. sporotrichioides
F. sambucinum

References: Lauren et al. (1987); Greenhalgh et al. (1989a,b);
Sanson et al. (1989).

Position	$^{13}C^a$	^1H
2	43.5	α: 1.74 (dd, 12.2, 2.2)
		β: 2.60 (ddd, 12.2, 6.2, 1.7)
3	72.8	4.28 (dddd, 2.2, 6.2, 1.3, 10.2)
4	44.5	α: 2.13 (dd, 12.9, 10.1)
		β: 1.67 (ddd, 12.9, 1.3, 1.7)
5	52.6	
6	44.7	
7	29.6	α: 1.37 (ddd, 13.1, 2.7, 5.6)
		β: 1.54 (dd, 13.1, 9.8)
8	27.7	α: 1.93 (m)
		β: 2.03 (m)
9	135.5	
10	121.4	5.52 (dd, 2.5, 1.3)
11	81.4	4.15 (dd, 2.5, 2.3)
12	92.6	
13	63.4	α: 3.17 (t, 11.3)
		β: 3.71 (dd, 11.3, 1.9)
14	19.4	0.92 (s)
15	17.8	0.99 (s)
16	22.6	1.63 (dd, 1.3, 2.3)

[a]: spectrum obtained on a ^{13}C-enriched sample.

3β, 13-DIHYDROXY-EPIAPOTRICHOTHECENE

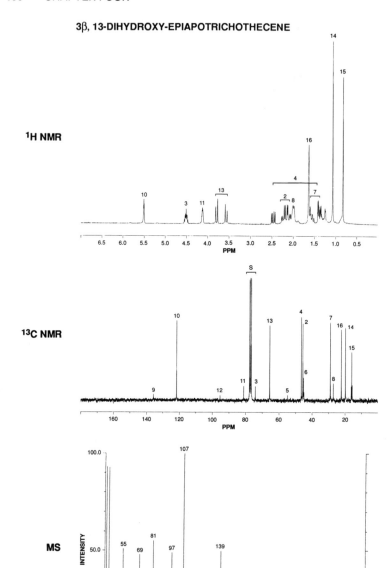

3β,13-DIHYDROXY-EPIAPOTRICHOTHECENE
(3β,13-DIHYDROXY-11-EPIAPOTRICHOTHEC-9-ENE)

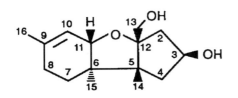

$C_{15}H_{24}O_3$

M.W. 252.1725

Species:

F. crookwellense
F. graminearum
F. culmorum
F. sporotrichioides
F. sambucinum

References: Lauren et al. (1987); Greenhalgh et al. (1989a,b);
Sanson et al. (1989).

Position	$^{13}C^a$	^1H
2	44.4	α: 2.23 (dd, 14.9, 6.3)
		β: 2.10 (ddd, 14.9, 4.1, 1.3)
3	73.4	4.51 (m, 6.3)
4	45.7	α: 2.45 (dd, 14.6, 6.4)
		β: 1.38 (ddd, 14.6, 1.3, 5.8)
5	54.9	
6	45.1	
7	29.1	α: 1.36 (m, 13.5)
		β: 1.57 (m, 13.5)
8	27.4	1.99 (m, 1.8, 5.4)
9	135.2	
10	121.1	5.50 (d, 1.6)
11	81.1	4.11 (d, 1.6)
12	95.1	
13	65.0	α: 3.56 (d, 10.9)
		β: 3.78 (d, 10.9)
14	20.0	1.07 (s)
15	16.2	0.92 (s)
16	22.5	1.63 (d, 1.6)

[a]: spectrum obtained on a ^{13}C-enriched sample.

Spectra not provided.

3-KETOAPOTRICHOTHECENE

$C_{15}H_{22}O_3$

M.W. 250.1569

Species:

F. sambucinum

References: Sanson et al. (1989).

Position	^{13}C	^1H
2	49.6	2.60 (dd, 20.0, 2.2)
		2.79 (dd, 20.0, 1.4)
3	215.5	
4	49.6	2.14 (dd, 19.2, 1.4)
		2.68 (dd, 19.2, 2.2)
5	51.3	
6	46.2	
7	28.8	1.48 (m)
		1.63 (m)
8	26.8	2.05 (m)
		2.15 (m)
9	135.7	
10	210.4	5.55 (m)
11		
11	80.1	4.23 (m)
12	90.8	
13	65.0	3.53 (d, 11.3)
		3.83 (d, 11.3)
14	19.2	1.16 (s)
15	15.3	0.76 (s)
16	22.6	1.68 (s)

3α, 11-OXY-13-HYDROXY-APOTRICHOTHECENE

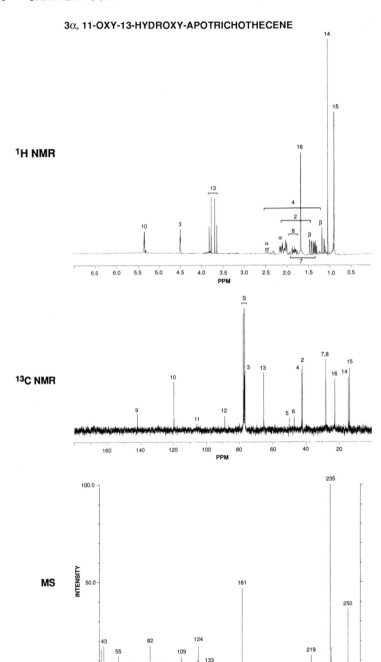

3α,11-OXY-13-HYDROXY-APOTRICHOTHECENE
(3α,11-EPOXY-13-HYDROXYAPOTRICHOTHEC-9-ENE)

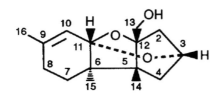

$C_{15}H_{22}O_3$

M.W. 250.1569

Species:

F. culmorum
F. sporotrichioides

References: Greenhalgh et al. (1989a,b).

Position	^{13}C	^1H
2	42.2	α: 2.13 (dd, 11.5, 2.0)
		β: 1.44 (dd, 11.5, 3.8)
3	76.3	4.51 (ddd, 1,8, 3.4, 1.8)
4	42.7	α: 2.46 (ddd, 13.5, 2.0, 3.4)
		β: 1.16 (br d, 13.5)
5	50.0	
6	47.1	
7	28.3	α: 1.34 (ddd, 12.9, 1.6, 6.6)
		β: 1.86 (dddd, 12.9, 5.6, 4.4, 1.0)
8	28.2	α: 1.80 (m, 7.2)
		β: 2.03 (m)
9	141.7	
10	119.5	5.35 (d, 1.6)
11	105.5	
12	88.7	
13	65.3	α: 3.66 (d, 12.1)
		β: 3.79 (d, 12.1)
14	14.7	1.04 (s)
15	13.9	0.90 (d, 0.8)
16	22.9	1.67 (d, 1.6)

2,13-DIACETOXY-8-HYDROXY-11-EPIAPOTRICHOTHECENE

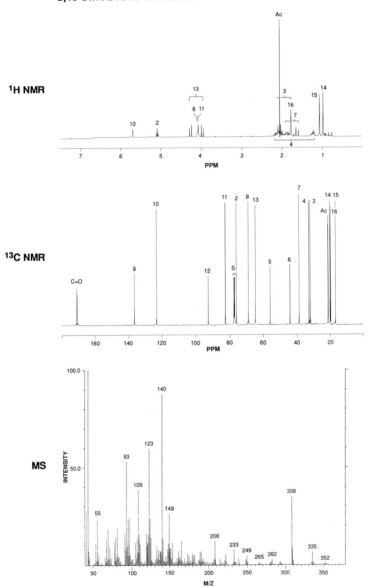

2,13-DIACETOXY-8-HYDROXY-11-EPIAPOTRICHOTHECENE

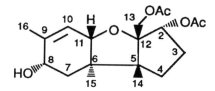

$C_{19}H_{28}O_6$

M.W. 352.1886

Species:

F. sporotrichioides

References: Greenhalgh et al. (1990).

Position	^{13}C	^1H
2	72.1	β: 5.07 (dd, 12.1, 6.3)
3	31.7	α: 1.85 (m)
		β: 2.07 (m)
4	32.6	α: 2.17 (br d, 13.1)
		β: 1.24 (dt, 13.1, 6.8)
5	55.3	
6	43.5	
7	38.5	α: 1.63 (br d, 14.2)
		β: 1.94 (dd, 14.2, 7.5)
8	68.8	β: 4.08 (d, 7.5)
9	136.2	
10	123.1	5.68 (br q, 1.1)
11	82.4	4.05 (br s)
12	95.0	
13	62.8	4.39 (d, 11.8)
		3.82 (d, 11.8)
14	19.4	0.97 (s)
15	16.7	1.06 (s)
16	19.0	1.77 (br s)
CH_3(Ac)	21.0	2.06 (s)
	21.1	2.06 (s)
C=O(Ac)	170.4	
	170.4	

2-ACETOXY-8,13-DIHYDROXY-11-EPIAPOTRICHOTHECENE

¹H NMR

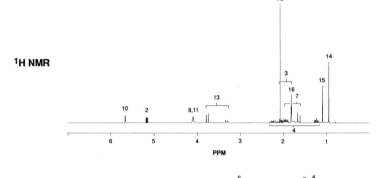

¹³C NMR

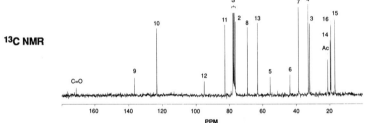

MS

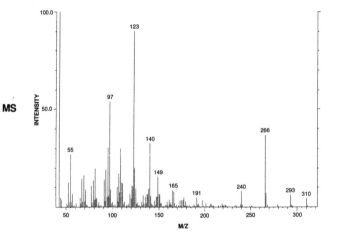

2-ACETOXY-8,13-DIHYDROXY-11-EPIAPOTRICHOTHECENE

$C_{17}H_{26}O_5$

M.W. 310.1780

Species:

F. sporotrichioides

References: Greenhalgh et al. (1990).

Position	^{13}C	^1H
2	75.7	β: 5.16 (m)
3	31.8	α: 1.95 (m)
		β: 2.05 (m)
4	32.6	α: 2.24 (dd, 13.3, 8.2)
		β: 1.22 (ddd, 13.3, 7.7, 5.3)
5	55.7	
6	43.9	
7	38.5	α: 1.63 (d, 14.1)
		β: 1.91 (m)
8	68.8	β: 4.10 (m)
9	136.1	
10	122.7	5.66 (dq, 3.3, 1.5)
11	82.3	4.09 (br s)
12	92.4	
13	64.4	3.76 (d, 11.6)
		3.31 (d, 11.6)
14	19.9	0.94 (s)
15	16.5	1.08 (br s)
16	19.5	1.77 (d, 1.6)
CH_3(Ac)	21.0	2.05 (s)
C=O(Ac)	171.1	

2,13-DIHYDROXY-3,11-EPOXYAPOTRICHOTHECENE

MS

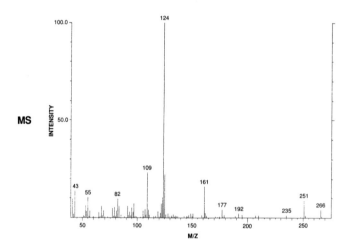

2,13-DIHYDROXY-3,11-EPOXYAPOTRICHOTHECENE

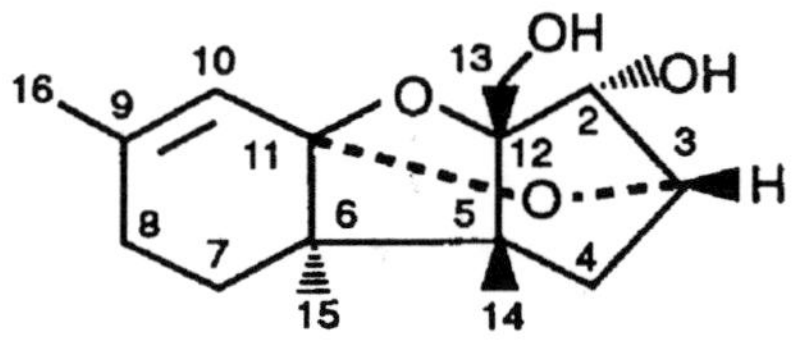

$C_{15}H_{22}O_4$

M.W. 266.1518

Species:

F. sporotrichioides

References: Greenhalgh et al. (1990).

Position	^{13}C	^1H
2	73.5	β: 3.53 (d, 2.1)
3	78.7	β: 4.16 (td, 2.1, 1.0)
4	34.8	α: 2.33 (dd, 13.3, 2.1)
		β: 1.07 (dd, 13.3, 1.0)
5	49.9	
6	46.5	
7	28.2	α: 1.90 (d, 13.2)
		β: 1.38 (ddq, 13.2, 6.0, 0.8)
8	28.0	α: 2.02 (m)
		β: 2.10 (m)
9	142.2	
10	118.6	5.39 (br q, 1.4)
11	105.6	
12	86.5	
13	62.5	3.86 (d, 12.2)
		3.74 (d, 12.2)
14	14.4	1.11 (s)
15	14.1	0.91 (d, 0.8)
16	23.0	1.77 (d, 1.4)

2,15-DIACETOXY-13-HYDROXY-11-EPIAPOTRICHOTHECENE

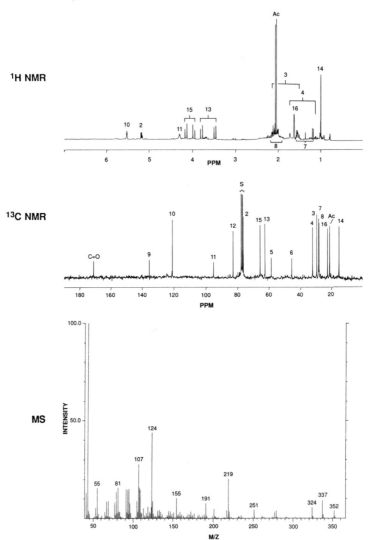

2,15-DIACETOXY-13-HYDROXY-11-EPIAPOTRICHOTHECENE

$C_{19}H_{28}O_6$

M.W. 352.1886

Species:

F. sporotrichioides

References: Greenhalgh et al. (1990).

Position	^{13}C	^1H
2	76.0	β: 5.18 (m)
3	29.3	α: 1.95 (m)
		β: 2.20 (m)
4	32.1	α: 2.20 (m)
		β: 1.50 (m)
5	58.4	
6	45.4	
7	28.1	α: 1.17 (m)
		β: 2.90 (m)
8	27.8	α: 1.80 (m)
		β: 2.20 (m)
9	135.3	
10	120.8	5.53 (br s)
11	82.6	4.29 (br s)
12	95.0	
13	62.4	3.48 (d, 12.2)
		3.79 (d, 12.2)
14	15.0	0.99 (s)
15	65.5	3.97 (d, 11.8)
		4.15 (d, 11.8)
16	22.3	1.63 (br s)
CH_3(Ac)	20.9	2.05 (s)
	21.0	2.06 (s)
C=O(Ac)	170.4	
	170.3	

SAMBUCINOL

¹H NMR

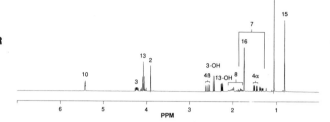

¹³C NMR

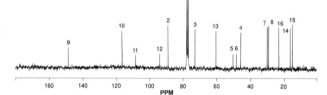

MS

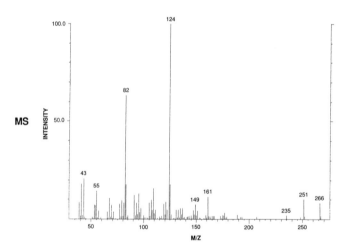

SAMBUCINOL

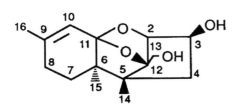

$C_{15}H_{22}O_4$

M.W. 266.1518

Species:

F. sambucinum
F. culmorum
F. graminearum
F. sporotrichioides
F. crookwellense

References: Mohr *et al.* (1984); Greenhalgh *et al.* (1986a, 1986b); Lauren *et al.* (1987).

Position	^{13}C	^{1}H
2	88.2	3.90 (br s)
3	72.5	4.22 (ddd, 8.1, 4.3, 6.5)
		2.40 (d, 3-OH, 6.5)
4	45.1	α: 1.48 (ddd, 15.1, 4.3, 0.9)
		β: 2.57 (dd, 15.1, 8.1)
5	51.0	
6	47.7	
7	29.7	α: 1.34 (dd, 13.0, 5.3)
		β: 1.93 (ddd, 13.0, 1.7, 0.5)
8	28.9	1.80 (m)
		2.02 (m)
9	146.0	
10	116.3	5.52 (m)
11	108.1	
12	93.5	
13	59.4	α: 4.03 (dd, 11.2, 6.9)
		β: 4.10 (dd, 11.2, 3.5)
		2.21 (d, 13-OH, 6.9, 3.5)
14	16.2	1.04 (s)
15	14.9	0.81 (d, 0.5)
16	22.6	1.73 (br s)

3,13-DIACETYL-SAMBUCINOL

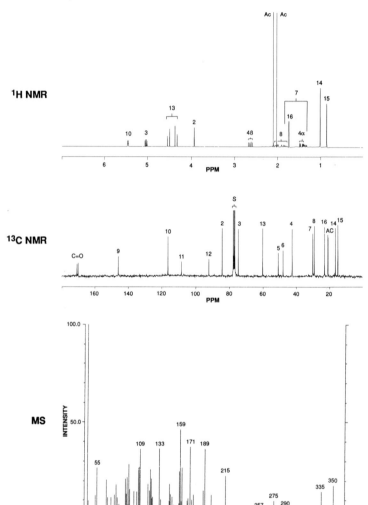

3,13-DIACETYL-SAMBUCINOL

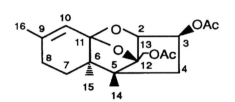

$C_{19}H_{26}O_6$

M.W. 350.1729

Species:

F. sporotrichioides

References: Fielder (1988)

Position	^{13}C	^1H
2	84.2	3.92 (br s)
3	74.6	5.01 (dd, 8.2, 4.9)
4	42.3	α: 1.50 (dd, 15.4, 8.2)
		β: 2.62 (dd, 15.4, 4.9)
5	50.7	
6	47.7	
7	30.3	1.40 (dd, 13.0, 5.5)
		2.08 (m)
8	29.0	1.80 (m)
		2.10 (m)
9	145.8	
10	116.1	5.45 (br s)
11	108.3	
12	92.2	
13	60.0	α: 4.34 (d, 12.6)
		β: 4.51 (d, 12.6)
14	16.3	1.00 (s)
15	14.8	0.85 (s)
16	22.8	1.73 (br s)
CH_3(Ac)	20.9	2.01 (s)
	21.7	2.09
C=O(Ac)	170.1	
	170.8	

3-DEOXYSAMBUCINOL

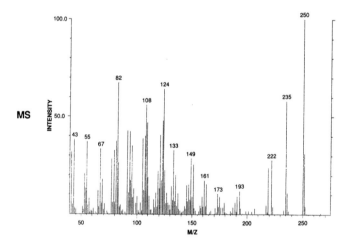

3-DEOXYSAMBUCINOL

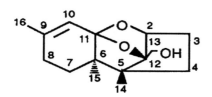

$C_{15}H_{22}O_3$

M.W. 250.1569

Species:

F. culmorum

References: unpublished.

Position	^{13}C	^1H
2	81.8	4.22 (dd, 2.6)
3	33.6	1.84-1.86 (m)
4	30.4	α: 1.48 (m)
		β: 1.95 (dd, 9.2, 4.3)
5	50.0	
6	48.2	
7	29.3	α: 1.85 (m, 13.0)
		β: 1.40 (ddd, 13.0, 5.5, 1.5)
8	27.7	2.05 (m)
		1.90 (m)
9	145.5	
10	116.6	5.49 (t, 1.2)
11	107.0	
12	95.6	
13	59.0	4.01 (d, 12.4)
		3.63 (d, 12.4)
14	16.4	0.94 (s)
15	14.5	0.86 (d, 1.5)
16	23.0	1.75 (br s)

SAMBUCOIN

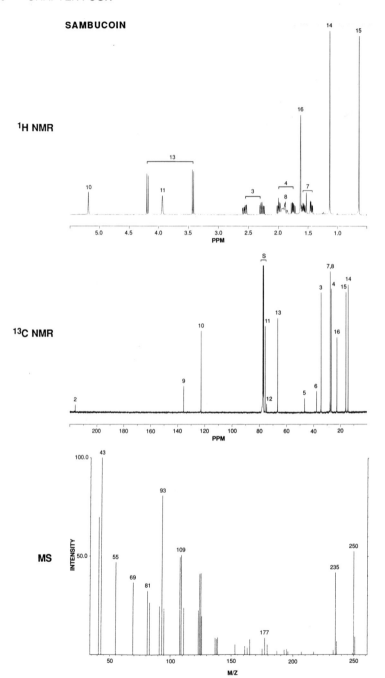

SAMBUCOIN

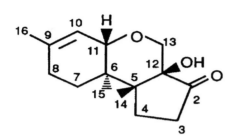

$C_{15}H_{22}O_3$

M.W. 250.1569

Species:

F. sambucinum
F. roseum
F. crookwellense
F. culmorum
F. sporotrichioides

References: Mohr et al. (1984); Lauren et al. (1987); Greenhalgh et al. (1986a, 1986b), Corley et al. (1987b).

Position	^{13}C	^1H
2	216.0	
3	34.3	α: 2.26 (ddd, 19.9, 7.9, 10.0)
		β: 2.55 (ddd, 20.0, 3.0, 11.8)
4	26.8	α: 1.74 (ddd, 13.8, 7.9, 11.7)
		β: 1.96 (ddd, 13.6, 3.2, 10.2)
5	46.4	
6	37.6	
7	27.5	1.44 (ddd, 13.3, 1.3, 5.4)
		1.58 (dd, 13.3, 6.5)
8	27.4	1.85 (m)
9	135.4	
10	122.7	5.18 (br s)
11	75.6	3.94 (br s)
12	74.6	
13	66.4	α: 3.42 (d, 11.2)
		β: 4.19 (d, 11.2)
14	14.1	1.13 (s)
15	15.8	0.64 (s)
16	22.4	1.62 (br s)

Spectra not provided.

8α-HYDROXY-SAMBUCOIN
8β-HYDROXY-SAMBUCOIN

$C_{15}H_{22}O_4$

M.W. 266.1518

Species:

F. sporotrichioides

References: Corley et al. (1987b).

Position	^{13}C	^1H α	β
2	216.1		
3	34.7	2.60 (ddd,3.3,11.5,19.6)	2.50 (ddd,2.8,11.4,19.9)
		2.32 (ddd,8.1,10.2,19.6)	2.28 (ddd,7.8,10.2,19.9)
4	27.2	2.07 (m)	2.02 (m)
		1.81 (m)	1.81 (m)
5	46.9		
6	38.0		
7	27.8	1.85 (m)	1.45 (dd, 9.6, 12.9)
		1.60 (m)	1.98 (dd, 5.9, 12.9)
8	27.7	4.08 (m)	4.13 (m)
9	135.2		
10	122.9	5.35 (br s)	5.33 (br s)
11	75.9	3.90 (br s)	4.03 (br s)
12	74.3		
13	67.4	3.42 (d, 11.3)	3.43 (d, 11.3)
		4.20 (d, 11.3)	4.21 (d, 11.3)
14	16.5	1.18 (s)	1.18 (s)
15	14.3	0.85 (s)	0.68 (s)
16	22.5	1.80 (br s)	1.76 (br s)

CULMORIN

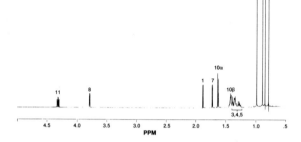

¹H NMR

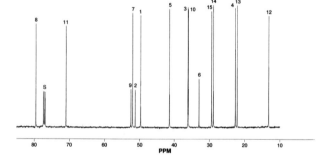

¹³C NMR

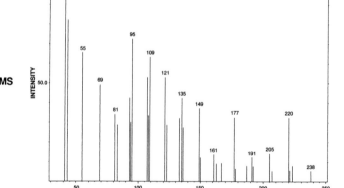

MS

CULMORIN

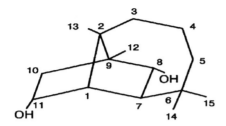

$C_{15}H_{26}O_2$

M.W. 238.1933

Species:

F. culmorum
F. graminearum
F. crookwellense

References: Ashley <u>et al</u>. (1937); Barton and Werstiuk (1967); Greenhalgh <u>et al.</u> (1984b); Lauren <u>et al.</u> (1987).

Position	^{13}C	^1H
1	49.5	1.88 (d, 4.5)
2	51.1	
3	36.0	1.25-1.40 (m)
4	22.6	1.25-1.40 (m)
5	41.3	1.25-1.40 (m)
6	32.9	
7	52.2	1.72 (d, 5.0)
8	79.5	3.78 (d, 5.0)
9	52.2	
10	36.0	α: 1.62 (dd, 13.7, 6.5)
		β: 1.42 (m, 13.7)
11	71.0	4.30 (ddd, 10.0, 6.5, 4.5)
12	13.1	0.84 (s)
13	22.0	0.78 (s)
14	28.8	0.88 (s)
15	29.2	0.98 (s)

5-HYDROXY-CULMORIN

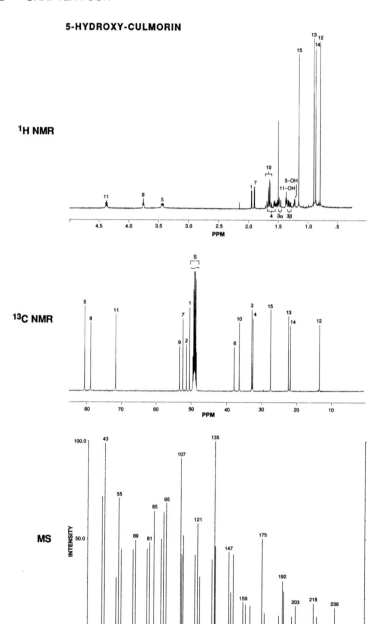

5-HYDROXY-CULMORIN

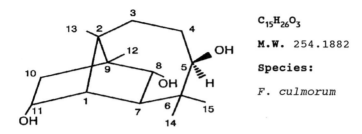

$C_{15}H_{26}O_3$

M.W. 254.1882

Species:

F. culmorum

References: Kasitu et al. (1992).

Position	$^{13}C^a$	^1H
1	50.5	1.94 (d, 4.5)
2	51.3	
3	32.8	α: 1.51 (m)
		β: 1.31 (ddd, 15.8, 6.9, 2.2)
4	32.5	α: 1.70-1.65 (m)
		β: 1.54 (dddd, 12.9, 7.0, 2.0, 2.0)
5	80.5	3.43 (ddd, 11.4, 5.7, 2.2)
6	37.8	
7	52.3	1.89 (d, 5.2)
8	78.8	3.75 (dd, 5.2, 1.3)
9	53.2	
10	36.4	α: 1.62 (dd, 13.7, 4.6)
		β: 1.67 (ddd, 13.7, 9.2, 1.3)
11	71.5	4.37 (ddd, 9.6, 4.5, 4.5)
12	13.4	0.79 (s)
13	22.3	0.90 (s)
14	21.8	0.86 (s)
15	27.4	1.15 (s)
		1.36 (t, 6.0, HO-11, HO-8)
		1.22 (d, 5.7, HO-5)

a: spectrum obtained in CD_3OD.

12-HYDROXY-CULMORIN

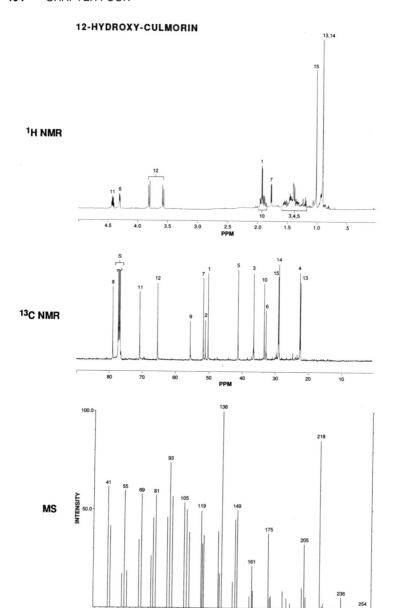

12-HYDROXY-CULMORIN

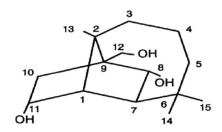

$C_{15}H_{26}O_3$

M.W. 254.1882

Species:

F. culmorum

References: Kasitu et al. (1992).

Position	^{13}C	^1H
1	50.3	1.92 (d, 4.4)
2	51.2	
3	36.6	1.54-1.18 (m)
4	22.5	1.54-1.18 (m)
5	41.2	1.54-1.18 (m)
6	32.8	
7	51.9	1.77 (d, 4.9)
8	79.0	4.30 (br dd, 4.9, 1.0)
9	55.6	
10	33.5	α: 1.94 (dd, 14.0, 4.2)
		β: 1.87 (ddd, 14.0, 9.6, 1.4)
11	70.8	4.42 (ddd, 9.7, 4.8, 4.8)
12	65.7	α: 3.81 (d, 10.4)
		β: 3.57 (d, 10.4)
13	22.4	0.90 (s)
14	28.9	0.90 (s)
15	29.0	1.01 (s)

15-HYDROXY-CULMORIN

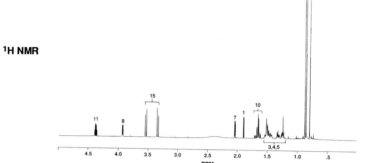

¹H NMR

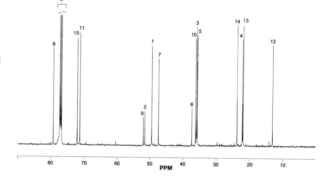

¹³C NMR

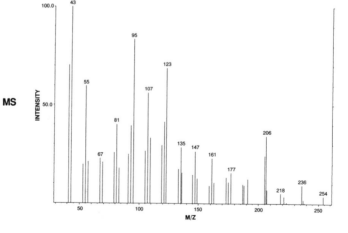

MS

15-HYDROXY-CULMORIN

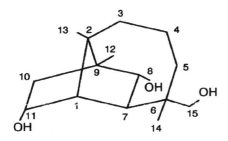

$C_{15}H_{26}O_3$

M.W. 254.1882

Species:

F. culmorum

References: Kasitu et al. (1992).

Position	^{13}C	^1H
1	49.4	1.89 (d, 4.4)
2	51.5	
3	35.8	1.53-1.43 (m)
4	22.2	1.53-1.43 (m)
5	35.6	1.53-1.43 (m)
6	37.3	
7	47.4	2.03 (d, 5.2)
8	79.2	3.93 (dd, 5.2, 1.5)
9	52.0	
10	36.1	α: 1.63 (dd, 13.9, 4.8)
		β: 1.67 (ddd, 13.9, 9.3, 1.7)
11	71.3	4.38 (ddd, 9.2, 4.6, 4.6)
12	13.1	0.80 (s)
13	22.0	0.89 (s)
14	23.7	0.86 (s)
15	72.0	3.53 (d, 11.2)
		3.34 (d, 11.2)

CULMORONE

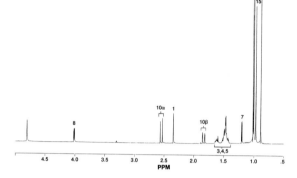

¹H NMR

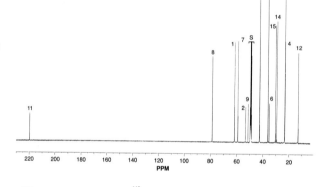

¹³C NMR

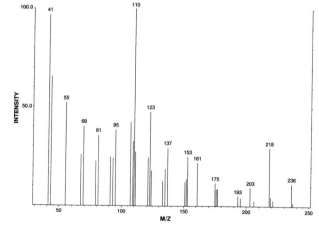

MS

CULMORONE

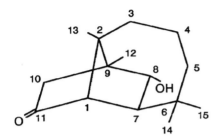

$C_{15}H_{24}O_2$

M.W. 236.1776

Species:

F. roseum
F. culmorum
F. crookwellense

References: Greenhalgh *et al.* (1986a, 1986b); Kasitu *et al.* (1992); Lauren *et al.* (1987).

Position	$^{13}C^a$	^1H
1	61.5	2.34 (s)
2	53.0	
3	35.9	1.62-1.42 (m)
4	22.9	1.62-1.42 (m)
5	42.3	1.62-1.42 (m)
6	35.0	
7	58.9	1.20 (d, 4.9)
8	78.4	4.01 (dd, 4.9, 2.0)
9	51.0	
10	42.2	α: 2.59 (d, 18.6)
		β: 1.83 (ddd, 18.6, 1.9, 1.2)
11	219.6	
12	12.6	1.02 (s)
13	23.0	0.89 (s)
14	29.0	1.00 (s)
15	30.0	0.97 (s)

^{a}: spectrum obtained in CD_3OD.

15-HYDROXY-CULMORONE

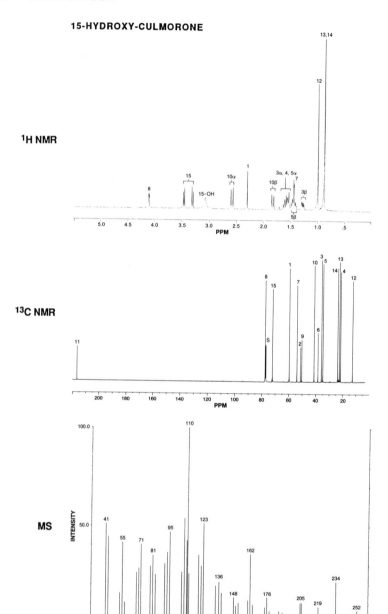

15-HYDROXY-CULMORONE

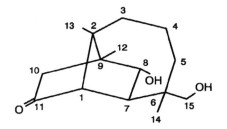

$C_{15}H_{24}O_3$

M.W. 252.1725

Species:

F. culmorum

References: Kasitu et al. (1992).

Position	^{13}C	^1H
1	59.5	2.30 (s)
2	51.5	
3	35.5	α: 1.59 (m)
		β: 1.27 (br dd, 14.2, 6.5)
4	21.5	1.51 (m)
5	34.5	α: 1.55 (m)
		β: 1.43 (m)
6	38.2	
7	53.9	1.43 (d, 3.5)
8	77.4	4.10 (d, 3.5)
9	50.2	
10	41.1	α: 2.58 (d, 18.5)
		β: 1.82 (d, 18.5)
11	216.2	
12	12.5	0.99 (s)
13	22.3	0.87 (s)
14	23.6	0.87 (s)
15	72.2	3.46 (d, 11.1)
		3.30 (d, 11.1)
		3.07 (s, OH)

CYCLONERODIOL

¹H NMR

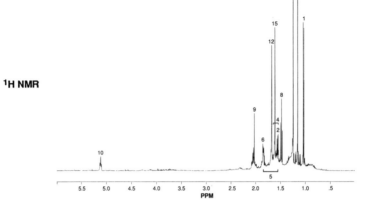

¹³C NMR

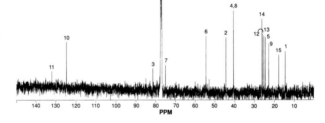

MS

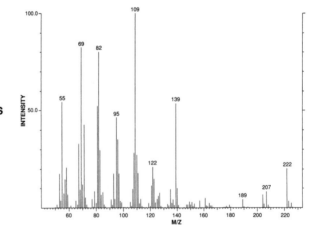

CYCLONERODIOL

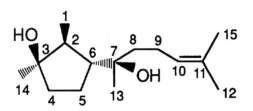

$C_{15}H_{28}O_2$

M.W. 240.2089

Species:

F. culmorum
F. moniliforme

References: Nozoe et al. (1970); Hanson et al. (1975); Cane et al. (1981).

Position	^{13}C	^1H
1	14.5	1.03 (d, 6.8)
2	44.2	1.58 (m)
3	81.3	
4	40.4	α: 1.59 (m)
		β: 1.67 (m)
5	24.2	α: 1.84 (m)
		β: 1.55 (m)
6	54.3	1.84 (m)
7	74.8	
8	40.4	1.47 (t, 8.3)
9	22.7	2.03 (dd, 8.3, 7.2)
10	124.5	5.10 (tq, 7.2, 0.7, 1.5)
11	131.8	
12	25.7	1.67 (d, 1.5)
13	25.1	1.15 (s)
14	26.1	1.24 (s)
15	17.7	1.61 (d, 0.7)

CYCLONEROTRIOL

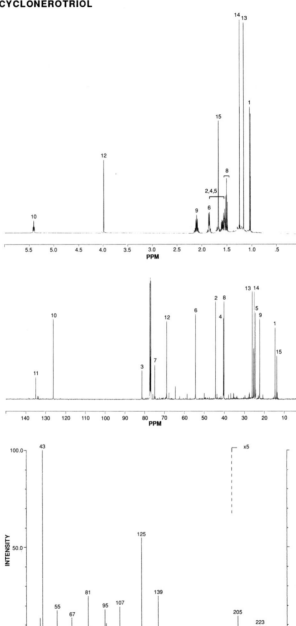

CYCLONEROTRIOL

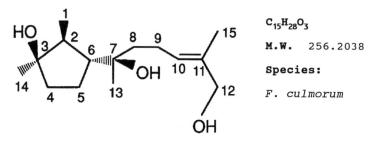

$C_{15}H_{28}O_3$

M.W.   256.2038

Species:

F. culmorum

References: Hanson *et al.* (1975); Kasitu *et al.* (1992)

Position	^{13}C	^1H
1	14.5	1.03 (d, 6.8)
2	44.3	1.60-1.53 (m)
3	81.3	
4	40.4	1.66 (m)
		1.60-1.53 (m)
5	24.4	1.85 (m)
		1.60-1.53 (m)
6	54.4	1.85 (m)
7	74.8	
8	40.1	1.50 (t, 8.3)
9	22.3	2.10 (m)
10	126.1	5.40 (t, 7.2)
11	135.0	
12	68.9	3.98 (s)
13	25.1	1.16 (s)
14	26.1	1.24 (s)
15	13.7	1.66 (s)

BUTENOLIDE

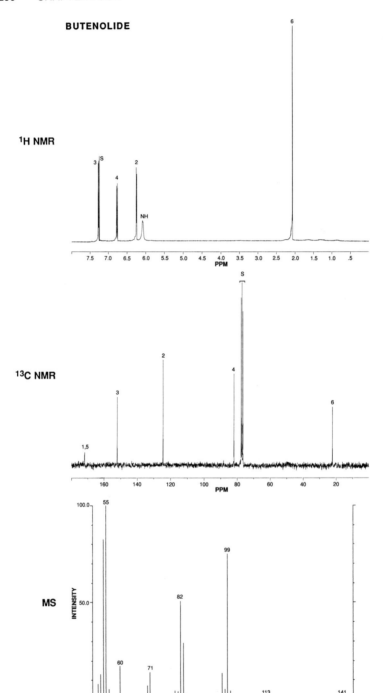

¹H NMR

¹³C NMR

MS

BUTENOLIDE
(4-Acetamido-4-hydroxy-2-butenoic acid γ-lactone)

$C_6H_7NO_3$

M.W. 141.0426

Species:

F. equiseti
F. tricinctum
F. graminearum
F. culmorum

References: Yates <u>et al.</u> (1967); Yates (1971); Vesonder <u>et al.</u> (1977).

Position	^{13}C	^1H
1	171.8	
2	124.1	6.24 (d, 5.5)
3	151.7	7.26 (dd, 5.5, 1.9)
4	81.9	6.76 (dd, 1.9, 9.0)
5	171.6	
6	22.3	2.06 (s)
NH		6.07 (br d, 9.0)

Spectra not provided.

MONILIFORMIN

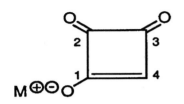

C$_4$HO$_3$(H, Na, or K)

M.W. 98.0004 (free acid)
         119.9824 (sodium salt)
         135.9563 (potassium salt)

Species:

F. moniliforme
F. subglutinans
F. sporotrichioides
F. culmorum
F. oxysporum

References: Abbas et al. (1990); Cole et al. (1973); Farber et al. (1988); Scott et al. (1987); Kriek et al. (1977); Thiel et al. (1982).

Position	$^{13}C^a$	$^1H^a$
1	202.1	
2	216.3	
3	202.1	
4	169.8	8.62 (s)

[a]: spectrum obtained in D$_2$O.

ZEARALENONE

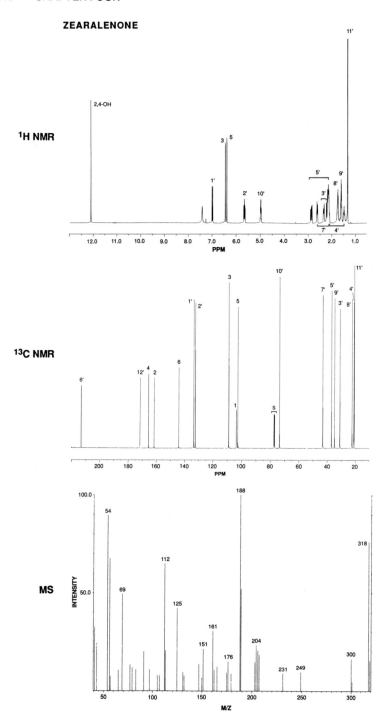

ZEARALENONE

$C_{18}H_{22}O_5$

M.W. 318.1467

Species:

F. graminearum
F. tricinctum
F. culmorum
F. crookwellense
F. equiseti

References: Christensen _et_ _al._ (1965); Hacking _et_ _al._ (1976);
Mirocha _et_ _al._ (1971); Golinski _et_ _al._ (1988);
Cordier _et_ _al._ (1990).

Position	^{13}C	^1H
1	103.7	
2	161.2	
3	108.7	6.42 (d, 2.3)
4	165.5	
5	102.6	6.36 (d, 2.3)
6	144.0	
1'	133.3	6.95 (d, 15.3)
2'	132.3	5.62 (ddd, 15.3, 10.3, 3.6)
3'	30.9	2.30, 2.21
4'	20.9	2.10, 1.46
5'	36.7	2.84, 2.15
6'	212.8	
7'	42.9	2.60, 2.10
8'	22.1	1.70
9'	34.7	1.59
10'	73.4	4.94 (dd, 6.1, 5.8)
11'	20.6	1.32 (d, 6.1)
12'	171.6	

Spectra not provided.

α-ZEARALENOL
β-ZEARALENOL

$C_{18}H_{24}O_5$

M.W. 320.1624

Species:

F. roseum
F. crookwellense

References: Urry et al. (1966); Stipanovic, R.D. and H.W.
Schroeder (1975); Hagler et al. (1979), Golinski et
al. (1988).

Position	^{13}C		^1H	
	α	β	α	β
1	103.8	103.8		
2	165.7	165.0		
3	102.5	102.6	6.56 (d, 2.5)	6.54 (d, 2.5)
4	160.5	160.8		
5	108.8	108.6	6.21 (d, 2.5)	6.34 (d, 2.5)
6	144.5	144.6		
1'	133.6	133.6	7.17 (d, 15)	6.90 (d, 15)
2'	132.9	131.7	5.89 (m)	5.98 (m)
3'	30.6	30.3	1.85 (m)	2.25 (m)
4'	22.8	21.6	1.73 (m)	1.90 (m)
5'	36.7	35.3	2.55 (m)	3.48 (m)
6'	66.9	67.0	3.85 (m)	3.76 (m)
7'	35.1	33.8	2.42 (m)	2.85 (m)
8'	21.7	18.7	1.51 (m)	1.62 (m)
9'	32.1	30.8	1.95 (m)	2.30 (m)
10'	73.8	72.6	5.08 (m)	5.10 (m)
11'	21.3	19.0	1.49 (d)	1.46 (d)
12'	171.7	170.3		

Spectra not provided.

8'-HYDROXY-ZEARALENONE

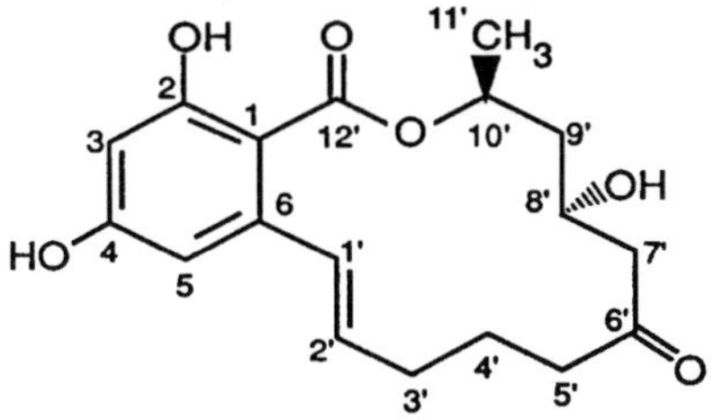

$C_{18}H_{22}O_6$

M.W. 334.1416

Species:

F. graminearum

References: Bolliger and Tamm (1972); Pathre and Mirocha (1979).

Position	^{13}C	^1H
1	102.8	
2	166.1	
3	102.1	7.06 (d, 2.5)
4	164.0	
5	108.4	6.82 (d, 2.5)
6	144.2	
1'		7.19 (d)
2'	131.0	6.30 (d)
3'	31.2	1.48 (m)
4'	23.1	1.34 (m)
5'	43.3	2.77 (m)
6'	212.1	
7'	51.7	2.35 (m)
8'	64.6	3.71 (m)
9'	39.1	1.99 (m)
10'	69.8	4.25 (m)
11'	21.0	1.04 (d)
12'	172.1	

Spectra not provided.

4-ACETYL-ZEARALENONE

$C_{20}H_{24}O_6$

M.W. 360.1573

Species:

F. graminearum

References: Munoz et al. (1989).

Position	^{13}C	^1H
1	108.3	
2	164.6	
3	113.7	6.66 (t, 2.7)
4	155.0	
5	109.6	6.66 (t, 2.7)
6	143.4	
1'	133.5	7.02 (dd, 1.6, 15.3)
2'	132.6	5.72 (m)
3'	31.0	2.04-2.42 (m)
4'	21.0	1.42-1.85 (m)
5'	36.6	2.82 (m)
6'	210.9	
7'	42.9	2.60 (m)
8'	22.0	1.42-1.85 (m)
9'	34.6	1.42-1.85 (m)
10'	74.0	5.03 (m)
11'	20.6	1.40 (d, 6.2)
12'	171.1	
CH_3(Ac)	21.0	2.30 (s)
C=O(Ac)	168.7	
OH		11.91 (s)

Spectra not provided.

4-ACETYL-CIS-ZEARALENONE

$C_{20}H_{24}O_6$

M.W. 360.1573

Species:

F. graminearum

References: Munoz et al. (1989).

Position	^{13}C	^1H
1	109.5	
2	164.7	
3	120.8	6.41 (d, 2.5)
4	154.6	
5	109.4	6.68 (d, 2.5)
6	141.6	
1'	131.1	6.63 (dd, 12.1, 2.1)
2'	131.0	5.46 (m)
3'	28.9	1.45-2.62 (m)
4'	21.2	1.45-2.62 (m)
5'	40.7	1.45-2.62 (m)
6'	210.3	
7'	41.5	1.45-2.62 (m)
8'	21.3	1.45-2.62 (m)
9'	30.0	1.45-2.62 (m)
10'	74.5	5.15 (m)
11'	21.0	1.36 (d, 6.3)
12'	170.7	
CH_3(Ac)	21.2	2.30 (s)
C=O(Ac)	168.6	

FUSARIC ACID

¹H NMR

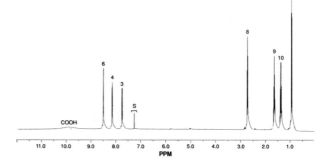

¹³C NMR

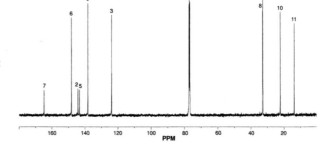

MS

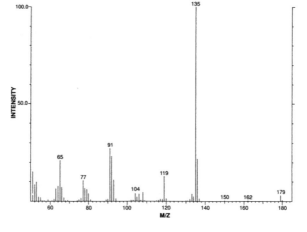

FUSARIC ACID

$C_{10}H_{13}NO_2$

M.W. 179.0946

Species:

F. moniliforme
F. oxysporum

References: Yabuta *et al.* (1934); Mutert *et al.* (1981); Burmeister *et al.* (1985).

Position	^{13}C	^1H
2	144.2	
3	123.7	7.73 (dd, 8.0, 1.1)
4	138.1	8.13 (d, 8.0)
5	143.3	
6	148.1	8.49 (br s)
7	164.6	
8	32.9	2.71 (t, 7.7)
9	32.8	1.63 (p, 7.6, 14.8)
10	22.2	1.37 (dq, 7.3, 14.8)
11	13.8	0.93 (t, 7.3)

ENNIATIN A

¹H NMR

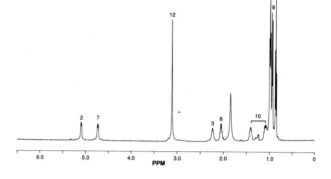

¹³C NMR

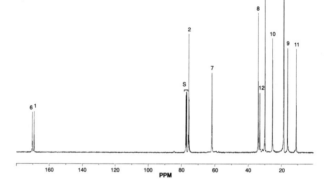

MS

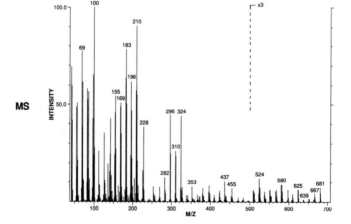

ENNIATIN A

$C_{36}H_{63}N_3O_9$

M.W. 681.4564

Species:

F. oxysporum
F. sambucinum
F. lateritium
F. avenaceum
F. fructigenum
F. compactum

References: Gaumann et al. (1947); Audhya and Russell (1973).

Position	^{13}C	^{1}H
1	169.3	
HyIv:		
2	75.6	5.10 (d, 7.8)
3	29.8	2.25 (m, 6.8)
4	18.3	0.92 (d, 6.8)
5	18.4	0.98 (d, 7.2)
6	170.4	
NMeIle:		
7	61.8	4.65 (d, 9.2)
8	34.0	2.04 (m, 9.4, 3.2)
9	16.2	1.00 (d, 8.3)
10	25.4	1.04 (m, 13.2, 7.5)
		1.42 (m, 13.2, 7.4, 3.2)
11	10.9	0.84 (t, 7.4)
12	33.0	3.09 (s)

ENNIATIN A1

¹H NMR

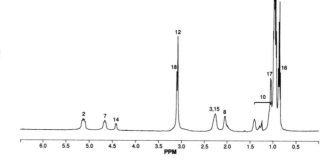

¹³C NMR

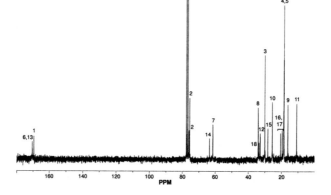

MS

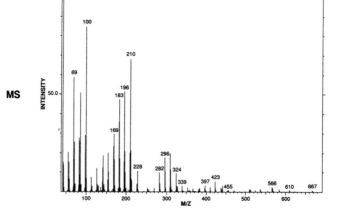

ENNIATIN A1

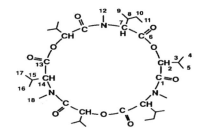

$C_{35}H_{61}N_3O_9$

M.W. 667.4407

Species:

F. avenaceum
F. compactum

References: Audhya and Russell (1973); Blais et al. (1992)

Position	^{13}C	^1H
1	169.5	
HyIv:		
2	75.6	5.10 (td, 8.8, 9.1, 7.0)
	76.0	
3	29.9	2.26 (m)
4	18.4-18.7	0.92-0.99 (d, 6.6)
5	18.4-18.7	0.92-0.99 (d, 6.6)
6	170.5	
NMeIle:		
7	61.4	4.62 (d, 9.2)
	61.4	4.66 (d, 9.2)
8	33.9	2.01 (m)
	34.0	2.01 (m)
9	16.2	0.92-0.99
10	25.5	1.04 (m)
		1.42 (m, 7.2)
11	10.9	0.83 (t, 7.5)
		0.85 (t, 7.1)
12	32.7	3.08 (s)
	32.8	3.08 (s)
NMeVal:		
13	170.4	
14	63.4	4.42 (d, 10.3)
15	28.1	2.28 (m)
16	19.5	0.87 (d)
17	20.5	1.03 (d, 6.2)
18	33.4	3.10 (s)

ENNIATIN A2

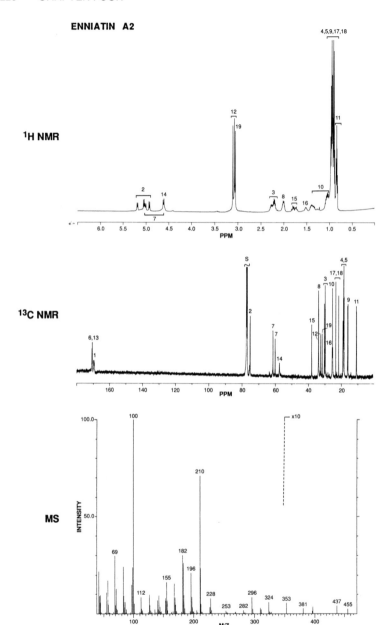

ENNIATIN A2

$C_{36}H_{63}N_3O_9$

M.W. 681.4564

Species:

F. avenaceum

References: Blais et al. (1992).

Position	^{13}C	^1H
1	169.4	
	169.4	
	169.7	
HyIv:		
2	77.2	4.92 (d, 8.9)
	75.0	5.05 (d, 8.2)
	75.2	5.18 (d, 8.2)
3	29.7	2.22 (m)
	29.8	2.22 (m)
	30.3	2.28 (m)
4,5	18.2-18.9	0.89, 0.90 (d, 5.5)
		0.93 (d, 7.4)
		0.94-0.96
6	170.5	
NMeIle:		
7	60.0	5.02 (d, 9.6)
	61.3	4.60 (m)
8	33.7	2.00 (m, 9.6, 3.1, 7.7)
	33.8	
9	15.9	0.94-0.96
	16.1	
10	25.1	1.02, 1.34 (m, 13.6, 7.5, 3.2)
	25.5	1.07, 1.40 (m, 13.6, 7.5, 3.3)
11	10.7	0.83 (t, 7.4)
	10.8	0.84 (t, 7.4)
12	32.5	3.09 (s)
	33.9	3.14 (s)
NMeLeu:		
13	170.8	
14	57.4	4.60 (m)
15	37.9	1.72, 1.81 (m, 14.2, 4.9, 9.6)
16	25.3	1.52 (m)
17,18	21.5, 23.2	0.90 (d, 7.1), 0.92 (d, 7.1)
19	31.6	3.05 (s)

ENNIATIN B

¹H NMR

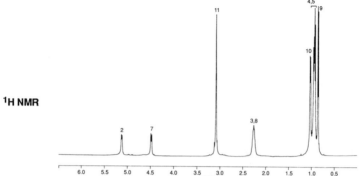

¹³C NMR

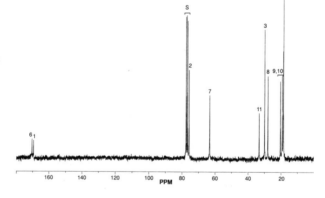

MS

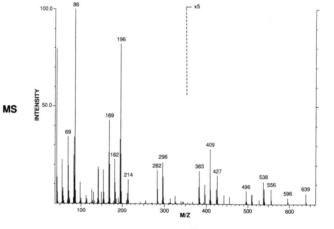

ENNIATIN B

$C_{33}H_{57}N_3O_9$

M.W. 639.4094

Species:

F. avenaceum
F. gibbosum
F. oxysporum
F. sambucinum
F. tricinctum
F. compactum

References: Nager (1948); Ovchinnikov et al. (1974).

Position	^{13}C	^1H
1	169.3	
HyIv:		
2	75.7	5.11 (d, 8.8)
3	29.9	2.27 (m)
4	18.5	0.93 (d, 6.7)
5	18.6	0.96 (d, 6.4)
6	170.3	
NMeVal:		
7	63.2	4.46 (d, 9.8)
8	27.9	2.27 (m)
9	19.3	0.86 (d, 6.8)
10	20.4	1.03 (d, 6.5)
11	33.2	3.09 (s)

ENNIATIN B1

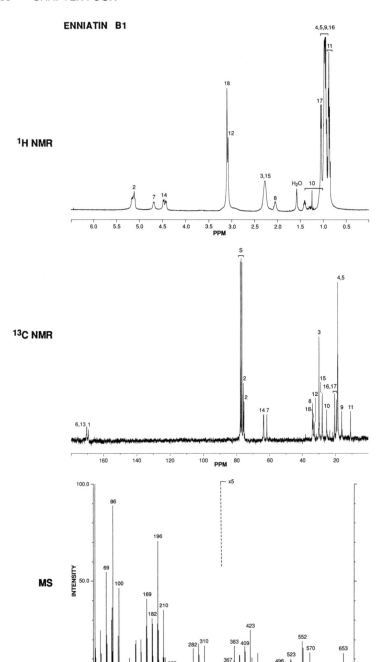

ENNIATIN B1

$C_{34}H_{59}N_3O_9$

M.W. 653.4251

Species:

F. avenaceum
F. compactum
F. oxysporum
F. tricinctum

References: Ovchinnikov et al. (1964); Audhya and Russell (1973); Burmeister and Plattner (1987); Blais et al. (1992).

Position	^{13}C	^1H
1	169.3	
HyIv:		
2	75.7	5.11 (m, 7.3, 9.1, 7.9)
	75.4	
3	29.8-30.2	2.24 (m)
4,5	18.4-18.7	0.92-0.98
6	170.4	
NMeIle:		
7	61.5	4.65 (d, 9.0)
8	33.8	2.04 (m)
9	16.1	0.92 (d, 7.7)
10	25.3	1.04, 1.42 (m, 13.8, 7.5, 8.7)
11	10.9	0.82 (t, 7.3)
12	32.7	3.08 (s)
NMeVal:		
13	170.4	
14	63.4	4.42 (d, 8.6)
	63.5	4.46 (d, 9.1)
15	27.9	2.27 (m)
16	19.4	0.87 (d, 6.5)
17	20.5	1.02 (d, 6.3)
18	33.4	3.10

ENNIATIN B2

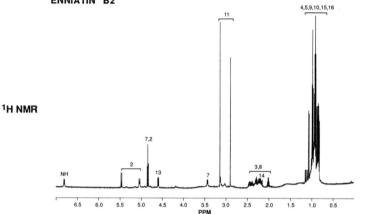

¹H NMR

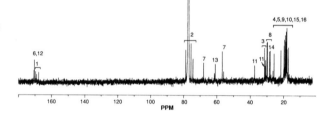

¹³C NMR

MS

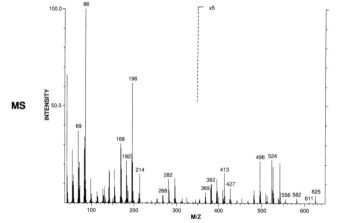

ENNIATIN B2

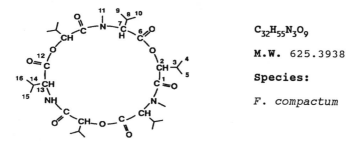

$C_{32}H_{55}N_3O_9$

M.W. 625.3938

Species:

F. compactum

References: Visconti et al. (1992).

Position	^{13}C	^1H
1	169.6	
	168.9	
	167.7	
HyIv:		
2	78.6	5.46 (d, 9.9)
	74.2	5.04 (d, 9.8)
	75.4	4.84 (d, 8.8)
3	31.4	2.35 (m)
	29.6	2.16 (m)
	29.9	2.01 (m)
4,5	17.0-25.6	0.83-1.01 (m)
6	170.4	
	170.4	
NMeVal:		
7	67.8	4.82 (d, 10.6)
	56.7	3.48 (d, 8.7)
8	30.7	2.38 (m)
	27.9	2.47 (m)
9,10	17.0-25.6	0.95-1.07 (m)
11	37.3	3.14 (s)
	31.0	2.88 (s)
Val:		
12	170.6	
13	60.9	4.60 (dd, 3.9, 7.9)
14	28.4	2.22 (m)
15,16	17.0-25.6	0.84 (d, 6.7), 0.94 (d, 6.7)
NH		6.80 (d, 8.0)

ENNIATIN B4

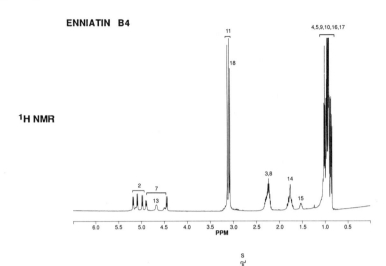

^{1}H NMR

^{13}C NMR

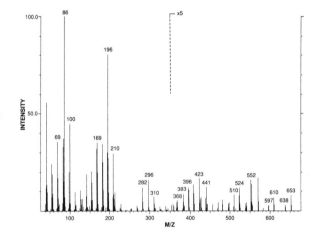

MS

ENNIATIN B4

$C_{34}H_{59}N_3O_9$

M.W. 653.4251

Species:

F. compactum

References: Visconti *et al*. (1992).

Position	^{13}C	^1H
1	169.0	
	169.3	
	169.6	
HyIv:		
2	74.9	4.98 (d, 9.0)
	75.2	5.09 (d, 8.5)
	75.6	5.18 (d, 8.4)
3	29.6	2.22–2.27(m)
	29.8	
	30.2	
4,5	18.3–18.9	0.93–0.97
6	170.3	
	170.4	
NMeVal:		
7	61.3	4.89 (d, 10.2)
	63.2	4.44 (d, 10.0)
8	27.5	2.23 (m)
	27.8	2.16 (m)
9,10	19.9,20.0	0.85 (d, 6.7), 0.87 (d, 6.8)
		1.02 (d, 6.2)
11	33.0	3.07 (s)
	33.8	3.10 (s)
NMeLeu:		
12	170.7	
13	57.2	4.66 (dd)
14	37.9	1.73, 1.81 (m, 13.5, 5.1, 10.1)
15	25.2	1.55 (m)
16,17	21.5, 23.3	0.91, 0.92 (dd, 6.6)
18	31.6	3.04 (s)

FUSARIN A

¹H NMR

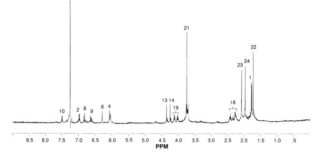

¹³C NMR

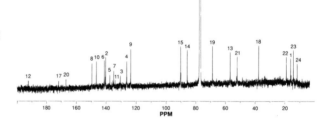

MS

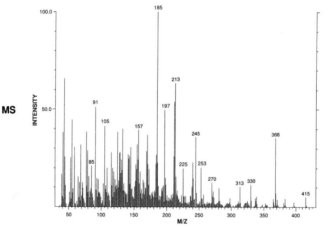

FUSARIN A

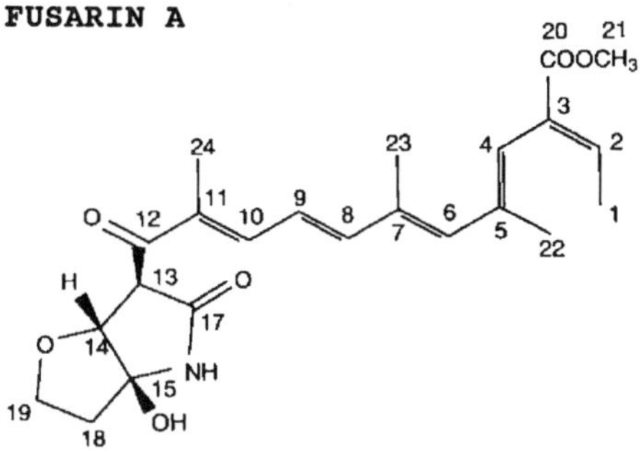

$C_{23}H_{29}NO_6$

M.W. 415.1995

Species:

F. moniliforme

References: Wiebe and Bjeldanes (1981); Steyn and Vleggaar (1985); Savard and Miller (1992).

Position	^{13}C	^1H
1	15.9	1.77 (dd, 7.2, 1.3)
2	140.1	6.97 (q, 7.2)
3	130.4	
4	126.2	6.06 (br s)
5	137.4	
6	140.9	6.28 (br s)
7	134.8	
8	149.2	6.83 (d, 15)
9	123.5	6.61 (dd, 15, 11)
10	146.2	7.44 (d, 11)
11	133.8	
12	192.5	
13	56.6	4.37 (s)
14	85.6	4.23 (d, 1.2)
15	90.0	
17	171.8	
18	37.5	2.35 (m)
		2.23 (m)
19	68.5	4.08 (m)
		3.91 (m)
20	166.4	
21	51.9	3.73 (s)
22	18.8	1.72 (d, 1.4)
23	14.1	2.07 (s)
24	11.6	1.96 (s)

FUSARIN C

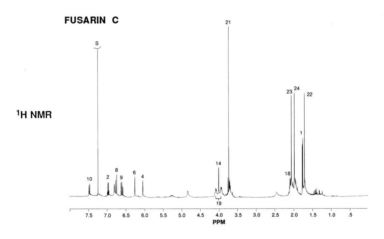

¹H NMR

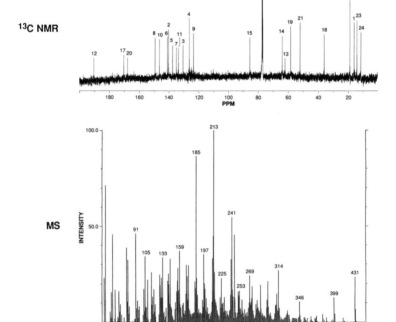

¹³C NMR

MS

FUSARIN C

20 21
COOCH₃

$C_{23}H_{29}NO_7$

M.W. 431.1944

Species:

F. moniliforme
F. sporotrichioides
F. graminearum
F. avenaceum
F. culmorum
F. sambucinum
F. crookwellense

References: Gelderblom et al. (1984); Farber and Sanders (1986); Mirocha et al. (1989); Golinski et al. (1988); Savard and Miller (1992).

Position	^{13}C	^1H
1	15.9	1.76 (dd, 7.3, 1.3)
2	140.1	6.97 (q, 7.3)
3	130.4	
4	126.2	6.05 (br s)
5	137.4	
6	140.9	6.25 (br s)
7	134.8	
8	149.3	6.74 (d, 15)
9	123.4	6.61 (dd, 15, 11)
10	146.3	7.44 (d, 11)
11	133.4	
12	190.2	
13	61.9	
14	63.7	3.99 (d, 2.5)
15	85.4	
17	170.1	
18	35.9	2.10 (m)
19	58.2	4.10 (m)
		3.96 (m)
20	167.6	
21	51.9	3.73 (s)
22	18.7	1.71 (d, 1.4)
23	14.1	2.06 (s)
24	11.4	1.99 (s)

Spectra not provided.

FUSARIN D

20 21
COOCH₃

$C_{23}H_{29}NO_7$

M.W. 431.1944

Species:

F. moniliforme

References: Wiebe and Bjeldanes (1981); Steyn and Vleggaar (1985). Vleggaar (pers. commun.)

Position	^{13}C	^1H
1	16.1	1.77 (dd, 7.2, 1.4)
2	140.2	6.95 (qd, 7.2, 1.1)
3	130.9	
4	126.3	6.05 (br s)
5	137.8	
6	140.0	6.27 (br s)
7	135.5	
8	147.7	6.74 (d, 15)
9	124.0	6.63 (dd, 15, 11)
10	144.0	7.36 (d, 11)
11	133.0	
12	197.1	
13	85.3	
14	91.8	4.31 (s)
15	95.6	
17	175.5	
18	38.4	2.28 (m)
19	69.2	3.97 (ddd, 8.9,6.7,4.8)
		3.80 (ddd, 8.9,8.9,6.5)
20	167.8	
21	52.0	3.71 (s)
22	17.0	1.72 (d, 1.4)
23	14.3	2.08 (d, 1.3)
24	12.7	1.98 (d, 1.2)

Spectra not provided.

FUSARIN E

$C_{23}H_{31}NO_7$

M.W. 433.2100

Species:

F. moniliforme

Reference: Sanson (1989).

Position	$^{13}C^a$	$^1H^a$
1	16.1	1.75 (dd, 7.2, 1.4)
2	140.0	6.93 (qd, 7.2, 1.1)
3	131.0	
4	125.0	5.97 (qd, 1.4, 1.1)
5	137.2	
6	142.0	6.1 (br s)
7	136.2	
8	137.9	6.45 (d, 15)
9	124.4	6.96 (dd, 15, 11)
10	135.6	7.16 (d, 6.0)
11	134.4	
12	190.2	
13	53.8	3.83 (d, 7)
14	84.6	4.50 (d, 7)
15	85.1	
17	170.3	
18	40.7	2.07 (ddd, 14, 8.3, 4)
		2.00 (ddd, 14, 5.8, 3.7)
19	58.7	4.02 (ddd, 11, 8.2, 3.8)
		3.93 (ddd, 11, 6, 4)
20	167.8	
21	52.1	3.72 (s)
22	19.0	1.67 (d, 1.4)
23	14.3	2.11 (d, 1.3)
24	19.3	1.96 (d, 1.3)

[a]:   spectra obtained in CD_2Cl_2

FUSARIN F

¹H NMR

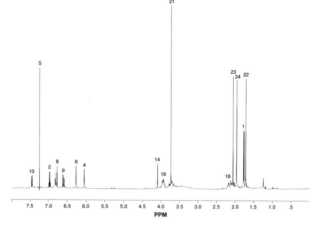

¹³C NMR

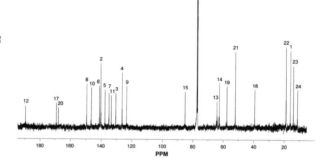

MS

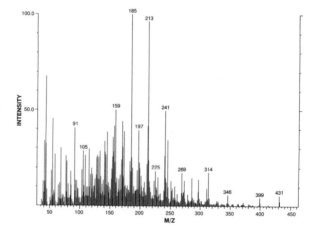

FUSARIN F

$C_{23}H_{29}NO_7$

M.W. 431.1944

Species:

F. moniliforme

References: Savard and Miller (1992).

Position	^{13}C	^1H
1	15.9	1.77 (dd, 7.2, 1.4)
2	140.1	6.96 (q, 7.2)
3	130.4	
4	126.2	6.06 (br s)
5	137.4	
6	140.9	6.27 (br s)
7	134.8	
8	149.2	6.76 (d, 15)
9	123.4	6.62 (dd, 15, 11.5)
10	146.5	7.47 (d, 11.7)
11	133.5	
12	189.5	
13	64.2	
14	62.4	4.11 (d, 2.4)
15	84.8	
17	168.2	
18	39.2	2.22 (m)
		2.09 (m)
19	57.6	3.99 (m)
		3.92 (m)
20	167.6	
21	51.9	3.73 (s)
22	18.7	1.72 (d, 1.4)
23	14.1	2.07 (s)
24	11.3	1.97 (s)

Spectra not provided.

SAMBUCINIC ACID
(5-HYDROXY-6,7,10-TRIMETHYLTRICYCLO[5.3.1.0]UNDEC-9-ENE-2-CARBOXYLIC ACID)

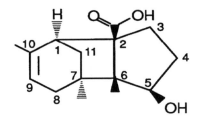

$C_{15}H_{22}O_3$

M.W. 250.1569

Species:

F. sambucinum

References: Rosslein et al. (1988).

Position	^{13}C	^1H
1	46.8	2.69 (d, 4.0)
2	70.9	
3	28.2	1.72 (td, 12, 6)
		2.35 (ddd, 12, 6.5, 2)
4	38.5	2.27 (dtd, 11, 5.5, 1.5)
		2.67 (m)
5	75.7	4.44 (dd, 10.5, 5.5)
6	60.3	
7	42.5	
8	41.3	2.10 (d quint, 18, 2.5)
		2.50 (dm, 18)
9	121.4	5.33 (m)
10	140.8	
11	46.1	1.82 (d, 11)
		2.90 (ddd, 10.5, 4, 1.3)
COOH	179.2	
CH_3(6)	16.1	1.55 (s)
(7)	25.4	1.15 (s)
(10)	25.0	1.63 (dt, 4, 2)

FUMONISIN B$_1$

^{1}H NMR

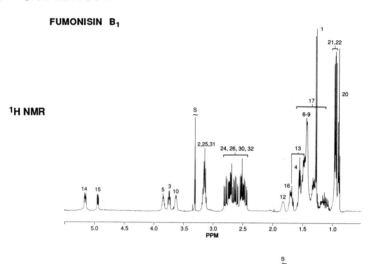

^{13}C NMR

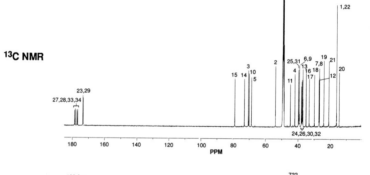

MS

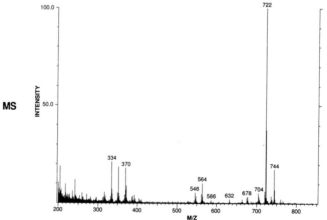

FUMONISIN B$_1$
(Macrofusine)

$C_{34}H_{59}NO_{15}$

M.W. 721.3884 (M+1: 722.3962)

Species:

F. moniliforme
F. proliferatum
F. nygamai

References: Bezuidenhout et al. (1988); Laurent et al. (1990); Ross et al. (1990); Thiel et al. (1991).

Position	$^{13}C^a$	$^1H^a$
1	16.0	1.27 (d, 6.7)
2	53.7	3.14 (dq, 6.7, 6.8)
3	70.3	3.74 (ddd, 9.6, 6.8, 3.2)
4	41.7	1.55 (m)
5	68.4	3.84 (m)
6	39.1	1.40-1.60
7	26.6	1.30-1.50
8	26.7	1.30-1.50
9	39.0	1.40-1.60
10	69.8	3.62 (m)
11	44.6	1.45 (m); 1.15 (m)
12	26.9	1.81 (m)
13	36.6	1.70 (m); 1.55 (m)
14	72.8	5.16 (ddd, 10.8, 3.7, 2.4)
15	78.7	4.94 (dd, 8.1, 3.7)
16	34.8	1.70 (m)
17	32.9	1.44 (m); 1.07 (m)
18	29.5	1.33 (m); 1.18 (m)
19	23.9	1.40 (m); 1.10 (m)
20	14.4	0.89 (t, 7.1)
21	20.8	0.96 (d, 6.7)
22	16.0	0.94 (d, 6.8)
23	173.5	
24	37.1	2.71 (dd, 16.7, 7.4)
		2.45 (dd, 16.7, 6.5)
25	39.6	3.10-3.20 (m)
26	37.8	2.66 (dd, 16.5, 8.3)
		2.50 (dd, 16.5, 5.2)
27	177.4	
28	178.5	
29	173.0	
30	36.9	2.79 (dd, 16.6, 7.2)
		2.60 (dd, 16.6, 6.4)
31	39.7	3.10-3.20 (m)
32	37.5	2.72 (dd, 16.5, 7.5)
		2.52 (dd, 16.5, 6.0)
33	177.0	
34	178.0	

[a]: spectra obtained in CD$_3$OD.

FUMONISIN B$_2$

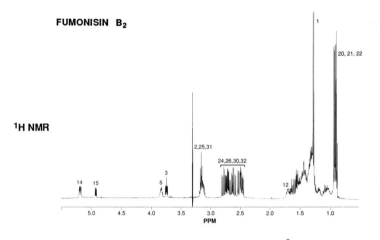

^{1}H NMR

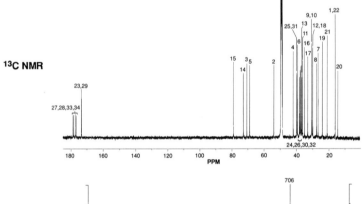

^{13}C NMR

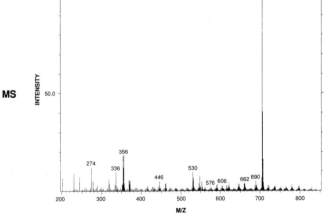

MS

FUMONISIN B$_2$

$C_{34}H_{59}NO_{14}$

M.W. 705.3935 (M+1: 706.4013)

Species:

F. moniliforme
F. proliferatum
F. nygamai

References: Bezuidenhout et al. (1988); Plattner et al. (1992).

Position	$^{13}C^a$	$^1H^a$
1	16.1	1.27 (d, 6.7)
2	53.7	3.16 (dq, 6.7, 6.8)
3	70.4	3.74 (ddd, 9.6, 6.8, 3.2)
4	41.8	1.55 (m)
5	68.7	3.82 (m)
6	39.1	1.30-1.50
7	26.5	1.30-1.50
8	27.4	1.40 (m)
9	30.4	1.10-1.20
10	30.5	1.30-1.50
11	36.1	1.45 (m)
12	29.8	1.68 (m)
13	36.5	1.60 (m); 1.45 (m)
14	72.7	5.18 (ddd, 10.8, 3.8, 2.4)
15	78.9	4.91 (dd, 8.0, 3.8)
16	34.9	1.53 (m)
17	32.9	1.30-1.50
18	29.8	1.30-1.50
19	23.9	1.25-1.35
20	14.4	0.89 (t, 7.1)
21	20.9	0.93 (d, 6.9)
22	16.1	0.91 (d, 6.6)
23	173.4	
24	37.1	2.72 (dd, 16.8, 6.9)
		2.46 (dd, 16.8, 6.8)
25	39.6	3.10-3.20 (m)
26	37.9	2.65 (dd, 16.5, 8.3)
		2.48 (dd, 16.5, 5.1)
27	176.6	
28	178.6	
29	173.2	
30	36.9	2.78 (dd, 16.7, 7.3)
		2.60 (dd, 16.7, 6.4)
31	39.7	3.10-3.20 (m)
32	37.6	2.71 (dd, 16.5, 7.5)
		2.51 (dd, 16.5, 6.0)
33	177.2	
34	178.1	

^{a}: spectra obtained in CD$_3$OD.

FUMONISIN B₃

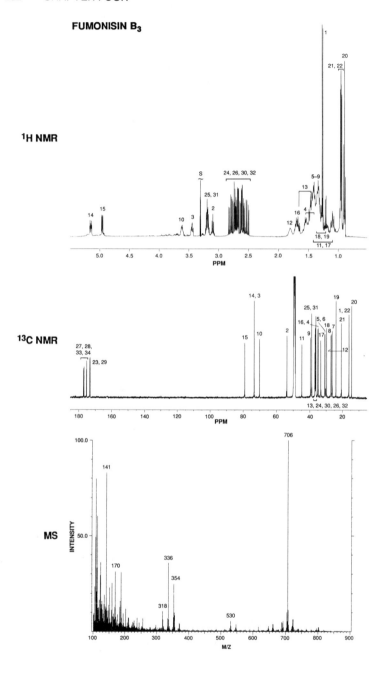

FUMONISIN B$_3$

$C_{34}H_{59}NO_{14}$

M.W. 705.3935 (M+1: 706.4013)

Species:

F. moniliforme
F. proliferatum
F. nygamai

References: Plattner *et al.* (1992).

Position	$^{13}C^a$	$^1H^a$
1	16.0	1.26 (d, 6.6)
2	53.5	3.08 (dq, 6.6, 6.8)
3	73.2	3.44 (ddd, 12.4, 6.8, 3.1)
4	34.7	1.44 (m); 1.66 (m)
5	30.6	1.30-1.40
6	30.7	1.30-1.40
7	26.2	1.52 (m); 1.45 (m)
8	26.8	1.46 (m); 1.34 (m)
9	39.3	1.36-1.44
10	69.9	3.61 (m)
11	44.6	1.46 (m); 1.09 (m)
12	27.0	1.80 (m)
13	36.7	1.48 (m); 1.68(m)
14	73.2	5.15 (ddd, 10.9, 2.6, 3.3)
15	78.8	4.95 (dd, 8.5, 3.3)
16	34.9	1.70 (m)
17	33.1	1.44 (m); 1.10 (m)
18	29.6	1.20 (m); 1.32 (m)
19	23.9	1.20 (m); 1.36 (m)
20	14.4	0.89 (t, 7.1)
21	20.6	0.95 (d, 6.8)
22	15.9	0.94 (d, 6.6)
23	173.2	
24	36.6	2.72 (dd, 16.8, 7.2)
		2.56 (dd, 16.8, 6.6)
25	38.6	3.18 (dddd, 7.2, 6.6, 5.8, 7.4)
26	36.2	2.69 (dd, 17.0, 7.4)
		2.59 (dd, 17.0, 5.8)
27	175.2	
28	177.0	
29	173.0	
30	36.4	2.80 (dd, 16.7, 7.1)
		2.65 (dd, 16.7, 6.2)
31	38.7	3.19 (dddd, 7.1, 6.2, 7.0, 6.3)
32	36.1	2.75 (dd, 16.9, 7.0)
		2.60 (dd, 16.9, 6.3)
33	175.0	
34	176.7	

^{a}: spectra obtained in CD$_3$OD

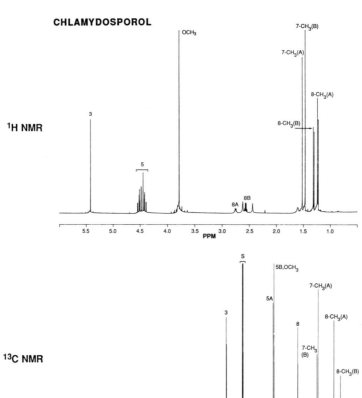

CHLAMYDOSPOROL

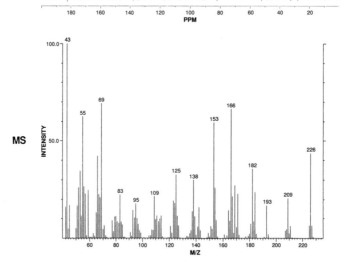

CHLAMYDOSPOROL

A   B

$C_{11}H_{14}O_5$

M.W. 226.0841

Species:

F. chlamydosporum
F. acuminatum

References: Savard et al. (1990); Grove and Hitchcock (1991).

Position		^{13}C	^{1}H
2	AB	168.49[ab]	
3	AB	88.05, 87.82	5.41 (s)
4	A	159.07[a]	
	B	157.68[b]	
4a	A	105.26	
	B	105.97	
5	A	56.56	4.39 (dd, 14.9, 1.9)
			4.53 (dd, 14.9, 3)
	B	56.11	4.41 (dd, 14.9, 0.7)
			4.50 (dd, 14.9, 1.8)
7	A	97.70	
	B	97.15	
8	A	40.08	2.74 (m, 7, 3, 1.9)
	B	39.95	2.56 (br q, 7, 1.8, 0.7)
8a	A	164.70[a]	
	B	164.64[b]	
7-CH$_3$	A	26.17	1.51 (s)
	B	26.96	1.46 (s)
8-CH$_3$	A	15.62	1.30 (d, 7)
	B	11.14	1.22 (d, 7)
OCH$_3$	AB	56.00	3.78 (s)

[a], [b]: may be interchanged

Spectra not provided.

EQUISETIN

$C_{22}H_{31}NO_4$

M.W. 373.2253

Species:

F. equiseti

References: Burmeister *et al.* (1974); Phillips *et al.* (1989).

Position	^{13}C	^1H
1	190.6	
2	48.4	
3	44.6	3.32 (br d, 9.0)
4	127.1[b]	5.40 (br s)
5	130.4[a]	5.40 (br s)
6	38.4	1.84 (br t, 12)
7	41.9	0.90 (q, 12)
		1.81 (br d, 12)
8	33.3	1.48 (m)
9	35.5	1.11 (q, 12)
		1.76 (br d, 12)
10	28.1	1.04 (q, 12)
		1.96 (br d, 12)
11	39.6	1.66 (br t, 11)
12	13.7	1.46 (s)
13	129.8[a]	5.17 (dd, 15, 9.0)
14	126.2[b]	5.25 (dq, 15, 6.5)
15	18.2	1.56 (d, 6.5)
16	22.5	0.92 (d, 6.5)
2'	176.7	
3'	99.8	
4'	198.9	
5'	66.4	3.67 (dd, 5.0, 3.5)
6'	60.0	4.07 (dd, 12, 3.5)
		3.90 (dd, 12, 5.0)
7'	27.2	3.07 (s)
enol		17.3 (s)

[a,b]: may be interchanged

Part Two
Aspergillus and *Penicillium*

Chapter 5

Penicillium and *Aspergillus* Toxins

P. M. Scott

Penicillium Toxins

The major *Penicillium* species occurring in Canadian grains are *P. verrucosum* var. *cyclopium* (*P. aurantiogriseum*), *P. chrysogenum*, and *P. simplicissimum* (Abramson et al 1983, 1985, 1990a,b). In addition, *P. expansum* is a common rot of apples in Canada (Harwig et al 1973), and *P. roquefortii* and *P. caseicola* (*P. camembertii*) are used as starter cultures in cheese manufacture (Scott 1981). The more important mycotoxins produced by these species are described in detail below.

OCHRATOXIN A

Studies on storage of Canadian grains have demonstrated an association of ochratoxin A production with strains of *P. verrucosum* var. *cyclopium* (*P. aurantiogriseum*) (Abramson et al 1983, 1985). According to Pitt (1987), ochratoxin A is a metabolite of *P. verrucosum* but not of *P. viridicatum,* a species name used in many previous studies (Harwig et al 1983, Kuiper-Goodman and Scott 1989). Natural occurrence of this carcinogenic mycotoxin in Canadian grains has been documented (Harwig et al 1983, Kuiper-Goodman and Scott 1989). Ochratoxin A is also an *Aspergillus* mycotoxin, originally isolated from *A. ochraceus* (Madhyastha et al 1990).

The chemistry of ochratoxin A (Fig. 5.1) has been reviewed (Harwig et al 1983, Steyn 1984, Kuiper-Goodman and Scott 1989), and where referencing below is incomplete, it may be found in these papers. The dechloro analogue of ochratoxin A, ochratoxin B, the methyl and ethyl esters of both compounds, and 4R-hydroxyochratoxin A are all fungal metabolites.

Ochratoxin A is a colorless, crystalline compound that is soluble in polar organic solvents and dilute sodium bicarbonate solution and slightly soluble in water. It has a melting point (mp) of about 90°C when crystallized from benzene as a solvate. Nonsolvated crystals of mp 169°C have been obtained from xylene; these were suitable for X-ray structural analysis (Bredenkamp et al 1989). Ochratoxin A is optically active. The ultraviolet (UV) absorption spectrum, which varies with pH and solvent polarity, shows maxima at 213 nm (ϵ 36,800) and 332 nm (ϵ 6,400) in ethanol. Fluorescence emission is maximum at 467 nm in 96% ethanol and 428 nm in absolute ethanol and increases in intensity with increasing pH and with protein binding.

The infrared spectrum of ochratoxin A in chloroform has peaks at 3,380, 2,988, 1,723, 1,674, 1,612, 1,528, 1,425, 1,381, 1,304, 1,260, 1,170, 1,140, 1,107, and 827 cm^{-1} (Steyn 1984). Infrared spectroscopy has demonstrated that in solution, ochratoxin A exists in the β-conformation, where the amide NH is hydrogen-bonded to the phenolic oxygen (Bredenkamp et al 1989). Support for this conformation in solution was provided by ^{13}C nuclear magnetic resonance (NMR) spectroscopy, with evidence for hydrogen bonding of the phenolic proton to the lactone carbonyl group (Bredenkamp et al 1989). Complete and partial ^{13}C NMR spectra of ochratoxin A have been published (de Jesus et al 1980, Pachter and Wessels 1980, Weisleder and Lillehoj 1980). The 100-MHz ^{1}H NMR spectrum (in CDCl$_3$) shows signals at δ 12.70, 10.80, and 8.55 (three exchangeable protons); 7.23 and 7.15 (six aromatic H); 4.71 and 5.07 (CH); 2.78 and 3.2 (CH$_2$); and 1.55 (CH$_3$) (de Jesus et al 1980). Pohland et al (1982) presented both 100 and 250 MHz ^{1}H NMR spectra. Major peaks in the electron impact (EI) mass spectrum are at m/z 239/241 and 255/257 with a low-intensity molecular ion at m/z 403 (Gallagher and Stahr 1981, Bauer and Gareis 1987), although a somewhat different spectrum that included major peaks at m/z 238, 239, 240, 241, 255, 256, and 257 was quoted by Steyn (1984) (see also Pohland et al 1982). Positive and negative chemical ionization (CI) mass spectra of ochratoxin A have base peaks at m/z 404 and 403, respectively (Abramson 1987, Bauer and Gareis 1987). The circular dichroism spectrum was recorded by Pohland et al (1982).

An important chemical reaction of ochratoxin A is esterification of the carboxyl group. The methyl ester, formed with boron trifluoride-methanol, is used for liquid chromatographic confirmation of identity of ochratoxin A (Bauer and Gareis 1987). Further methylation of ochratoxin A to the

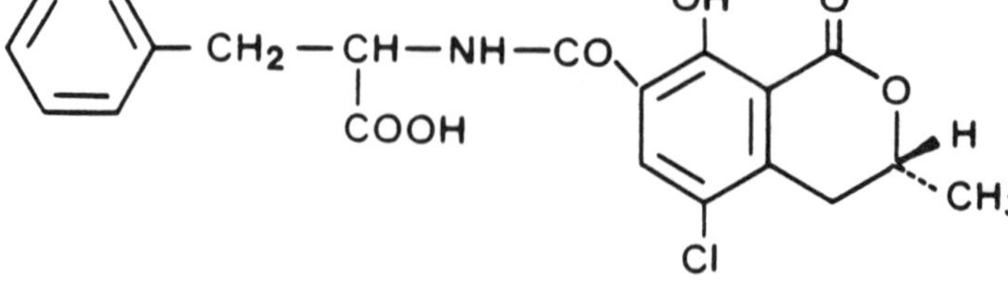

Figure 5.1. Ochratoxin A.

O-methyl ester has been achieved with diazomethane after 16 h reaction (van der Merwe et al 1965). Ochratoxin A is hydrolyzed by vigorous treatment with acid or by treatment with enzymes to yield L-β-phenylalanine and 7-carboxy-5-chloro-3,4-dihydro-8-hydroxy-3-methylisocoumarin (ochratoxin α). Detailed study of the acid hydrolysis indicated deviation from first-order kinetics and faster hydrolysis of ochratoxin A than ochratoxin B (Bredenkamp et al 1989).

Binding of ochratoxin A to bovine and human serum albumins is high in the pH range 4.0–6.0. Ochratoxin A also binds more specifically to unidentified plasma macromolecules with molecular weights of about 20,000.

Recently, ochratoxin A has been converted to ochratoxin B by catalytic dechlorination with palladium-charcoal and ammonium formate (Bredenkamp et al 1989). The Fe^{3+} complex of ochratoxin A was shown to produce hydroxyl radicals in the presence of an enzymatic reductase; it was proposed that these radicals may be partly responsible for the toxicity of ochratoxin A (Hasinoff et al 1990).

General stability of ochratoxin A is high, and solutions in ethanol can be stored in the refrigerator for a year or more; exposure to fluorescent light should be avoided. Heat stability is good as shown by studies with various foods (Harwig et al 1983, Kuiper-Goodman and Scott 1989). Gamma irradiation (up to 7.5 Mrad) caused no decomposition of ochratoxin A in methanol solution (Paster et al 1985).

There have been several chemical syntheses of ochratoxin A. The most recent, which led to ochratoxin α and hence formally to ochratoxin A, was reported by Sibi et al (1985). Tyrosine and 12 other amino acid analogues of ochratoxin A have also been synthesized. There have been syntheses of radioactive ochratoxin A labeled in the phenylalanine moiety with ^{14}C or ^3H (Rousseau et al 1984, Hult 1986). Previously, randomly labeled ochratoxin A had been prepared by ^3H exchange (Chang and Chu 1976).

CITRININ

There is good evidence for natural occurrence of citrinin and citrinin-producing strains of *Penicillium* in Canadian grains (Scott et al 1970, 1972; Abramson et al 1990a). Concentrations found are often several times higher than for accompanying ochratoxin A. Although citrinin is also a metabolite of some *Aspergillus* species (Scott 1977, Betina 1984), this is of no significance for its natural occurrence in Canada.

The extensive earlier chemical studies on citrinin, which was first described in 1931, have been briefly reviewed, together with summaries of its physico-chemical properties (Scott 1977, Betina 1984).

(−)-Citrinin is an optically active, yellow crystalline compound for which a 178–179°C mp (dec) has been reported. The first X-ray analysis of crystalline citrinin showed it to be a *p*-quinone methide (Fig. 5.2) with two intramolecular hydrogen bonds (Rodig et al 1971). More recently the alternative *o*-quinone methide structure has also been reported in the crystal (Sankawa et al 1983, Destro and Marsh 1984). In aqueous buffer solution (D_2O, pH 7.4) citrinin exists as two diastereoisomeric hydrates, demonstrated by ^1H and ^{13}C NMR spectroscopy (Barber et al 1987b), where a molecule of water has been added and a dihydric phenolic structure formed. Citrinin is in fact very sparingly soluble in water, but it is soluble in dilute sodium hydroxide, sodium carbonate, sodium bicarbonate (Nakazato et al 1981), sodium acetate, and in polar organic solvents.

The UV absorption spectrum of citrinin in 95% ethanol (7×10^{-5} *M*) has maxima at 222 (ϵ 22,280), 253 (ϵ 8,279), and 319 (ϵ 4,710) nm (Neely et al 1972); molar absorptivities were reported to vary with the concentration. Fluorescence excitation and emission maxima in this solvent are at 336 and 521 nm, respectively; in 40% methanol solution containing 0.2% $AlCl_3$ the emission maximum is at 465 nm (Nakazato et al 1981). The infrared spectrum (KBr) has maxima at 3,484, 2,985, 1,675, and 1,639 cm^{-1} (Scott 1977). The infrared spectrum in carbon tetrachloride indicated two hydrogen bonding conformations that are in equilibrium, one of which was supported by ^1H NMR spectroscopy. The 60 MHz spectrum in $CDCl_3$ has signals at δ 15.2 and 13.7 (OH); 8.3 (double bond CH); 4.84 and 3.04 (CH); and at 2.03, 1.38, and 1.25 (CH_3). The 100 and 250 MHz spectra were presented by Pohland et al (1982). ^{13}C NMR spectra in $CDCl_3$ and D_2O (pH 7.4) were published more recently (Barber et al 1987a,b) in addition to the ^2H NMR spectrum of labeled citrinin (Barber and Staunton 1980, Barber et al 1987a). The EI mass spectrum has a molecular ion at *m/z* 250 and prominent fragment ions at *m/z* 232, 217, 206, 191, 177, 91, 77, and 44 (Scott et al 1972, Pohland et al 1982).

Several syntheses or partial syntheses of citrinin have been carried out (Scott 1977, Barber et al 1986) as well as an asymmetric synthesis of the unnatural enantiomer (+)-citrinin (Regan and Staunton 1987).

Various derivatives of citrinin have been prepared, including a mono-acetate, diethyl derivative, methyl ester, and dihydro derivatives; it forms azo dyes by coupling with diazotized aromatic amines (Scott 1977). Citrinin

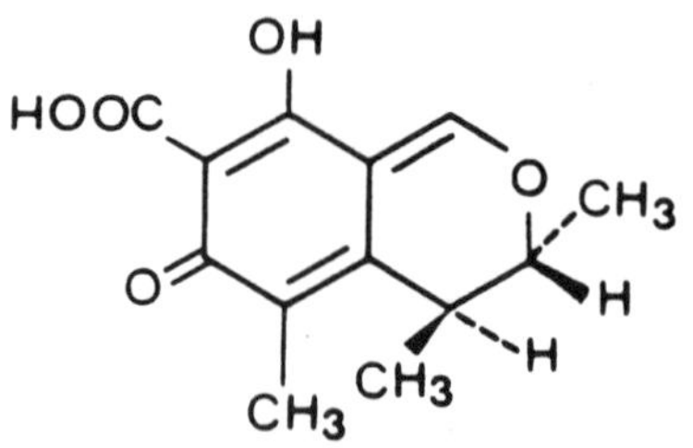

Figure 5.2. Citrinin.

also has metal complexing properties and forms 1:1 and 1:2 copper(II)-citrinin chelates (Ahles et al 1976).

Citrinin binds in vitro to human serum protein (Damodaran 1977). However, no evidence could be obtained of interaction with DNA (Barber et al 1987b). It is inactivated by cysteine.

Citrinin is fairly heat stable in solution in 95% ethanol or *n*-hexane but not in acid or alkaline solution. Losses have also been observed in heated grains (Scott 1991).

PATULIN

The most recent reviews on patulin are by Betina (1989) and the International Agency for Research on Cancer (IARC) (1986). Its formation by fungi is not limited to species of the genera *Penicillium* and *Aspergillus*, and recently many others were added to the list of patulin producers (Steiman et al 1989). The major source of patulin in the Canadian food supply is *P. expansum* growing on apples processed into juice.

The physical and chemical properties of patulin were summarized in two earlier reviews (Scott 1974, 1977). It is a colorless, crystalline compound with about 110°C mp and is soluble in water and polar organic solvents. The chemical structure of patulin, proposed in 1949, is shown in Figure 5.3. The crystal contains hydrogen-bonded pairs of D and L enantiomers present in a 1:1 ratio (Hubbard et al 1977, Lezi et al 1981).

Maximum UV absorption is at 276 nm (ϵ 14,450) in alcohol. The infrared absorption spectrum in chloroform solution includes peaks at 3,580, 3,340, 1,782, and 1,755 cm^{-1}. The 100 MHz ^{1}H NMR spectrum (in CDCl$_3$) has signals at δ 5.97 (OCH(OH), double bond CH), 4.73 (CH–O), 4.40 (CH–O), and 3.46 (OH). The 250 MHz spectrum and additional data on the ^{1}H NMR spectrum of patulin in other solvents have been published, together with ^{13}C NMR spectra (Pohland et al 1982, Chernyshev et al 1988, Seijas et al 1989). The EI mass spectrum shows prominent ions at m/z 154 (M$^+$), 136, 126, 110, 97, 82, 71, 69, 55, and 53 (Scott 1977, Pohland et al 1982, Chernyshev et al 1988, Seijas et al 1989).

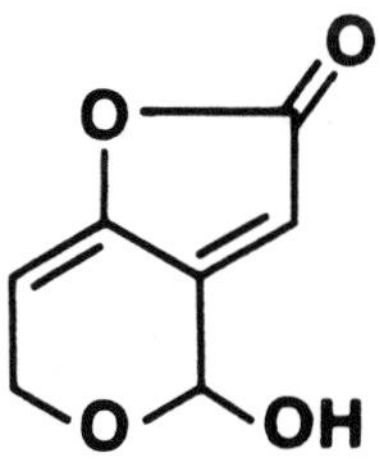

Figure 5.3. Patulin.

Synthetic work on patulin has been limited until recently, when total synthesis was achieved from arabinose (Gill et al 1988, Seijas et al 1989, Bennett et al 1991).

The chemistry of patulin has centered mainly on reactions of the secondary alcohol group and the potential carbonyl group in the hemiacetal function. Thus patulin forms an acetate, cinnamate, and benzoate; in addition to the acetate, the chloroacetate, trimethylsilyl ether, and heptafluorobutyrate have been used as derivatives for gas chromatographic (GC) determination of patulin (Tarter and Scott 1991). The phenylhydrazone, semicarbazone, and oxime derivatives have been made also.

Patulin hemisuccinate has been prepared and then coupled with bovine serum albumin by the carbodiimide method to form an immunogen (Mehl et al 1986).

Patulin yields ascladiol on reduction with sodium borohydride whereas hydrogenation gives desoxypatulinic acid, β-n-propyl-α-butyrolactone or other products, depending on the catalyst and conditions. Treatment of patulin with sodium hypoiodite yields 4-hydroxy-2H-pyran-2-one, with loss of two carbons from the lactone ring (Rao and Reddy 1989).

Patulin is unstable in alkaline solution but is stable under acidic conditions at room temperature. Boiling with 2N sulfuric acid forms tetrahydro-γ-pyrone-2-carboxylic acid. Slow decomposition takes place in distilled water, and a detailed study of the kinetics of patulin decomposition in buffers of pH 6.0 to 8.0 at 25°C was undertaken by Brackett and Marth (1979).

Patulin is of particular interest to the food scientist because of its stability in foodstuffs and in the presence of their constituents as well as its reaction with food additives (Scott 1984a). Patulin is appreciably stable in apple and grape juices. The rate of disappearance in apple juice increases with addition of vitamin C. On the other hand, sucrose protects patulin during heat treatment in berry jams, although patulin is unstable in these jams over a six-month period. Stability of patulin in ground cereals depends on the type of cereal and decreases with increasing moisture content. At high concentrations (up to 2,000 μg/L) sulfur dioxide reacts with patulin. Reactivity of patulin with sulfhydryl compounds such as cysteine, thioglycolic acid, glutathione, and dimercaptopropanol has been well documented, although no pure reaction products have been isolated. Covalent interaction with amino groups in an enzyme was also demonstrated. Gamma irradiation of patulin in apple juice causes loss of patulin; the dose required to cause 50% decomposition was 0.35 kGy (Żegota et al 1988).

PENICILLIC ACID

Strains of *P. cyclopium* (*P. verrucosum* var. *cyclopium*, syn. *P. aurantiogriseum*) that produce penicillic acid have been isolated from

Canadian grains and feeds (Scott et al 1972), although the mycotoxin itself has not been detected in Canadian agricultural commodities.

Physical and chemical properties of penicillic acid were reviewed by Scott (1977). Penicillic acid is an optically inactive, colorless, crystalline compound that forms a monohydrate, mp 64–65° C, from water, and anhydrous needles of mp 87° C from light petroleum. Solubility in cold water is moderate (2%) and in polar organic solvents is high.

The UV absorption maximum at 225 (ϵ 10,500) nm and infrared absorption (in chloroform) at 3,335, 1,757, and 1,645 cm^{-1} support the lactone structure (Fig. 5.4) rather than the open-chain form, although it appears that a pH-dependent equilibrium between these two tautomeric forms exists in aqueous solution. The 60 MHz ^{1}H NMR spectrum (in CDCl$_3$) shows chemical shifts (δ) at 5.48 (double bond CH), 5.20 and 5.15 (OH, double bond CH), 3.92 (OCH$_3$), and 1.75 (CH$_3$). Complete assignments for the unsaturated methylene protons were determined from europium-induced shifts in the 90 MHz spectrum (Elvidge et al 1977). The 100 and 250 MHz ^{1}H NMR spectra were presented by Pohland et al (1982). The ^{13}C NMR spectrum of ^{13}C-labeled penicillic acid was published by Seto et al (1974a), and the ^{3}H NMR spectrum of ^{3}H-labeled penicillic acid has also been reported (Al-Rawi et al 1974; Elvidge et al 1977). The EI mass spectrum has ions at m/z 170 (M$^+$), 129, 101, 69, and 41 (Yeh et al 1978, Pohland et al 1982). In addition to the earlier 12-step synthesis of penicillic acid by Raphael (1948), a convenient four-step synthesis from tartaric acid was reported by Yeh et al (1978).

Birkinshaw et al (1936) described several chemical reactions of penicillic acid. Penicillic acid reacts rapidly with excess diazomethane to form a pyrazoline derivative. Phillips et al (1981) showed that there is an initial methyl esterification at the carboxyl group of the ring-opened form of penicillic acid, with subsequent addition of diazomethane at the methylene double bond. The reaction of phenylhydrazine with penicillic acid also yields a pyrazoline, with incorporation of two molecules of phenylhydrazine. Hydroxylamine forms a dioxime, hydrogenation in the presence of palladium-charcoal yields dihydropenicillic acid, bromine gives penicillic acid dibromide, and acetylation with acetic anhydride and sodium acetate yields a monoacetate. Ozonolysis and acid and alkaline hydrolysis were also investigated by Birkinshaw et al (1936).

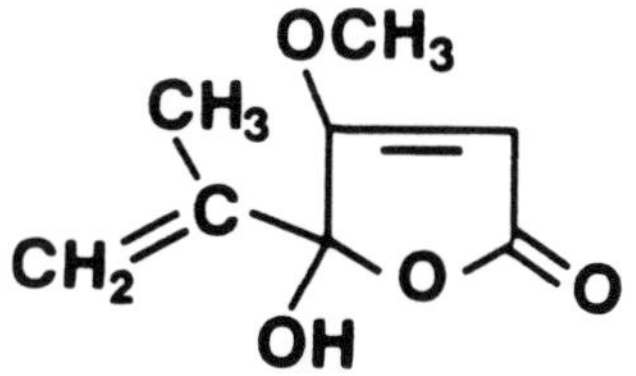

Figure 5.4. Penicillic acid.

For GC analysis of penicillic acid, trimethylsilyl, and trifluoroacetate derivatives have been prepared. Two epimeric forms of penicillic acid trimethylsilyl derivative were separated by capillary GC (Keromnes and Thouvenot 1985).

Penicillic acid is inactivated by sulfhydryl compounds, and the site of addition of the sulfhydryl group was shown to be the isolated rather than the conjugated double bond. The product formed with cysteine in aqueous solution is derived from the open-chain form of penicillic acid according to Black (1966) but from the lactone form according to Ciegler et al (1972). Penicillic acid also reacts with amines and other amino acids but reaction products have not been identified. As might be expected, penicillic acid is unstable in foods containing reactive amino acids.

ROQUEFORTINE (ROQUEFORTINE C)

The indole alkaloid roquefortine, originally named roquefortine C by Japanese workers (Ohmomo et al 1975), is a metabolite of *P. roquefortii* and a number of other *Penicillium* species (Scott 1984b, Frisvad and Filtenborg 1989, Reshetilova and Kozlovsky 1990). The chemical structure is shown in Figure 5.5, where the 3,12 double bond in the dehydrohistidine side chain has the *E* configuration (Scott et al 1979). The remaining stereochemistry was established recently (Yamaguchi et al 1991). Roquefortine has been detected in blue cheese (containing *P. roquefortii*) obtained from nine countries including Canada, where it is made in a Quebec monastery (Ohmomo et al 1975, Scott and Kennedy 1976, Ware et al 1980). Roquefortine has not been detected in Canadian grains or feeds, although it was recently isolated from feed grain in Sweden (Häggblom 1990).

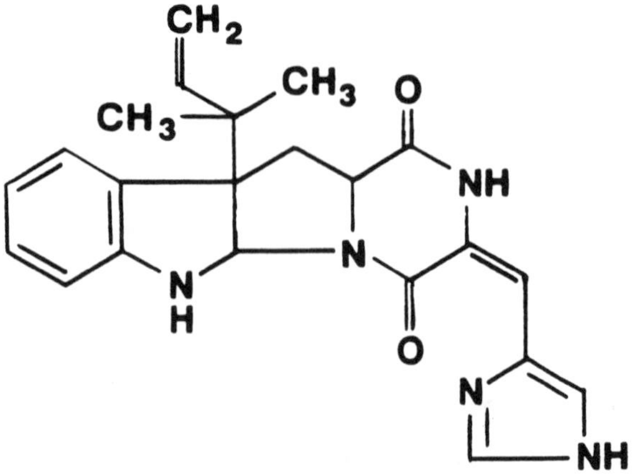

Figure 5.5. Roquefortine.

Recently, roquefortine was found in the stomach contents of six dogs suffering from strychninelike poisoning in Alberta (Lowes et al 1992).

Spectral and chemical properties of roquefortine were reviewed by Scott (1984b). It crystallizes from methanol-water as solvated colorless needles, which are soluble in polar organic solvents but poorly soluble in water and hexane. The characteristic UV absorption spectrum has maxima (in 95% ethanol) at 209, 240, and 328 nm (log ϵ 4.47, 4.21, and 4.43, respectively). Infrared absorption occurs at 3,430, 3,380, 3,190, 1,685, 1,665 and 1,608 cm^{-1} (in chloroform). The 90 MHz ^1H NMR spectrum (in $CDCl_3$) has chemical shifts (δ) at 11.81 and 11.02 (NH), 7.66 and 7.12 (both imidazole CH), 6.57–7.20 (aromatic and double bond CH), 6.03 (double bond CH), 5.66 (CH), 5.10 (double bond CH_2), 5.00 (NH), 4.09 (CH), 2.50 (CH_2), and at 1.14 and 1.02 (CH_3) (Scott et al 1976). The ^{13}C NMR spectrum was published by Scott et al (1976), Vleggaar and Wessels (1980), and for ^{13}C-enriched roquefortine, by Gorst-Allman et al (1982). The ^2H NMR spectrum of roquefortine enriched by incorporation of $^2H_5$-tryptophan has been recorded by Bhat et al (1990).

The EI mass spectrum of roquefortine is characterized by ions at m/z 389 (M^+), 320 (base peak), 198, 192, 157, 130, and 108.

An important chemical reaction of roquefortine is photoisomerization at the 3,12 double bond linking the diketopiperazine and imidazolyl moieties. Comparison of the spectral properties of the isomer (iso-roquefortine) with those of roquefortine established the stereochemistry of this double bond. Reduction of roquefortine with zinc in acetic acid at 70°C yields a mixture of two stereoisomeric 3,12-dihydroroquefortines, whereas catalytic reduction in acetic acid with hydrogen and platinum reduces the double bond in the dimethylallyl side chain. When roquefortine is heated in deuteromethanol and deuterium oxide, two deuterium atoms are incorporated as C–D bonds. Roquefortine undergoes an unusual fragmentation when refluxed in dilute methanolic hydrochloric acid, with loss of the dimethylallyl side chain and ring opening between the indole and diketopiperazine (at the nitrogen atom) moieties.

α-CYCLOPIAZONIC ACID

α-Cyclopiazonic acid is a metabolite of several Penicillium and Aspergillus species and is of Canadian interest from two viewpoints. First, one of the important producers, P. aurantiogriseum (formerly P. cyclopium) (Pitt et al 1986), commonly occurs in stored Canadian grains (Mills 1990), although the mycotoxin itself has not been detected or even looked for. A recent report of the natural occurrence of cyclopiazonic acid in sunflower seed screenings in North Dakota (Ross et al 1991) raises the distinct possibility of Canadian occurrence. Secondly, α-cyclopiazonic acid is a metabolite of P. camembertii, which is used in the manufacture of mold-ripened white cheeses, and it can be found in the crust of these cheeses (Le Bars et al 1988).

Pure α-cyclopiazonic acid (colorless crystals from methanol) has an mp of 245–246°C (Holzapfel 1968). Its chemical structure is shown in Figure 5.6. UV absorption (in methanol) is at 225, 253, 275 (sh), 284, and 292 (sh) nm (log ϵ 4.60, 4.22, 4.28, 4.31 and 4.24, respectively) according to Holzapfel (1968), and (in ethanol) at 224, 282, and 290 nm (log ϵ 4.51, 4.25, and 4.15, respectively) according to Ohmomo et al (1973); the absorption curve is illustrated in the compilation of Pohland et al (1982). The infrared absorption spectrum (in chloroform) shows maxima at 3,478, 3,200–2,600, 1,708, and 1,618 cm^{-1} (Holzapfel 1968); spectra recorded in KBr disks were presented by Ohmomo et al (1973) and Pohland et al (1982). The 100 MHz ^{1}H NMR spectrum (in CDCl$_3$) has signals at δ 11.75 (OH), 8.20 (NH), 7.14–7.19 (indole CH), 6.91 (indole CH), 4.06 (CHN), 3.65 (CH), 3.05 (CH$_2$), 2.62 (CH), 2.44 (CH$_3$-C=), and at 1.67 and 1.63 (CH$_3$) (Yokota et al 1981). Both 100 MHz and 250 MHz ^{1}H NMR spectra were given by Pohland et al (1982). Detailed analysis of the ^{13}C NMR spectrum was published by Yokota et al (1981). Prominent ions in the EI mass spectrum occur at m/z 336 (M$^+$) 196, 182, 181, 155, and 154 (Holzapfel 1968, Pohland et al 1982), and circular diochroism curves have been recorded (Holzapfel 1968, Luk et al 1977, Pohland et al 1982).

α-Cyclopiazonic acid was synthesized by Kozikowski et al (1984); an unusual zinc-mediated replacement of a phenylthio group by a methyl group was used to form one of the *gem*-dimethyls. The final product of the synthesis was a 2.5:1 mixture of α-cyclopiazonic acid and *iso*-α-cyclopiazonic acid, which were separated by chromatography.

The chemistry of α-cyclopiazonic acid was investigated by Holzapfel (1968). Methylation with diazomethane gives an amorphous *O*-methyl derivative and an epoxide. Treatment with 0.5N sodium hydroxide solution causes isomerization at the CH attached to the amide nitrogen, forming *iso*-α-cyclopiazonic acid. Tetradeutero-*iso*-α-cyclopiazonic acid and tetradeutero-

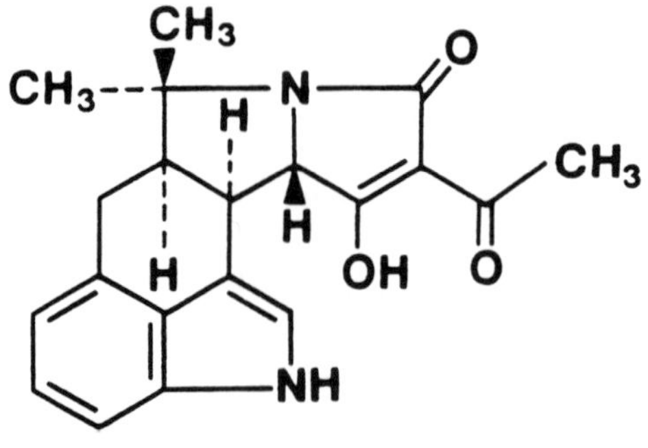

Figure 5.6. α-Cyclopiazonic acid.

α-cyclopiazonic acid have been prepared by heating α-cyclopiazonic acid with sodium and D_2O. Hydrolysis by refluxing with $0.05N$ hydrochloric acid yields deacetylcyclopiazonic acid. In other reactions, α-cyclopiazonic acid forms a hydrazone when treated with 85% hydrazine and forms a noncrystalline acetate with acetic anhydride and pyridine. Peterson et al (1989) observed by mass spectrometry that cyclopiazonic acid was binding with sodium ions from a soda-lime glass vial. Cyclopiazonic acid was not reduced by catalytic hydrogenation over 10% palladium on charcoal (Luk et al 1977).

VERRUCOSIDIN

Another mycotoxin produced by *P. aurantiogriseum* (*P. verrucosum* var. *cyclopium*) is verrucosidin, formerly known as S-toxin. It is a potent neurotoxin.

Verrucosidin has the structure shown in Figure 5.7, which was established by spectroscopic, X-ray crystallographic, chemical, and synthetic studies and is closely related to that of citreoviridin (Fig. 5.8) (Burka et al 1983, Ganguli et al 1984, Nishiyama et al 1986, Hatakeyama et al 1988). Verrucosidin crystallizes from ether as colorless plates, mp 90–91°C, $[\alpha]_D^{26}$ + 92.4° (c = 0.25 g/100 mL, methanol). The UV spectrum has maxima

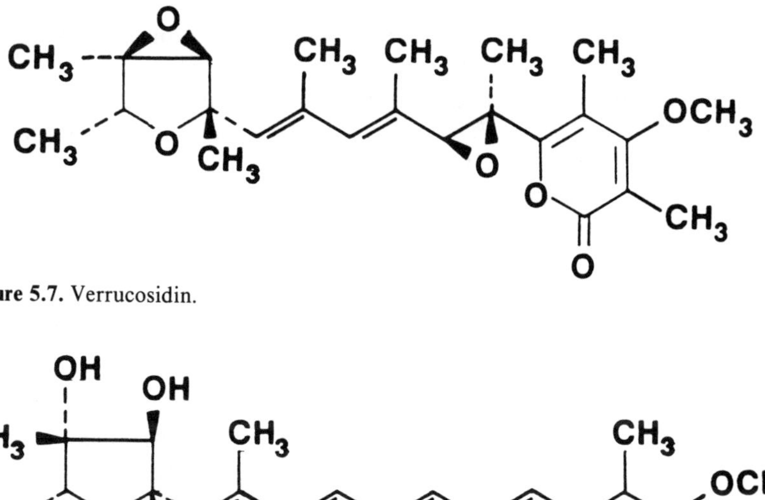

Figure 5.7. Verrucosidin.

Figure 5.8. Citreoviridin.

at 294 (ϵ 13,000) and 241 (ϵ 21,000) nm. Infrared absorption (in CDCl$_3$) is at 2,970, 1,700, 1,640, and 1,575 cm^{-1}. The ^{1}H NMR spectrum (90 MHz, in CDCl$_3$) shows signals at δ 5.88 and 5.50 (double bond CH); 4.12 (CH.O); 3.85 (OCH$_3$); 3.50 and 3.45 (epoxide CH); 2.05, 2.05, 1.98, and 1.93 (CH$_3$-CH$=$); 1.48, 1.44, and 1.42 (CH$_3$-epoxide); and 1.21 (CH$_3$). The ^{13}C NMR spectrum has 24 resonances and indicated an α-pyrone unit. The EI mass spectrum shows ion at m/z 416 (M$^+$), 127, and 99, whereas the CI (CH$_4$) mass spectrum has prominent ions at 417 (M + H$^+$), 399, 345, 249, 197, 183, and 127.

Verrucosidin is hydrolyzed with methanolic potassium hydroxide to give an aldehyde, containing all but one of the carbon atoms in the side chain and with loss of the α-pyrone moiety, together with a rearranged cyclopenten-2-one carboxylic acid, whose methyl ester was synthesized (Ganguli et al 1984).

Total synthesis of (+)-verrucosidin was achieved by Hatakeyama et al (1988).

OTHER *PENICILLIUM* MYCOTOXINS

Numerous other mycotoxins are known to be produced by *Penicillium* species. Citreoviridin has already been mentioned. Others include rubratoxins A and B, PR toxin, luteoskyrin, cyclochlorotine, rugulosin, erythroskyrine, secalonic acid D, gliotoxin, viridicatumtoxin, kojic acid, β-nitropropionic acid, xanthomegnin, viomellein, rugulovasines A and B, chlororugulovasines A and B, verruculotoxin, emodin, chaetoglobosin C, penitrem A and related compounds, tryptoquivaline and related compounds, verruculogen, paxilline, fumitremorgens A and B, and mycophenolic acid (Cole and Cox 1981). None of these have been found in Canadian foodstuffs. Recently a Canadian isolate of *P. aurantiogriseum* var. *aurantiogriseum* was identified as a producer of the naphthoquinones xanthomegnin and viomellein (Mills et al 1992).

Aspergillus Toxins

Several of the mycotoxins described under *Penicillium*, e.g., ochratoxin A, citrinin, patulin, penicillic acid, and α-cyclopiazonic acid, are also *Aspergillus* toxins, although this genus is not expected to be their source in Canadian agricultural commodities. Among other *Aspergillus* mycotoxins, the aflatoxins are rarely found, except in imported foodstuffs such as nuts. During storage of grains in Canada, the predominant *Aspergillus* toxin is sterigmatocystin associated with *A. versicolor* and *Eurotium* species, and it has occasionally been detected in feeds (Mills 1990). Since *Eurotium* species are members of the *A. glaucus* group and can produce mycotoxins, two of these (echinulin and flavoglaucin) are also described here. This group of fungi also produces toxic and mutagenic anthraquinones and anthraquinone derivatives—physcion, erythroglaucin, physcion-anthrone B_1, and physcion-dianthrone (Bachmann et al 1979).

STERIGMATOCYSTIN

Sterigmatocystin is a pale yellow crystalline compound, mp 247–248°C (from amyl acetate), which can be sublimed in vacuo. It is strongly levorotatory, with an $[\alpha]_D^{26}$ of −398° (c = 1 g/100 mL, chloroform) (Birkinshaw and Hammady 1957). It is sparingly soluble in most organic solvents in the cold but is readily soluble in chloroform. It is insoluble in water.

The UV absorption spectrum of sterigmatocystin has maxima (in ethanol) at 205, 233, 246, and 325 nm (log ϵ 4.40, 4.49, 4.53, and 4.21, respectively) (Davies et al 1960). Sterigmatocystin is weakly fluorescent and can be observed as a brick red spot on a thin-layer plate under long-wave UV light. At pH 7.0 the fluorescence emission maximum is at 570 nm (340 nm excitation) (Maness et al 1976). Enhanced fluorescence is observed in sulfuric acid solutions. Infrared absorption (KBr) is at 3,450, 3,099, 2,995, 2,975, 2,920, 1,650, 1,627, and 1,610 cm^{-1} (Davies et al 1960). The chemical structure (Fig. 5.9) was proposed by Bullock et al (1962); X-ray crystallographic analyses have since been carried out (Tanaka et al 1970, Fukuyama et al 1976).

The ^{1}H NMR spectra of sterigmatocystin and other 7,8-dihydrofuro[2,3-b]furans were originally published by Bullock et al (1962) and further discussed by Rodricks (1969); 100 and 250 MHz spectra have been published (Pachler et al 1976, Pohland et al 1982). The most recent ^{1}H NMR spectrum of sterigmatocystin was recorded at 300 MHz in CDCl$_3$: δ 7.50, 6.83, 6.76, and 6.44 (aromatic H); 6.83 (CH–O); 6.50 and 5.45 (CH=); and 3.99 (OCH$_3$).

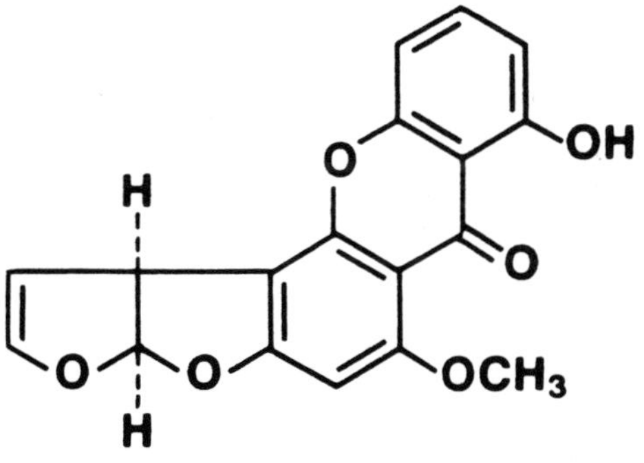

Figure 5.9. Sterigmatocystin.

^{13}C NMR spectra (in CDCl$_3$) were published by Pachler et al (1976), Cox and Cole (1977), and Maes and Steyn (1984) in addition to ^{13}C NMR spectra in other solvents using ^{13}C-enriched sterigmatocystin (Tanabe et al 1970, Seto et al 1974b). Where another isotopically labeled element, ^{2}H or ^{18}O, was also incorporated into the sterigmatocystin molecule, this can be detected in the ^{13}C NMR spectrum by the isotope shifts (Sankawa et al 1982, Simpson and Stenzel 1982, Nakashima and Vederas 1982). The ^{2}H NMR spectrum has also been analyzed (Simpson and Stenzel 1982). The EI mass spectrum of sterigmatocystin shows prominent ions at m/z 324 (M$^+$, base peak), 306, 295, 278, 265, and 237 (Koller and Orth 1974, Cole and Cox 1981).

The chemistry of sterigmatocystin has been extensively investigated and briefly reviewed by van der Watt (1974). With hot ethanolic potassium hydroxide, sterigmatocystin is converted to *iso*-sterigmatocystin with opening of the dihydrofuran ring attached to the xanthone moiety. Sterigmatocystin can be methylated with dimethyl sulfate or methyl iodide but not with diazomethane. Dihydrosterigmatocystin is formed by catalytic hydrogenation in acetic acid over palladium-charcoal. Oxidation of sterigmatocystin with potassium permanganate in refluxing acetone yields γ-resorcylic acid or, with a smaller proportion of potassium permanganate, 3,8-dihydroxy-1-methoxyxanthone-4-carboxylic acid. Aluminum chloride and sterigmatocystin heated in chlorobenzene produce 1,3,8-trihydroxyxanthone (Davies et al 1960). Reaction with aluminum chloride has important analytical applications to form a yellow fluorescent spot on thin-layer chromatography and for postcolumn fluorescence enhancement in reversed-phase liquid chromatography (Neely and Emerson 1990).

Acetylation of sterigmatocystin with acetic anhydride in pyridine yields *O*-acetylsterigmatocystin, which can also be used in liquid chromatographic analysis (Abramson and Thorsteinson 1989). Acetylation under more rigorous conditions gives *O*-acetyldihydro-acetoxysterigmatocystin (Davies et al 1960). Sterigmatocystin also forms an *O*-benzoate. Addition of acetic acid across the vinyl ether double bond of sterigmatocystin without *O*-acetylation is accomplished by refluxing sterigmatocystin in glacial acetic acid for 30 min (Engebrecht and Altenkirk 1972); both α- and β-isomers are formed. Methanol and ethanol also add to this double bond in acidic solution (Bullock et al 1962). By analogy, the acid-catalyzed addition of water to the vinyl ether double bond gives α- and β-hemiacetals, whose spectral properties were described by Orti et al (1989) and by Ashley et al (1987). The ratio of α- to β-isomers varies with reaction and storage conditions. Sterigmatocystin hemiacetal is conjugated to proteins to give immunogens used for making antibodies with specificity for sterigmatocystin, its hemiacetal, and *O*-methyl sterigmatocystin (Li and Chu 1984, Chung et al 1989). Another approach to immunogen preparation is to first treat sterigmatocystin with sebacoyl dichloride then add bovine serum albumin to give sterigmatocystin hemiacetal-7-decane-1,8-dioic acid amide protein conjugate (Morgan et al 1986).

Sterigmatocystin has been converted into dihydroaspertoxin acetate via O-methyldihydrosterigmatocystin, opening of the furan ring with alkali, acetylation, and oxidation with lead acetate (Hutchison and Holzapfel 1971). The trimethylsilyl derivative of sterigmatocystin has been used for its GC determination (Suzuki et al 1976).

Sterigmatocystin binds covalently to calf thymus DNA by incubation in the presence of phenobarbital-induced rat liver microsomes. The ^{1}H NMR spectrum and field desorption mass spectrum of the major adduct following hydrolysis, 1,2-dihydro-2-(N^7-guanyl)-1-hydroxysterigmatocystin, are published (Essigmann et al 1979). This compound, together with 1,2-dihydro-1,2-dihydroxysterigmatocystin, can also be prepared nonenzymatically by benzil-sensitized photochemical epoxidation of sterigmatocystin in the presence of 2'-deoxyguanosine-3',5'-dibutyrate and oxygen, then acid hydrolysis of the reaction mixture (Büchi et al 1982). More recently, it was prepared from sterigmatocystin epoxide added to calf thymus DNA in phosphate buffer pH 7.0, followed by acid hydrolysis (Baertschi et al 1989). The epoxide itself is readily formed by treating sterigmatocystin with dimethyldioxirane. Sterigmatocystin also binds with the albumin fraction of human serum (Meerarani and Shanmugasundaram 1987).

Gamma irradiation requires a dose of 50 Mrad for destruction of sterigmatocystin in the dry state (Kume et al 1983).

ECHINULIN

The indole alkaloid echinulin may now be regarded as a mycotoxin (Ali et al 1989) and causes mice to refuse drinking water (Vesonder et al 1988). Echinulin is a colorless, crystalline compound of mp 242–243°C, which is soluble in certain polar organic solvents, including chloroform (Quilico and Panizzi 1943). The chemical structure of echinulin (Fig. 5.10) was established by Casnati et al (1962). The UV absorption maxima (in ethanol) are at 230 nm (log ϵ 4.60), 279 nm (log ϵ 3.98), and 286 nm (log ϵ 3.96) (Casnati et al 1960). The infrared spectrum, exhibiting a major peak at 1,660 cm^{-1}, was published by Casnati et al (1960) (nujol) and by Inoue et al (1977a) (KBr). Various authors have recorded the ^{1}H NMR spectrum, including Cole and Cox (1981) (100 MHz, in CDCl$_3$ + DMSO), Inoue et al (1977a) (in CDCl$_3$), and Podojil et al (1979) (in deuterated DMSO); the latter had signals at δ 9.48 (tryptophan NH); 8.15 and 7.18 (NH); 7.58 and 6.65 (aromatic H); 6.29, 5.37, 5.15, and 4.90 (olefinic H); 3.99 and 3.84 (CH); 3.56 (CH$_2$); 1.67 and 1.62 (olefinic CH$_3$); and at 1.54 and 1.39 (CH$_3$). ^{13}C NMR assignments have also been made (Cardillo et al 1974, Nagasawa et al 1979). Prominent peaks in the EI mass spectrum occur at m/z 461 (M$^+$), 335, 334 (base peak), 319, 278, 266, and 69 (Inoue et al

1977a, Selva and Traldi 1977, Podojil et al 1979, Cole and Cox 1981). The major fragmentation to tryptophanyl and diketopiperazine moieties and its diagnostic value was discussed by Selva and Traldi (1977). Echinulin undergoes catalytic hydrogenation (platinum black) to hexahydroechinulin (Quilico and Panizzi 1943). Alcoholic potassium hydroxide at 200°C gives an indolic amino acid, isoechinin, with loss of alanine (Casnati et al 1960). Sodium ethylate in alcohol racemizes echinulin (Casnati et al 1960).

Total synthesis of optically active echinulin was achieved by Inoue et al (1977a).

FLAVOGLAUCIN

Flavoglaucin was first isolated in 1934 (Gould and Raistrick 1934), but evidence for its being a mycotoxin is relatively recent (Nazar et al 1984). Yellow crystals of pure flavoglaucin melt at 108–109°C after recrystallization from *n*-hexane (Inoue et al 1977b). It is readily soluble in most organic solvents (except light petroleum) and insoluble in water (Raistrick et al 1937). Maximum UV absorption (in methanol) at 270 nm (log ϵ 3.85) and 394 nm (log ϵ 3.66) and infrared absorption at 1,690 and 1,630 cm^{-1} (KBr) were recorded by Inoue et al (1977b). An additional UV absorption maximum at 236 nm was reported by Podojil et al (1979). The ^{1}H NMR spectrum (in CDCl$_3$) has signals at δ 11.93 (OH), 10.28 (CHO), 6.93 (aromatic H), 5.32 (olefinic H), 4.74 (OH), 3.32 (ArCH$_2$-C=), 2.91 (ArCH$_2$), 1.76 and 1.71 (CH$_3$-C=), 1.10–1.68 (CH$_2$), and 0.88 (CH$_3$) (Inoue et al 1977b), consistent with the structure shown in Figure 5.11 (Quilico et al 1953).

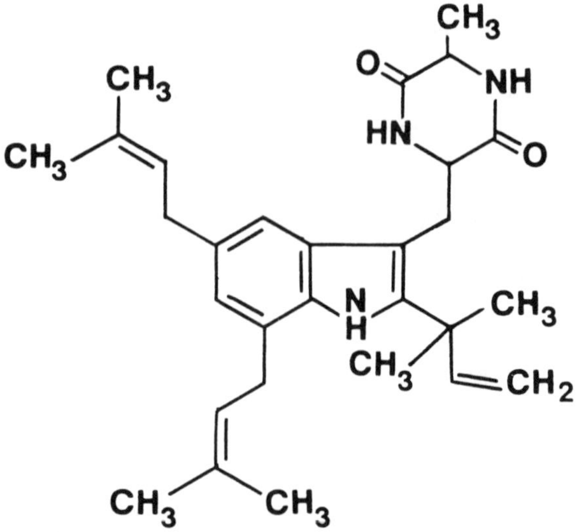

Figure 5.10. Echinulin.

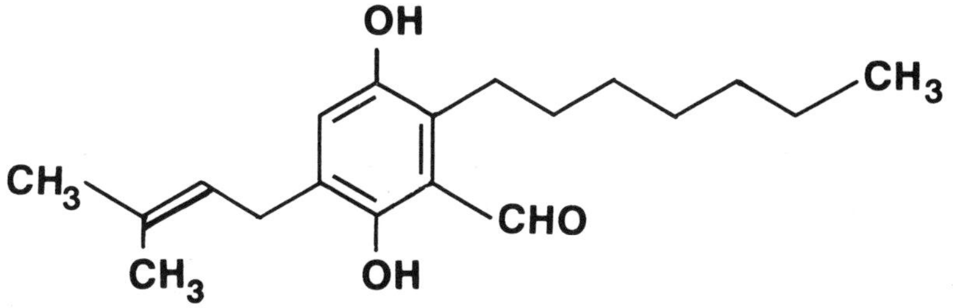

Figure 5.11. Flavoglaucin.

Slightly different shifts were reported by Podojil et al (1979) for the isoprenyl methyl groups. The EI mass spectrum includes ions at m/z 304 (M^+), 249, 230, 215, and 163 (Inoue et al 1977b, Podojil et al 1979).

The chemistry of flavoglaucin consists mainly of reactions of the carbonyl, hydroxyl, and olefinic groups. Thus a Schiff's base with o-phenylene diamine, a phenylhydrazone, and a 2,4-dinitrophenylhydrazone are readily formed (Raistrick et al 1937); ferric chloride oxidizes flavoglaucin to a quinone (Quilico et al 1949); and reduction with hydrogen and palladium-strontium carbonate yields dihydroflavoglaucin (Cruickshank et al 1938). On hydrogenation with Raney nickel, flavoglaucin is converted to a cyclic glycol (Quilico et al 1949). Oxidation of flavoglaucin with hydrogen peroxide and sodium hydroxide yields n-octanoic acid and two other acids, including iso-amylsuccinic acid (Quilico et al 1949, 1953). Under slightly different conditions (Quilico et al 1953) a hydroxybenzoquinone is formed.

OTHER ASPERGILLUS MYCOTOXINS

In addition to those previously mentioned, there are many other known *Aspergillus* toxins. These include versicolorin A, austamide, maltoryzine, viriditoxin, aspergillic acid, paspaline, aflatrem, cytochalasin E, fumigaclavine C, fumagillin, terreic acid, malformin A_1, and nigragillin (Cole and Cox 1981) (this is not a complete list). They have not been detected in Canadian foodstuffs.

Literature Cited

Abramson, D. 1987. Measurement of ochratoxin A in barley extracts by liquid chromatography-mass spectrometry. J. Chromatogr. 391:315-320.

Abramson, D., and Thorsteinson, T. 1989. Determination of sterigmatocystin in barley by acetylation and liquid chromatography. J. Assoc. Off. Anal. Chem. 72:342-344.

Abramson, D., Sinha, R. N., and Mills, J. T. 1983. Mycotoxin and odor formation in barley stored at 16 and 20% moisture in Manitoba. Cereal Chem. 60:350-355.

Abramson, D., Sinha, R. N., and Mills, J. T. 1985. Mycotoxin formation and quality changes in granary-stored corn at 16 and 21% moisture content. Sci. Aliment. 5:653-663.

Abramson, D., Mills, J. T., and Sinha, R. N. 1990a. Mycotoxin production in amber durum wheat stored at 15 and 19% moisture content. Food Addit. Contam. 7:617-627.

Abramson, D., Sinha, R. N., and Mills, J. T. 1990b. Mycotoxin formation in HY-320 wheat during granary storage at 15 and 19% moisture content. Mycopathologia 111:181-189.

Ahles, L. A., Fernando, Q., Rodig, O. R., and Quante, J. M. 1976. Interaction of copper (II) with the fungal metabolite, citrinin. Bioinorg. Chem. 5:361-366.

Ali, M., Mohammed, N., Alnaqeeb, M. A., Hassan, R. A. H., and Ahmad, H. S. A. 1989. Toxicity of echinulin from *Aspergillus chevalieri* in rabbits. Toxicol. Lett. 48:235-241.

Al-Rawi, J. M. A., Elvidge, J. A., Jaiswal, D. K., Jones, J. R., and Thomas, R. 1974. Use of tritium nuclear magnetic resonance for the direct location of ^{3}H in biosynthetically-labelled penicillic acid. J. Chem. Soc., Chem. Commun. pp. 220-221.

Ashley, D. L., Orti, D. L., and Hill, R. H., Jr. 1987. Proton nuclear magnetic resonance evidence for two configurations of the hemiacetals of aflatoxin B_1 and sterigmatocystin. J. Agric. Food Chem. 35:782-785.

Bachmann, M., Lüthy, J., and Schlatter, C. 1979. Toxicity and mutagenicity of molds of the *Aspergillus glaucus* group. Identification of physcion and three related anthraquinones as main toxic constituents from *Aspergillus chevalieri*. J. Agric. Food Chem. 27:1342-1347.

Baertschi, S. W., Raney, K. D., Shimada, T., Harris, T. M., and Guengerich, F. P. 1989. Comparison of rates of enzymatic oxidation of aflatoxin B_1, aflatoxin G_1, and sterigmatocystin and activities of the epoxides in forming guanyl-N^7 adducts and inducing different genetic responses. Chem. Res. Toxicol. 2:114-122.

Barber, J., and Staunton, J. 1980. Biosynthesis of citrinin in *Penicillium citrinum*: Advanced precursor studies using ^{2}H nuclear magnetic resonance spectroscopy. J. Chem. Soc., Chem. Commun. pp. 552-554.

Barber, J. A., Staunton, J., and Wilkinson, M. R. 1986. A diastereoselective synthesis of the polyketide antibiotic citrinin using toluate anion chemistry. J. Chem. Soc., Perkin Trans. 1, pp. 2101-2109.

Barber, J., Chapman, A. C., and Howard, T. D. 1987a. Methods of tracing hydrogen in polyketide biosynthesis: High field NMR spectroscopy of citrinin produced in a D_2O based medium. Mycotoxin Res. 31:13-18.

Barber, J., Cornford, J. L., Howard, T. D., and Sharples, D. 1987b. The structure of citrinin *in vivo*. J. Chem. Soc., Perkin Trans. 1, pp. 2743-2744.

Bauer, J., and Gareis, M. 1987. Ochratoxin A in der Nahrungsmittelkette. J. Vet. Med. Ser. B 34:613-627.

Bennett, M., Gill, G. B., Pattenden, G., Shuker, A. J., and Stapleton, A. 1991. Ylidenebutenolide mycotoxins—Concise syntheses of patulin and neopatulin from carbohydrate precursors. J. Chem. Soc., Perkin Trans. 1, pp. 929-937.

Betina, V. 1984. Citrinin and related substances. Pages 217-236 in: Mycotoxins—Production, Isolation, Separation and Purification. V. Betina, ed. Elsevier, Amsterdam.

Betina, V. 1989. Patulin and other small lactones. Pages 242-270 in: Mycotoxins—Chemical, Biological and Environmental Aspects. Elsevier, Amsterdam.

Bhat, B., Harrison, D. M., and Lamont, H. M. 1990. The biosynthesis of the tryptophan-derived mould metabolites roquefortine and aszonalenin. J. Chem. Soc., Chem. Commun. pp. 1518-1519.

Birkinshaw, J. H., and Hammady, I. M. M. 1957. Studies in the biochemistry of microorganisms. 99. Metabolic products of *Aspergillus versicolor* (Vuillemin) Tiraboschi. Biochem. J. 65:162-166.

Birkinshaw, J. H., Oxford, A. E., and Raistrick, H. 1936. LXIV. Studies in the biochemistry of micro-organisms. XLVIII. Penicillic acid, a metabolic product of *Penicillium puberulum* Bainier and *P. cyclopium* Westling. Biochem. J. 30:394-411.

Black, D. K. 1966. The addition of L-cysteine to unsaturated lactones and related compounds. J. Chem. Soc. C, pp. 1123-1127.

Brackett, R. E., and Marth, E. H. 1979. Stability of patulin at pH 6.0-8.0 and 25°C. Z. Lebensm. Unters. Forsch. 169:92-94.

Bredenkamp, M. W., Dillen, J. L. M., van Rooyen, P. H., and Steyn, P. S. 1989. Crystal structures and conformational analysis of ochratoxin A and B: Probing the chemical structure causing toxicity. J. Chem. Soc., Perkin Trans. 2, pp. 1835-1839.

Büchi, G., Fowler, K. W., and Nadzan, A. M. 1982. Photochemical epoxidation of aflatoxin B_1 and sterigmatocystin: Synthesis of guanine-containing adducts. J. Am. Chem. Soc. 104:544-547.

Bullock, E., Roberts, J. C., and Underwood, J. G. 1962. Studies in mycological chemistry. Part XI. The structure of isosterigmatocystin and an amended structure for sterigmatocystin. J. Chem. Soc. pp. 4179-4183.

Burka, L. T., Ganguli, M., and Wilson, B. J. 1983. Verrucosidin, a tremorgen from *Penicillium verrucosum* var. *cyclopium*. J. Chem. Soc., Chem. Commun. pp. 544-545.

Cardillo, R., Fuganti, C., Gatti, G., Ghiringhelli, D., and Grasselli, P. 1974. Molecular structure of cryptoechinuline A, a new metabolite of *Aspergillus amstelodami,* isolated during investigations on echinuline biosynthesis. Tetrahedron Lett. pp. 3163-3166.

Casnati, G., Piozzi, F., Quilico, A., and Cardani, C. 1960. Echinulina. Nota X. (XVI di ricerche chimiche nel gruppo dell'*Aspergillus glaucus*). Gazz. Chim. Ital. 90:476-500.

Casnati, G., Cavalleri, R., Piozzi, F., and Quilico, A. 1962. Echinulina. Nota XI (XVII di ricerche chimiche nel gruppo dell'*Aspergillus glaucus*). Gazz. Chim. Ital. 92:105-128.

Chang, F. C. C., and Chu, F. S. 1976. Preparation of ^{3}H-labelled ochratoxins. J. Labelled Compd. Radiopharm. 12:231-238.

Chernyshev, A. I., Shorshnev, S. V., Klyuev, N. A., Kudinova, M. K., Murenets, N. V., Anisova, L. N., Krasilnikova, O. L., and Bartoshevich, Yu. E. 1988. Study of patulin by nmr and mass spectroscopy. (In Russian.) Antibiot. Khimoterapia 33:335-338.

Chung, D.-H., Abouzied, M. M., and Pestka, J. J. 1989. Immunochemical assay applied to mycotoxin biosynthesis: ELISA comparison of sterigmatocystin production by *Aspergillus versicolor* and *Aspergillus nidulans*. Mycopathologia 107:93-100.

Ciegler, A., Mintzlaff, H.-J., Weisleder, D., and Leistner, L. 1972. Potential production and detoxification of penicillic acid in mold-fermented sausage (salami). Appl. Microbiol. 24:114-119.

Cole, R. J., and Cox, R. H. 1981. Handbook of Toxic Fungal Metabolites. Academic Press, New York.

Cox, R. H., and Cole, R. J. 1977. Carbon-13 nuclear magnetic resonance studies of fungal metabolites, aflatoxins, and sterigmatocystins. J. Org. Chem. 42:112-114.

Cruickshank, J. H., Raistrick, H., and Robinson, R. 1938. The chemistry of *Aspergillus* colouring matters. Part II. J. Chem. Soc. pp. 2056-2064.

Damodaran, C. 1977. In vitro binding of citrinin to serum protein. Experientia 33:598-599.

Davies, J. E., Kirkaldy, D., and Roberts, J. C. 1960. Studies in mycological chemistry. Part VII. Sterigmatocystin, a metabolite of *Aspergillus versicolor* (Vuillemin) Tiraboschi. J. Chem. Soc. pp. 2169-2178.

de Jesus, A. E., Steyn, P. S., Vleggaar, R., and Wessels, P. L. 1980. Carbon-13 nuclear magnetic resonance assignments and biosynthesis of the mycotoxin ochratoxin A. J. Chem. Soc., Perkin Trans. 1, pp. 52-54.

Destro, R., and Marsh, R. E. 1984. Temperature dependence of tautomeric equilibria in the solid state: The case of citrinin. J. Am. Chem. Soc. 106:7269-7271.

Elvidge, J. A., Jaiswal, D. K., Jones, J. R., and Thomas, R. 1977. Tritium nuclear magnetic resonance spectroscopy. Part 7. New information from the tritium distribution in biosynthetically labelled penicillic acid. J. Chem. Soc., Perkin Trans. 1, pp. 1080-1083.

Engebrecht, J. C., and Altenkirk, B. 1972. Comparison of some biological effects of sterigmatocystin and aflatoxin analogues on primary cell cultures. J. Natl. Cancer Inst. 48:1647-1655.

Essigmann, J. M., Barker, L. J., Fowler, K. W., Francisco, M. A., Reinhold, V. N., and Wogan, G. N. 1979. Sterigmatocystin-DNA interactions: identification of a major adduct formed after metabolic activation in vitro. Proc. Natl. Acad. Sci. 76:179-183.

Frisvad, J. C., and Filtenborg, O. 1989. Terverticillate penicillia: Chemotaxonomy and mycotoxin production. Mycologia 81:837-861.

Fukuyama, K., Hamada, K., Tsukihara, T., Katsube, Y., Hamasaki, T., and Hatsuda, Y. 1976. Crystal structures of sterigmatocystin and O-methylsterigmatocystin, metabolites of genus *Aspergillus*. Bull. Chem. Soc. Jpn. 49:1153-1154.

Gallagher, R. T., and Stahr, H. M. 1981. Mass spectral confirmation of ochratoxin A. Appl. Spectrosc. 35:131-132.

Ganguli, M., Burka, L. T., and Harris, T. M. 1984. Structural studies of the mycotoxin verrucosidin. J. Org. Chem. 49:3762-3766.

Gill, G. B., Pattenden, G., and Stapleton, A. 1988. A concise synthesis of patulin from arabinose. Tetrahedron Lett. 29:2875-2878.

Gorst-Allman, C. P., Steyn, P. S., and Vleggaar, R. 1982. The biosynthesis of roquefortine. An investigation of acetate and mevalonate incorporation using high field n.m.r. spectroscopy. J. Chem. Soc., Chem. Commun. pp. 652-653.

Gould, B. S., and Raistrick, H. 1934. CCXVIII. Studies in the biochemistry of microorganisms. XL. The crystalline pigments of species in the *Aspergillus glaucus* series. Biochem. J. 28:1640-1656.

Häggblom, P. 1990. Isolation of roquefortine C from feed grain. Appl. Environ. Microbiol. 56:2924-2926.

Harwig, J., Chen, Y-K., Kennedy, B. P. C., and Scott, P. M. 1973. Occurrence of patulin and patulin-producing strains of *Penicillium expansum* in natural rots of apple in Canada. Can. Inst. Food Sci. Technol. J. 6:22-25.

Harwig, J., Kuiper-Goodman, T., and Scott, P. M. 1983. Microbial food toxicants: Ochratoxins. Pages 193-238 in: CRC Handbook of Foodborne Diseases of Biological Origin. M. Rechcigl, Jr., ed. CRC Press, Boca Raton, FL.

Hasinoff, B. B., Rahimtula, A. D., and Omar, R. F. 1990. NADPH-cytochrome-P-450 reductase promoted hydroxyl radical production by the iron(III)-ochratoxin A complex. Biochim. Biophys. Acta 1036:78-81.

Hatakeyama, S., Sakurai, K., Numata, H., Ochi, N., and Takano, S. 1988. A novel chiral route to substituted tetrahydrofurans. Total synthesis of (+)-verrucosidin and formal synthesis of (-)-citreoviridin. J. Am. Chem. Soc. 110:5201-5203.

Holzapfel, C. W. 1968. The isolation and structure of cyclopiazonic acid, a toxic metabolite of *Penicillium cyclopium* Westling. Tetrahedron 24:2101-2119.

Hubbard, C. R., Mighell, A. D., and Ware, G. M. 1977. 4-Hydroxy-4H-furo[3,2c]pyran-2(6H)-one (DL-patulin). Acta Crystallogr. B 33:928-931.

Hult, K. 1986. Ochratoxin A labelled with ^{14}C or ^{3}H in the phenylalanine moiety. J. Labelled Compd. Radiopharm. 23:801-805.

Hutchison, R. D., and Holzapfel, C. W. 1971. The conversion of sterigmatocystin into dihydroaspertoxin. Tetrahedron 27:425-431.

IARC. 1986. Patulin. Pages 83-98 in: IARC Monographs on the Evaluation of the Carcinogenic Risks of Chemicals to Humans. Vol. 40, Some Naturally Occurring and Synthetic Food Components, Furocoumarins and Ultraviolet Radiation. IARC, Lyon, France.

Inoue, S., Takamatsu, N., and Kishi, Y. 1977a. Synthetic studies on echinulin and related natural products. II. A total synthesis of echinulin. (In Japanese.) Yakugaku Zasshi 97:558-563.

Inoue, S., Hashizume, K., Takamatsu, N., Nagano, H., and Kishi, Y. 1977b. Synthetic studies on echinulin and related natural products. IV. Isolation, structure and synthesis of flavoglaucin-auroglaucin type natural products isolated from *Aspergillus amstelodami*. (In Japanese.) Yakugaku Zasshi 97:569-575.

Keromnes, J., and Thouvenot, D. 1985. Role of penicillic acid in the phytotoxicity of *Penicillium cyclopium* and *Penicillium canescens* to the germination of corn seeds. Appl. Environ. Microbiol. 49:660-663.

Koller, W. D., and Orth, R. 1975. Quantitative Bestimmung von Sterigmatocystin mit Hilfe des Massenspektrometrie. Lebensmittelchem. Gerichtl. Chem. 29:153-157.

Kozikowski, A. P., Greco, M. N., and Springer, J. P. 1984. Total synthesis of the unique mycotoxin α-cyclopiazonic acid (αCA): An unusual dimethylzinc-mediated replacement of a phenylthio substituent by a methyl group and a contrathermodynamic Raney nickel desulfurization reaction. J. Am. Chem. Soc. 106:6873-6874.

Kuiper-Goodman, T., and Scott, P. M. 1989. Risk assessment of the mycotoxin ochratoxin A. Biomed. Environ. Sci. 2:179-248.

Kume, T., Ito, H., Iizuka, H., and Takehisa, M. 1983. Radiosensitivity of *Aspergillus versicolor* isolated from animal feeds and destruction of sterigmatocystin by gamma-irradiation. Agric. Biol. Chem. 47:1065-1069.

Le Bars, J., Gripon, J. C., Vassal, L., and Le Bars, P. 1988. Production d'acide cyclopiazonique dans des fromages en fonction des souches de *Penicillium camembertii* et des conditions de maturation. Microbiol. Aliments Nutr. 6:337-343.

Lezi, S. Z., Thonart, P., Wathelet, J.-P., Blanpain, P., and Bechet, J. 1981. La patuline: Production, purification, dégradation et dosage par CPG. Rev. Ferment. Ind. Aliment. 36:122-129.

Li, Y., and Chu, F. S. 1984. Production and characterization of antibody against sterigmatocystin. J. Food Saf. 6:119-127.

Lowes, N. R., Smith, R. A., and Beck, B. E. 1992. Roquefortine in the stomach contents of dogs suspected of strychnine poisoning in Alberta. Can. Vet. J. 33:535-538.

Luk, K. C., Kobbe, B., and Townsend, J. M. 1977. Production of cyclopiazonic acid by *Aspergillus flavus* Link. Appl. Environ. Microbiol. 33:211-212.

Madhyastha, S. M., Marquardt, R. R., Frohlich, A. A., Platford, G., and Abramson, D. 1990. Effects of different cereal and oilseed substrates on the growth and production of toxins by *Aspergillus alutaceus* and *Penicillium verrucosum*. J. Agric. Food Chem. 38:1506-1510.

Maes, C. M., and Steyn, P. S. 1984. Polyketide-derived fungal metabolites from *Biopolaris sorokiniana* and their significance in the biosynthesis of sterigmatocystin and aflatoxin B_1. J. Chem. Soc., Perkin Trans. 1, pp. 1137-1140.

Maness, D. D., Schneider, L. W., Sullivan, G., Yakatan, G. J., and Scholler, J. 1976. Fluorescence behavior of sterigmatocystin. J. Agric. Food Chem. 24:961-963.

Meerarani, S., and Shanmugasundaram, E. R. B. 1987. Preparation of ^{14}C-sterigmatocystin and its in vivo and in vitro binding with macromolecules. Indian J. Exp. Biol. 25:122-123.

Mehl, M., Starke, P., Jacobi, H. D., Schleinitz, K. D., and Wasicki, P. 1986. Immunologischer Nachweis des Mykotoxins Patulin. Pharmazie 41:147-148.

Mills, J. T. 1990. Mycotoxins and toxigenic fungi on cereal grains in western Canada. Can. J. Physiol. Pharmacol. 68:982-986.

Mills, J. T., Siefert, K. A., Frisvad, J. C., Abramson, D., and Louis-Seize, G. 1992. Relative importance of *Penicillium* spp. producing nephrotoxins in Canadian grain products. Ext. Abstr. Int. Symp. Stored Grain Ecosystems, pp. 29-31.

Morgan, M. R. A., Kang, A. S., and Chan, H. W.-S. 1986. Production of antisera against sterigmatocystin hemiacetal and its potential for use in an enzyme-linked immunosorbent assay for sterigmatocystin in barley. J. Sci. Food Agric. 37:873-880.

Nagasawa, H., Isogai, A., Suzuki, A., and Tamura, S. 1979. ^{13}C-NMR spectra and stereochemistry of isoechinulins A, B and C. Agric. Biol. Chem. 43:1759-1763.

Nakashima, T. T., and Vederas, J. C. 1982. Biosynthesis of the aflatoxin precursor sterigmatocystin by *Aspergillus versicolor*; spin-echo resolution of ^{18}O isotope shifts in ^{13}C-n.m.r. spectroscopy. J. Chem. Soc., Chem. Commun. pp. 206-208.

Nakazato, M., Kanmuri, M., Nakazawa, K., Ariga, T., Fujinuma, K., Nishijima, M., and Naoi, Y. 1981. Fluorometric determination of citrinin in cereals (Studies on mycotoxins in Food. XV). (In Japanese.) J. Food Hyg. Soc. Jpn. 22:391-396.

Nazar, M., Ali, M., Fatima, T., and Gubler, C. J. 1984. Toxicity of flavoglaucin from *Aspergillus chevalieri* in rabbits. Toxicol. Lett. 23:233-237.

Neely, F. L., and Emerson, C. S. 1990. Determination of sterigmatocystin in fermentation broths by reversed-phase high-performance liquid chromatography using post-column fluorescence enhancement. J. Chromatogr. 523:305-311.

Neely, W. C., Ellis, S. P., Davis, N. D., and Diener, U. L. 1972. Spectroanalytical parameters of fungal metabolites. I. Citrinin. J. Assoc. Off. Anal. Chem. 55:1122-1127.

Nishiyama, S., Shizuri, Y., Shigemori, H., and Yamamura, S. 1986. Synthetic study on verrucosidin and its absolute configuration. Tetrahedron Lett. 27:723-726.

Ohmomo, S., Sugita, M., and Abe, M. 1973. Isolation of cyclopiazonic acid, cyclopiazonic acid imine and bissecodehydrocyclopiazonic acid from the cultures of *Aspergillus versicolor* (Vuill.) Tiraboschi (Production of alkaloids and related substances by fungi Part XI). (In Japanese.) J. Agric. Chem. Soc. Jpn. 47:57-63.

Ohmomo, S., Sato, T., Utagawa, T., and Abe, M. 1975. Isolation of festuclavine and three new indole alkaloids, roquefortine A, B and C from the cultures of *Penicillium roqueforti* (Production of alkaloids and related substances by fungi Part XII). (In Japanese.) J. Agric. Chem. Soc. Jpn. 49:615-623.

Orti, D. L., Grainger, J., Ashley, D. L., and Hill, R. H., Jr. 1989. Chromatographic and spectroscopic properties of hemiacetals of aflatoxin and sterigmatocystin metabolites. J. Chromatogr. 462:269-279.

Pachler, K. G. R., Steyn, P. S., Vleggaar, R., Wessels, P. L., and Scott, de B. 1976. Carbon-13 nuclear magnetic resonance assignments and biosynthesis of aflatoxin B$_1$ and sterigmatocystin. J. Chem. Soc., Perkin Trans. 1, pp. 1182-1189.

Pachter, R., and Wessels, P. L. 1980. Detection of partially resolved carbon-13-proton couplings with selective population inversion (SPI). Org. Magn. Reson. 14:374-378.

Paster, N., Barkai-Golan, R., and Padova, R. 1985. Effect of gamma radiation on ochratoxin production by the fungus *Aspergillus ochraceus*. J. Sci. Food Agric. 36:445-449.

Peterson, R. E., Shannon, G. M., and Shotwell, O. L. 1989. Purification of cyclopiazonic acid by liquid chromatography. J. Assoc. Off. Anal. Chem. 72:332-335.

Phillips, T. D., Ivie, G. W., Heidelbaugh, N. D., Kubena, L. F., Cysewski, S. J., Hayes, A. W., and Witzel, D. A. 1981. Confirmation of penicillic acid by high pressure liquid and gas-liquid chromatography. J. Assoc. Off. Anal. Chem. 64:162-165.

Pitt, J. I. 1987. *Penicillium viridicatum, Penicillium verrucosum*, and production of ochratoxin A. Appl. Environ. Microbiol. 53:266-269.

Pitt, J. I., Cruickshank, R. H., and Leistner, L. 1986. *Penicillium commune, P. camembertii*, the origin of white cheese moulds, and the production of cyclopiazonic acid. Food Microbiol. 3:363-371.

Podojil, M., Sedmera, P., Vokoun, J., Betina, V., Baráthová, H., Ďuráková, Z., Horáková, K., and Nemec, P. 1979. *Eurotium (Aspergillus) repens* metabolites and their biological activity. Folia Microbiol. 23:438-443.

Pohland, A. E., Schuller, P. L., Steyn, P. S., and van Egmond, H. P. 1982. Physiochemical data for some selected mycotoxins. Pure Appl. Chem. 54:2219-2284.

Quilico, A., and Panizzi, L. 1943. Chemische Untersuchungen über *Aspergillus echinulatus*, I. Mitteilung. Ber. Dtsch. Chem. Ges. 76:348-358.

Quilico, A. Panizzi, L., and Mugnaini, E. 1949. Structure of flavoglaucin and auroglaucin. Nature 164:26-27.

Quilico, A., Cardani, C., and d'Alcontres, G. S. 1953. Sulla struttura della flavoglaucina: Prodotti di degradazione con H_2O_2 e alcali. Nota VIII di ricerche chimiche nel gruppo dell'*Aspergillus glaucus*. Gazz. Chim. Ital. 83:754-773.

Raistrick, H., Robinson, R., and Todd, A. R. 1937. The chemistry of *Aspergillus* colouring matters. Part I. J. Chem. Soc. pp. 80-88.

Rao, K. V., and Reddy, G. C. S. 1989. A new reaction of patulin. J. Nat. Prod. 52:1376-1378.

Raphael, R. A. 1948. Compounds related to penicillic acid. Part III. Synthesis of penicillic acid. J. Chem. Soc. pp. 1508-1512.

Regan, A. C., and Staunton, J. 1987. Asymmetric synthesis of (+)-citrinin using an ortho-toluate carbanion generated by a chiral base. J. Chem. Soc., Chem. Commun. pp. 520-521.

Reshetilova, T. A., and Kozlovsky, A. G. 1990. Synthesis and metabolism of roquefortine in *Penicillium* species. J. Basic Microbiol. 30:109-114.

Rodig, O. R., Shiro, M., and Fernando, Q. 1971. The crystal and molecular structure of citrinin. Chem. Commun. pp. 1553-1554.

Rodricks, J. V. 1969. Fungal metabolites which contain substituted 7:8-dihydrofuro[2,3-*b*]furans (DHFF) and 2,3,7,8-tetrahydrofuro[2,3,-*b*]furans (THFF). J. Agric. Food Chem. 17:457-461.

Ross, P. F., Rice, L. G., Casper, H. H., Crenshaw, J. D., and Richard, J. L. 1991. Novel occurrence of cyclopiazonic acid in sunflower seeds. Vet. Hum. Toxicol. 33:284-285.

Rousseau, D., Slegers, G., Van Peteghem, C., and Claeys, A. 1984. Synthesis of ^{14}C-ochratoxin A. J. Labelled Compd. Radiopharm. 21:429-440.

Sankawa, U., Shimada, H., Kobayashi, T., Ebizuka, Y., Yamamoto, Y., Noguchi, H., and Seto, H. 1982. Biosynthesis of averufin and sterigmatocystin in *Aspergillus parasiticus* and *A. versicolor*. Application of multiple labelled acetate in polyketide biosynthesis. Heterocycles 19:1053-1058.

Sankawa, U., Ebizuka, Y., Noguchi, H., Isikawa, Y., Kitaghawa, S., Yamamoto, Y., Kobayashi, T., and Iitak, Y. 1983. Biosynthesis of citrinin in *Aspergillus terreus*. Incorporation studies with $[2\text{-}^{13}C, 2\text{-}^2H_3]$, $[1\text{-}^{13}C, {}^{18}O_2]$ and $[1\text{-}^{13}C, {}^{17}O]$-acetate. Tetrahedron 39:3583-3591.

Scott, P. M. 1974. Patulin. Pages 383-403 in: Mycotoxins. I. F. H. Purchase, ed., Elsevier, Amsterdam.

Scott, P. M. 1977. Penicillium mycotoxins. Pages 283-356 in: Mycotoxic Fungi,

Mycotoxins, Mycotoxicoses. An Encyclopedic Handbook, vol. 1. T. D. Wyllie and L. G. Morehouse, eds. Marcel Dekker, New York.

Scott, P. M. 1981. Toxins of *Penicillium* species used in cheese manufacture. J. Food Prot. 44:702-710.

Scott, P. M. 1984a. Effects of food processing on mycotoxins. J. Food Prot. 47:489-499.

Scott, P. M. 1984b. Roquefortine. Pages 463-468 in: Mycotoxins—Production, Isolation, Separation and Purification. V. Betina, ed. Elsevier, Amsterdam.

Scott, P. M. 1991. Possibilities of reduction or elimination of mycotoxins present in cereal grains. Pages 529-572 in: Cereal Grain. Mycotoxins, Fungi and Quality in Drying and Storage. J. Chelkowski, ed. Elsevier, Amsterdam.

Scott, P. M., and Kennedy, B. P. C. 1976. Analysis of blue cheese for roquefortine and other alkaloids from *Penicillium roqueforti*. J. Agric. Food Chem. 24:865-868.

Scott, P. M., van Walbeek, W., Harwig, J., and Fennell, D. I. 1970. Occurrence of a mycotoxin, ochratoxin A, in wheat and isolation of ochratoxin A and citrinin producing strains of *Penicillium viridicatum*. Can J. Plant Sci. 50:583-585.

Scott, P. M., van Walbeek, W., Kennedy, B., and Anyeti, D. 1972. Mycotoxins (ochratoxin A, citrinin, and sterigmatocystin) and toxigenic fungi in grains and other agricultural products. J. Agric. Food Chem. 20:1103-1109.

Scott, P. M., Merrien, M.-A., and Polonsky, J. 1976. Roquefortine and isofumigaclavine A, metabolites from *Penicillium roqueforti*. Experientia 32:140-142.

Scott, P. M., Polonsky, J., and Merrien, M.-A. 1979. Configuration of the 3,12 double bond of roquefortine. J. Agric. Food Chem. 27:201-202.

Seijas, J. A., Vázquez Tato, M. P., Estévez, R., Castedo, L., and Riguera, R. 1989. New total synthesis of patulin. Heterocycles 29:181-184.

Selva, A., and Traldi, P. 1977. The electron impact induced fragmentation of *Aspergillus amstelodami* alkaloids and derivatives. Biomed. Mass Spectrom. 4:143-145.

Seto, H., Cary, L. W., and Tanabe, M. 1974a. Utilization of ^{13}C-^{13}C coupling in structural and biosynthetic studies. IV. Penicillic acid. J. Antibiot. 27:558-559.

Seto, H., Cary, L. W., and Tanabe, M. 1974b. Utilization of ^{13}C-^{13}C coupling in structural and biosynthetic studies. V. The ^{13}C FT nmr spectrum of sterigmatocystin. Tetrahedron Lett. pp. 4491-4494.

Sibi, M. P., Chattopadhyay, S., Dankwardt, J. W., and Snieckus, V. 1985. Combinational O-aryl carbamate and benzamide directed ortho metalation reactions. Synthesis of ochratoxin A and ochratoxin B. J. Am. Chem. Soc. 107:6312-6315.

Simpson, T. J., and Stenzel, D. J. 1982. Biosynthesis of aflatoxins. Incorporation of $[1,2-^{13}C_2]$ acetate, $[^2H_3]$ acetate, and $[1-^{13}C, {}^2H_3]$ acetate into sterigmatocystin in *Aspergillus versicolor*. J. Chem. Soc., Chem. Commun. pp. 890-892.

Steiman, R., Seigle-Murandi, F., Sage, L., and Krivobok, S. 1989. Production of patulin by Micromycetes. Mycopathologia 105:129-133.

Steyn, P. S. 1984. Ochratoxins and related dihydroisocoumarins. Pages 183-216 in: Mycotoxins—Production, Isolation, Separation and Purification. V. Betina, ed. Elsevier, Amsterdam.

Suzuki, T., Hoshino, Y., Kikuchi, Y., Nose, N., and Watanabe, A. 1976. Determination of sterigmatocystin by gas-liquid chromatography with an electron-capture detector. (In Japanese.) J. Food Hyg. Soc. Jpn. 17:253-257.

Tanabe, M., Hamasaki, T., Seto, H., and Johnson, L. 1970. Biosynthetic studies with carbon-13: ^{13}C nuclear magnetic resonance spectra of the metabolite sterigmatocystin. Chem. Commun. pp. 1539-1540.

Tanaka, N., Katsube, Y., Hatsuda, Y., Hamasaki, T., and Ishida, M. 1970. Structure analysis of *p*-bromobenzoate of sterigmatocystin by X-ray diffraction method. Bull. Chem. Soc. Jpn. 43:3635.

Tarter, E. J., and Scott, P. M. 1991. Determination of patulin by capillary gas chromatography of the heptafluorobutyrate derivative. J. Chromatogr. 538:441-446.

van der Merwe, K. J., Steyn, P. S., and Fourie, L. 1965. Mycotoxins. Part II. The constitutions of ochratoxins A, B, and C, metabolites of *Aspergillus ochraceus* Wilh. J. Chem. Soc. pp. 7083-7088.

van der Watt, J. J. 1974. Sterigmatocystin. Pages 369-382 in: Mycotoxins. I. F. H. Purchase, ed. Elsevier, Amsterdam.

Vesonder, R. F., Lambert, R., Wicklow, D. T., and Biehl, M. L. 1988. *Eurotium* spp., and echinulin in feed refused by swine. Appl. Environ. Microbiol. 54:830-831.

Vleggaar, R., and Wessels, P. L. 1980. Stereochemistry of the dehydrogenation of (2*S*)-histidine in the biosynthesis of roquefortine and oxaline. J. Chem. Soc., Chem. Commun. pp. 160-162.

Ware, G. M., Thorpe, C. W., and Pohland, A. E. 1980. Determination of roquefortine in blue cheese and blue cheese dressing by high pressure liquid chromatography with ultraviolet and electrochemical detectors. J. Assoc. Off. Anal. Chem. 63:637-641.

Weisleder, D., and Lillehoj, E. 1980. Carbon-13 nuclear magnetic resonance assignments and biosynthesis of ochratoxin A. Tetrahedron Lett. 21:993-996.

Yamaguchi, T., Nozawa, K., Nakajima, S., Kawai, K., and Udagawa, S. 1991. Absolute configuration of roquefortine C, a tremorgenic mycotoxin. Proc. Jpn. Assoc. Mycotoxicol. pp. 29-32.

Yeh, C.-L., Colwell, W. T., and DeGraw, J. I. 1978. A convenient synthesis of penicillic acid. Tetrahedron Lett. pp. 3987-3988.

Yokota, T., Sakurai, A., Iriuchijima, S., and Takahashi, N. 1981. Isolation and ^{13}C NMR study of cyclopyazonic acid, a toxic alkaloid produced by muscardine fungi *Aspergillus flavus* and *A. oryzae*. Agric. Biol. Chem. 45:53-56.

Żegota, H., Żegota, A., and Bachmann, S. 1988. Effect of irradiation and storage on patulin disappearance and some chemical constituents of apple juice concentrate. Z. Lebensm.-Unters. Forsch. 187:321-324.

Chapter 6

Ecological Aspects of Growth and Mycotoxin Production by Storage Fungi

K. H. Ominski, R. R. Marquardt, R. N. Sinha, and D. Abramson

Cereal grains, oilseeds, and their products are the main dietary component for both humankind and livestock. These commodities are stored mainly to ensure an even food supply throughout the year, to facilitate transport of surplus food to a deficit area, to serve as a reserve during times of food scarcity, and to use as seeds for the next growing season (Salunkhe et al 1985). During storage, the cereal crop commonly undergoes quality loss, which is characterized by increased susceptibility to infection by fungi, insects, and mites which directly or indirectly affect grain quality. A decrease in the nutritional content of the commodity may also occur with a loss of carbohydrates, proteins, amino acids, and vitamins and with increased fat acidity. Fungi are a major cause of spoilage in stored grains and seeds and rank second only to insects as a cause of deterioration and loss (CAST 1989). Fungal infection may result in a decrease in quality, grade, and market value of cereal grains and their products. Fungal growth does not necessarily denote the presence of mycotoxins because not all species and strains of fungi produce mycotoxins. Conditions favoring the development of mycotoxins in stored cereals are not well understood but are of considerable importance to grain-exporting countries like Canada, which are concerned with ensuring high standards of quality in their produce.

Storage Fungi

Fungal infection, development, and toxin production can occur in the field, in storage, or both. Thus, stored crops may contain mycotoxins arising from field infection and/or toxins formed during storage (Sauer 1978, Abramson 1991). Dormant conidia arising from soil particles, plant debris, and residues from harvest equipment may occur on seed surfaces (Flannigan 1978). Occasionally, dormant mycelium may occur under the seed pericarp (Christensen and Kaufmann 1969). Field fungi that are present at the onset of storage are succeeded by storage fungi as the duration of storage increases (Sinha et al 1969; Abramson et al 1990a,b). Thus, the species composition and relative dominance of individual species and the production of secondary metabolites vary according to the developmental stage of the stored-grain ecosystem (Sinha 1992). Field fungi such as *Alternaria, Fusarium,* and *Cladosporium* require high relative humidities and water contents and are not competitive in storage systems (Sauer 1978), which become dominated by species of storage fungi, particularly *Aspergillus* and *Penicillium*.

NOMENCLATURE OF STORAGE FUNGI

Storage fungi generally include several genera of seedborne fungi including *Aspergillus, Penicillium, Rhizopus, Mucor,* and *Wallemia*. Some species of *Aspergillus* and *Penicillium* have adapted to a life without free water and can produce a range of fungal metabolites including those known as mycotoxins. Accurate identification is essential when working with toxin-producing species and describing their toxins. Failure to do so can result in conflicting reports of the abilities of species to produce particular toxins. This problem can be seen particularly with reports of toxin production by *Penicillium* species and results from misidentification of species, the use of different systems of nomenclature, and synonymy of fungal names (El-Banna et al 1987, Frisvad 1989).

New approaches to fungal taxonomy, in conjunction with traditional taxonomy, seem to have clarified the taxonomy of many fungal species. *P. viridicatum* Westling, for example, is widely reported to produce ochratoxin A. Pitt (1987), however, demonstrated that *P. viridicatum* produces xanthomegnin, viomellin, or brevianamide A but does not produce ochratoxins or citrinin. *P. verrucosum* Dierckx chemotypes I and II were cited as the only food- and feed-related species in the genus *Penicillium* to be confirmed as producers ochratoxin A (Pitt 1987, Frisvad and Filtenborg 1989), although it is possible that other species of *Penicillium* may produce this toxin. For instance, ochratoxin A production by *P. chrysogenum* Thom has been reported (Mills et al 1989, Mills and Abramson 1982), although Frisvad (1989) was unable to confirm this. Table 6.1 provides a list of *Penicillium* species that have been reidentified using modern taxonomic tools. *Aspergillus ochraceus* K. Wilh. has been renamed *A. alutaceus* var. *alutaceus* Berkley & Curtis 1875 (Kozakiewicz 1989). The names of fungal species in this review are cited as they appeared in the original research paper.

MYCOLOGY

Traditionally, fungal identification was based primarily on morphology. In recent years, however, new techniques for identification have been explored; when these techniques are utilized concurrently with morphology-based taxonomy, they provide a more comprehensive identification of fungal species. The techniques include chemotaxonomy, electrophoresis, genetic and ultrastructure studies, and immunological techniques (Pitt and Samson 1990). Chemotaxonomy involves the characterization of secondary metabolite production by a given fungal species (Frisvad and Filtenborg 1983). A rapid thin-layer chromatography technique for screening toxigenic fungi was developed by Filtenborg and Frisvad and associates (Filtenborg and Frisvad 1980, Filtenborg et al 1983). Another tool utilized in fungal taxonomy is electrophoresis, in which patterns of electrophoretically separated enzymes or total soluble proteins provide information about the fungal genome. Cruikshank and Pitt (1987) studied amylase and ribonuclease isozymes to differentiate species in the subgenus *Penicillium*. They demonstrated a strong correlation between this technique and morphological identification. Recently, genetic studies have also been used for fungal identification and classification. The techniques currently used include determination of guanine and cytosine molar percentage, DNA-DNA complementarity, rRNA sequence comparisons, and restriction fragment length polymorphism (Samson et al 1991). Immunological techniques include enzyme-linked immunosorbency assays (ELISA), immunofluorescent and latex agglutination

TABLE 6.1
Recent Nomenclatural Changes for Common *Penicillium* Names[a]

Old Name	Current Name
P. candidum	P. camembertii Thom
P. caseicola	P. camembertii Thom
P. claviforme	P. vulpinum (Cooke & Massee) Seifert & Samson
P. corymbiferum	P. hirsutum Dierckx
P. cyclopium	P. aurantiogriseum Fierckx
P. frequentans	P. glabrum (Wehmer) Westling
P. glaucum	Doubtful name
P. granulatum	P. glandicola (Oudem.) Seifert & Samson
P. janthinellum	P. simplicissimum (Oudem.) Thom, pro parte
P. martensii	P. aurantiogriseum Dierckx
P. nigricans	P. janczewskii Zaleski
P. notatum	P. chrysogenum Westling[b]
P. palitans	P. commune Thom
P. patulum	P. griseofulvum Dierckx
P. puberulum	P. aurantiogriseum Dierckx
P. urticae	P. griseofulvum Dierckx
P. verrucosum var. cyclopium	P. aurantiogriseum Dierckx

[a]Source: Samson et al (1991); reprinted by courtesy of Marcel Dekker, Inc.
[b]Referred to as *P. chrysogenum* Thom in the *Catalogue of Filamentous Fungi* (Jong and Edwards 1991).

tests, radioimmunoassay, immunoelectron microscopy, and immunoprecipitation (Samson et al 1991). The use of these tools in fungal taxonomy has increased the knowledge base in this field considerably. Bridge et al (1992) have developed an interactive, computer-assisted, probabilistic identification scheme for terverticillate penicillia known as PENIMAT (*Penicillium* identification matrix), which utilizes morphological and scanning electron microscopy features as well as physiological and biochemical test criteria. One may expect further advancements in this field as new techniques are developed for the identification of fungi.

Penicillium

Although both *Aspergillus* and *Penicillium* species are of great economic concern as causes of food spoilage, *Penicillium* is more diverse in terms of the number of species and the range of habitats in which it can survive (Pitt and Hocking 1985). It has been estimated that 70–80% of the penicillia have the potential to produce mycotoxins (Pohland and Wood 1987). Mycotoxins produced by *Penicillium* species include brevianamide A, citreoviridin, citrinin, cyclopiazonic acid, fumitremorgin B, griseofulvin, luteoskyrin, ochratoxin A, patulin, penicillic acid, penitrem A, PR-toxin, roquefortine, rugulosin, verrucosidin, verruculogen, viridicatumtoxin, and xanthomegnin (El-Banna et al 1987).

Mold-ripened foods such as cheese and meat are often fermented with potentially toxigenic strains of penicillia (Leistner 1984). Toxicological screening, sensory evaluation, and mycotoxin analysis are employed to ensure the molds are neither pathogenic nor toxigenic (Leistner 1990).

The main toxins produced by *Penicillium* species on stored cereals in western Canada are ochratoxin A and citrinin (Scott et al 1972). *P. aurantiogriseum* Dierckx (*P. verrucosum* var. *cyclopium*), *P. chrysogenum*, and *P. verrucosum* were identified as the main producers of ochratoxin A and citrinin on cereals in western Canada (Scott et al 1972, Mills 1989, Mills et al 1989), but several researchers (Pitt 1987, Frisvad and Filtenborg 1989) have now presented evidence that *P. verrucosum* is the only *Penicillium* species that produces ochratoxin A and citrinin. Mills et al (1992) recently isolated *Penicillium* species from 533 samples of wheat, barley, oats, and rye obtained from Manitoba, Saskatchewan, and Alberta. *P. aurantiogriseum* var. *aurantiogriseum* was the most commonly isolated taxon, followed by *P. aurantiogriseum* var. *viridicatum* (Westling) Frisvad & Filtenborg and *P. aurantiogriseum* var. *polonicum*. *P. verrucosum, P. chrysogenum, P. rubrum* Stoll, *P. griseofulvum* Dierckx, and *P. oxalicum* Currie & Thom occurred at lower frequencies. The occurrence of fungal species in Canada and the factors affecting the variability of these species have not been fully explored. Predominant *Penicillium* species identified to date in Canada, both in terms of frequency of occurrence and abundance, and the mycotoxins which may be associated with these species are listed in Table 6.2.

Aspergillus

Although the genus *Aspergillus* occurs throughout the world, it is more abundant in tropical countries (Krogh 1987). This genus is important because it contains many species capable of growing at low water activity (Moss 1987). Mycotoxins associated with *Aspergillus* species include aflatoxin, sterigmatocystin, versicolorins, and ochratoxins (Frisvad and Samson 1991). The main toxin produced by *Aspergillus* species on cereal grains in western Canada is sterigmatocystin (Scott et al 1972; Abramson et al 1983a,b). The main fungal species responsible for the production of this toxin is *A. versicolor* (Vuill.) Tiraboschi. The sterigmatocystin-producing ability of *A. versicolor* isolates from western Canadian barley and canola was examined by Mills and Abramson (1986). A large portion of the isolates produced sterigmatocystin under favorable substrate and environmental conditions.

DETECTION OF FUNGAL DETERIORATION

To protect grain from serious quality loss, the presence of fungi in stored cereals must be detected early while deterioration is at a minimum and intervention is still logistically and economically feasible. Often, when the

TABLE 6.2
Mycotoxin Production by Important Foodborne and Feedborne Filamentous Fungi[a,b]

Penicillium Species	Mycotoxins
P. aurantiogriseum var. *aurantiogriseum*	Auranthine, aurantiamine, penicillic acid, puberulic acid, terrestric acid (few isolates), verrucofortin, verrucosidin, viomellein, vioxanthin, viridicatins,[c] xanthomegnin, nephrotoxin glycopeptide
P. aurantiogriseum var. *polonicum*	Penicillic acid, verrucofortine, verrucosidin, viridicatins[c]
P. aurantiogriseum var. *viridicatum*	Brevianamide A, penicillic acid, rubrosulphin, verrucofortine, viomellen, viopurpurin, vioxanthim, viridamin, viridicatins,[c] viridic acid, xanthomegnin, xanthoviridicatin D and G
P. chrysogenum	Emodic acid, meleagrin, penicillin, roquefortine C
P. griseofulvum	Cyclopiazonic acid, griseofulvin, patulin, roquefortine C
P. oxalicum	Oxaline, roquefortine C, secalonic acid D
P. rubrum	Rubratoxins, rugulovasins
P. verrucosum	Citrinin, ochratoxin A

[a] Source: Frisvad and Samson (1991); reprinted by courtesy of Marcel Dekker, Inc.
[b] In Canada, *P. aurantiogriseum* Dierkx has been identified as a common species. This species can be further subdivided as indicated above.
[c] Viridicatins: cyclopenol, cyclopenin, viridicatin, viridicatol, cyclopeptin, dehydrocyclopeptin, 3-*O*-methylviridicatin.

signs of fungal presence are visible, irreversible quality loss has already occurred, and as a result the grain may be downgraded considerably. Thus, the use of a specific biochemical indicator to detect early spoilage and thereby prevent severe quality losses would be extremely useful. The biochemical indicators investigated to date are chemical by-products of the moldy substrate and include fungal volatiles (Kaminski et al 1974, Richard-Molard et al 1976, Abramson et al 1980, Sinha et al 1988) and carbon dioxide (White et al 1982a,b). Unlike carbon dioxide, which is a measure of the total respiration from all organisms in the bin, volatile compounds are more characteristic of the organism from which they are produced. Volatile compounds produced by microflora can be distinguished from those produced by the dormant seed (see Sinha et al 1988 for a review). Several researchers have demonstrated that certain volatile odor components such as 3-methyl-1-butanol, 1-octen-3-ol, and 3-octanone are commonly produced by storage fungi (Kaminski et al 1974; Richard-Molard et al 1976; Abramson et al 1980, 1983a; Sinha et al 1988) and may be used to detect early microbial spoilage on farms and in commercial stores. Tuma et al (1990) demonstrated that tridecane, neral, and geranial are associated with mite-infested grain. These volatile metabolites differ from those observed in mitefree, fungus-infected grain. Thus causes of infestation and spoilage of stored grain by two very different types of organisms can be differentiated by means of their characteristic odors. Fungal volatiles may be used to predict the rate of metabolism of the fungi and the potential toxins that may be produced, but they do not reflect the fungal biomass that is present. Biomass may be estimated using alternative chemical techniques, which include the ergosterol assay (Seitz et al 1979) or analysis of chitin (Ride and Drysdale 1972).

Ergosterol, a predominant sterol component of fungal cell membranes, is specific to fungi and is seldom present in higher plants or animals (Tothill et al 1992). Several researchers have demonstrated that fungal biomass and ergosterol yield are correlated (Matcham et al 1985, Tothill et al 1992). However, it was observed that substrate composition, extent of aeration, growth phase of the mycelium and water activity influence ergosterol content (Nout et al 1987, Tothill et al 1992). Börjesson et al (1990) observed a high correlation between production of volatile metabolites and CO_2, a moderate correlation between volatiles and ergosterol, and no correlation between volatiles and colony-forming units of *P. aurantiogriseum*.

Chitin is a fungal cell-wall polymer, which may be quantified as the polymer using near-infrared reflectance spectroscopy (Roberts et al 1991) or as the monomer *N*-acetyl-D-glucosamine (Ride and Drysdale 1972). Acid hydrolysis of chitin yields glucosamine which may be quantified using ion-exchange chromatography (Rotter et al 1989). Analysis of chitin via glucosamine to quantify fungal biomass has been criticized because glucosamine is also found in insect exoskeletons (Hicks and Newell 1983) and seed glycoproteins (Pusztai 1964). Furthermore, Sharma et al (1977) demonstrated that chitin content varies with age of mycelium, nutrient content of the mycelium, and oxygen tension. Roberts et al (1991), however, indicated that glucosamine from nonfungal sources does not interfere with fungal glucosamine. Neither the ergosterol nor the chitin assay are completely reliable predictors of fungal biomass. A more extensive discussion of the use of ergosterol and other measures of fungal biomass can be found in Newell (1992).

Ecological Parameters

GRAIN STORAGE

Many organisms, from microorganisms to small mammals, interact with grain at the various stages of storage, as do the various components of the environment such as climate and granary structure (Sinha 1973). Factors that are likely to affect mycotoxin formation include water activity, temperature, time, damage to the seed, oxygen and carbon dioxide levels, composition of the substrate, fungal abundance, prevalence of toxigenic strains, spore load, microbial interactions, and invertebrate vectors (Hesseltine 1976). Although spoilage, fungal growth, and mycotoxin formation result from the complex interaction of these factors, an understanding of each factor involved is essential for comprehension of the overall process and may facilitate the prediction and prevention of mycotoxin development. Thus, these factors will be discussed both individually and collectively as they interact under natural conditions. A review of the literature shows that in only a few cases have several ecological parameters been studied simultaneously, in a holistic manner under natural conditions. Nevertheless, inductive studies that are summarized in this chapter are important to establish cause and effect relationships and are useful in the interpretation of multivariate interactions in a holistic study. Figure 6.1 provides a schema of these ecological parameters.

Water Activity and Water Content

In temperate climates, fungi that invade cereal grains can be categorized into three groups that differ in their moisture requirements: field fungi, storage fungi, and advanced decay fungi (Christensen 1965). Field and advanced decay fungi that require 20–25% water content to grow will be discussed in other sections of this book. This categorization does not apply in tropical and subtropical regions or during periods of drought stress and insect damage when *Aspergillus* species may occur in the field (Moss 1987).

Storage fungi, primarily *Aspergillus* and *Penicillium* species, can grow at 13–18% water content (Lillehoj and Elling 1983). Availability of water for the growth of fungi may be expressed as moisture content, as given above, or may alternatively be expressed as water activity (a_w). The a_w of a product is defined as the ratio of the vapor pressure of the product to that of pure water (Pitt and Hocking 1985). The lower the a_w, the less water that is available to the fungi. All microorganisms including toxigenic fungi have a minimum, optimum, and maximum a_w for growth. The minimum a_w for most species colonizing stored cereal grains is about 0.70 (Lacey 1989). Adequate water for fungal growth in grain may result from inadequate drying before storage, penetration of rain- or windblown snow into storage structures, or moisture migration because of temperature gradients within a bin (Christensen and Kaufmann 1974, Sauer 1978, Lacey 1989). Differences in temperature between different portions of a grain bulk can result in a fairly rapid transfer of water vapor from warm to cool areas. Thus, grain may gain water by absorption from warm moist air until it reaches equilibrium, and condensation may occur if the moist air cools below its dew point (Anderson et al 1943, Lacey 1989). Insect infestations may also increase grain temperature through their metabolic activity driving water to other parts of the grain bulk (Christensen and Kaufmann 1974).

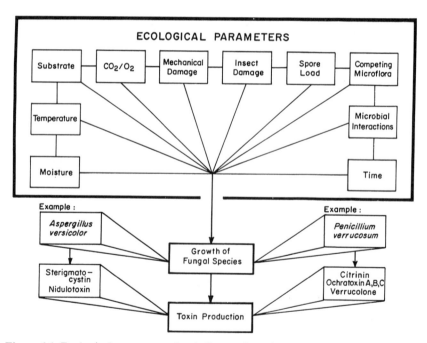

Figure 6.1. Ecological parameters that influence fungal growth and toxin production, with examples of two fungal species and associated mycotoxins that occur on feedstuffs stored in western Canada.

Several studies have investigated the interactive effect of a_w and other ecological parameters such as temperature, substrate, and pH on the growth of selected fungus species (Ayerst 1969; Northolt et al 1976, 1977, 1978, 1979a,b; Magan and Lacey 1984a,b,c).

Northolt and his colleagues investigated the effects of both temperature and a_w on the production of several mycotoxins by several species of fungi (Northolt et al 1976, 1977, 1978, 1979a,b). *P. expansum* Link, *P. patulum* Bainier, and *A. clavatus* Desmaz. were able to grow down to 0.83–0.85 a_w. However, the limits for patulin production were higher: 0.99 a_w for *P. expansum* and *A. clavatus* and 0.95 a_w for *P. patulum* (Northolt et al 1978). The production of penicillic acid by *A. ochraceus* and *P. cyclopium* Westling also followed a similar pattern with limits for fungal growth and   . toxin production of 0.85–0.88 a_w and 0.97–0.99 a_w, respectively (Northolt et al 1979a). In general, these studies demonstrated that the rate of fungal growth and production of the associated mycotoxin may be influenced by temperature as well as substrate (Northolt et al 1976, 1977, 1978, 1979a,b).

Magan and Lacey (1984a) studied the effect of a_w, temperature, and pH on fungal growth. A decrease in substrate pH from 6.5 to 4 resulted an increase in the minimum a_w for germination by 0.2 a_w at optimum temperature and a 0.5 a_w at suboptimal temperature. The effects of a_w, temperature, and gas composition on fungal growth have also been studied (Magan and Lacey 1984b). As a_w and temperature were decreased, *Aspergillus* became more sensitive to O_2. *Penicillium* species, however, were tolerant of low O_2 at lower temperatures (14°C). Growth of some species of fungi was stimulated at CO_2 concentrations of 5–10% and 0.98 a_w but was decreased at the same concentrations of CO_2 and low a_w. Magan and Lacey also investigated the effect of a_w, temperature, and substrate on interactions between fungi (1984c). Species domination varied with a_w, temperature, and substrate.

The interactive effects of a_w, temperature, and fungal interactions on mycotoxin production have also been studied (Cuero et al 1988). The growth of *A. flavus* Link in paired cultures at 25°C was greater at all a_w values (0.98, 0.95, and 0.90) than the other competing species—*Fusarium graminearum* Schwabe, *P. viridicatum, A. niger* Tiegh., and *A. oryzae* (Ahlburg) Cohn. *P. viridicatum* grew better than *A. flavus* at all a_w at 16°C. Aflatoxin production in paired cultures was decreased at high water activity (0.98 a_w) but was enhanced when water activity was low (0.90 a_w). Lacey et al (1992) studied the interactive effects of a_w, temperature, and fungal interactions on fungal growth and mycotoxin formation in barley. Interactions between paired species (*A. flavus, P. verrucosum, Fusarium poae* (Peck) Woolenweber, *Hyphophichia burtonii*) differed depending on the combination of species and appeared to be influenced by growth rate and branching rather than inhibitory metabolites. Mycotoxin production was determined by species, environmental conditions, and period of incubation. Enhancement, partial or complete inhibition, and inhibition after initial enhancement of mycotoxin production were observed. The interactive effects of water activity and other ecological parameters have been reviewed by Lacey and Magan (1991).

These studies concluded that water requirements are influenced by species, temperature, pH, and substrate, and furthermore, that mycotoxin production usually occurs within a more restricted range than fungal growth. The latter observation is illustrated in Figure 6.2, where a higher water activity is necessary for toxin production than that required for growth of selected fungal species. Frisvad and Samson (1991) summarized the minimum water activities for germination and toxin production for several species of fungi. In general, mycotoxin formation may occur if the water content of stored products increases above 13–16%, and maximum production may occur at 20–25% water content (Bullerman et al 1984).

Temperature

Temperature is another important factor affecting mycotoxin formation. As with water, each fungal species has characteristic minimum, optimum, and maximum temperature requirements for growth. Some may have minima close to or below 0°C, whereas others have maxima up to 55–60°C (Lacey 1989). Temperature has considerable influence on water requirements, because the minimum requirements for growth are different at different temperatures and on different substrates. The minimum a_w for growth can be lowest at the optimum temperature and highest near minimum and maximum growth temperatures (Bullerman et al 1984). Northolt et al (1978) demonstrated that patulin production occurred at temperatures ranging

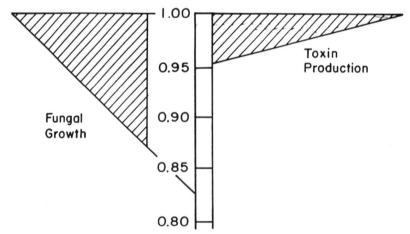

Figure 6.2. Fungal growth and mycotoxin production of selected fungal species as effected by a_w. (Data from Northolt et al 1978, 1979a.)

between 0 and 24°C for *P. expansum* and 4 and 31°C for *P. patulum* at 0.99 a_w; however, *A. clavatus* did not produce patulin below 12°C even at a high a_w. A similar pattern for ochratoxin production by *P. cyclopium, P. viridicatum*, and *A. ochraceus* was also observed (Northolt et al 1979b). Such studies demonstrate that *Penicillium* species can grow and produce secondary metabolites over a broader range of temperatures than *Aspergillus* species. *Penicillium* species are more abundant in temperate climatic regions, e.g., Canada and Scandinavia, whereas *Aspergillus* species tend to predominate in warmer climates, e.g., Yugoslavia and Australia (Krogh 1987).

Substrate

Mycotoxigenic fungi are capable of growing on various substrates. Substrates differ in their ability to support mycotoxin production (Madhyastha et al 1990), affecting both the nature and amount of metabolic toxin produced. Madhyastha et al (1990) showed that peanut and soybean were much better substrates than rapeseed, wheat, or corn for production of ochratoxin by *A. alutaceus,* whereas wheat was a better substrate for production of ochratoxin A by *P. verrucosum* than corn or the oilseed crops. Citrinin was produced only by *P. verrucosum* on wheat and corn but not on oilseeds, and ochratoxin B was only produced by *A. alutaceus* on oilseeds and not on cereal crops. Abramson and co-workers (1980, 1983a, 1985, 1987, 1990a,b) concluded that the risk of mycotoxin development by molds associated with particular crops stored under prairie granary conditions were affected by the species or type of cereal. The relative vulnerability of crops to mycotoxin formation were minimal for oats, low for hard red spring wheat and two-row barley, moderate for six-row barley and corn, and high for amber durum wheat.

Differences in fungal growth and to a greater extent differences in toxin production among different substrates may be attributed to differences in the physical and chemical characteristics of the substrate (Le Bars 1982, Lacey 1989). Physical characteristics of foodstuffs play an important role in toxin formation. These include parameters such as available water (solvent water) at a given water activity, which differs according to species; mechanical resistance to packing, which determines residual air and thus oxygen availability; and thermal conductivity, which influences temperature migration in the commodity (Le Bars 1982). Chemical characteristics such as fat and protein content and trace mineral, amino acid, and fatty acid composition may influence toxin production.

Gupta and Venkitasubramanian (1975) observed increased aflatoxin production in soybeans when zinc was added to the culture. They hypothesized that soybeans contain only a small amount of zinc that is bound to phytic acid. They also observed increased aflatoxin production when soybeans were autoclaved and suggested that this process may breakdown

phytic acid. Subsequent additions of phytic acid resulted in decreased aflatoxin synthesis.

In addition to the presence of essential trace elements, the presence of
certain amino acids may also affect mycotoxin production. Payne and Hagler
(1983) demonstrated that proline could stimulate aflatoxin production by
A. parasiticus Speare and *A. flavus.* Häggblom and Ghosh (1985) demonstrated that proline, as well as glutamic acid, promoted ochratoxin A
production by *A. alutaceus* and *P. verrucosum* in liquid culture. They showed
a positive correlation between the protein concentration of the barley and
the production of ochratoxin A by both *A. alutaceus* and *P. verrucosum*
at 12 and 25°C.

Although some substrates contain compounds that appear to stimulate
toxin production, some foods may contain compounds that are inhibitory
to fungal growth and mycotoxin production. Hitokoto et al (1978, 1980)
demonstrated that cloves, allspice, and cinnamon bark completely inhibited
mold growth of *A. flavus, A. ochraceus*, and *A. versicolor*, whereas mustard
powder, green garlic powder, powdered green tea leaves, and various other
spices including thyme, sage, dill seeds, turmeric, and oregano inhibited
only toxin production.

Oxygen and Carbon Dioxide Levels

Most fungi causing grain deterioration, like other filamentous fungi, have
an absolute requirement for oxygen (Pitt and Hocking 1985). Fungal growth
and toxin production have been observed at various levels of O_2, CO_2,
and N_2. Wilson and Jay (1975) reported that *A. flavus* was able to grow
in atmospheres containing 61.7% CO_2, 8.7% O_2, and 29.7% N_2. Although
little visible mold was apparent under storage at 99.7% N_2 and 0.3% O_2
or at 13.5% CO_2, 0.5% O_2, and 84.4% N_2, a fermentation-like odor was
evident. Peterson et al (1956) reported that *A. glaucus* (L.) Link grew in
wheat aspired with N_2 ($O_2 = 0.2\%$). Thus, it is apparent that O_2 requirements
for fungal growth are very low and may in some cases be overestimated
(Lacey 1989). Landers et al (1967) demonstrated that growth and aflatoxin
production by *A. flavus* occurred at 1% O_2 and 99% N_2 and at 1% O_2,
79% N_2, and 20% CO_2, but was inhibited by 1% O_2, 19% N_2, and 80%
CO_2. Thus they concluded that fungal growth and toxin production are
more sensitive to high CO_2 concentrations than to high N_2 and low O_2
concentrations. Paster et al (1983) reported that ochratoxin A production
by *A. ochraceus* could be completely inhibited at 30% CO_2 regardless of
the O_2 level but that complete inhibition of fungal growth was noted only
when the levels of CO_2 exceeded 80%. When the inhibitory gas concentrations
were terminated, ochratoxin production returned to normal, indicating that
the modified atmosphere must be maintained throughout the storage period,
which negates its practical application. Furthermore, high levels of CO_2
may not be sufficient to inhibit mold growth and toxin production in all

cases, because fungal growth has been reported in beer (Cole et al 1983). As reported by Magan and Lacey (1984c), the sensitivity of fungi to gas composition is influenced by a_w and temperature.

It appears that toxin production is more sensitive than fungal growth to levels of atmospheric gases. Low O_2 concentration ($<1.0\%$) and/or increased concentrations of CO_2 may be effective in preventing mold development and inhibiting mycotoxin production in most cases (Paster and Bullerman 1988).

Microbial Interactions

The presence of other microorganisms, either bacteria or other fungi, may alter fungal growth and mycotoxin production. Bacteria flourish near neutral pH, a pH at which fungi cannot compete unless some other factor, such as low temperature, creates an unfavorable environment for the bacteria (Pitt and Hocking 1985). Marth and his co-workers demonstrated that when *A. parasiticus* was grown in the presence of *Streptococcus lactis* (Wiseman and Marth 1981) and *Lactobacillus casei* (El-Gendy and Marth 1981), aflatoxin production was decreased. This decrease in aflatoxin production in the presence of *S. lactis* was apparent when *S. lactis* was grown first and *A. parasiticus* added three days later or when *A. parasiticus* was grown first and *S. lactis* added later. Simultaneous addition of the two organisms resulted in an increase in aflatoxin production. The rapid growth of bacteria on fresh meat is thought to be responsible for the scarcity of fungi on such a commodity (Northolt and Bullerman 1982).

As with bacteria, competition for available nutrients also occurs between different species of fungi and thus may influence mycotoxin production. Mislivec et al (1988) reported that when either *A. parasiticus* or *A. ochraceus* were grown together with *A. flavus* on the same substrate, there was no apparent effect on aflatoxin production. *Penicillium* species (*P. citrinum* Thom, *P. cyclopium,* and *P. urticae* Bainier), however, substantially lowered aflatoxin production. The inhibition was attributed to the presence of a heat stable metabolite produced by the penicillia. Chelack et al (1991) demonstrated that a reduction in competing microbial flora by irradiation or by chemical treatment resulted in enhanced mycotoxin production. It has been shown that *A. flavus* does not produce aflatoxin in the presence of *A. niger* or *Trichoderma viride* Pers.:Fr. when grown on sterile corn (Wicklow et al 1980). Horn and Wicklow (1983) observed an inhibition of aflatoxin production when *A. niger* was cultured simultaneously with *A. flavus.* This inhibition was attributed to the production of compounds that inhibited aflatoxin formation and lowered the pH of the medium. Similarly, Wicklow et al (1988) demonstrated that when *A. flavus* was grown on wound-infected kernel tissue of commercial hybrid corn, high levels of aflatoxin were produced, even when competing fungi were introduced simultaneously. There was, however, a 12-fold difference in the aflatoxin levels between kernels simultaneously inoculated with *A. flavus* and *Fusarium moniliforme* J. Sheld. and kernels inoculated only with *A. flavus*. It was hypothesized that the competing species reduced the available resource pool or in some way interfered with toxin production. Seitz et al (1982) did not observe the same pattern of inhibition. When stored corn was naturally contaminated with *A. flavus* and *A. niger*, aflatoxin production by *A. flavus* was not prevented. Serafini et al (1991) investigated the effect of ochratoxin A (produced by *A. ochraceus*) on aflatoxin production by *A. parasiticus*. At low concentrations (<0.5 mg/L), the ochratoxin had no effect on fungal growth or aflatoxin production. At high concentrations (>2.5 mg/L), however, aflatoxin production was increased substantially. As ochratoxin A inhibits coupled respiration (Harwig 1974), it has been hypothesized (Serafini et al 1988) that the uncoupling action of ochratoxin may be important for aflatoxin production as a result of the accumulation of acetate units (Shih and Marth 1974). Thus, it appears that the presence of competing fungi and their effects on toxin production are unpredictable. In general, growth and toxin production of storage fungi appear to be considerably reduced in the presence of other competing microorganisms. Wicklow et al (1980) suggested that successful establishment of the fungi and subsequent toxin production may depend on the sequence in which the inoculum reaches the commodity in relation to their colonization by other fungi and the biological properties of the co-invading fungal partner.

Mechanical Damage and Insect Infestation

Mechanical damage resulting from harvesting or from the action of insects disrupts the seed coat and facilitates the penetration of inoculum into the interior of the grain (Le Bars 1982). Wicklow et al (1988) demonstrated that silk inoculation of uninjured ears of corn with *A. flavus* resulted in less than 2% of kernels infected and no detectable aflatoxin. Sterile toothpick-wounding, however, of several kernels on ears that were similarly silk-inoculated promoted *A. flavus* infestation and aflatoxin production on the wounded kernels. Certain kinds of stored-grain insects, such as the grain weevil, in which larvae and pupae develop within the infested kernel, carry numerous spores of storage fungi; a developing infestation is likely to provide both the temperature and water to promote rapid growth of storage fungi (Christensen and Kaufmann 1974, Abramson and Mills 1985). Sinha (1961) found that in farm granaries the water content of grain in hot spots infested by insects and of grain above the hot spots was considerably higher than that of noninfested grain in the same bin. In addition to providing higher water contents, the nitrogenous waste products from insects provide a more favorable carbon-nitrogen ratio and thereby facilitate further microbial infection (Sinha 1973). Griffiths et al (1959) observed that mites developing in moldy grain picked up spores of storage fungi and carried the spores on the outside of their bodies, in their digestive tracts, and in their feces. Therefore, inoculation of clean grain with fungal spores occurred when the mites came in contact with it. It appears that the relationship between storage fungi and insects is unpredictable. Although some species of storage insect spread storage fungi, others are capable of decreasing fungal growth. In addition, some storage fungi attract storage insects, leading to increases in their populations, whereas others are detrimental to them (Dunkel 1988).

Spore Load

When sterilized barley was inoculated with 1.4×10^3 and 1.4×10^5 spores, respectively, from ochratoxin A-producing strains of *A. ochraceus* and *P. viridicatum*, the decrease in the amount of inoculum had no effect on ochratoxin A production by *P. viridicatum* but resulted in the production of considerably less ochratoxin A by *A. ochraceus* (Häggblom 1982). All fungi produced the same amount of biomass irrespective of initial inoculum size. Chelack et al (1991) observed that postirradiation inoculation of barley with *A. alutaceus* resulted in enhancement of ochratoxin A production; they noted greater enhancement if the barley was inoculated with 10^2 conidia/gram than if inoculated with 10^5 conidia/gram. Karunaratne and Bullerman (1990) investigated the relationship between initial spore load of *A. flavus* subsp. *parasiticus* (formerly *A. parasiticus*) on subsequent growth and aflatoxin production. They reported that high amounts of aflatoxin were produced at 28°C when when 10^3 spores were inoculated into rice, while lower and higher spore levels produced comparatively lower amounts of aflatoxin. However, at 35°C the lowest spore level ($<10^1$ spores) resulted in the highest amounts of aflatoxin. Odamtten et al (1987) showed the production of aflatoxin increased three- to 12-fold when the amount of *A. flavus* inoculum was reduced. In both of these studies, vegetative growth was independent of inoculum size. It is possible that aflatoxin production in a medium may be associated with mycelial branching and differentiation (Sharma et al 1980). With smaller populations, more lateral branching and less hyphal fusion during mycelial growth may result in higher yields of toxin (Odamtten et al 1987).

Time

Madhyastha et al (1990) investigated the ability of cereal substrates and oilseeds to support the production of fungal growth and toxin production by *A. alutaceus* and *P. verrucosum* after 7, 15, and 30 days of incubation. They observed little toxin production by the two fungi prior to day 5 of incubation. Considerable growth and toxin production were evident, however, by day 7. It was hypothesized that fungal growth and toxin production might have occurred in a logarithmic manner between days

5 and 7. Times of maximal fungal growth and toxin production differed for a given fungus species depending on the substrate. When incubated with *A. alutaceus*, soybeans produced most ochratoxin A after day 26. Henceforth, toxin production began to decline, suggesting that the toxin might have been metabolized so that additional nutrients could be provided in the contained environment, which was approaching nutrient deficiency. Toxin production continued to increase in all other substrates (wheat, corn, rapeseed, and peanuts) to day 30, at which point a plateau had not been reached.

ENSILED COMMODITIES

The ecological parameters have thus far been reviewed with reference to conventionally stored grain. The conditions in an ensiled commodity differ from those of traditionally stored grain and thus will be considered separately. Silage is the product formed when grass or other material of sufficiently high moisture content (e.g., forage legumes and forage corn, high-moisture grain) and susceptible to aerobic microorganisms, is stored anaerobically. During the ensiling process, the commodity undergoes acid fermentation in which bacteria produce lactic, acetic, and butyric acids from sugar in the raw material. In addition, the term *ensiled* is also used to describe fodder or grain in which fermentation has been restricted by the addition of preservatives, usually organic acids such as propionic acid.

Although there have been several suspected cases of mycotoxicosis due to spoiled silage, it is very difficult to link a fungus and its respective toxin to the illness in question (Mirocha and Christensen 1975). Scientific research in this area is also limited. Because filamentous fungi have an absolute requirement for oxygen (Pitt and Hocking 1985), their presence should be of little significance in the ensiling process. Although fungi can grow at 0.5% oxygen and survive at low pH, their ability to compete with anaerobic bacteria and yeasts is limited. Mirocha and Christensen (1975) packed fermented alfalfa silage into mason jars and inoculated it with many *A. flavus* spores. Neither *A. flavus* nor any other fungus developed, and within 2–3 days the *A. flavus* died. Most of the problems involved with silage occur under improper (aerobic) conditions, such as when cracks in the wall or floor allow the entrance of air and subsequent fungal growth. Many different genera and species of fungi may colonize silage because it is an ideal substrate. In silage samples observed by Mirocha and Christensen (1975), the fungal population consisted of a mixture of numerous genera and species. Occasionally, *Aspergillus* (*A. glaucus* or *A. versicolor*), *Penicillium*, or *Scopulariopsis* species predominated, but never was a single species observed in isolation. Lacey (1971) identified the following species in moist barley grain in an unsealed silo: *Absidia* spp., *A. flavus, A. fumigatus* Fresen., *Penicillium* spp., *Mucor pusillus* Lindt, *Humicola lanuginosa*

(Griffon & Maublanc) Bunce, *Micropolyspora faeni* Cross, Maciver, & Lacey, and *Thermoactinomyces vulgaris* Tsiklinsky. A succession of fungal species in the silo was apparent. In 1985–1986, stored commodities including high-moisture barley and silage were collected from various locations in Manitoba (Frohlich et al 1991). Forty-four percent of the high-moisture barley contained detectable levels of ochratoxin A, and 12% of the silage samples contained detectable levels of ochratoxin upon moisturization and incubation. The presence of toxic fungi in a given sample does not necessitate the presence of toxin. Hacking and Rosser (1981) isolated 26 *Paecilomyces* species from silage, of which 21 were thought to be capable of producing patulin; however, patulin was not detected in any of the silage samples. Cole et al (1977) and Smith and Lynch (1973) isolated *A. fumigatus* from moldy silage samples but were unable to detect the presence of any mycotoxins.

Organic acids such as acetic and propionic acid are often used to preserve high-moisture commodities. The effectiveness of propionic acid has been demonstrated in damp cereal grains (Huitson 1968). A series of experiments was performed by Lord and Lacey and co-workers (Lacey and Lord 1977; Lord and Lacey 1978; Lacey et al 1978; Lord et al 1981a,b) to determine the effectiveness of propionic acid for the preservation of hay. Hay was incompletely preserved as a result of poor distribution of acid in the bale. This uneven distribution permitted fungi tolerant to propionic acid to grow. *Paecilomyces* and species of the *A. glaucus* group could not only tolerate propionic acid, but could also metabolize it. Al-Hilli and Smith (1979) showed that sublethal concentrations of propionic acid may stimulate aflatoxin production in cultures of *A. flavus*. Nørregaard et al (1984) also showed that several organic acids are capable of stimulating penicillin production in cultures of *P. chrysogenum*. More recent evidence indicates that the distribution of preservatives in the bale can be improved; however, the widespread use of propionic acid for hay preservation is limited as a result of the cost (K. Wittenberg, *personal communication*).

The above research suggests that under the correct ensiling conditions, the chance of silage being contaminated with mycotoxins is slight. Nevertheless, the question of the risks imposed by mycotoxins in ensiled commodities remains unresolved. Most of the investigations were conducted with fungal isolates from silage under artificial conditions. Thus, the potential for mycotoxin production and persistence in silage and hay under natural conditions requires further investigation.

Multivariate Analysis of Ecological Parameters

Laboratory research on the aforementioned ecological parameters has contributed significantly to the understanding of both fungal infection and toxin production. Under natural conditions, however, these parameters interact concurrently, intermittently, or sequentially. Thus, it is logical to study stored grain ecosystems using a holistic approach with both descriptive and multivariate analysis (Sinha 1979). One method of assessing the interaction of the variables responsible for growth and toxin production is to study the stored product under natural conditions, that is, in a grain bin as illustrated in Figure 6.3. Several such studies have been carried out (Seitz et al 1982; Abramson et al 1980, 1983a, 1985, 1987, 1990a,b). One tool that aids the assessment of quality changes that occur in such granary studies is principal component analysis (PCA) (Sinha et al 1986). A PCA statistical procedure is advantageous in assessing interrelations among variables, because no predetermined assumptions about the underlying structure of the variables are required. Usually, an optimum linear combination of variables is produced, which better accounts for the variance in the data than do other statistical treatments. An example of such an analysis is demonstrated in the recent work by Abramson et al (1990a). In this 60-week study, HY-320 wheat, a cultivar of the new Canada Prairie Spring class, was kept at 15 and 19% initial moisture contents in simulated storage in a Manitoba farm granary. In early storage, (0–28 weeks) the five factors describing the variability of the system were the concentration of ochratoxin A, the presence of *A. versicolor* and *Alternaria* (mainly *Alternaria alternata* (Fr.:Fr.) Keissler), total fungal propagule count, and germination. Thus, ochratoxin A production appeared to be associated

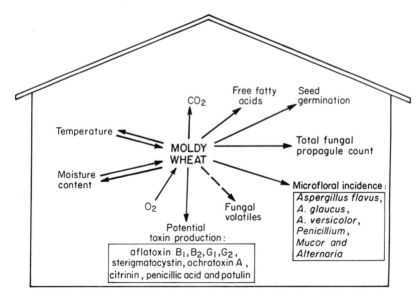

Figure 6.3. The holistic approach used to assess the interaction of factors responsible for fungal growth and toxin production in a grain bin. (Data from Abramson 1991.)

with decreasing seed germination and decreasing *Alternaria* population and a corresponding increase in fungal propagules and *A. versicolor*. In the later data set (32–60 weeks), ochratoxin A reproduction appeared to increase in a multidimensional niche characterized by increased water content, CO_2, fungal propagule count, and *Penicillium* and with decreased *Alternaria* and seed germination. Other parameters that did not affect the variability of the system were temperature, O_2, fat acidity, and the presence of *A. flavus* and *A. glaucus*. This analysis indicates that ochratoxin A is produced in a niche containing moist decaying seed with actively growing populations of storage fungi, particularly *Penicillium*. This type of study also indicates that medium-protein wheat initially stored at 19% moisture is a suitable substrate for ochratoxin A formation.

Conclusions

It is apparent from the literature that many factors are involved in fungal growth and toxin production. Studies performed in both the laboratory and in actual storage environments have greatly contributed to our knowledge in this area. Further studies of this nature are required to more fully understand the overall processes of grain spoilage, fungal infection, and mycotoxin production. Such an understanding will aid in early identification of such conditions and will provide a sound basis for the development of more economically feasible and effective control procedures. One possible avenue to pursue in this regard is the identification of fungal volatiles, which may be characteristic of specific mycotoxin-producing fungi.

Literature Cited

Abramson, D. 1991. Development of molds, mycotoxins and odors in moist cereals during storage. Pages 119-147 in: Cereal Grain—Mycotoxins, Fungi and Quality in Drying and Storage. J. Chelkowski, ed. Elsevier, Amsterdam.

Abramson, D., and Mills, J. T. 1985. Mycotoxin production during storage. Pages 69-79 in: P. M. Scott, H. L. Trenholm, and M. D. Sutton, eds. Mycotoxins: A Canadian Perspective. NRCC Publications, Ottawa.

Abramson, D., Sinha, R. N., and Mills, J. T. 1980. Mycotoxin and odor formation in moist cereal grain during granary storage. Cereal Chem. 57:346-351.

Abramson, D., Sinha, R. N., and Mills, J. T. 1983a. Mycotoxin and odor formation in barley stored at 16 and 20% moisture in Manitoba. Cereal Chem. 60:350-355.

Abramson, D., Mills, J. T., and Boycott, B. R. 1983b. Mycotoxins and mycoflora in animal feedstuffs in western Canada. Can. J. Comp. Med. 47:23-26.

Abramson, D., Mills, J. T., and Sinha, R. N. 1985. Mycotoxin formation and quality changes in granary-stored corn at 16 and 21% moisture content. Sci. Aliment. 5:653-663.

Abramson, D., Mills, J. T., and Sinha, R. N. 1987. Mycotoxin formation in moist 2-row and 6-row barley during granary storage. Mycopathologia 97:179-185.

Abramson, D., Sinha, R. N., and Mills, J. T. 1990a. Mycotoxin formation in HY-320 wheat during granary storage at 15 and 19% moisture content. Mycopathologia 111:181-189.

Abramson, D., Mills, J. T., and Sinha, R. N. 1990b. Mycotoxin production in amber durum wheat stored at 15 and 19% moisture content. Food Addit. Contam. 7:617-627.

Al-Hilli, A. L., and Smith, J. E. 1979. Influence of propionic acid on growth and aflatoxin production by *Aspergillus flavus*. FEMS Microbiol. Lett. 6:367-370.

Anderson, J. A., Babbitt, J. D., and Meredith, W. O. S. 1943. The effect of temperature differential on moisture content of stored wheat. Can. J. Res. Sect. C Bot. Sci. 21:297-306.

Ayerst, G. 1969. The effects of moisture and temperature on growth and spore germination in some fungi. J. Stored Prod. Res. 5:127-141.

Börjesson, T., Stöllman, U., and Schnürer, J. 1990. Volatile metabolites and other indicators of *Penicillium aurantiogriseum* growth on different substrates. Appl. Environ. Microbiol. 56:3705-3710.

Bridge, P. D., Kozakiewicz, Z., and Paterson, R. R. M. 1992. PENIMAT: A computerized identification scheme for terverticillate *Penicillium* species. Pages 55-56 in: Extended Abstracts. International Symposium on Stored-Grain Ecosystems. D. S. Jayas, N. D. G. White, W. E. Muir, and R. N. Sinha, eds. Department of Agricultural Engineering, University of Manitoba, Winnipeg, Canada.

Bullerman, L. B., Schroeder, L. L., and Park, K.-Y. 1984. Formation and control of mycotoxins in food. J. Food Prot. 47:637-646.

CAST. 1989. Mycotoxins. Economic and Health Risks. Task Force Rep. 116. Council for Agricultural Science and Technology, Ames, IA.

Chelack, W. S., Borsa, J., Marquardt, R. R., and Frohlich, A. A. 1991. Role of competitive microbial flora in the radiation-induced enhancement of ochratoxin production by *Aspergillus alutaceus* var. *alutaceus* NRRL 3174. Appl. Environ. Microbiol. 57:2492-2496.

Christensen, C. M. 1965. Fungi in cereal grains and their products. Pages 9-14 in: Mycotoxins in Foodstuffs. G. N. Wogan, ed. MIT Press, Cambridge, MA.

Christensen, G. M., and Kaufmann, H. H. 1969. Grain Storage: The Role of Fungi in Quality Loss. University of Minnesota Press, Minneapolis.

Christensen, C. M., and Kaufmann, H. H. 1974. Microflora. Pages 158-192 in: Storage of Cereal Grains and Their Products. American Association of Cereal Chemists, St. Paul, MN.

Cole, R. J., Dorner, J. W., Cox, R. H., and Raymond, L. W. 1983. Two classes of alkaloid mycotoxins produced by *Penicillium crustosum* Thom isolated from contaminated beer. J. Agric. Food Chem. 31:655-657.

Cole, R. J., Kirksey, J. W., Dorner, J. W., Wilson, D. M., Johnson, Jr., J. C., Johnson, A. N., Bedell, D. M., Springer, J. P., Chexal, K. K., Clardy, J. C., and Cox, R. H. 1977. Mycotoxins produced by *Aspergillus fumigatus* species isolated from molded silage. J. Agric. Food Chem. 25:826-830.

Cruikshank, R. H., and Pitt, J. I. 1987. Identification of species in *Penicillium* subgenus *Penicillium* by enzyme electrophoresis. Mycologia 79:614-620.

Cuero, R., Smith, J. E., and Lacey, J. 1988. Mycotoxin formation by *Aspergillus flavus* and *Fusarium graminearum* in irradiated maize grains in the presence of other fungi. J. Food Prot. 51:452-456.

Dunkel, F. V. 1988. The relationship of insects to the deterioration of stored grain by fungi. Int. J. Food. Microbiol. 7:227-244.

El-Banna, A. A., Pitt, J. I., and Leistner, L. 1987. Production of mycotoxins by *Penicillium* species. Syst. Appl. Microbiol. 10:42-46.

El-Gendy, S. M., and Marth, E. H. 1981. Growth and aflatoxin production by *Aspergillus parasiticus* in the presence of *Lactobacillus casei*. J. Food Prot. 44:211-212.

Filtenborg, O., and Frisvad, J. C. 1980. A simple screening method for toxigenic moulds in pure cultures. Lebensm. Wiss. Technol. 13:128-130.

Filtenborg, O., Frisvad, J. C., and Svendsen, J. A. 1983. Simple screening method for molds producing intracellular mycotoxins in pure cultures. Appl. Environ. Microbiol. 45:581-585.

Flannigan, B. 1978. Primary contamination of barley and wheat grain by storage fungi. Trans. Br. Mycol. Soc. 71:37-42.

Frisvad, J. C. 1989. The connection between the penicillia and aspergilli and mycotoxins with special emphasis on misidentified isolates. Arch. Environ. Contam. Toxicol. 18:452-467.

Frisvad, J. C., and Filtenborg, O. 1983. Classification of terverticillate penicillia based on profiles of mycotoxins and other secondary metabolites. Appl. Environ. Microbiol. 46:1301-1310.

Frisvad, J. C., and Filtenborg, O. 1989. Terverticillate penicillia: chemotaxonomy and mycotoxin production. Mycologia 81:837-861.

Frisvad, J. C., and Samson, R. A. 1991. Filamentous fungi in foods and feeds: Ecology, spoilage, and mycotoxin production. Pages 31-68 in: D. K. Arora, K. G. Mukerii, and E. H. Marth, eds. Handbook of Applied Mycology. Vol. 3, Foods and Feeds. Marcel Dekker, New York.

Frohlich, A. A., Marquardt, R. R., and Ominski, K. H. 1991. Ochratoxin A as a contaminant in the human food chain: A Canadian perspective. Pages 139-143 in: M. Castegnaro, R. Plestina, G. Dirheimer, I. Chernozemsky, and H. Bartsch, eds. Mycotoxins: Nephropathy and Urinary Tract Tumors. International Agency for Research on Cancer, Lyons, France.

Griffiths, D. A., Hodson, A. C., and Christensen, G. M. 1959. Grain storage fungi associated with mites. J. Econ. Entomol. 52:514-518.

Gupta, S. K., and Venkitasubramanian, T. A. 1975. Production of aflatoxin on soybeans. Appl. Microbiol. 29:834-836.

Hacking, A., and Rosser, W. R. 1981. Patulin production by *Paecilomyces* species isolated from silage in the United Kingdom. J. Sci. Food Agric. 32:620-623.

Häggblom, P. E. 1982. Production of ochratoxin A in barley by *Aspergillus ochraceus* and *Penicillium viridicatum*: Effect of fungal growth, time, temperature, and inoculum size. Appl. Environ. Microbiol. 43:1205-1207.

Häggblom, P. E., and Ghosh, J. 1985. Postharvest production of ochratoxin A by *Aspergillus ochraceus* and *Penicillium viridicatum* in barley with different protein levels. Appl. Environ. Microbiol. 49:787-790.

Harwig, J. 1974. Ochratoxin A and related metabolites. Pages 345-367 in: I. F. H. Purchase, ed. Mycotoxins. Elsevier Scientific Publishing, Amsterdam.

Hesseltine, C. W. 1976. Conditions leading to mycotoxin contamination of foods and feeds. Pages 1-22 in: Mycotoxins and Other Fungal Related Food Problems. J. V. Rodericks, ed. American Chemical Society, Washington, DC.

Hicks, R. G., and Newell, S. Y. 1983. An improved gas chromatographic method for measuring glucosamine and muramic acid concentrations. Anal. Biochem. 128:438-445.

Hitokoto, H., Morozumi, S., Wauke, T., Sakai, S., and Ueno, I. 1978. Inhibitory effects of condiments and herbal drugs on the growth and toxin production of toxigenic fungi. Mycopathologia 66:161-167.

Hitokoto, H., Morozumi, S., Wauke, T., Sakai, S., and Kurata, H. 1980. Inhibitory effects of spices on growth and toxin production of toxigenic fungi. Appl. Environ. Microbiol. 39:818-822.

Horn, B. W., and Wicklow, D. T. 1983. Factors influencing the inhibition of aflatoxin production in corn by *Aspergillus niger*. Can. J. Microbiol. 29:1087-1091.

Huitson, J. J. 1968. Cereals preservation with propionic acid. Process Biochem. 3:31-32.

Jong, S. C., and Edwards, M. J., eds. 1991. American Type Culture Collection Catalogue of Filamentous Fungi. 18th ed. American Type Culture Collection, Rockville, MD.

Kaminski, E., Stawicki, S., and Wasowicz, E. 1974. Volatile flavor compounds produced by molds of *Aspergillus, Penicillium* and Fungi Imperfecti. Appl. Microbiol. 27:1001-1004.

Karunaratne, A. E., and Bullerman, L. B. 1990. Interactive effects of spore load and temperature on aflatoxin production. J. Food Prot. 53:227-229.

Kozakiewicz, Z. 1989. *Aspergillus* species on stored products. Mycol. Pap. 161:1-188.

Krogh, P. 1987. Ochratoxins in Food. Pages 97-121 in: Mycotoxins in Food. P. Krogh, ed. Academic Press, London.

Lacey, J. 1971. The microbiology of moist barley stored in unsealed silos. Ann. Appl. Biol. 69:187-212.

Lacey, J. 1989. Prevention of mold growth and mycotoxin production through control of environmental factors. Pages 161-168 in: Mycotoxins and Phycotoxins '88. S. Natori, K. Hashimoto, and Y. Ueno, eds. Elsevier, Amsterdam.

Lacey, J., and Lord, K. A. 1977. Methods for testing chemical additives to prevent moulding of hay. Ann. Appl. Biol. 87:327-335.

Lacey, J., and Magan, N. 1991. Fungi in cereal grains: Their ocurrence and water and temperature relations. Pages 77-118 in: Cereal Grain—Mycotoxins, Fungi and Quality in Drying and Storage. J. Chelkowski, ed. Elsevier, Amsterdam.

Lacey, J., Lord, K. A., King, H. G. C., and Manlove, R. 1978. Preservation of baled hay with propionic and formic acids and a proprietary additive. Ann. Appl. Biol. 88:65-73.

Lacey, J., Ramakrishna, N., and Smith, J. 1992. Interactions between water activity, temperature, and different species on colonization of grain and mycotoxin formation. Pages 37-38 in: Extended Abstracts. International Symposium on Stored-Grain Ecosystems. D. S. Jayas, N. D. G. White, W. E. Muir, and R. N. Sinha, eds. Department of Agricultural Engineering, University of Manitoba, Winnipeg, Canada.

Landers, K. E., Davis, N. D., and Diener, U. L. 1967. Influence of atmospheric gases on aflatoxin production by *Aspergillus flavus* in peanuts. Phytopathology 57:1086-1090.

Le Bars, J. 1982. Toxigenesis as a function of the ecological conditions of the grain/microorganisms systems. Pages 347-366 in: Preservation and Storage of Grains. Seeds and Their By-Products. J. L. Multon, ed. Lavoisier Publishing, New York.

Leistner, L. 1984. Toxigenic penicillia occurring in food and feeds: a Review. Food Technol. Aust. 36:404-409,413.

Leistner, L. 1990. Mould-fermented foods: Recent developments. Food Biotechnol. 4:433-441.

Lillehoj, E. B., and Elling, F. 1983. Environmental conditions that facilitate ochratoxin contamination of agricultural commodities. Acta Agric. Scand. 33:113-128.

Lord, K. A., and Lacey, J. 1978. Chemicals to prevent the moulding of hay and other crops. J. Sci. Food Agric. 29:574-575.

Lord, K. A., Cayley, G. R., and Lacey, J. 1981a. Laboratory application of preservatives to hay and the effects of irregular distribution on mould development. Anim. Feed Sci. Technol. 6:73-82.

Lord, K. A., Lacey, J., Cayley, G. R., and Manlove, R. 1981b. Fatty acids as substrates and inhibitors of fungi from propionic acid treated hay. Trans. Br. Mycol. Soc. 77:41-45.

Madhyastha, S. M., Marquardt, R. R., Frohlich, A. A., Platford, G., and Abramson, D. 1990. Effects of different cereal and oilseed substrates on the growth and production of toxins by *Aspergillus alutaceus* and *Penicillium verrucosum*. J. Agric. Food Chem. 38:1506-1510.

Magan, N., and Lacey, J. 1984a. The effect of temperature and pH on the water relations of field and storage fungi. Trans. Br. Mycol. Soc. 82:71-81

Magan, N., and Lacey, J. 1984b. Effects of gas composition and water activity on growth of field and storage fungi and their interactions. Trans. Br. Mycol. Soc. 82:305-314.

Magan, N., and Lacey, J. 1984c. Effect of water activity, temperature and substrate on interactions between field and storage fungi. Trans. Br. Mycol. Soc. 82:83-93.

Matcham, S. E., Jordan, B. R., and Wood, D. A. 1985. Estimation of fungal biomass in a solid substrate by three independent methods. Appl. Microbiol. Biotechnol. 21:108-112.

Mills, J. T. 1989. Ecology of toxigenic fungi associated with grains in Manitoba, Canada. Pages 13-20 in: S. Natori, K. Hashimoto, and Y. Ueno, eds. Mycotoxins and Phycotoxins '88. Elsevier, Amsterdam.

Mills, J. T., and Abramson, D. 1982. Ochratoxigenic potential of *Penicillium* spp. isolated from stored rapeseed and cereals in western Canada. Can. J. Plant Pathol. 4:37-41

Mills, J. T., and Abramson, D. 1986. Production of sterigmatosystin by isolates of *Aspergillus versicolor* from western Canadian stored barley and rapeseed/canola. Can. J. Plant Pathol. 8:151-153.

Mills, J. T., Abramson, D., Frohlich, A. A., and Marquardt, R. R. 1989. Citrinin and ochratoxin A produced by *Penicillium* spp. from stored durum wheat. Can. J. Plant Pathol. 11:357-360.

Mills, J. T., Seifert, K. A., Frisvad, J. C., Abramson, D., and Louis-Seize, G. 1992. Relative importance of *Penicillium* species producing nephrotoxins in Canadian grain products. Pages 29-31 in: Extended Abstracts. International Symposium on Stored-Grain Ecosystems. D. S. Jayas, N. D. G. White, W. E. Muir, and R. N. Sinha, eds. Department of Agricultural Engineering, University of Manitoba, Winnipeg, Canada.

Mirocha, C. J., and Christensen, C. M. 1975. Mycotoxins in stored grain. Pages 253-274 in: Technological Papers Presented at the 2nd International Silage Conference. National Silo Association, Cedar Falls, IA.

Mislivec, P. B., Trucksess, M. W., and Stoloff, L. 1988. Effect of other toxigenic mold species on aflatoxin production by *Aspergillus flavus* in sterile broth shake culture. J. Food Prot. 51:449-451.

Moss, M. O. 1987. Food mycology. Pages 3-34 in: Mycotoxins in Food. P. Krogh, ed. Academic Press, London.

Newell, S. Y. 1992. Estimating fungal biomass and productivity in decomposing litter. Pages 521-526 in: The Fungal Community, Its Organization and Role in the Ecosystem. 2nd ed. G. C. Carroll and D. T. Wicklow, eds. Marcel Dekker, New York.

Nørregaard, P., Holm, H., and Emborg, C. 1984. Organic acids and penicillin production. Appl. Microbiol. Biotechnol. 20:221-224.

Northolt, M. D., and Bullerman, L. B. 1982. Prevention of mold growth and toxin production through control of environmental conditions. J. Food Prot. 45:519-526.

Northolt, M. D., Verhulsdonk, C. A. H., Soentoro, P. S. S., and Paulsch, W. E. 1976. Effect of water activity and temperature on aflatoxin production by *Aspergillus parasiticus*. J. Milk Food Technol. 39:170-174.

Northolt, M. D., van Egmond, H. P., and Paulsch, W. E. 1977. Differences between *Aspergillus flavus* strains in growth and aflatoxin B₁ production in relation to water activity and temperature. J. Food Prot. 40:778-781.

Northolt, M. D., van Egmond, H. P., and Paulsch, W. E. 1978. Patulin production by some fungal species in relation to water activity and temperature. J. Food Prot. 41:885-890.

Northolt, M. D., van Egmond, H. P., and Paulsch, W. E. 1979a. Penicillic acid production by some fungal species in relation to water activity and temperature. J. Food Prot. 42:476-484.

Northolt, M. D., van Egmond, H. P., and Paulsch, W. E. 1979b. Ochratoxin A production by some fungal species in relation to water activity and temperature. J. Food Prot. 42:485-490

Nout, M. J. R., Bonants-van Laarhoven, T. M. G., de Jongh, P., and de Koster, P. G. 1987. Ergosterol content of *Rhizopus oligosporus* NRRL 5905 grown in liquid and solid substrates. Appl. Microbiol. Biotechnol. 26:456-461.

Odamtten, G. T., Appiah, V., and Langerak, D. I. 1987. Influence of inoculum size of *Aspergillus flavus* Link on the production of aflatoxin B₁ in maize medium before and after exposure to combination treatment of heat and gamma radiation. Int. J. Food Microbiol. 4:119-127.

Paster, N., and Bullerman, L. B. 1988. Mould spoilage and mycotoxin formation in grains as controlled by physical means. Int. J. Food Microbiol. 7:257-265.

Paster, N., Lisker, N., and Chet, I. 1983. Ochratoxin A production by *Aspergillus ochraceus* Wilhelm, grown under controlled atmospheres. Appl. Environ. Microbiol. 45:1136-1139.

Payne, G. A., and Hagler, Jr., W. M. 1983. Effect of specific amino acids on *Aspergillus flavus* in defined media. Appl. Environ. Microbiol. 46:805-812.

Peterson, A., Schlegel, V., Hummel, B., Cuendet, L. S., Geddes, W. F., and Christensen, C. M. 1956. Grain storage studies. XXII. Influence of oxygen and carbon dioxide concentrations on mold growth and grain deterioration. Cereal Chem. 33:53-66.

Pitt, J. I. 1987. *Penicillium viridicatum, Penicillium verrucosum,* and production of ochratoxin A. Appl. Environ. Microbiol. 53:266-269.

Pitt, J. I., and Hocking, A. D. 1985. Fungi and Food Spoilage. Academic Press, Sydney, Australia.

Pitt, J. I., and Samson, R. A. 1990. Approaches to *Penicillium* and *Aspergillus* Systematics. Stud. Mycol. 32:77-90.

Pohland, A. E., and Wood, G. E. 1987. Occurrence of mycotoxins in food. Pages 35-64 in: Mycotoxins in Food. P. Krogh, ed. Academic Press, London.

Pusztai, A. 1964. Hexosamines in the seeds of higher plants (spermatophytes). Nature 201:1328-1329.

Richard-Mollard, D., Cahagnier, B., Poisson, J., Drapron, R., and Desserme, C. 1976. Comparative evolution of volatile constituents and microflora in maize stored under different conditions of temperature and humidity. (In French.) Ann. Technol. Agric. 25:29-44

Ride, J. P., and Drysdale, R. B. 1972. A rapid method for the chemical estimation of filamentous fungi in plant tissue. Physiol. Plant Pathol. 2:7-15.

Roberts, C. A., Marquardt, R. R., Frohlich, A. A., McGraw, R. L., Rotter, R. G., and Henning, J. C. 1991. Chemical and spectral quantification of mold in contaminated barley. Cereal Chem. 68:272-275.

Rotter, R. G., Frohlich, A. A., Marquardt, R. R., and Mills, P. A. 1989. Estimation of fungal contamination of cereal grains as determined by measuring glucosamine concentration. Can. J. Anim. Sci. 69:235-245.

Salunkhe, D. K., Chavan, J. K., and Kadam, S. S. 1985. Postharvest Biotechnology of Cereals. CRC Press, Boca Raton, FL.

Samson, R. A., Frisvad, J. C., and Arora, D. K. 1991. Taxonomy of filamentous fungi in foods and feeds. Pages 1-29 in: Handbook of Applied Mycology. Vol. 3, Foods and Feeds. D. K. Arora, K. G. Mukerji, and E. H. Marth, eds. Marcel Dekker, New York.

Sauer, D. B. 1978. Contamination by mycotoxins: When it occurs and how to prevent it. Pages 147-158 in: Mycotoxic Fungi, Mycotoxins, Mycotoxicoses of Man and Plants–An Encyclopedic Handbook. Vol. 3, Mycotoxin Control and Regulatory Practices. T. D. Wyllie and L. D. Morehouse, eds. Marcel Dekker, New York.

Scott, P. M., van Walbeek, W., Kennedy, B., and Anyeti, D. 1972. Mycotoxins (ochratoxin A, citrinin, and sterigmatocystin) and toxigenic fungi in grains and other agricultural products. J. Agric. Food Chem. 20:1103-1109.

Seitz, L. M., Sauer, D. B., Burroughs, R., Mohr, M. E., and Hubbard, J. D. 1979. Ergosterol as a measure of fungal growth. Phytopathology 69:1202-1203.

Seitz, L. M., Sauer, D. B., Mohr, H. E., and Aldis, D. F. 1982. Fungal growth and dry matter loss during bin storage of high-moisture corn. Cereal Chem. 59:9-14.

Serafini, M., Foddai, S., Pieretti, S., Tomassini, L., and Nicoletti, M. 1991. Effect of ochratoxin A on aflatoxin production by *Aspergillus parasiticus*. Can. J. Bot. 69:16-17

Sharma, P. D., Fisher, P. J., and Webster, J. 1977. Critique of the chitin assay technique for estimation of fungal biomass. Trans. Br. Mycol. Soc. 69:479-483.

Sharma, A., Behere, A. G., Padwal-Desai, S. R., and Nadkarni, G. B. 1980. Influence of inoculum size on *Aspergillus parasiticus* spores on aflatoxin production. Appl. Environ. Microbiol. 40:989-993.

Shih, C. N., and Marth, E. H. 1974. Aflatoxin formation, lipid synthesis and glucose metabolism by *Aspergillus parasiticus* during incubation with and without agitation. Biochim. Biophys. Acta 338:286-296.

Sinha, R. N. 1961. Insects and mites associated with hot spots in farm stored grain. Can. Entomol. 93:609-621.

Sinha, R. N. 1973. Ecology of storage. Ann. Technol. Agric. 22:351-369.

Sinha, R. N. 1979. Ecology of microflora in stored grain. Ann. Technol. Agric. 28:191-209.

Sinha, R. N. 1992. The fungal community in the stored grain ecosystem. Pages 797-815 in: The Fungal Community: Its Organization and Role in the Ecosystem. Marcel Dekker, New York.

Sinha, R. N., Wallace, H. A. H., and Chebib, F. S. 1969. Principal-component analysis of interrelations among fungi, mites, and insects in grain bulk ecosystems. Ecology 50:536-547.

Sinha, R. N., Abramson, D., and Mills, J. T. 1986. Interrelations among ecological variables in stored cereals and associations with mycotoxin production in the climatic zones of western Canada. J. Food Prot. 49:608-614.

Sinha, R. N., Tuma, D., Abramson, D., and Muir, W. E. 1988. Fungal volatiles associated with moldy grain in ventilated and non-ventilated bin-stored wheat. Mycopathologia 101:53-60

Smith, D. F., and Lynch, G. P. 1973. *Aspergillus fumigatus* in samples of moldy silage. J. Dairy Sci. 56:828-829.

Tothill, I. E., Harris, D., and Magan, N. 1992. The relationship between fungal growth and ergosterol content of wheat grain. Mycol. Res. 96:965-970.

Tuma, D., Sinha, R. N., Muir, W. E., and Abramson, D. 1990. Odor volatiles associated with mite-infested bin-stored wheat. J. Chem. Ecol. 16:713-724.

White, N. D. G., Sinha, R. N., and Muir, W. E. 1982a. Intergranular carbon dioxide as an indicator of biological activity associated with the spoilage of stored wheat. Can. Agric. Eng. 24:35-42.

White, N. D. G., Sinha, R. N., and Muir, W. E. 1982b. Intergranular carbon dioxide as an indicator of deterioration in rapeseed. Can. Agric. Eng. 24:43-49.

Wicklow, D. T., Hesseltine, C. W., Shotwell, O. L., and Adams, G. L. 1980. Interference competition and aflatoxin levels in corn. Phytopathology 70:761-764.

Wicklow, D. T., Horn, B. T., Shotwell, O. L., Hesseltine, C. W., and Caldwell, R. W. 1988. Fungal interference with *Aspergillus flavus* infection and aflatoxin contamination of maize grown in a controlled environment. Phytopathology 78:68-74.

Wilson, D. M., and Jay, E. 1975. Influence of modified atmosphere storage on aflatoxin production in high moisture corn. Appl. Microbiol. 29:224-228.

Wiseman, D. W., and Marth, E. H. 1981. Growth and aflatoxin production by *Aspergillus parasiticus* when in the presence of *Streptococcus lactis*. Mycopathologia 73:49-56.

Part Three
Alternaria

Chapter 7

Alternaria Toxins

A. Visconti and A. Sibilia

Molds of the genus *Alternaria* are widely distributed in many crops as well as in soil, household dust, and decaying organic material. They include both plant pathogenic and saprophytic species that may cause extensive spoilage of crops in the field and during transport and storage. *Alternaria* species require a relatively high moisture content (28–34% moisture) for growth. In grains or seeds, *Alternaria* infection usually occurs before harvest when the seed moisture content is still high. The infection becomes extensive enough to produce secondary metabolites only when drying in the field is delayed considerably by rainfall and high humidity. Fruits and vegetables are subject to *Alternaria* infection both in the field and postharvest. Fruit imperfection, overripening, cold stress, and surface physical damages are factors promoting fungal infection. Since *Alternaria* can grow at low temperatures, they are often involved in the spoilage of refrigerated products. *Alternaria* species produce about 70 secondary metabolites belonging to several classes of chemicals, including dibenzopyrones, perylenequinones, tetramic acids, lactones, anthraquinones, cyclic peptides, and others (Montemurro and Visconti 1992). Various biological activities—phytotoxic, cytotoxic, insecticidal, antiviral, antifungal, antibacterial, mutagenic, teratogenic, fetotoxic, and others—have been shown for different *Alternaria* metabolites (Pero et al 1973, Harvan and Pero 1976, King and Schade 1984, Chelkowski and Visconti 1992). Many *Alternaria* pathotypes are host-specific plant pathogens and produce host-specific toxins that play an important role in plant pathogenesis (Nishimura and Kohomoto 1983, Graniti et al 1989, Chelkowski and Visconti 1992).

In this chapter, the physical and chemical properties, analytical methods, toxicity, and natural occurrence of some major *Alternaria* mycotoxins are reviewed. These toxins have been considered on the basis of their toxicity, their possible involvement in food contamination, and their potential hazard to humans or animals. They include alternariols and altenuenes (alternariol, alternariol methyl ether, altenuene, isoaltenuene); altertoxins (altertoxin I, altertoxin II, altertoxin III, stemphyltoxin III); tenuazonic acid (TeA); and *Alternaria alternata* f. sp. *lycopersici* toxins (AAL toxins). Their chemical structures are presented in Figures 7.1–7.3.

Physical and Chemical Properties

ALTERNARIOLS AND ALTENUENES

Alternariol (AOH, molecular weight [MW] = 258.0528) and alternariol monomethyl ether (AME, MW = 272.0684) crystallize from ethanol as colorless needles: melting point (mp) (AOH) 350°C (decomposes), mp (AME) 267°C (decomposes). They sublime without decomposition in a high vacuum at 250°C (AOH) and 180–200°C (AME), and both show optical activity: $[\alpha]_D^{20}$ 0° (c = 0.8 in ethanol). They give blue fluorescence under ultraviolet (UV) light and a purple color after reaction with ethanolic iron(III) chloride (Raistrick et al 1953, Freeman 1965, Pero et al 1971a, Pero and Harvan 1973). They show the same principal peaks in the infrared (IR) spectrum: ν_{max} at 3,460, 3,100, 1,672, 1,590, 1,430, 1,365, 1,200, and 850 cm^{-1} (Pero et al 1971a). UV data, λ_{max}, AOH (EtOH): 218, 258, 302, and 330 nm (Freeman 1965); AME (MeOH): 230, 257, 290, 301, 335–342 nm (broad) (Seitz et al 1975b). They show the parent ion in the mass spectrum at m/z 258 and 272, respectively. Other fragments are present

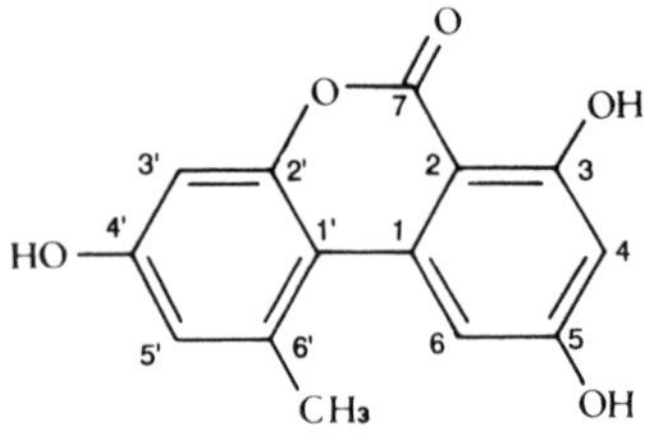

ALTERNARIOL

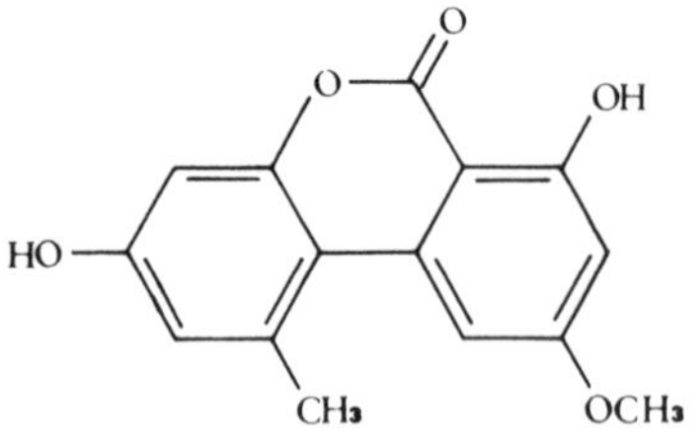

ALTERNARIOL METHYLETHER

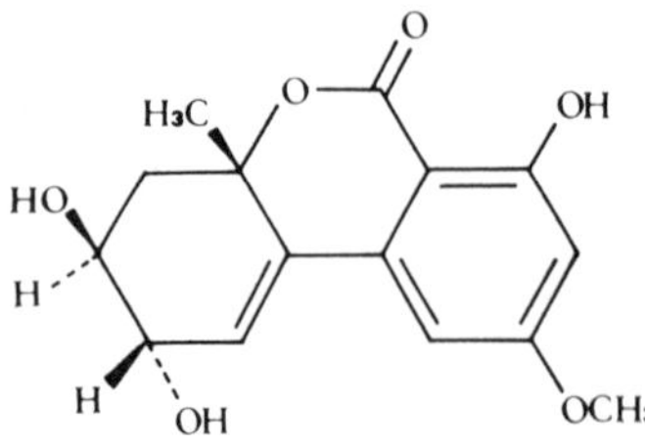

ALTENUENE

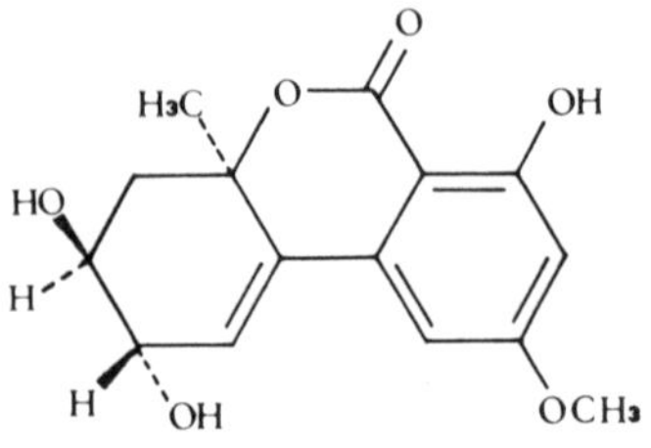

ISOALTENUENE

Figure 7.1. Alternariols and altenuenes.

at m/z 230, 213, 187, 115, 69, and 51 for AOH, and at 243, 215, 201, 185, 172, 155, 145, 139, 136, 127, 115, and 102 for AME (Seitz et al 1975b, Dusold et al 1978). The ^{1}H NMR and ^{13}C NMR show the following chemical shifts. ^{1}H NMR, AOH (CDCl$_3$/DMSO): H4, 6.30; H6, 7.14; H3', 7.65; H5', 6.58; H8', 2.66. ^{1}H NMR, AME (CDCl$_3$/DMSO): H4, 6.29; H6, 7.12; H8, 3.84; H3', 7.60; H5', 6.60; H8', 2.68. ^{13}C NMR, AOH (CDCl$_3$/DMSO): C2, 97.6; C4, 100.9; C6, 104.2; C7, 165.0; C3', 101.7; C5', 117.3; C6', 137.8; C8', 25.4, 128.6; 138.2; 152.6; 158.0; 164.3; 165.0. ^{13}C NMR, AME (CDCl$_3$/ DMSO): C2, 98.7; C4, 98.5; C6, 103.4; C7, 165.0; C8, 55.4; C3', 101.9; C5', 117.5; C6', 137.8; C8', 25.4; 137.9; 146.7; 152.6; 158.2; 164.4; 166.0 (Seitz et al 1975b). The AME spectra show additional signals at 3.84 in the ^{1}H NMR and 55.4 in the ^{13}C NMR (due to an aromatic methoxy group) which are absent in the AOH spectra.

Altenuene (ALT, MW = 292.0946) crystallizes from acetone-hexane as colorless prisms: mp 190–191°C (Pero et al 1971a). Its structure was elucidated by McPhail et al (1973). Isoaltenuene (isoALT, MW = 292.0946) was isolated recently by Visconti et al (1989a). ALT and isoALT give yellow fluorescence under UV light. They show the same peaks in the IR spectrum: ν_{max} at 3,394, 1,662, 1,619, 1,578, 1,493, 1,457, 1,438, 1,060, and 1,038 cm^{-1} (Pero et al 1971a, Visconti et al 1989a). UV data, λ_{max} ALT (EtOH): 240, 278, and 319 nm (Pero et al 1971a). isoALT: 243.6, 281.4, and 325.7 nm (Visconti et al 1989a). They show the parent ion in the mass spectrum at m/z 292; other fragments are present at m/z 256, 248, 246, 228, 219, 206, 204, 191, and 177 (Dusold et al 1978, Visconti et al 1989a). The ^{1}H NMR and ^{13}C NMR show the following chemical shifts. ^{1}H NMR, ALT (CDCl$_3$/DMSO): H4, 6.15; H6, 6.53; H8, 3.87; H3'a, 1.78; H3'b, 2.54; H4', 3.78; H5', 4.12; H6', 6.41; H7', 1.51. ^{1}H NMR, isoALT: H4, 6.40; H6, 6.50; H8, 3.79; H3'a, 2.22; H3'b, 2.32; H4', 3.79; H5', 4.33; H6', 6.06; H7', 1.53. ^{13}C NMR, ALT (CDCl$_3$/DMSO): C4, 100.4; C6, 102.4; C7, 165.7; C8, 55.5; C2', 81.0; C3', 40.2; C4', 69.6; C5', 72.4; C6', 132.5; C7', 27.8; 130.9; 139.1; 143.6; 165.7; 168.7. ^{13}C NMR, isoALT (CDCl$_3$): C1, 127.28; C2, 100.58; C3, 166.33; C4, 101.54; C5, 164.46; C6, 103.07; C7, 168.10; C8, 55.77; C1', 137.40; C2', 82.09; C3', 43.07; C4', 71.83; C5', 73.95; C6', 134.27; C7', 26.00. The difference of the chemical shifts of H3'a and H3'b in the ^{1}H NMR spectra of the two isomers showed the change of configuration at the C2' position, the chemical shifts and coupling constants of H4' and H5' being identical for both compounds (Visconti et al 1989a).

ALTERTOXINS

Altertoxin I (ATX-I, MW = 352.0946) was first reported by Pero et al (1973), although the structure was unknown. Stinson et al (1982) first established a structure that was later corrected by Stack et al (1986). At the same time Okuno et al (1983) isolated a new compound from *Alternaria* spp., named dihydroalterperylenol, which had the same structure as ATX-I. ATX-I is a yellow amorphous solid: mp >180°C (Stack et al 1986). It fluoresces bright yellow under UV light and has $[\alpha]_D$ +484° (c = 0.002 in chloroform) (Stack et al 1986). The IR spectrum shows ν_{max} (nujol) at 3,450, 1,641, 1,598, 1,489, 1,465, 1,368, 1,337, 1,234, 1,170, 1,062, 951, and 829 cm^{-1} (Stack et al 1986). UV data, λ_{max} (MeOH): 215, 256, 285, 296, and 356 nm. ATX-I shows the parent ion in the mass spectrum at m/z 352; other fragments are present at m/z 334, 316, and 314. The ^{1}H NMR and ^{13}C NMR show the following chemical shifts. ^{1}H NMR: H1, 7.87; H2, 7.01; H5eq, 2.65; H5ax, 3.17; H6ax, 2.43; H6eq, 3.17; H6b, 3.09; H7, 4.78; H8ax, 2.94; H8eq, 3.07; H11, 6.92; H12, 7.82; 12.7, 12.4. ^{13}C NMR: C1, 132.7; C2, 119.5; C3, 162.3; C3a, 116.9; C4, 205; C5eq, 34.0; C6ax, 34.5; C6a, 69.2; C6b, 51.9; C7, 66.1; C8ax, 47.7; C9, 202; C9a, 117.4; C9b, 139.1; C10, 162.0; C11, 117.5; C12, 132.4; C12a, 124.1; C12b, 122.7; C12c, 135.5 (Stack et al 1986).

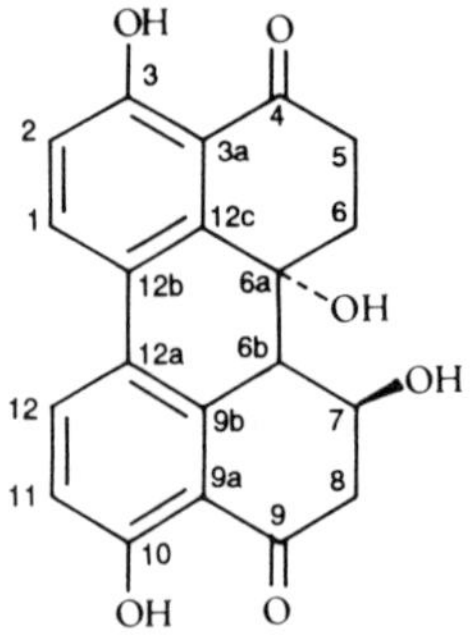

ATX-I

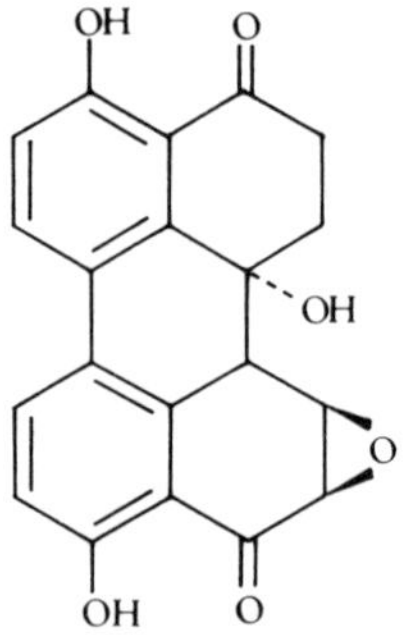

ATX-II

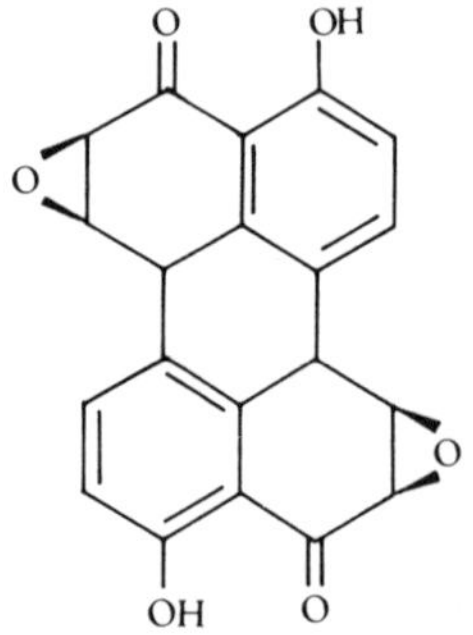

ATX-III

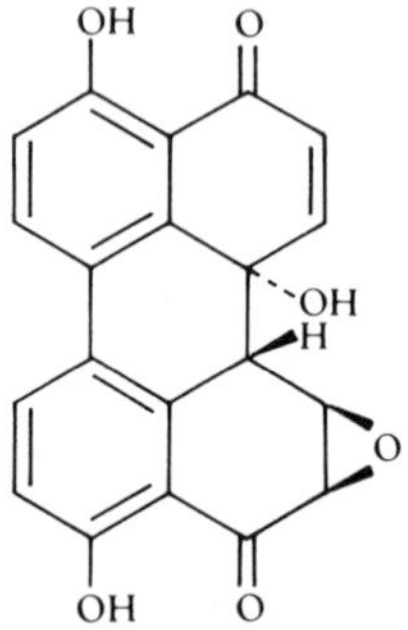

STEMPHYLTOXIN-III

Figure 7.2. Altertoxins.

Altertoxin II (ATX-II, MW = 350.0790) was reported by Pero et al (1973) together with ATX-I. Its structure was elucidated later by Stack et al (1986). ATX-II is an orange crystalline solid: mp 245–250°C. It fluoresces orange-yellow under UV light and has $[\alpha]_D$ +636° (c = 0.001 in chloroform) (Stack et al 1986). The IR spectrum shows ν_{max} (KBr) at 3,462, 1,643, 1,598, 1,485, 1,384, 1,342, 1,321, 1,296, 1,223, 1,184, 1,138, 1,103, 1,095, 1,061, 1,041, 952, 833, and 774 cm^{-1} (Stack et al 1986). UV data, λ_{max} (MeOH): 215, 258, 286, 297, and 358 nm. ATX-II shows the parent ion at m/z 350 in the mass spectrum; other fragments are present at m/z 332, 321, 305, and 291. The ^{1}H NMR and ^{13}C NMR show the following chemical shifts. ^{1}H NMR (CDCl$_3$): H1, 7.91; H2, 7.12; H5eq, 2.83; H5ax, 3.26; H6ax, 2.41; H6eq, 2.89; H8ax, 3.55; H11, 7.07; H12, 7.86; 12.7, 12.1. ^{13}C NMR (CDCl$_3$): C1, 133.0; C2, 119.9; C3, 163.3; C3a, 113.5; C4, 204.1; C5, 32.1; C6, 33.3; C6a, 68.3; C6b, 45.1; C7, 55.7; C8, 52.8; C9 196.6; C9a, 114.6; C9b, 138.8; C10, 162.6; C11, 118.0; C12, 132.6; C12a, 124.0; C12b, 122-4; C12c, 133.5 (Stack et al 1986).

Altertoxin III (ATX-III, MW = 348.0633) has mp 175–230°C (sublimes) and $[\alpha]_D$ +845° (c = 4 × 10^{-4} in CHCl$_3$) (Stack et al 1986). The IR spectrum shows ν_{max} (KBr) at 3,452, 1,650, 1,620, 1,473, 1,440, 1,382, 1,342, 1,249, 1,234, 1,198, 1,045, 1,003, 806, 746, and 715 cm^{-1}. UV data, λ_{max} (MeOH): 210, 265, and 352 nm. ATX-III shows the parent ion at m/z 348 in the mass spectrum; other fragments are present at m/z 346, 331, 329, 319, 314, 291, 263, and 250. The ^{1}H NMR and ^{13}C NMR show the following chemical shifts. ^{1}H NMR (CDCl$_3$): H1, 4.60; H2, 3.86; H5, 6.90; H6, 7.60; H6b, 4.20; H7, 4.60; H8, 3.86; H11, 6.90; H12, 7.60; H12b, 4.2; 11.5; 11.5. ^{13}C NMR (DMSO-d6): C1, 56.0; C2, 53.6; C3 196.8; C3a, 112.3; C4, 159.6; C5, 114.5; C6, 132.1; C6a, 128.8; C6b, 37.5; C7, 56.0; C8, 53.6; C9 196.8; C9a, 112.3; C9b, 143.0; C10, 159.6; C11, 114.5; C12, 132.1; C12a, 128.8; C12b, 37.5; C12c, 143.0 (Stack et al 1986).

Stemphyltoxin III (MW = 348.0636) was first isolated and chemically characterized by Arnone et al (1986) from cultures of *Stemphylium botryosum* var. *lactucum*, from which the name derived. Afterwards Stack and Mazzola (1989) isolated it from *Alternaria*. The compound, whose structure corresponds to reduced altertoxin II, is a brown solid: mp >300°C (decomposes) and $[\alpha]_D$ +688.5° (c = 0.13 in MeOH) (Arnone et al 1986). The IR spectrum shows ν_{max} (KBr) at 3,439, 3,047, 3,028, 1,658, 1,644, 1,607, 1,457, 1,335, 1,228, 1,172, 843, and 827 cm^{-1} (Stack and Mazzola 1989). UV data, λ_{max} (MeOH): 215, 269, 287, 300, and 374 nm. It shows the parent ion at m/z 348 and other fragments at m/z 330, 319, 314, and 302. The ^{1}H NMR and ^{13}C NMR show the following chemical shifts. ^{1}H NMR (Me$_2$CO-d6): H8, 3.77; H6b, 3.9; H7, 4.60; OH6a, 5.15; H5, 6.55; H11, 7.02; H2, 7.09; H6, 7.86; H1 and H12, 8.14; OH3, 12.2; OH19, 12.4. ^{13}C NMR (Me$_2$CO-d6): C6b, 43.4; C8, 53.5; C7, 57.2; C6a, 66.7; C5, 129.3; C112c, 136.5; C9b, 140.6; C6, 147.8; C4. 191.0; C9 198.4; 113.7; 115.2;

117.4; 118.8; 125.0; 125.8; 132.7; 133.7; 161.8; 163.6 (Stack and Mazzola 1989).

TENUAZONIC ACID

Tenuazonic acid (TeA, MW $= 197.1051$) isolated by Rosett et al (1957) and its structure was elucidated by Stickings (1959). TeA is a colorless, viscous oil. TeA is a monobasic acid with pK_a 3.5. It exhibits dark quenching spots on fluorescent thin-layer chromatograpy plates irradiated with UV light at 254 nm; $[\alpha]_D^{20}$ $-121°$ (c $= 2.0$ in chloroform) (Steyn and Rabie 1976). On standing for a long period, heating, or treatment with a base, the optical activity of TeA is lost, and crystallization may occur as a result of the formation of isotenuazonic acid (Rosett et al 1957). TeA turns red-brown with iron(III) chloride, yellow with 2,4-dinitrophenylhydrazine, and gray brown with 1% $Ce(SO_4)_2$ in 6N H_2SO_4. The IR spectrum shows ν_{max} (KBr) at 3,310, 2,950, 2,910, 1,734, 1,710, 1,620, 1,460, 1,300, and 1,180 cm^{-1} (Meronuck et al 1972, Davis et al 1977). UV data, λ_{max} (MeOH): 218 and 277.5 nm in acidic methanol, and 240 and 280 nm in neutral and basic methanol (Rosett et al 1957, Stickings 1959, Steyn and Rabie 1976). TeA shows the parent ion at m/z 197 in the mass spectrum; other fragments are present at m/z 182, 168, 154, 141, 123, 111, 99, and 84 (Umetsu et al 1973). The ^1H NMR and ^{13}C NMR show the following chemical

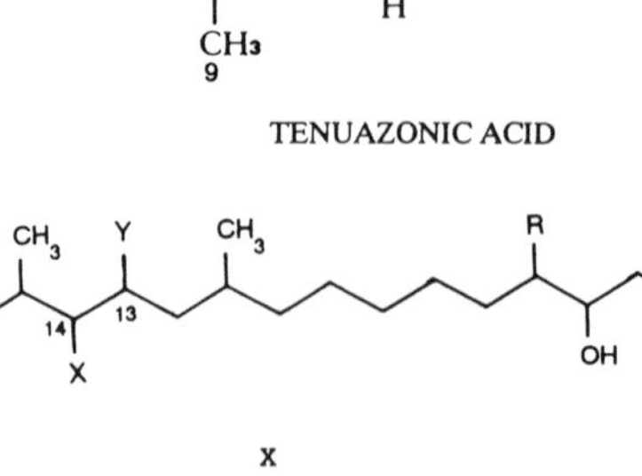

TENUAZONIC ACID

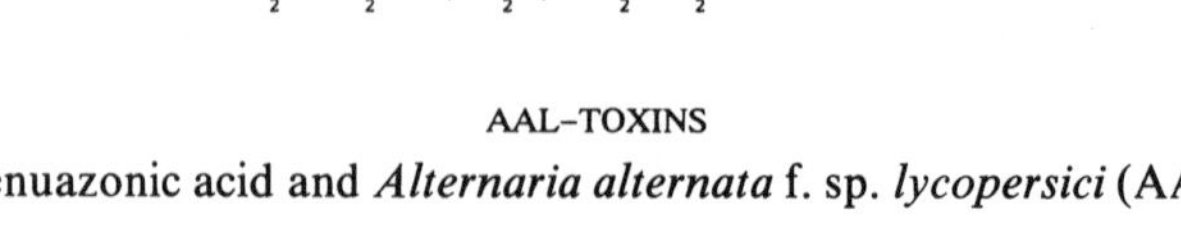

AAL-TOXIN	R	X	Y
T_A a	OH	OH	$-O_2C-CH_2-CH(CO_2H)-CH_2-CO_2H$
b	OH	$-O_2C-CH_2-CH(CO_2H)-CH_2-CO_2H$	OH

AAL-TOXINS

Figure 7.3. Tenuazonic acid and *Alternaria alternata* f. sp. *lycopersici* (AAL) toxins.

shifts. ^{1}H NMR (CDCl$_3$): H8, 0.9; H9, 0.9; H7, 1.3; H6 2.0; H11, 2.5; H5, 3.7; NH, 6.3; OH, 9.3 (Davis et al 1977). ^{13}C NMR (methanol-d^4):12.1, 15.9, 20.0, 24.8, 38.2, 67.2, 103.9, 175.2, 187.0, 198.1 (Cole and Cox 1981). TeA forms covalent complexes with Ca^{2+}, Mg^{2+}, Cu^{2+}, Fe^{3+}, and Ni^{2+} (Steyn and Rabie 1976, Lebrun et al 1985). Furthermore, the copper salt is quite stable and convenient for storage (Rosett et al 1957, Stickings 1959).

AAL-TOXINS

AAL-toxins include two fractions: T$_A$ and T$_B$ (Gilchrist and Grogan 1976, Bottini et al 1981a,b). T$_A$ (C$_{13}$H$_{53}$NO$_{15}$, MW = 679) consists of esters (C$_{13}$ or C$_{14}$) of propane-1,2,3-tricarboxylic acid and 1-amino-11,15-dimethyl-heptadeca-2,4,5,13,14-pentol. T$_B$ (C$_{13}$H$_{53}$NO$_{13}$, MW = 647) consists of esters (C$_{13}$ or C$_{14}$) of propane-1,2,3-tricarboxylic acid and 1-amino-11,15-di-methylheptadeca-2,4,13,14-tetrol. T$_A$ has $[\alpha]_D$ 22° (c = 2.7 in water). Bottini et al (1981b) reported the following NMR data (D$_2$O) for the two esters. T$_A$ (C$_{13}$) ^{1}H NMR: 0.853, 0.903, 0.878. ^{13}C NMR: 45.28, 64.97, 35.74, 70.46, 74.67, 31.49, 25.70, 28.92, 25.02, 37.02, 10.14, 20.16, 28.62, 33.82, 74.46, 76.30, 36.35, 24.27, 14.48, 34.73, 42.07, 40.35. T$_A$ (C$_{14}$) ^{1}H NMR: 0.833, 0.9273, 0.857. ^{13}C NMR: 45.28, 64.97, 35.74, 70.46, 74.70, 31.49, 25.70, 29.15, 25.16, 37.82, 10.48, 20.28, 28.37, 36.78, 69.16, 81.72, 34.89, 24.70, 14.80, 34.73, 42.07, 40.35. The structure of T$_B$ esters, having R = H instead of R = OH, was derived by comparison with these data (Bottini et al 1981b).

Analysis

ALTERNARIOLS, ALTENUENES, AND ALTERTOXINS

Alternariols, altenuenes, and altertoxins have been analyzed by thin-layer chromatography (TLC) with a variety of solvent systems and visualized by irradiation with short (254 nm) or long (360 nm) wavelength UV light (Schade and King 1984, Seitz 1984, Palmisano and Visconti 1988). The most popular solvent systems are chloroform-acetone (88:12) and toluene-ethyl acetate-formic acid (6:3:1) (Visconti et al 1986).

Pero et al (1971b) applied gas chromatography (GC) to analyze AOH, AME, and ALT trimethylsilyl derivatives with flame ionization detection (FID). By comparing different columns, they established that glass column packed with 3% OV-17 on Gas-Chrom Q gave the best results. The detection limit was about 0.1 μg for each toxin. Pero and Harvan (1973) obtained a good separation of the three toxins with OV-11 and OV-101 columns, although they failed to separate two additional *Alternaria* metabolites. Visconti et al (1989a) used a wide-bore HP-1 methyl silicone gum column and flame ionization detector for the GC determination of altenuenes as trimethylsilyl derivatives. A detection limit of 20 μg/kg was obtained by Kellert et al (1984) with a DB-1 fused silica capillary column and a mass spectrometer with multiple-ion detection.

High-performance liquid chromatography (HPLC) possesses a highly desirable advantage over GC in that no derivatization step is required. Most HPLC applications have been performed on reversed-phase C_{18} columns with conventional UV detection (Heisler et al 1980, Wittkowski et al 1983, Stack et al 1985, Visconti et al 1986, Ozcelik et al 1990). The best detection limits for AME, AOH, and ALT were in the low-nanogram range or from 30 to 100 ng/g of sample (Stack et al 1985, Visconti et al 1986). Silica columns were also used by Seitz and Mohr (1976) and Ozcelik et al (1990) with detection limits higher than 60 ng per injection.

Palmisano et al (1989a) developed an HPLC method for the simultaneous detection of ATX-I, ATX-II, and isoALT in addition to ALT, AOH, and AME. They used a reversed-phase C_{18} column with a gradient elution system (50–85% methanol in acidic water, phosphoric acid at pH 3) and a UV-diode array detector. Such a detector can scan a wide range of wavelengths and acquire full UV spectra within a few seconds. Chromatograms could be retrieved at selected wavelengths in order to increase the sensitivity or selectivity for each mycotoxin. The best sensitivity was obtained at 240 nm for altenuenes and at 257 nm for alternariols and altertoxins. The best selectivity was at 340 nm. Toxin identification was confirmed by comparison with the retention time and UV spectrum of the pure standard, previously recorded in the detector library. The detection limit of these toxins ranged from 3 to 10 ng. Extracts of a naturally contaminated sunflower seed sample were successfully analyzed with this method. AOH and AME at respective levels of 360 and 130 ng/g were detected in the naturally contaminated sample.

Because of the phenolic moiety present in the structure of these toxins, HPLC procedures with electrochemical detection were investigated in order to improve sensitivity (Palmisano et al 1989b). The optimum working potential was established by hydrodynamic voltammograms. Four of these mycotoxins (alternariols and altertoxins) were found to be anodically electroactive in a range useful for HPLC analysis, whereas the oxidation of ALT and isoALT could only occur at potentials higher than the detector anodic limit. Although the detection limit in isocratic elution was quite low (about 10–50 pg on column), separations were unsatisfactory for real samples. A suitable gradient elution system was defined to overcome the limited compatibility of electrochemical detection and gradient elution. The gradient solvent composition (essentially methanol in phosphate buffer) was adjusted in order to maintain optimum pH and ionic strength during the run (Palmisano et al 1989b).

Trace analysis of altertoxins in natural samples may be compromised because of co-occurring AOH, which is often produced at higher levels by *Alternaria* (Visconti et al 1986, Logrieco et al 1988). With the procedure of Palmisano et al (1989b), AOH was eluted between ATX-I and ATX-II. Visconti et al (1991) overcame this problem by using a dual in-series electrode detection in the "redox" mode. Analytes were electrochemically oxidized at the upstream electrode (generator), and the oxidation products were reduced at the downstream electrode (collector). In this way the selectivity for altertoxins was consistently improved due to their 10-fold higher collection efficiency (downstream/upstream peak current ratio) compared to AOH. The method was rapid and used a reversed-phase C_{18} column and isocratic elution with methanol-water containing $0.1M$ sodium nitrate and 1 mM nitric acid (60:40). Detection limits were 30 pg for ATX-I and 100 pg for ATX-II (Visconti et al 1991).

Indirect electrochemical determination of altenuenes was performed by postcolumn derivatization with bromine, using the same mobile phase (Visconti et al 1991) with addition of $0.01M$ potassium bromide (Palmisano et al 1990). Bromine, which was formed on-line by reduction of bromide at a coulometric cell, was passed through a reactor (0.5 mm i.d. coiled Teflon tube) and detected downstream at an analytical amperometric cell. The bromine signal decreased when ALT or isoALT eluted through the reactor, and toxin concentrations were determined from negative signals. Although the sensitivity of this method was similar to that of UV-diode array detection (Palmisano et al 1989a), the selectivity was considerably improved.

HPLC with fluorescence detection might also be used for the analysis of these toxins because of their native fluorescence. A preliminary approach was reported by Seitz and Mohr (1976) for AOH, AME, and ALT.

TENUAZONIC ACID

Several solvent systems have been used for TLC of TeA (Schade and King 1984). The most popular was toluene-ethyl acetate-formic acid (5:4:1 or 6:3:1). A smaller quantity (0.5 µg) could be detected by its quenching of fluorescence under short-wavelength UV light (Scott and Kanhere 1980, Visconti et al 1986).

TeA has also been analyzed by GC with FID as its trimethylsilyl derivative. Harvan and Pero (1974) found a detection limit of 0.1 µg using a glass column packed with 3% OV-17 on Gas Chrom Q and a GC oven temperature programmed from 100 to 250°C at 8°C/min. Sauer et al (1978) found a detection limit of 1 µg/g in sorghum grain using dual columns, one with 3% OV-17 and the other with 3% OV-101. Higher sensitivity (10 µg/kg) was obtained with mass spectrometric detection in single-ion-monitoring mode (Harwig et al 1979, Scott and Kanhere 1980). Confirmation of TeA at parts per billion levels was made in tomato samples by chemical ionization gas chromatography/mass spectrometry (GC/MS) (Stack et al 1985).

Scott and Kanhere (1980) studied six different HPLC conditions, including reversed-phase and ion-pair chromatography. Elution with methanol-water containing 1% ammonium acetate, 0.025% 4-dodecyldiethylenetriamine (C_{12}-diene) and 1 mM zinc sulfate (60:40) resulted the best eluent with the reversed-phase C_{18} column. UV detection was performed at 280 nm with a 5-ng detection limit. By using an anion exchange column with acetonitrile-water containing $0.05M$ KH_2PO_4 and $0.1M$ KCl (2.5:97.5) as the mobile phase, a good detection limit (1 ng) was obtained, although the calibration curve was not linear. Broad peaks, with a detection limit of 50 ng, were obtained by Heisler et al (1980), who used a reversed-phase C_{18} column eluted with methanol-water (90:10) and UV detection at 278 nm. Stack et al (1985) improved this method by using methanol-water containing 300 mg/L of $ZnSO_4 \cdot 7H_2O$ (85:15) as the mobile phase. In this way TeA formed a strong complex with Zn^{2+}, which was eluted as a sharp peak, with a 1-ng detection limit. Three chromatographic systems (ion-pair, anion-exchange, and ligand-exchange) were compared by Lebrun et al (1989). The best results were obtained with ion-pair chromatography and the ion-pairing reagent centrimide and water-methanol (45:55) buffered with potassium phosphate as the mobile phase.

AAL-TOXINS

Siler and Gilchrist (1982) analyzed AAL-Toxins by TLC using ethyl acetate-acetic acid-water (6:3:1) as the solvent system, and ninhydrin as the spray reagent for visualization. They also reported an HPLC method to determine and separate AAL-toxins. Maleyl-derivatized toxins, prepared by reaction with maleic anhydride, with an absorbance maximum at 250 nm, were detected for 0.5–10 nmol following chromatogrphy on a C_{18} bonded-phase column using a binary gradient system.

Mirocha et al (*in press*) analyzed AAL toxins by continuous flow fast atom bombardment and ionspray mass spectrometry, which were sensitive to picomole and femtomole ranges, respectively. The pseudomolecular ion (522) and sodium adduct (544) were observed. The hydrolysis product (pentolamine) of AAL toxins could be analyzed by gas GC/MS as trifluoroacetate derivatives in negative chemical ionization with methane (Mirocha et al, *in press*). Fluorescamine or *o*-phtaldialdehyde could be candidate reagents for fluorescent compounds easily detectable after HPLC separation. These derivatizing agents have been successfully applied to fumonisins, which are *Fusarium* toxins structurally similar to the AAL toxins (Shephard et al 1990, Sydenham et al 1990).

Toxicity

ALTERNARIOLS AND ALTENUENES

AME, AOH, and ALT, administered intraperitoneally as a single dose, were weakly toxic to mice (Pero et al 1973). Approximate LD_{50} values were higher than 400 mg/kg of body weight for AME and AOH and about 75–100 mg/kg body weight for ALT. Mice given either toxin were sedated within a few minutes. The eyes of mice recovering were dull; there were occasional stomach spasms and periodic panting. When administered subcutaneously, AOH and AME in combination (1:1) and AOH alone showed fetotoxic and teratogenic effects in mice, and a synergistic effect was observed when the combination of toxins (25 mg/kg of each toxin) was used (Pero et al 1973). AME was maternally toxic and fetotoxic to Syrian golden hamsters when given intraperitoneally at 200 mg/kg of body weight on day 8 of gestation (Pollock et al 1982). Cultures of *Alternaria* strains producing AME, AOH and ALT, but no other known toxins, were not toxic when fed to chicks and rats as half the diet (Sauer et al 1978). No evidence of toxicity was observed in rats or chicks when AME, AOH, and ALT were fed for 21 days at levels up to 24, 39, and 10 μg/g, respectively. The lack of toxicity of AME in poultry was confirmed by Griffin and Chu (1983) after feeding pure toxin in the standard diet at levels of up to 100 mg/kg of feed for four weeks. By injecting the toxin through the yolk sac of seven-day-old chicken embryos, no mortality or teratogenic effect was observed for AME, AOH, and ALT at respective doses of up to 0.5, 1.0, and 1.0 mg/egg (Griffin and Chu 1983).

AOH was not mutagenic to *Salmonella typhimurium* strain TA 98, AME was weakly mutagenic, and no synergism of the two toxins was observed (Scott and Stoltz 1980). An et al (1989) found AME highly mutagenic to *Escherichia coli* strain ND-160 without S9 activation. The level of revertants caused by 50 and 100 μg of AME/plate were six- and 10-fold greater than spontaneous revertants.

AME, AOH, and ALT showed cytotoxic activity to bacterial and mammalian cells (Spalding et al 1970, Pero et al 1973). The respective ID_{50} values of AME and AOH to HeLa cells were 8–14 and 6 μg/ml, and the ID_{25} for ALT was 28 μg/ml. At a dose of 25μg/ml, cytotoxicity to the human erythroleukemia cell line K-562 was observed for AOH (50% cell survival) and ALT (65% cell survival), whereas AME was not toxic (Minervini and Visconti, *unpublished*). The growth of *Bacillus mycoides* was inhibited by AOH, ALT, and AME at respective doses of 60, 125, and 500 μg/disk. AOH suppressed the growth of *Staphylococcus aureus* and *E. coli* at 25 and 50 μg/ml, respectively (Raistrick et al 1953).

The only biological activity reported for isoaltenuene, a novel member of this group, is phytotoxicity to tomato leaves at 20 μg/spot, similar to ALT (Visconti et al 1989a,b).

ALTERTOXINS

One aspect of *Alternaria* toxicity that is of some concern, mutagenicity, is ascribed mainly to altertoxins. The risks are somewhat ameliorated because much smaller amounts of altertoxins are produced by *A. alternaria* in comparison with the toxins already discussed (Logrieco et al 1990a; Visconti et al 1986, 1992).

ATX-I and ATX-II were lethal to mice at the dose of 200 mg/kg (Pero et al 1973). Toxic effects were represented by inactivity, subendocardial and subarachnoid hemorrhages, and blood in the cerebral ventricles. They were cytotoxic to HeLa cells with respective ID_{50}s of 20 and 0.5 $\mu g/ml$ for ATX-I and ATX-II (Pero et al 1973). The 100-fold difference in cytotoxic activity of these toxins was also found with Chinese hamster lung fibroblast (V79) cells (Boutin et al 1989). The noncytotoxic levels were found to be less than 5 and 0.02 $\mu g/ml$ for ATX-I and ATX-II, respectively, and 0.2 $\mu g/ml$ for ATX-III. The least cytotoxic, ATX-I, weakly disrupted metabolic communication at two concentrations, 4 and 5 $\mu g/ml$, whereas ATX-II and ATX-III did not significantly inhibit gap junction communication more than the weak tumor promoter 4-*O*-methyl ether tetracanoylphorbol 13-acetate (Boutin et al 1989). Altertoxins were highly mutagenic to *S. typhimurium* TA98, TA100, and TA1537 with and without metabolic activation (Stack and Prival 1986). Of the three compounds, ATX-III was the most mutagenic, followed by ATX-II and ATX-I. ATX-III gave a maximum response of 0.7 revertants per picomole with TA100; the maximum responses for ATX-II and ATX-I were respectively 0.5 and <0.03 revertants/pmol (Stack and Prival 1986). Stemphyltoxin III, whose chemical structure differs from that of ATX-II only by a double bond, was highly mutagenic for *S. typhimurium* TA98 and TA1537 and marginally mutagenic for TA100, both with and without metabolic activation (Davis and Stack 1991). The minimum dose of stemphyltoxin at which a doubling of revertants per plate occurred was, in general, higher than those of ATX-II and ATX-III but was about the same as that of ATX-I (Davis and Stack 1991).

Osborne et al (1988) examined the activity of ATX-I and ATX-III in the Raji cell Epstein-Barr virus early antigen (EBV-EA) induction system and in the C3H/10T(1/2) murine fibroblast cell transformation system to determine whether these toxins might function as carcinogens or tumor promoters in vitro. Both ATX-I and ATX-III significantly enhanced fibroblast transformation and activated EBV-EA expression by eight- to 10-fold. Since EBV-EA in Raji cells has been positively correlated with tumor promoters, these results suggest that altertoxins are not just mutagens but have a potential role in cell transformation (Osborne et al 1988).

Dong et al (1987) reported that extracts of *A. alternata* isolated from grains in Linxian, China, induced mutagenesis and transformation in mammalian cells, and they suggested that grain contamination with *Alternaria* metabolites may contribute to the high incidence of esophageal cancer in China. Considering the above reports (Stack and Prival 1986, Osborne et al 1988, Davis and Stack 1989) altertoxins and stemphyltoxin III may be the *Alternaria* metabolites responsible for such activities.

TENUAZONIC ACID

TeA is the most studied *Alternaria* metabolite because of its interesting biological activities and the availability of convenient procedures for its production and isolation from fungal cultures.

TeA exhibited antitumor activity against human adenocarcinoma-1 (Kaczka et al 1964, Gitterman et al 1964, Gitterman 1965). Antiviral activity of TeA was shown against measles virus, enteroviruses, respiratory viruses, vaccinia, herpes simplex HF, poliovirus, and B virus (Miller et al 1963). However, a number of adverse side effects were observed in many test animals, including guinea pigs, mice, rabbits, dogs, and rhesus monkeys when the toxin was administered to them as sodium and N,N'-dibenzyl-ethylenediamine salt (Smith et al 1968). They induced salivation, emesis, anorexia, erythema, gastrointestinal hemorrhages, convulsions, increase in packed cell volume, and many other effects (Smith et al 1968). The single oral dose LD_{50} of the sodium salt of TeA was 174 mg/kg in rats. Dogs administered repeated daily oral doses of 10 mg/kg died in eight or nine days (Smith et al 1968). TeA was toxic to young chickens, which showed an LD_{50} of 37.5 mg/kg by single-dose esophageal intubation of the toxin. At doses of 1.25 mg/kg in the diet, TeA caused suppression of weight gain and pathological changes in the spleen and gizzard. Birds receiving the toxin by intubation showed hemorrhages in subcutaneous tissue, on the surface of the heart, and in the intestinal lumen (Giambrone et al 1978). TeA was toxic to chick embryos; LD_{50}s 10 and 18 days after treatment were 50 and 5 mg/kg, respectively (Davis et al 1977). The LD_{50} of TeA by injection into the embryo yolk sac was 0.55 mg/egg (Griffin and Chu 1983).

TeA and cogeneric tetramic acids were reported to have antibacterial activity against *Bacillus subtilis, Bacillus megaterium, Alcaligenes faecalis, Brucella brachiseptica, E. coli, Salmonella gallinarum, Staphylococcus aureus, Vibrio percolans,* and *Xanthomonas vesicatoria* (Gitterman 1965, Harris et al 1965). Antifungal activity towards *Geotrichum candidum* has been shown for TeA at 20 µg/disk (Visconti et al 1989b). TeA insecticidal activity has been reported by Cole and Rolinson (1972). Several authors have studied TeA as a major nonspecific phytotoxin of *A. alternata* (Umetsu et al 1972, 1974; Janardhanan and Hussain 1984; Visconti et al 1987, 1989b).

Alternaria cultures containing TeA were toxic when they constituted half the diet for rats and chicks; the lethal effects of such cultures were apparently due to TeA (Sauer et al 1978). TeA has been found in cultures of *Phoma sorghina* strains isolated in South West Africa from millet and sorghum implicated in the etiology of onyalai, a human hematologic disease (Rabie et al 1975, Steyn and Rabie 1976).

The principal mode of action of TeA appeared to be inhibition of protein synthesis by suppression of the release of newly formed proteins from the ribomosomes into the supernatant fluid (Shigeura and Gordon 1963). Further investigations on protein synthesis inhibition showed that TeA acted at the peptide bond formation step in which the 60S unit, site of the peptidyltransferase center, appeared to be involved (Carrasco and Vasquez 1973). The toxicity of TeA also might be related in part to its ability to complex selectivity in vivo with trace metals (Steyn and Rabie 1976).

AAL-TOXINS

The toxicity of AAL-toxins has been mostly assessed for plants, in particular for susceptible tomato cultivar Earlypak-7, on which foliar interveinal necrosis was induced at concentrations as low as 10 ng/ml (Bottini et al 1981a). Following the great interest of mycotoxicologists in the *Fusarium* toxins, fumonisins, which are structurally related to AAL-toxins, AAL-toxins were also tested for toxicity to animals and cell cultures (Shier et al 1991; Mirocha et al, 1992). Rice cultures of an isolate of *A. alternata* producing AAL-toxins caused death of four out of four rats fed for five days as a 1:1 dilution of their regular diet (Mirocha et al, 1992). Nevertheless the gastric intubation of the pure toxin at levels of about 100 mg/kg of body weight did not induce any toxic effects in rats, leading to the conclusion that toxins other than AAL-toxin were responsible for the death of rats caused by the culture of *Alternaria* (Mirocha et al, *in press*). A series of 26 mammalian cell lines was examined by Shier et al (1991) for sensitivity to the cytotoxic effect of AAL toxin. Complete cell killing by AAL toxin was observed only for some rat hepatoma cell lines (six of nine) and in the MDCK dog kidney cell line. The approximate IC_{50} values for the sensitive hepatoma lines ranged from 10 to 50 μg/ml; for the MDCK kidney line the ID_{50} was 5 μg/ml. The nonsensitive cell types included fibroblast, cervical carcinoma, ovary, neuroblastoma, muscle, myeloma, connective tissue, adrenal pheochromocytoma, and some hepatoma and kidney lines.

Natural Occurrence

ALTERNARIOLS AND ALTENUENES

AOH and AME have been found at parts per million levels in samples of weathered and discolored sorghum, and their concentrations were correlated with the degree of discoloration and rainy days during plant growth (Seitz et al 1975a, Sauer et al 1978). Good quality sorghum was not contaminated by these toxins (Seitz et al 1975a). ALT was found at concentrations ranging from 0.1 to 1.5 $\mu g/g$ in three of 12 samples of heavily weathered grain sorghum (Sauer et al 1978). Ansari and Shrivastava (1990) detected AME (up to 1.8 $\mu g/g$) and ALT (up to 0.7 $\mu g/g$) in nine out of 20 sorghum samples collected in India. They also found these toxins in two samples of ragi (*Eleusine coracana* Gaertner). AOH and AME were detected, but not quantified, in discolored pecans by Schroeder and Cole (1977). Grabarkiewicz-Szczesna et al (1989) detected AOH in four (out of 21) samples of wheat chaff and in one (out of 10) samples of rye chaff at levels ranging from 0.27 to 1.7 $\mu g/g$. They also detected AOH (0.59 $\mu g/g$) in a sample of wheat kernels and AME (0.51 $\mu g/g$) in a sample of wheat chaff. No other *Alternaria* mycotoxins were detected in the same samples. AOH and AME were found, up to 1.3 and 0.3 $\mu g/g$, respectively, in tomato fruit naturally infected by *Alternaria* (Stinson et al 1981, Visconti et al 1987). Wittkowski et al (1983) analyzed about 50 commercial products and 20 moldy fruits and found AOH and AME at lower than parts per million levels in two apple samples. The natural occurrence of these toxins and ALT in apples had been previously reported by Stinson et al (1981). Four out of 13 samples of olives examined by Visconti et al (1986) contained alternariols (up to 2.9 $\mu g/g$ of AME and 2.3 $\mu g/g$ of AOH); one of these also contained 1.4 $\mu g/g$ of ALT. Although *Alternaria* species were widespread on all the examined samples, mycotoxins were detected only on physically damaged olives. The preparation of an oil sample from the most contaminated olive sample showed that the amount of toxin transferred into the oil was 4% for AME, 1.8% for AOH, and none for the other toxins (ALT and TeA) present in the original sample (Visconti et al 1986). Alternariols were detected at parts per million levels in mandarin oranges naturally infected by *A. alternata*. Black and grey heart rot were observed in naturally infected mandarins, and alternariols were found mainly associated with black rot (Logrieco et al 1990b). AOH and AME were also found in sunflower seeds and peppers, while only AME was detected in melons naturally infected with *Alternaria* (Logrieco et al 1988). No mycotoxins were detected in a number of oilseed rape samples collected in southern Italy, despite the high incidence of toxigenic *A. alternata* in them (Visconti et al 1992). Alternariols and altenuene were formed on wheat and whole wheat bread, olives, ragi, apples, oranges, mandarins, lemons, blueberries, and tomatoes inoculated with toxigenic strains of *A. alternata* (Harwig et al 1979; Stinson et al 1980, 1981; Reiss 1983; Visconti et al 1986; Ansari and Shrivastava 1990; Logrieco et al 1990a,b; Ozcelik et al 1990). The toxigenic potential of *Alternaria* strains was better expressed on rice cultures than on the substrate of origin (Visconti et al 1986, Ansari and Shrivastava 1990, Logrieco et al 1990a).

ALTERTOXINS

The only reports of the natural occurrence of altertoxins are referred to the finding of trace levels of ATX-I in three (out of 12) samples of weathered sorghum (Sauer et al 1978) and the presence of ATX-I (not quantified) in five (out of eight) samples of apples (Stinson et al 1981). Nevertheless the ability to produce these toxins has been shown by many *Alternaria* strains isolated from various plants, including mandarin oranges, olives, oilseed rape, tomatoes, blueberries, wheat, ragi, and sorghum (Sauer et al 1978; Stinson et al 1980, 1981; Visconti et al 1986, 1987, 1992; Ansari and Shrivastava 1990; Logrieco et al 1990a,b). The lack of data may be related to the limited number of laboratories searching for these toxins in plant products, the lack of reference standards and their relative instability, and the lower levels of toxin produced by most toxigenic isolates of *Alternaria* compared with production of TeA and alternariols. Fortunately, some exceptions exist in *Alternaria* strains yielding relatively high amounts of ATX-I (up to 250 mg/kg culture) (Visconti et al 1992). Considering that this group of toxins generates the most concern among the Alternaria mycotoxins, because of their mutagenicity and the potential cancer promoting activity, more investigations on their natural occurrence in food are recommended.

TENUAZONIC ACID

Up to 7.2 mg/kg of TeA was detected in tomato fruits affected by black mold in the field (Visconti et al 1987). The occurrence of TeA in commercial tomato products has been reported at trace levels in eight tomato paste samples (Scott and Kanhere 1980) and in 73 of 142 tomato samples collected from commercial processing lines (Stack et al 1985). The toxin was found in two olive samples at up to 0.26 µg/g (Visconti et al 1986), in mandarin oranges affected by grey and black rot at up to 174 µg/g (Logrieco et al 1990b), and in peppers and melons at levels of 54 and 80 ng/g, respectively (Logrieco et al 1988). Natural occurrence of TeA in sorghum and ragi at levels lower than 6 µg/g was reported by Ansari and Shrivastava (1990). Large amounts of TeA could be produced (at grams per kilogram levels) on rice culture by several strains of *A. alternata* isolated from tomatoes (Visconti et al 1987), oilseed rape (Visconti et al 1992), wheat (Logrieco et al 1990a), and mandarins (Logrieco et al 1990b). The same strains produced the toxin to a lesser extent when grown on their substrate of origin. Stinson et al (1981) found TeA in samples of tomatoes, apples, oranges, and mandarins infected with *Alternaria*. No TeA was detected in several samples of oilseed rape (Visconti et al 1992) or wheat kernels and chaff (Grabarkiewicz-Szczesna et al 1989) naturally infected by *Alternaria*. TeA was found in blast diseased rice plants (Umetsu et al 1973) and in brown spot tissue of tobacco plants (Mikami et al 1971).

AAL-TOXINS

No reports are available on the natural occurrence of these toxins in plant products, probably because *A. alternata* f. sp. *lycopersici*, the toxin-producing fungus, is a rarely occurring pathotype of *A. alternata*. More data are needed on the production of these toxins by other fungal species.

Literature Cited

An, Y., Zheng, T., Miao, J., Liu, G., Zheng, Y., Xu, Y., and Van Etten, R. L. 1989. Isolation, identification, and mutagenicity of alternariol monomethyl ether. J. Agric. Food Chem. 37:1341-1343.

Ansari, A. A., and Shrivastava, A. K. 1990. Natural occurrence of *Alternaria* mycotoxins in sorghum and ragi from Noth Bihar, India. Food Addit. Contam. 7:815-820.

Arnone, A., Nasini, G., Merlini L., and Assante G. 1986. Secondary mould metabolites. Part 16. Stemphyltoxins. New reduced perylenequinone metabolites from *Stemphylium botryosum* var. *lactucum*. J. Chem. Soc., Perkin Trans. 1, 525-530.

Bottini, A. T., and Gilchrist, D. G. 1981a. Phytotoxins. I. A 1-aminodimethyl-heptadecapentol from *Alternaria alternata* f. sp. *lycopersici*. Tetrahedron Lett. pp. 2719-2722.

Bottini, A. T., Bowen, J. R., and Gilchrist, D. J. 1981b. Phytotoxins. II. Characterization of a phytotoxic fraction from Alternaria alternata f. sp. lycopersici. Tetrahedron Lett. pp. 2723-2726.

Boutin, B. K., Peeler, J. T., and Twedt, R. M. 1989. Effect of purified altertoxin I, II, and III in the metabolic communication V79 system. J. Toxicol. Environ. Health 26:75-81.

Carrasco, L., and Vasquez D. 1973. Differences in eukaryotic ribosomes detected by selective action of an antibiotic. Biochem. Byophys. Acta 319:209-215.

Chelkowski, J., and Visconti, A., eds. 1992. *Alternaria*—Biology, Plant Diseases and Metabolites. Elsevier, Amsterdam.

Cole, M., and Rolinson, G. H. 1972. Microbial metabolites with insecticidal properties. Appl. Microbiol. 24:660-662.

Cole, R. J., and Cox, R. H., eds. 1981. pages 488-496 in: Handbook of Toxic Metabolites. Academic Press, New York.

Davis, N. D., Diener, U. L., and Morgan-Jones, G. 1977. Tenuazonic acid production by *Alternaria alternata* and *Alternaria tenuissima* isolated from cotton. Appl. Environ. Microbiol. 34:155-157.

Davis, V. M., and Stack, M. E. 1991. Mutagenicity of Stemphyltoxin III, a metabolite of Alternaria alternata. Appl. Environ. Microbiol. 57:180-182.

Dong, Z., Liu, G., Dong, Z., Qian, Y., An, Y., Miao, J., and Zhen, Y. 1987. Induction of mutagenesis and trasformation by the extract of *Alternaria alternaria* isolated from grains in Linxian, China. Carcinogenesis 8:989-991.

Dusold, L. R., Dreifuss, R. A., Pohland, A., and Sphon, J. A., eds. 1978. In: Mycotoxins Mass Spectral Data Bank. Association of Official Analytical Chemists, Washington, DC.

Freeman, G. G. 1965. Isolation of alternariol and alternariol monomethyl ether from *Alternaria dauci* (Kuhn) Groves and Skolko. Phytochemistry 5:719-725.

Giambrone, J. J., Davis, N. D., and Diener, U. L. 1978. Effect of tenuazonic acid on young chickens. Poult. Sci. 57:1554-1558.

Gilchrist, D. G., and Grogan, R. G. 1976. Production and nature of a host-specific toxin from *Alternaria alternata* f. sp. *lycopersici*. Phytopathology 66:165-171.

Gitterman, C. O. 1965. Antitumor, cytotoxic, and antibacterial activities of tenuazonic acid and cogeneric tetramic acids. J. Med. Chem. 8:483-486.

Gitterman, C. O., Dulaney, E. L., Kaczka, E. A., Campbell, G. W., Hendlin, D., and Woodruff, H. B. 1964. The human tumor-egg host system III. Tumor-inhibitory properties of tenuazonic acid. Cancer Res. 24:440-443.

Grabarkiewicz-Szczesna, J., Chelkowski, J., and Zajkowski, P. 1989. Natural occurrence of *Alternaria* mycotoxins in the grain and chaff of cereals. Mycotoxin Res. 5:77-80.

Graniti, A., Durbin, R. D., and Ballio A., eds. 1989. Phytotoxins and Plant Pathogenesis. NATO ASI Ser. Ser. H 27. Springer-Verlag, Berlin.

Griffin, G. F., and Chu, F. S. 1983. Toxicity of the *Alternaria* metabolites alternariol, alternariol methyl ether, altenuene, and tenuazonic acid in the chicken embryo assay. Appl. Environ. Microbiol. 46:1420-1422.

Harris, S. A., Fisher, L. V., and Folkers, K. 1965. The synthesis of tenuazonic acid and congeneric tetramic acids. J. Med. Chem. 8:483-486.

Harvan, D. J., and Pero, R. W. 1974. Gas chromatographic analysis of the *Alternaria* metabolite, tenuazonic acid. J. Chromatogr. 101:222-224.

Harvan, D. J., and Pero, R. W. 1976. The structure and toxicity of the *Alternaria* metabolites. Pages 344-355 in: Mycotoxins and Other Fungal Related Food Problems. J. V. Rodricks, ed. Adv. Chem. Ser. 149.

Harwig, J., Scott, P. M., Stoltz, D. R., and Blanchfield, B. J. 1979. Toxins of mould from decaying tomato fruit. Appl. Environ. Microbiol. 38:267-274.

Heisler, E. G., Siciliano, J., Stinson, E. E., Osman, S. F., and Bills, D. D. 1980. High-performance liquid chromatographic determination of major mycotoxins produced by *Alternaria* molds. J. Chromatogr. 194:89-94.

Janardhanan, K. K., and Hussain, A. 1984. Phytotoxic activity of tenuazonic acid isolated from *Alternaria alternata* (Fr.) Keisser causing leaf blight of *Datura innoxia* Mill. and its effect on host metabolism. Phytopathol. Z. 111:305-311.

Kaczka, E. A., Gitterman, C. O., Dulaney, E. L., Smith, M. C., Hendlin, D., Woodruff, H. B., and Folkers, K. 1964. Discovery of inhibitory activity of tenuazonic acid for growth of human adenocarcinoma-1. Biochem. Biophys. Res. Commun. 14:54-57.

Kellert, M., Blaas, W., and Wittkowski, M. 1984. Determination of *Alternaria* toxins in fruit and vegetable products by gas chromatography/mass spectrometry. Fresenius Z. Anal. Chem. 318:419-424.

King, A. D., and Schade, J. E. 1984. *Alternaria* toxins and their importance in food. J. Food Prot. 47:886-901.

Lebrun, M. H., Duvert, P., Gaudemer, F., Gaudemer, A., Deballon, C., and Boucly, P. 1985. Complexation of the fungal metabolite tenuazonic acid with copper(II), iron(III), nickel(II), and magnesium(II) ions. J. Inorgan. Biochem. 24:167-181.

Lebrun, M. H., Gaudemer, F., Bouter, M., Nicolas, L., and Gaudemer, A. 1989. Ion-pair, anion-exchange, and ligand-exchange high-performance liquid chromatography of tenuazonic acid and 3-acetyl 5-substituted pyrrolidine-2,4-diones. J. Chromatogr. 464:307-322.

Logrieco, A., Bottalico, A., Visconti, A., and Vurro, M. 1988. Natural occurrence of *Alternaria* mycotoxins in some plant products. Microbiol. Aliment. Nutr. 6:13-17.

Logrieco, A., Bottalico, A., Solfrizzo, M., and Mulé, G. 1990a. Incidence of *Alternaria* species in grains from Mediterranean countries and their ability to produce mycotoxins. Mycologia 82:501-505.

Logrieco, A., Visconti, A., and Bottalico, A. 1990b. Mandarin fruit rot caused by

Alternaria alternata and associated mycotoxins. Plant Dis. 74:415-417.

McPhail, A. T., Miller, R. W., Harvan, D., and Pero, R. W. 1973. X-ray crystal structure revision for the fungal metabolite (±)-altenuene. J. Chem. Soc. Chem. Commun. p. 682.

Meronuck, R. A., Steele, J. A., Mirocha, C. J., and Christensen, C. M. 1972. Tenuazonic acid, a toxin produced by *Alternaria alternata*. Appl. Microbiol. 23:613-617.

Mikami, Y., Nishijima, Y., Iimura, H., Suzuki, A., and Tamura, S. 1971. Chemical studies on brown-spot disease of tobacco plants. Part1. Tenuazonic acid as a vivotoxin of *Alternaria longipes*. Agric. Biol. Chem. 35:611-618.

Miller F. A., Rightsel, W. A., Sloan, B. J., Ehrlich, J., French, J. C., and Bartz, Q. R. 1963. Antiviral activity of tenuazonic acid. Nature 200:1338-1339.

Mirocha, C. J., Gilchrist, D. G., Shier, W. T., Abbas, H. K., Wen, Y., and Vesonder, R. F. 1992. AAL toxins, fumonisins (biology and chemistry) and host-specificity concepts. Mycopathologia 117:47-56.

Montemurro, N., and Visconti, A. 1992. Alternaria metabolites—Chemical and biological data. Pages 451-560 in: *Alternaria*—Biology, Plant Diseases and Metabolites. J. Chelkowski and A. Visconti, eds. Elsevier, Amsterdam.

Nishimura, S., and Kohomoto, K. 1983. Host-specific toxins and chemical structures from *Alternaria* species. Annu. Rev. Phytopathol. 21:87-116.

Okuno, T., Natsume, I., Sawai, K., Furusaki, A., and Matsumoto, T. 1983. Structure of antifungal and phytotoxic pigments produced by *Alternaria* spp. Tetrahedron Lett. 24:5653-5656.

Osborne, L. C., Jones, V. I., Peeler, J. T., and Larkin, E. P. 1988. Transformation of $C3H/10T_{1/2}$ cells and induction of EBV-early antigen in raji cells by altertoxins I and III. Toxicol. In Vitro 2:97-102.

Ozcelik, S., Ozcelik, N., and Beuchat, L. R. 1990. Toxin production by *Alternaria alternata* in tomatoes and apples stored under various conditions and quantitation of the toxins by high-performance liquid chromatography. Int. J. Food Microbiol. 11:187-194.

Palmisano, F., and Visconti A. 1988. *Alternaria* mycotoxins in foodstuffs: Natural occurrence and analytical aspects. Exedra 1(Prod. Chim. Suppl.):29-33.

Palmisano, F., Zambonin, P. G., Visconti, A., and Bottalico, A. 1989a. Profiling of *Alternaria* mycotoxins in foodstuffs by high-performance liquid chromatography with diode-array ultraviolet detection. J. Chromatogr. 465:305-313.

Palmisano, F., Zambonin, P. G., Visconti, A., and Bottalico, A. 1989b. Determination of *Alternaria* mycotoxins in foodstuffs by gradient elution liquid chromatography with electrochemical detection. Chromatographia 27:425-430.

Palmisano, F., Sibilia, A., and Visconti, A. 1990. Determination of altenuene and isoaltenuene by liquid chromatography—Electrochemical detection with on-line generated bromine. Chromatographia 29:333-337.

Pero, R. W., and Harvan, D. 1973. Simultaneous detection of metabolites from several toxigenic fungi. J. Chromatogr. 80:255-258.

Pero, R. W., Owens, R. G., Dale, S. W., and Harvan, D. 1971a. Isolation and identification of a new toxin, altenuene, from the fungus *Alternaria tenuis*. Biochem. Byophys. Acta 230:170-179.

Pero, R. W., Owens, R. G., and Harvan, D. 1971b. Gas and thin layer chromatographic methods for analysis of the mycotoxins altenuene, alternariol, and alternariol monomethyl ether. Anal. Biochem. 43:80-88.

Pero, R. W., Posner, H., Blois, M., Harvan, D., and Spalding, J. W. 1973. Toxicity of metabolites produced by the "*Alternaria*". EHP, Environ. Health Perspect., June, pp. 87-94.

Pollock, G. A., DiSabatino, C. E., Heimsch, R. C., and Hilbelink, D. R. 1982. The subchronic toxicity and teratogenicity of alternariol monomethyl ether produced by *Alternaria solani.* Food Chem. Toxicol. 20:899-902.

Rabie, C. J., Van Rensberg, S. J., Van Der Watt, J. J., and Lubben, A. 1975. Onyalai— The possible involvement of a mycotoxin produced by *Phoma sorghina* in the aetiology. S. Afr. Med. J. (Nutr. Suppl.) 49:1647-1650.

Raistrick, H., Sticking, C. E., and Thomas, R. 1953. Studies in the biochemistry of micro-organisms. 90. Alternariol and alternariol monomethyl ether, metabolic products of *Alternaria tenuis.* Biochem. J. 55:421-433.

Reiss, J. 1983. *Alternaria* mycotoxin in grain and bread. XVII. Mycotoxins in foodstuffs. Z. Lebensm. Unters. Forsch. 176:36-39.

Rosett, T., Sankhala, R. H., Stickings, C. E., Taylor, M. E. U., and Thomas, R. 1957. Studies on the biochemistry of micro-organisms. 103. Metabolites of *Alternaria tenuis.* Auct.: Culture filtrate products. Biochem. J. 67:390-400.

Sauer, D. B., Seitz, L. M., Burroughs, R., Mohr, H. E., West, J. L., Milleret, R. J., and Anthony, H. D. 1978. Toxicity of *Alternaria* metabolites found in weathered sorghum grain at harvest. J. Agric. Food Chem. 26:1380-1383.

Schade, J. E., and King, A. D., Jr. 1984. Analysis of the major *Alternaria* toxins. J. Food Prot. 47:978-995.

Schroeder, H. W., and Cole, R. J. 1977. Natural occurrence of alternariols in discolored pecans. J. Agric. Food Chem. 25:204-206.

Scott, P. M., and Kanhere, S. R. 1980. Liquid chromatographic determination of tenuazonic acid in tomato paste. J. Assoc. Off. Anal. Chem. 63:612-621.

Scott, P. M., and Stoltz, D. R. 1980. Mutagens produced by *Alternaria alternata.* Mutat. Res. 78:33-40.

Seitz, L. M. 1984. *Alternaria* metabolites. Pages 443-455 in: V. Betina, eds. Mycotoxin— Production, Isolation, Separation and Purification. Elsevier Science Publishers, Amsterdam.

Seitz, L. M., and Mohr, H. E. 1976. Analysis of *Alternaria* metabolites by high-pressure liquid chromatography. Anal. Biochem. 70:224-230.

Seitz, L. M., Sauer, D. B., Mohr, H. E., and Burroughs, R. 1975a. Weathered grain sorghum: Natural occurrence of alternariols and storability of the grain. Phytopathology 65:1259-1263.

Seitz, L. M., Sauer, D. B., Mohr, H. E., Burroughs, R., and Paukstelis, J. V. 1975b. Metabolites of *Alternaria* in grain sorghum. Compounds which could be mistaken for zearalenone and aflatoxin. J. Agric. Food Chem. 23:1-4.

Shephard, G. S., Sydenham, E. W., Thiel, P. G., and Gelderblom, W. C. A. 1990. Quantitative determination of fumonisins B1 and B2 by high-performance liquid chromatography. J. Liq. Chromatogr. 13:2077-2087.

Shier, W. T., Abbas, H. K., and Mirocha, C. J. 1991. Toxicity of the mycotoxins fumonisins B1 and B2 and *Alternaria alternata* f. sp. *lycopersici* toxin (AAL) in cultured mammalian cells. Mycopathologia 116:97-104.

Shigeura, H. T., and Gordon, C. N. 1963. The biological activity of tenuazonic acid. Biochemistry 2:1132-1137.

Siler, D. J., and Gilchrist, D. G. 1982. Determination of host-selective phytotoxins from *Alternaria alternata* f. sp. lycopersici as their maleyl derivatives by high-performance liquid chromatography. J. Chromatogr. 238:167-173.

Smith, E. R., Fredrickson, T. N., and Hadidian, Z. 1968. Toxic effects of the sodium and the *N,N'*-dibenzylethylenediamine salts of tenuazonic acid (NSC-525816 and NSC-82260). Cancer Chemother. Rep. 52:579-585.

Spalding, J. W., Pero, W., and Owens, R. G. 1970. Inhibition of the G2 phase of the mammalian cell cycle by the mycotoxin alternariol. (Abst.) J. Cell Biol. 47(2, Part 2):199a.

Stack, M. E., and Prival, M. J. 1986. Mutagenicity of the *Alternaria* metabolites altertoxin I, II, and III. Appl. Environ. Microbiol. 52:718-722.

Stack, M. E., and Mazzola, E. P. 1989. Stemphyltoxin III from *Alternaria alternata*. J. Nat. Prod. 52:426-427.

Stack, M. E., Mislivec, P. B., Roach, J. A. G., and Pohland, A. E. 1985. Liquid chromatographic determination of tenuazonic acid and alternariol methyl ether in tomatoes and tomato products. J. Assoc. Off. Anal. Chem. 68:640-642.

Stack, M. E., Mazzola, E. P., Page, S. W., Pohland, A. E., Highet, R. J., Tempesta, M. S., and Corley, D. G. 1986. Mutagenic perylequinone metabolites of *Alternaria alternata*: Altertoxin I, II, III. J. Nat. Prod. 49:866-871.

Steyn, P. S., and Rabie, C. J. 1976. Characterization of magnesium and calcium tenuazonate from *Phoma sorghina*. Phytochemistry 15:1977-1979.

Stickings, C. E. 1959. Studies in the biochemistry of micro-organisms. 106. Metabolites of *Alternaria tenuis* Auct.: The structure of tenuazonic acid. Biochem. J. 72:332-340.

Stinson, E. E., Bills, D. D., Osman, S. F., Siciliano, J., Ceponis, M. J., and Heisler, E. G. 1980. Mycotoxin production by *Alternaria* species grown on apples, tomatoes, and blueberries. J. Agric. Food Chem. 28:960-963.

Stinson, E. E., Osman, S. F., Heisler, E. G., Siciliano, J., and Bills, D. D. 1981. Mycotoxin production in whole tomatoes, apples, oranges, and lemons. J. Agric. Food Chem. 29:790-792.

Stinson, E. E., Osman, S. F., and Pfeffer, P. E. 1982. Structure of Altertoxin I, a mycotoxin from *Alternaria*. J. Org. Chem. 47:4110-4113.

Sydenham, E. W., Thiel, P. G., and Marasas, W. F. O. 1988. Occurrence and chemical determination of zearalenone and altrnariol monomethyl ether in sorghum-based mixed feeds associated with an outbreak of suspected hyperestrogenism in swine. J. Agric. Food Chem. 36:621-625.

Sydenham, E. W., Gelderblom, W. C. A., Thiel, P. G., and Marasas, W. F. O. 1990. Evidence for the natural occurrence of fumonisin B1, A mycotoxin produced by *Fusarium moniliforme,* in corn. J. Agric. Food Chem. 38:285-290.

Umetsu, N., Kaji, J., and Tamari, K. 1972. Investigation on the toxin production by several blast fungus strains and isolation of tenuazonic acid as a novel toxin. Agric. Biol. Chem. 36:859-866.

Umetsu, N., Kaji, J., and Tamari, K. 1973. Isolation of tenuazonic acid from blast-diseased rice plants. Agric. Biol. Chem. 37:451-452.

Umetsu, N., Muramatsu, T., Honda, H., and Tamari, K. 1974. Studies of the effect of tenuazonic acid on plant cells and seedlings. Agric. Biol. Chem. 38:791-799.

Visconti, A., Logrieco, A., and Bottalico, A. 1986. Natural occurrence of *Alternaria* mycotoxins in olives—Their production and possible transfer into the oil. Food Addit. Contam. 3:323-330.

Visconti, A., Logrieco, A., Vurro, M., and Bottalico, A. 1987. Tenuazonic acid in blackmold tomatoes: Occurrence, production by associated *Alternaria* species, and phytotoxic properties. Phytopathol. Mediterr. 26:125-128.

Visconti, A., Bottalico, A., Solfrizzo, M., and Palmisano, F. 1989a. Isolation and structure elucidation of isoaltenuene, a new metabolite of *Alternaria alternata*. Mycotoxin Res. 5:69-76.

Visconti, A., Bottalico, A., and Solfrizzo, M. 1989b. Activity of *Alternaria alternata* metabolites on tomato leaves and *Geotrichum candidum*. Pages 457-459 in: Phytotoxins and Plant Pathogenesis. A. Graniti, R. D. Durbin, and A. Ballio, eds. NATO ASI Ser. Ser. H 27. Springer-Verlag, Berlin.

Visconti, A., Sibilia, A., and Palmisano, F. 1991. Selective determination of altertoxins by high-performance liquid chromatography with electrochemical detection with dual "in-series" electrodes. J. Chromatogr. 540:376-382.

Visconti, A., Sibilia, A., and Sabia, C. 1992. *Alternaria alternata* from oilseed rape: Mycotoxin production and toxicity to *Artemia salina* larvae and rape seedlings. Mycotoxin Res. 8:9-16.

Wittkowski M., Baltes, W., Kronert, W., and Weber, R. 1983. Determination of *Alternaria* toxins in fruit and vegetable products. Z. Lebensm. Unters. Forsch. 177:447-453.

Part Four
Effects of Mycotoxins on Animals

Chapter 8

Immunotoxic Effects of Mycotoxins

James J. Pestka and Genevieve S. Bondy

Immunotoxicology addresses two general types of adverse effects that can be elicited when a toxicant or its metabolites alter the delicate balance of the immune system. One type occurs when a chemical suppresses one or more functions of the immune system, and the net result is increased susceptibility to infection or neoplasia. In the second type, the chemical stimulates immune function, and this results in hypersensitivity or an auto-immunelike disorder. Veterinary clinicians have long recognized immuno-suppression in food-producing animals that have ingested mycotoxin-contaminated grains (Richard et al 1978). There is substantial evidence that mycotoxins can be immunotoxic in experimental animals. Seminal studies in this area have been reviewed by Corrier 1991, Otokawa 1983, Pestka and Bondy 1990, Pier 1981, Thurston et al 1986, and Vidal 1990.

The potential impact of mycotoxin-induced modulation of animal and human immune function is important for several reasons. It is theoretically possible that altered immune function mechanistically contributes to symptoms of certain animal mycotoxicoses. Mycotoxins may predispose livestock to infectious disease, and this might result in feed refusal and decreased productivity. Increased infections in food-producing animals might also result in increased animal-to-human transmission of pathogens such as *Salmonella* and *Listeria*. Ingestion of mycotoxins by humans might also contribute to decreased resistance to infectious agents and neoplasms, and these compounds may function as unrecognized etiologic agents of immune dysfunction diseases. From the perspective of economic and health risks, feeding studies are most relevant in assessing immunotoxicity of mycotoxins because this is the natural exposure route. However, discussion of in vivo studies using other toxin exposure routes and of in vitro studies is important when oral studies are lacking and to provide additional insight into the mechanistic basis of mycotoxin-induced immune alteration. The potential immunotoxicity of several mycotoxins produced by the *Fusarium* and the *Aspergillus-Penicillium* groups is specifically discussed in this review.

Fusarium Toxins

TRICHOTHECENES

The capacity of trichothecene mycotoxins to be potent protein synthesis inhibitors and to interact with the cell membrane apparently contributes to their potential to modulate immune function. Acute exposure to trichothecenes results in severe damage to actively dividing cells in tissues such as bone marrow, lymph nodes, spleen, thymus, and intestinal mucosa. General effects on function of immunocompetent cells, host resistance, and immunoglobulin (Ig) production at lower doses using oral and other routes of exposure have been reported and reviewed extensively (Otokawa 1983, Thurston et al 1986, Vidal 1990).

One general means of assessing effects of toxicants on immunocompetent cells is to measure the proliferation of lymphocytes after a mitogenic stimulus. In vitro trichothecene exposure impairs, or at lower doses enhances, mitogen-induced lymphocyte proliferation (Atkinson and Miller 1984; Bondy et al 1991; Cooray 1984; Hughes et al 1988, 1990; Miller and Atkinson 1986, 1987; Tomar et al 1986, 1987, 1988). Dose-dependent decreases or increases in B- and T-cell mitogen responses can be found in lymphocytes obtained from animals exposed to T-2 toxin, vomitoxin, or various macrocyclic trichothecenes (Friend et al 1983; Hughes et al 1988, 1989; Lafarge-Frayssinet et al 1979; Paucod et al 1990; Robbana-Barnat et al 1988; Taylor et al 1985, 1987; Tryphonas et al 1986). Dietary deoxynivalenol (DON or vomitoxin) exposures as low as 2 ppm for five weeks (Tryphonas et al 1986) and 5 ppm for one week (Robbana-Barnat et al 1988) also depress mitogen responses. When the structural effects of trichothecenes are evaluated in mitogen-induced lymphocyte proliferation assays, in vitro concentrations required for inhibition are approximately 10-fold lower than those required to inhibit cellfree protein synthesis (Forsell et al 1985, Forsell and Pestka 1985, Pestka and Forsell 1988). In rank order the macrocyclic group is more potent than type A group, which is more potent than type B group (Forsell et al 1985, Forsell and Pestka 1985, Mekhancha-Dahel et al 1990, Pestka and Forsell 1988, Thompson and Wannemacher 1986, Tomar et al 1988). Metabolites and precursors related to the trichothecene biosynthetic pathway can also inhibit murine lymphocyte proliferation (Bondy et al 1991). These studies suggest that in vitro lymphotoxicity decreases upon substitution of acyl groups at the C-8 with keto or hydroxy moieties and is also dependent on the nature of substitutions at the C-3, C-4, and C-15 positions. Thus metabolism that results in deacylation might be expected to reduce immunotoxicity. Differences in sensitivities exist among lymphocyte cell types, and this may be related to differences in both uptake and metabolism to less toxic compounds (Porcher et al 1988).

Repeated exposure of experimental animals to T-2 toxin or diacetoxy-scirpenol results in markedly increased susceptibility to *Candida* (Salazar et al 1980), *Cryptococcus* (Fromentin et al 1981), *Listeria* (Corrier and Ziprin 1987), *Salmonella* (Boonchuvit et al 1975; Tai and Pestka 1988a, 1990; Vidal and Mavet 1989), *Mycobacterium* (Kanai and Kondo 1984), and herpes simplex virus type 1 (Friend et al 1983). There is a marked increase in susceptibility to lipopolysaccharide (LPS) when T-2 toxin is coadministered, suggesting that impaired resistance to LPS might be one mechanism for increased susceptibility to gram-negative bacteria (Tai and Pestka 1988b). Taylor et al (1989, 1991) also hypothesized that T-2 toxin treatment enhances LPS absorption and that this produces some of the pathologic sequelae associated with T-2 toxicosis.

Corrier et al (1987b) determined that short-term intraperitoneal preinoculation treatment with T-2 toxin can be immunostimulatory and can actually enhance resistance to *Listeria*, whereas postinoculation T-2 toxin treatment is markedly immunosuppressive. Cooray and Jonsson (1990) noted a similarly enhanced resistance to mastitis pathogens by preinoculation gavage of T-2 toxin in mice. Ingestion of DON decreases time-to-death intervals and splenic counts in mice challenged with *Listeria monocytogenes* (Pestka et al 1987, Tryphonas et al 1986). Evaluation by assays such as allograft rejection time, neutrophil migration, macrophage phagocytosis, and delayed-type hypersensitivity indicates that depression of host resistance by trichothecenes involves suppression of several cellular functions (Buening et al 1982; Corrier et al 1987a,b; Gerberick et al 1984; Gerberick and Sorenson 1983; Niyo et al 1988; Paucod et al 1990; Pestka et al 1987; Rosenstein et al 1979; Sorenson et al 1986).

Trichothecenes can also cause both suppression and stimulation of immunoglobulin production. Repeated exposure to trichothecenes impairs murine antibody responses to challenge with sheep red blood cells (Rosenstein et al 1979, Pestka et al 1987, Robbana-Barnat et al 1988) and other antigens (Otokawa 1983). Dietary DON causes marked elevation of serum IgA in the mouse with concurrent decreases in IgM and IgG (Forsell et al 1986). The IgA is polyspecific and reacts with bacterial and self antigens (Rasooly and Pestka 1993). Concurrent with serum IgA elevation are immunopathologic effects that include increased polymeric IgA and IgA immune complexes, kidney mesangial IgA accumulation, and hematuria. These effects are persistent (Dong and Pestka 1993) and are highly analogous to the common human glomerulonephritis, IgA nephropathy (Dong et al 1991). Interestingly, there is a progressive increase in serum IgE after withdrawal of DON from the diet (Pestka and Dong 1992). IgA nephropathy occurs to a greater extent in males than in females. Recently Greene et al (1993) determined that in the B6C3F1 mouse, males exhibit a predilection for DON-induced IgA nephropathy in terms of threshold toxin dose and response onset and magnitude. Another common feature of the human disease and the murine model is the involvement of polyspecific "natural" IgA, which may be associated with immune complex formation and the subsequent glomerulonephritis (Rasooly and Pestka 1992, Rasooly et al 1993). Although the etiology of IgA nephropathy is still unknown, grain-containing diets have been shown to play a role (Coppo et al 1990). The potential for trichothecenes to induce IgA-mediated immune complex glomerulonephritis is perhaps one of the most intriguing aspects of myco-toxin-induced immune alteration that has been observed to date.

The suppressive effects of trichothecenes on immune function are explain-able by their potent capacity to inhibit protein synthesis, but their immuno-stimulatory effects at low doses are problematic. Superinduction of cytokines such as interleukins 1 (IL-1) and 2 (IL-2) have been demonstrated with trichothecenes and other protein synthesis inhibitors (Efrat et al 1984; Holt et al 1988; Miller and Atkinson 1986, 1987). Proposed mechanisms for these effects include interference with synthesis of high-turnover proteins that limit transcription or half-life of interleukin mRNA; analogous mecha-nisms could be proposed at the level of translation. Dysregulation of IgA production is apparently brought about by the T-cell–mediated polyclonal differentiation of B cells to IgA secretion at the level of the Peyer's patch (Bondy and Pestka 1991; Pestka et al 1990a,b).

Studies on the effect of trichothecenes on cytokine production are warranted. Recently, it was shown that in vitro exposure of T cells to low levels of DON resulted in superinduction of IL-5 mRNA and enhanced secretion of IL-4 and IL-5 (Dong and Pestka, *unpublished data*). One hypothetical mechanism for trichothecene hyperelevation of IgA production is via superinduction of T helper 2 (TH2) cytokines. The following temporal order of peaked events upon trichothecene exposure in a lymph node might be predicted: 1) protein synthesis inhibition, 2) enhanced TH cytokine mRNA expression, 3) elevated TH cytokine production after metabolism or removal of vomitoxin and cessation of protein synthesis inhibition, and 4) terminal differentiation of immunoglobulin-secreting cells. The Peyer's patch may be particularly prone to this type of dysregulation, since it would be exposed to concentrated levels of vomitoxin in food and after enterohepatic recirculation.

ZEARALENONE

There have been few investigations of potential zearalenone immuno-toxicity, but other estrogens have been shown to decrease host resistance to pathogens, increase tumor susceptibility, and impair cell-mediated im-munity and inflammatory responses (Luster et al 1985). Zearalenone and zearalenol can induce thymic atrophy and macrophage activation (Luster et al 1984). Zearalenone and its analogues are capable of inhibiting mitogen-stimulated lymphocyte proliferation (Forsell and Pestka 1985). Dietary exposure to 10 ppm of zearalenone for two weeks decreases resistance to *Listeria* but has no effects on antibody or delayed type hypersensitivity responses (Pestka et al 1987). The same study indicated that zearalenone and vomitoxin apparently do not act synergistically when the same immune functions are monitored.

FUMONISIN

Produced by *Fusarium moniliforme,* fumonisin B_1 (FB_1) is the etiologic agent of equine leukoencephalomalacia and porcine pulmonary edema and is hepatotoxic and carcinogenic to rats (Colvin and Harrison 1992, Gelderblom et al 1991, Kellerman et al 1990). To date there is relatively little evidence to suggest that FB_1 or any of the structurally related fumonisin toxins can modulate immune function in vivo. Dombrink-Kurtzman et al (1992c) noted reduced lymphocyte viability in chickens fed diets containing 62 ppm of FB_1 and 15 ppm of FB_2. There is an apparent correlation between "mystery swine disease" and feeding corn contaminated with FB_1 and FB_2 at concentrations of 20 mg/kg or higher (Bane et al 1992). Concurrently with pyrexia, anorexia, and problems with gestation and farrowing, affected swine exhibit increased infection rates, indicating the potential involvement of an immunosuppressive agent (Moore et al 1990). Haschek et al (1992) observed that pigs dosed intravenously with fumonisin over several days (total 4.6–7.9 mg/kg) or fed 48–166 ppm of the toxin exhibited membranous material in pulmonary macrophages.

FB_1 is cytotoxic in vitro to the MDCK dog kidney cell line and to several rat hepatoma cell lines (Shier et al 1991). FB_1 and FB_2 are cytotoxic to turkey lymphocytes; 50% cytotoxic doses range between 0.3 and 2 μg/ml (Dombrink-Kurtzman et al 1992a,b). Chicken peritoneal macrophages treated with FB_1 (10–100 μg) show morphological alterations, small but significant reductions in cell viability, and reduced phagocytic potential (Qureshi and Hagler 1992). Tumoricidal factor production in a chicken macrophage cell line was unaffected (Qureshi and Hagler 1992). Besides investigation of the immunomodulatory effects, there is a clear need for in vivo data to confirm in vitro suppression of lymphocyte and macrophage function, as well as for evaluation studies of the effects of FB_1 on the immune mechanisms associated with tumor resistance.

FUSAROCHROMANONE

Fusarochromanone (TDP-1) is a mycotoxin produced by *Fusarium equiseti* that has been associated with tibial dyschondroplasia (TDP) in poultry. The compound is both inhibitory and stimulatory (at lower levels) in the mitogen-induced lymphocyte proliferation assay (Minervini et al 1992). Deacetylation to its metabolite, TDP-2, strongly reduces these immunotoxic effects. Additional studies are warranted to verify whether immunotoxicity is similarly evident in vivo.

Penicillium and *Aspergillus* Toxins

OCHRATOXIN A

Ochratoxin A (OTA) immunomodulation was reviewed by Thurston et al (1986). General indicators of immunosuppression following OTA ingestion include lymphocytopenia and depletion of lymphoid cells, particularly in the thymus, bursa of Fabricius, and Peyer's patches (Boorman et al 1984; Chang et al 1979; Dwivedi and Burns 1984, 1985; Gupta et al 1979). The effects of OTA on lymphoid cells have been examined in several in vitro systems. OTA is more toxic to porcine blood lymphocytes than to porcine blood granulocytes, as determined by a colorimetric tetrazolium cleavage test as an indicator of cytotoxicity (Reubel et al 1987). Lea et al (1989) found in in vitro studies that OTA inhibits human peripheral B and T lymphocyte proliferation and abolishes IL-2 and IL-2 receptor production.

Although several studies indicate that intraperitoneal injection of OTA suppresses antibody responses to sheep red blood cells (SRBC) or *Brucella abortus* (Creppy et al 1988, Haubeck et al 1981, Prior and Sisodia 1982), depressed T-independent or T-dependent antigen responses are not seen when OTA is administered in the diet (Campbell et al 1983, Prior and Sisodia 1982, Richard et al 1975). Since phenylalanine has been shown to attenuate OTA-induced immunosuppression in mice and to reduce mortality in broiler chicks receiving OTA-contaminated diets, it is possible that dietary factors account for some of the discrepancies in studies of OTA immunotoxicity using different administration routes (Gibson et al 1990, Haubeck et al 1981). The complement system is unaffected by dietary OTA (0.45 mg/ml) in guinea pigs (Richard et al 1975); however, complement activity is significantly reduced in chickens fed OTA (2 ppm) (Campbell et al 1983). It remains to be determined whether species or other experimental variables are responsible for these differences.

OTA is reportedly a renal carcinogen (Bendele et al 1985) and it is suggested to contribute to human renal and urinary tumors (Austwick 1981, Radovanovic et al 1984). When Manolova et al (1990) tested in vitro the ability of OTA to induce chromosomal aberrations in human peripheral lymphocytes in the presence or absence of a kidney microsomal metabolic activation system, OTA induced X chromosome aberrations that were similar to those previously detected in lymphocytes from patients suffering from endemic nephropathy. The interaction between OTA-induced immunosuppression and carcinogenesis was examined by Luster et al (1987). Suppression of natural killer cell activity by oral OTA appears to be the result of inhibition of endogenous interferon levels. The ability of OTA to modulate natural killer cell activity may contribute to its capacity to induce renal and hepatic carcinomas, because 1) the dose for tumor induction is similar to that required for immunosuppression, 2) immunosuppression precedes development of detectable tumors, and 3) immune alterations are consistent with those involved in tumor immunity and correlate with changes in resistance to transplantable syngeneic tumor cells. One problem with this hypothesis identified by the aforementioned authors is that existing evidence suggests that most immunodeficiency tumors are lymphomas and leukemias rather than solid tumors as occurs with OTA. A second concern is that the above studies were conducted with very high levels of OTA that may not be encountered naturally.

AFLATOXINS

Although they are primarily known as hepatotoxins and hepatocarcinogens, aflatoxins have numerous immunosuppressive effects. Aflatoxin-induced immune modulation has been discussed in great detail in the following reviews: Corrier 1991, Pier 1986, Pier et al 1986, Pier and McLoughlin 1985, and Richard et al 1978. In general, experimental evidence supports the contention that aflatoxin B_1 (AFB_1) particularly suppresses the cell-mediated immune response. There is dose-related suppression (0.145 and 0.7 mg/kg of body weight [bw]) of delayed type hypersensitivity to keyhole limpet hemocyanin in CD-1 mice (Reddy et al 1987). Significant decreases in splenic CD4 (TH) cells at 0.75 mg/kg bw of AFB_1 and in IL-2 production by splenic lymphocytes at 0.15 and 0.75 mg/kg bw of AFB_1 have been observed in C57Bl/6 mice (Hatori et al 1991). There is a reduced response to intradermal phytohemagglutinin (PHA) at 0.045 mg/kg bw of AFB_1 and a reduction in delayed cutaneous hypersensitivity to *Nocardia asteroides* at 0.042 mg/kg bw of AFB_1 in guinea pigs receiving oral aflatoxin (Pier et al 1986, 1989).

Because aflatoxin is an economic threat to the poultry industry, there has been extensive study of the effects of aflatoxin on avian immunity. Cell-mediated responses again appear to be particularly sensitive, as reflected by reduced thymus weight and reduced peripheral T lymphocyte numbers in mice fed AFB_1 (Ghosh et al 1991, Ghosh et al 1990, Virdi et al 1989). Graft versus host response is suppressed in chickens receiving 300 ppm of AFB_1 (Kadian et al 1988). Delayed hypersensitivity response to dinitrofluorobenzene is reduced in broiler chicks receiving 1 ppm AFB_1 feed (Ghosh et al 1991). Oral administration of AFB_1 to chicks at 0.1 and 0.5 mg/kg bw lowers peripheral blood lymphocyte proliferation responses to the T cell mitogen concanavalin A (ConA) (Scott et al 1991).

In vitro data also suggest that T-cell–mediated responses are sensitive to AFB_1. Yang et al (1986) found that the mitogenic response to ConA was more sensitive to AFB_1 than LPS- or phytohemagglutinin-stimulated murine spleen lymphocytes cocultured with rat hepatocytes (Yang et al 1986).

Modulatory effects of AFB_1 on humoral immunity are not as apparent as effects on cell-mediated immunity (CMI), particularly upon comparison across species barriers. In swine receiving up to 500 ppm of AFB_1 in feed and inoculated with *Erysipelothrix rhusiopathiae* bacteria, there is no significant difference in antibody titers compared with those of inoculated swine receiving uncontaminated feed (Panangala et al 1986). The ability of guinea pigs to produce antibodies to *Brucella abortus* is unaffected by aflatoxin (0.045 mg/kg bw) administered orally (Pier et al 1989). In rabbits receiving aflatoxin (approximately 24 ppm in feed), the antibody-forming response to SRBC, a T-cell–dependent antigen, is unchanged compared with those of animals receiving AFB_1-free diets (Singh et al 1987). In contrast, bursal weight loss is significant in chicks fed AFB_1 but is comparatively less than thymus weight loss (Virdi et al 1989). Furthermore, C57B1/6 mice receiving up to 0.75 mg/kg bw of AFB_1 by gavage exhibit a dose-related decrease in plasma IgM titer to SRBC (Hatori et al 1991). Oral AFB_1 exposure (0.145 to 0.7 mg/kg bw) reduced the number of splenic IgM-producing cells specific for SRBC but did not alter the number of antibody-producing cells specific for the T-cell–independent antigen LPS in CD-1 mice (Reddy et al 1987). This suggests that AFB_1-induced suppression of B cell function is T-cell mediated.

The effects of oral AFB_1 administration to mice on spontaneous and mitogen-stimulated lymphocyte blastogenesis do not clearly indicate specific suppression of either CMI or humoral immunity. Blastogenesis responses to AFB_1 appear to be strain-dependent in mice. In C57B1/6 mice LPS and pokeweed mitogen-stimulated spleen lymphocyte blastogenesis is suppressed to a greater extent than ConA and phytohemagglutinin-stimulated lymphocytes, suggesting specific effects of AFB_1 ingestion on splenic B cell function (Hatori et al 1991). In CD-1 mice receiving AFB_1 orally (0.03 to 0.7 mg/kg bw), spontaneous spleen lymphocyte blastogenesis is more sensitive than mitogen-stimulated blastogenesis, whereas in BALB/c mice mitogen-stimulated blastogenesis is suppressed without regard for mitogen specificity (Reddy et al 1987, Reddy and Sharma 1989).

Phagocytic functions of macrophages and of the reticuloendothelial system are consistently inhibited by AFB_1. In chicks, the percentage of nitro-blue-tetrazolium–positive cells in spleen tissue is depressed by AFB_1 (0.3 to 1.0 ppm), indicating depressed macrophage function (Ghosh et al 1991). Oral administration depresses both macrophage number and function in rats of AFB1 (0.35 to 0.7 mg/kg bw) (Raisuddin et al 1990). Clearance of circulating colloidal carbon is reduced in chicks receiving AFB_1 (0.3 mg/kg of feed), suggesting that there is reduced phagocytic status of the reticuloendothelial system (Kadian et al 1988). In vitro data also indicate that there is suppression of phagocytic activity seen in vivo in chickens and rats. Phagocytosis, intracellular killing of *Candida albicans*, and spontaneous O^{2-} production are suppressed in rat peritoneal macrophages exposed to aflatoxins in vitro (Cusumano et al 1990). Activation of AFB_1 by mixed-function oxidases is apparently required for macrophage toxicity (Neldon-Ortiz and Qureshi 1991, 1992).

Aspects of innate immunity are suppressed to various degrees by AFB_1. Natural killer cell-mediated cytolysis of YAC-1 target cells is suppressed in BALB/c mice receiving oral AFB_1 at concentrations of 0.03 to 0.7 mg/ kg bw (Reddy and Sharma 1989) but is unaffected in C57B1/6 mice (0.03 to 0.75 mg/kg bw [Hatori et al 1991]) and in rabbits (24 ppm in feed [Singh et al 1987]). Serum complement activity is reduced in pigs receiving feed containing 500 ppm of AFB_1, whereas complement activity is unaffected in pigs receiving 300 ppm AFB_1 in feed and in rabbits receiving 95 ppm AFB_1 in feed (Panangala et al 1986, Singh et al 1987).

In summary, AFB_1 suppresses CMI to a greater extent than humoral immunity. Some aspects of innate immunity, especially phagocytic responses, are also inhibited by AFB_1. It is clear that aflatoxins in the low parts per million range would be immunomodulatory. These might be most likely encountered in animal feed that is not inspected for interregional or international commerce. In contrast, human food is regulated at the low parts per billion range in Canada, the United States, and most developed countries because of potent hepatocarcinogenicity of aflatoxins. Thus, vigilant monitoring should minimize the potential for aflatoxin-induced immunosuppression in humans.

PATULIN

Recent efforts have addressed the immunosuppressive action of patulin, which is known for producing ulceration, congestion, and hemorrhagic lesions, particularly in the gastrointestinal tract (McKinley et al 1982, Speijers et al 1988). Several studies indicate that patulin inhibits multiple aspects of macrophage function in vitro. Sorenson et al (1985) found that protein synthesis is inhibited in rat alveolar macrophages exposed to patulin in vitro and that cell membrane function is compromised. In rat and rabbit peritoneal macrophages, in vitro patulin suppresses the oxidative burst (Escoula et al 1988b). In mouse peritoneal macrophages, patulin significantly reduces O^{2-} production, phagosome-lysosome fusion, and lysosomal enzyme activity (Bourdiol et al 1990).

Oral administration of patulin (10 mg/kg for four days) decreases the mortality of mice experimentally infected with *Candida albicans* from 80% in control mice to 50% in mice treated with patulin (Escoula et al 1988a). Patulin administration also results in increased circulating neutrophils, which could contribute to increased resistance. McKinley et al (1982) attribute neutrophilia in rats administered patulin orally to gastrointestinal inflammation.

CITRININ

Citrinin is primarily known as a nephrotoxin, and few studies have addressed its potential for immunotoxicity. Campbell et al (1981) administered citrinin (125–500 ppm for three weeks) in feed to broiler chicks and found that cellular and humoral immune responses are not affected. In mice receiving citrinin intraperitoneally (0.12-3.0 mg/kg bw), decreased peripheral leukocytes are observed (Reddy et al 1988). Antibody response to SRBC is depressed at the lowest citrinin dose but is stimulated at higher doses. Splenic lymphocyte blastogenesis with or without mitogens is stimulated by citrinin administration. From this study, citrinin appears to be immunostimulatory rather than immunosuppressive; however, it remains to be determined whether ingestion of feed contaminated with low levels of citrinin produces responses similar to intraperitoneal exposure.

WORTMANNIN

Wortmannin is a steroid metabolite first isolated from *Penicillium wortmanni* and later found to be produced by other fungi including *Myrothecium roridum, Fusarium oxysporum,* and *F. sambucinum* (Abbas and Mirocha 1988). The potential for immunosuppression by wortmannin was suggested by the anti-inflammatory activity of structurally related fungal metabolites. 11-Desacetoxy-wortmannin inhibits carrageenin-induced rat paw edema and reduces symptoms of arthritis (Wiesinger et al 1974). Exposure of human neutrophils or mouse bone marrow–derived macrophages to wortmannin or its 11-desacetoxy analogue in vitro inhibits induction of the respiratory burst during phagocytosis (Baggiolini et al 1987). Gastric and myocardial hemorrhage are prominent in rats receiving crude wortmannin in feed or purified wortmannin by gavage (Gunther et al 1989). Accompanying the hemorrhagic lesions are necrosis of the spleen, lymph nodes, and gut-associated lymphoid tissue. Rats receiving crude wortmannin in feed display depletion and mild necrosis of thymus lymphoid tissue, whereas in wortmannin-gavaged rats there is severe necrosis of the thymus cortex accompanied by hemorrhage (Gunther et al 1989). There is a possibility that wortmannin contributes to the mycotoxicosis known as alimentary toxic aleukia; however, this has not been proven experimentally (Mirocha and Abbas 1989).

CYCLOPIAZONIC ACID

Cyclopiazonic acid (CPA) has been reported to be produced by several species of *Penicillium* and *Aspergillus* (Dorner et al 1983). Subchronic toxicity studies indicate that CPA is toxic to the gastrointestinal tract, liver, heart, kidneys, and skeletal muscle (Voss et al 1990) and is not primarily an immunosuppressive agent. Dorner et al (1983) found lymphoid cell depletion and reduced follicle size in the bursa of Fabricius in chickens receiving CPA in feed. However, in guinea pigs receiving oral CPA (2.2 mg/kg) there were no detectable effects on responses to *Brucella abortus,* complement titers, delayed cutaneous hypersensitivity, or proliferative responses to intracutaneous phytohemagglutinin injections (Pier et al 1989). In another study, oral CPA administration to guinea pigs did not affect cutaneous hypersensitivity to *Mycobacterium tuberculosis* or complement activity (Richard et al 1986). In rats receiving CPA intraperitoneally (0.1–5.0 mg/kg bw), there is a transient depression of the antibody response to SRBC, but at day 14 after SRBC injection antibody production is higher in treatment animals than in controls (Hill et al 1986). From the available evidence, it appears that although low doses of CPA administered intraperitoneally are transiently immunosuppressive, there is insufficient evidence to conclude that oral exposure to CPA would have similar effects.

SECALONIC ACID D

Secalonic acid D (SAD) is one of a group of toxic pigments first found to be produced by *Claviceps purpurea* and also produced by *Penicillium oxalicum* and *Aspergillus aculeatus* (Anderson et al 1977, Steyn 1984). In mice subjected to influenza virus infection, intraperitoneal injection of SAD (35 mg/kg bw) results in significantly increased mortality. Mice administered SAD also have higher virus titers than control mice, and anti-influenza virus antibody titers are significantly lower (Fleishhacker et al 1986). Eldeib and Reddy (1990) found neutrophilia and lymphocytopenia in mice receiving SAD intraperitoneally (15–45 mg/kg bw). To date there is insufficient evidence to suggest that oral exposure to low levels of SAD in food or feed is significantly immunosuppressive.

Conclusions

Some generalizations can be made upon examination of studies on mycotoxin-induced immune modulation. Clearly, mycotoxins can experimentally suppress or stimulate immune function depending on the timing of exposure and dose. While immune alteration may be detectable following multiple high-dose injections, examination of the dietary effects of mycotoxins, particularly at levels found in human food and animal feed, are most desirable. It is important to note that nutritional effects associated with feed refusal may also contribute to observed immune alterations. It is equally critical to consider that mycotoxin mixtures are likely to occur naturally and these may alter immunity in an additive or synergistic manner. Additional investigation of the immune effects of inhaled mycotoxins is also desirable because of the potential for environmental exposure via grain dust or mold-contaminated air supplies (Sorenson et al 1987, Miller 1992). Finally, while systemic immunity is the focus of most investigations, it is very likely that mycotoxins have their greatest effect on mucosal lymphoid tissue (particularly gut and bronchial) before they are absorbed and metabolized. This possibility is verified by the observation that vomitoxin ingestion results in IgA hyperelevation and IgA-mediated glomerulonephritis.

There is as yet insufficient data to extrapolate specific immune dysfunction in humans or animals that have ingested citrinin, cyclopiazonic acid, patulin, secalonic acid, wortmannin, fumonisin, fusarochromanone, or zearalenone. Studies on the aflatoxins, trichothecenes, and ochratoxin A might indicate, however, a possible role for toxin-induced immune modulation in human and animal health problems. Despite experimental evidence for immunotoxicity by these mycotoxins, supportive epidemiologic data for the above conjectures are poor. Extrapolation of immunotoxicity studies performed in genetically defined animals to actual disease in livestock and humans must be done with extreme care. Although in vivo comparison and validation with livestock is feasible, similar studies with humans raise ethical concerns. Thus, wherever possible the cellular and molecular basis for in vivo observations should be determined, reconstituted in vitro, and potential species differences compared. Further emphasis should be placed on the effects of mycotoxins on lymphocyte profile as measured by flow cytometry, lymphocyte growth and differentiation, interleukin production, and receptor function in immunocompetent cells from selected target species.

Literature Cited

Abbas, H. K., and Mirocha, C. J. 1988. Isolation and purification of a hemorrhagic factor (wortmannin) from *Fusarium oxysporum* (N17B). Appl. Environ. Microbiol. 54:1268-1274.

Anderson, R., Buchi, G., Kobbe, B., and Demain, A. L. 1977. Secalonic acids D and F are toxic metabolites of *Aspergillus aculeatus*. J. Org. Chem. 42:352-353.

Atkinson, H. A. C., and Miller, K. 1984. Inhibitory effect of deoxynivalenol, 3-acetyldeoxynivalenol and zearalenone on induction of rat and human lymphocyte proliferation. Toxicol. Lett. 23:215-221.

Austwick, P. C. 1981. Balkan nephropathy. Practitioner 225:1031-1038.

Baggiolini, M., Dewald, B., Schnyder, J., Ruch, W., Cooper, P. H., and Payne, T. G. 1987. Inhibition of the phagocytosis-induced respiratory burst by the fungal metabolite wortmannin and some analogues. Exp. Cell Res. 169:408-418.

Bane, D. P., Neumann, E. J., Hall, W. F., Harlin, K. S., and Slife, R. L. N. 1992. Relationship between fumonisin contamination of feed and mystery swine disease. Mycopathologia 117:121-124.

Bendele, A. M., Carlton, W. W., Krogh, P., and Lillehoj, E. B. 1985. Ochratoxin A carcinogenesis in the (C57BL/6J × C3H) F_1 mouse. J. Natl. Cancer Inst. 75:733-742.

Bondy, G. S., and Pestka, J. J. 1991. Dietary exposure to the trichothecene vomitoxin (deoxynivalenol) stimulates terminal differentiation of Peyer's patch B cells to IgA secreting plasma cells. Toxicol. Appl. Pharmacol. 108:520-530.

Bondy, G. S., Beremand, M. N., McCormick, S. P., and Pestka, J. J. 1991. Murine lymphocyte proliferation impaired by substituted neosolaniols and calonectrins—*Fusarium* metabolites associated with trichothecene biosynthesis. Toxicon 29:1107-1113.

Boonchuvit, G., Hamilton, P. B., and Burmeister, H. R. 1975. Interaction of T-2 toxin with *Salmonella* infections in chickens. Poult. Sci. 54:1693-1696.

Boorman, G. A., Hong, H. L., Dieter, M. P., Hayes, H. T., Pohland, A. E., Stack, M., and Luster, M. I. 1984. Myelotoxicity and macrophage alteration in mice exposed to ochratoxin A. Toxicol. Appl. Pharmacol. 72:304-312.

Bourdiol, D., Escoula, L., and Salvayre, R. 1990. Effect of patulin on microbicidal activity of mouse peritoneal macrophages. Food Chem. Toxicol. 28:29-33.

Buening, G. M., Mann, D. D., Hook, B., and Osweiler, G. D. 1982. The effect of T-2 toxin on the bovine immune system: Cellular factors. Vet. Immunol. Immunopathol. 3:411-417.

Campbell, M. L., Doerr, J. A., and Wyatt, R. D. 1981. Immune status in broiler chickens during citrinin toxicosis. Poult. Sci. 60:1634.

Campbell, M. L., May, J. D., Huff, W. E., and Doerr, J. A. 1983. Evaluation of immunity of young broiler chickens during simultaneous aflatoxicosis and ochratoxicosis. Poult. Sci. 62:2138-2144.

Chang, C. F., Huff, W. E., and Hamilton, P. B. 1979. A leukocytopenia induced in chickens by dietary ochratoxin A. Poult. Sci. 58:555-558.

Colvin, B. M., and Harrison, L. R. 1992. Fumonisin-induced pulmonary edema and hydrothorax in swine. Mycopathologia 117:79-82.

Cooray, R. 1984. Effects of some mycotoxins on mitogen-induced blastogenesis and SCE frequency in human lymphocytes. Food Chem. Toxicol. 22:529-534.

Cooray, R., and Jonsson, P. 1990. Modulation of resistance to mastitis pathogens by pretreatment of mice with T-2 toxin. Food Chem. Toxicol. 28:687-692.

Coppo, R., Roccatello, D., Amore, A., Quattrochio, G., Molino, A., Giangolo, B., Amoroso, A., Bajardi, P., and Piccoli G. 1990. Effects of gluten-free diet in primary IgA nephropathy. Nephrology 33:72-86.

Corrier, D. E. 1991. Mycotoxicosis: Mechanisms of immunosuppression. Vet. Immunol. Immunopharmacol. 30:73-87.

Corrier, D. E., and Ziprin, R. L. 1987. Immunotoxic effects of T-2 mycotoxin on cell-mediated resistance to *Listeria monocytogenes* infection. Vet. Immunol. Immunopathol. 14:11-21.

Corrier, D. E. Holt, P. S., and Mollenhauer, H. H. 1987a. Regulation of murine macrophage phagocytosis of sheep erythrocytes by T-2 toxin. Am. J. Vet. Res. 48:1304-1307.

Corrier, D. E., Ziprin, R. L., and Mollenhauer, H. H. 1987b. Modulation of cell-mediated resistance to listeriosis in mice given T-2 toxin. Toxicol. Appl. Pharmacol. 89:323-331.

Creppy, E. E., Stormer, F. C., Roschenthaler, R., and Dirheimer, G. 1988. Effects of two metabolites of ochratoxin A, (4R)-4-hydroxyochratoxin A and ochratoxin, on immune response in mice. Infect. Immun. 39:1015-1018.

Cusumano, V., Costa, G. B., and Seminara, S. 1990. Effect of aflatoxins on rat peritoneal macrophages. Appl. Environ. Microbiol. 56:3482-3484.

Dombrink-Kurtzman, M. A., Bennett, G. A., and Richard, J. L. 1992a. Avian lymphocytes as in vitro models to predict fumonisin cytotoxicity. (Abstr.) FASEB (Fed. Am. Soc. Exp. Biol.) J. 6:A2007.

Dombrink-Kurtzman, M. A., Bennett, G. A., and Richard, J. L. 1992b. Cytotoxicity of fumonisins in avian lymphocytes. (Abstr.) Assoc. Off. Anal. Chem. Ann. Meet. 106, p. 145.

Dombrink-Kurtzman, M. A., Javed, T., Bennett, G. A., Richard, J. L., Cote, L. M., and Buck, W. B. 1992c. Lymphocyte cytotoxicity and erythrocyte abnormalities induced in broiler chicks by fumonisins B$_1$ and B$_2$ and moniliformin from *Fusarium proliferatum*. (Abstr.) Assoc. Off. Anal. Chem. Ann. Meet. 106, p. 233.

Dong, W., and Pestka, J. J. 1993. Persistent dysregulation of IgA production and IgA nephropathy in the B6C3F1 mouse following withdrawal of dietary vomitoxin (deoxynivalenol). Fundam. Appl. Toxicol. 20:38-47.

Dong, W., Sell, J. E., and Pestka, J. J. 1991. Quantitative assessment of mesangial immunoglobulin A (IgA) accumulation, elevated, circulating IgA immune complexes, and hematuria during vomitoxin-induced IgA nephropathy. Fundam. Appl. Toxicol. 17:197-207.

Dorner, J. W., Cole, R. J., Lomax, L. G., Gosser, H. S., and Diener, U. L. 1983. Cyclopiazonic acid production by *Aspergillus flavus* and its effects on broiler chickens. Appl. Environ. Microbiol. 46:698-703.

Dwivedi, P., and Burns, R. B. 1984. Pathology of ochratoxicosis A in young broiler chicks. Res. Vet Sci. 36:92-103.

Dwivedi, P., and Burns, R. B. 1985. Immunosuppressive effects of ochratoxin A in young turkeys. Avian Pathol. 14:213-225.

Efrat, S., Zelig, S., Yagen, B., and Kaempfer, R. 1984. Superinduction of human interleukin-2 messenger RNA by inhibitor of translation. Biochem. Biophys. Res. Commun. 123:842-848.

Eldeib, M. M. R., and Reddy, C. S. 1990. Toxic effects of secalonic acid D in mice and protection by dimethylsulfoxide. Bull. Environ. Contam. Toxicol. 44:692-699.

Escoula, L., Bourdiol, D., Linas, M. S., Recco, P., and Seguela, J. P. 1988a. Enhancing resistance and modulation of humoral immune response to experimental *Candida albicans* infection by patulin. Mycopathologia 103:153-156.

Escoula, L., Thomsen, M., Bourdiol, D., Pipy, B., Peuriere, S., and Roubinet, F. 1988b. Patulin immunotoxicology: Effect on phagocyte activation and the cellular and humoral immune system of mice and rabbits. Int. J. Immunopharmacol. 10:983-989.

Fleischhacker, D. S., Akers, T. G., Katz, S. P., and Palmgren, M. S. 1986. Enhancement of influenza virus infections by secalonic acid D. Environ. Health Perspect. 66:113-118.

Forsell, J. H., and Pestka, J. J. 1985. Relation of 8-ketotrichothecene and zearalenone analogue structure to inhibition of mitogen-induced human lymphocyte blastogenesis. Appl. Environ. Microbiol. 50:1304-13009.

Forsell, J., Kately, J. R., Yoshizawa, T., and Pestka, J. J. 1985. Inhibition of mitogen-induced blastogenesis in human lymphocytes by T-2 toxin and its metabolites. Appl. Environ. Microbiol. 49:1523-1526.

Forsell, J. H., Witt, M. F., Tai, J.-H., Jensen, R., and Pestka, J. J. 1986. Effects of 8-week exposure of the B6C3F1 mouse to dietary deoxynivalenol (vomitoxin) and zearalenone. Food Chem. Toxicol. 24:213-219.

Friend, S. C. E., Babiuk, L. A., and Schiefer, H. B. 1983. The effects of T-2 toxin on the immunological function and herpes simplex reactivation in Swiss mice. Toxicol. Appl. Pharmacol. 69:234-244.

Fromentin, H., Salazar-Mejicanos, S., and Mariat, F. 1981. Experimental Cryptococcoses in mice treated with diacetoxyscirpenol, a mycotoxin of *Fusarium*. Sabouraudia 19:311-313.

Gelderblom, W. C. A., Kriek, N. P. J., Marasas, W. F. O., and Thiel P. G. 1991. Toxicity and carcinogenicity of the *Fusarium moniliforme* metabolite, fumonisin B_1, in rats. Carcinogenesis 12:1247-1251.

Gerberick, G. F., and Sorenson, W. G. 1983. Toxicity of T-2 toxin, a *Fusarium* mycotoxin, to alveolar macrophages in vitro. Environ. Res. 32:269-285.

Gerberick, G. F., Sorenson, W. G., and Lewis, D. M. 1984. The effects of T-2 toxin on alveolar macrophage function in vitro. Environ. Res. 33:246-260.

Ghosh, R. C., Chauhan, H. V. S., Roy, S. 1990. Immunosuppression in broilers under experimental aflatoxicosis. Br. Vet. J. 146:457-462.

Ghosh, R. C., Chauhan, H. V. S., and Jha, G. J. 1991. Suppression of cell-mediated immunity by purified aflatoxin B_1 in broiler chicks. Vet. Immunol. Immunopathol. 28:165-172.

Gibson, R. M., Bailey, C. A., Kubena, L. F., Huff, W. E., and Harvey, R. B. 1990. Impact of L-phenylalanine supplementation on the performance of three-week-old broilers fed diets containing ochratoxin A.1. Effects on body weight, feed conversion, relative organ weight, and mortality. Poult. Sci. 69:414-419.

Greene, D., Azcona-Olivera, J., Warner, R., and Pestka, J. J. 1993. Male predilection for vomitoxin-induced IgA nephropathy in the B6C3F1 mouse. (Abstr.) Soc. Toxicol. Ann. Meet. 1993. p. 105.

Gunther, R., Abbas, H. K., and Mirocha, C. J. 1989. Acute pathological effects on rats of orally administered wortmannin-containing preparations and purified wortmannin from *Fusarium oxysporum*. Food Chem. Toxicol. 27:173-179.

Gupta, M., Bandopadhyay, S., Paul, B., and Majumder, S. K. 1979. Hematological changes produced in mice by ochratoxin A. Toxicology 14:95-98.

Haschek, W. M., Motelin, G., Ness, D. K., Harlin, K. S., Hall, W. F., Vesonder, R. F., Peterson, R. E., and Beasley, V. R. 1992. Characterization of fumonisin toxicity in orally and intravenously dosed swine. Mycopathologia 117:83-96.

Hatori, Y., Sharma, R. P., and Warren, R. P. 1991. Resistance of C57B$_1$6 mice to immunosuppressive effects of aflatoxin B_1 and relationship with neuroendocrine mechanisms. Immunopharmacology 22:127-136.

Haubeck, H.-D., Lorkowski, G., Kolsch, E., and Roschenthaler, R. 1981. Immunosuppression by ochratoxin A and its prevention by phenylalanine. Appl. Environ. Microbiol. 41:1040-1042.

Hill, J. E., Lomax, L. G., Cole, R. J., and Dorner, J. W. 1986. Toxicologic and immunologic effects of sublethal doses of cyclopiazonic acid in rats. Am. J. Vet. Res. 47:1174-1177.

Holt, P. S., Corrier, D. E., and DeLoach, J. R. 1988. Suppressive and enhancing effect of T-2 toxin on murine lymphocyte activation and interleukin 2 production. Immunopharmacol. Immunotoxicol. 10:365-385.

Hughes, B. J., Taylor, M. J., and Sharma, R. P. 1988. Effects of verrucarin A and roridin A, macrocyclic trichothecene mycotoxins, on the murine immune system. Immunopharmacology 16:79-87.

Hughes, B. J., Hsieh, G. C., Jarvis, B. B., and Sharma, R. P. 1989. Effects of macrocyclic trichothecene mycotoxins on the murine immune system. Arch. Environ. Contam. Toxicol. 18:388-395.

Hughes, B. J., Jarvis, B. B., and Sharma, R. P. 1990. Effects of macrocyclic trichothecene congeners on the viability and mitogenesis of murine splenic lymphocytes. Toxicol. Lett. 50:57-67.

Kadian, S. K., Monga, D. R., and Goel, M. C. 1988. Effect of aflatoxin B_1 on the delayed type hypersensitivity and phagocytic activity of reticuloendo-thelial system in chickens. Mycopathologia 104:33-36.

Kanai, K., and Kondo, E. 1984. Decreased resistance to *Mycobacterium* infection in mice fed a trichothecene compound. Jpn. J. Med. Sci. Biol. 37:97-104.

Kellerman, T. S., Marasas, W. F. O., Thiel, P. G., Gelderblom, W. C. A., Cawood, M., and Coetzer, J. A. W. 1990. Leucoencephalomalacia in two horses induced by oral dosing of fumonisin B_1. Onderstepoort J. Vet. Res. 57:269-275.

Lafarge-Frayssinet, C., Lespinaats, G., LaFont, P., Loisillier, F., Mousset, S., Rosenstein, Y., and Frayssinet, C. 1979. Immunosuppressive effects of *Fusarium* extracts and trichothecenes: Blastogenic response of murine splenic and thymic cells or mitogen. Proc. Soc. Exp. Biol Med. 160:302-311.

Lea, T., Steien, K., and Stormer, F. A. 1989. Mechanism of ochratoxin A-induced immunosuppression. Mycopathologia 107:153-159.

Luster, M. I., Boorman, G. A., Korach, K. S., Dieter, M. P., and Hong, L. 1984. Myelotoxicity toxicity resulting from exogenous estrogens evidence for bimodal mechanism of action. Int. J. Immunopharmacol. 6:287-297.

Luster, M. I., Pfeifer, R. W., and Tucker, A. N. 1985. The immunotoxicity of natural and environmental estrogens. Pages 315-326 in: Immunotoxicology and Pharmacology. J. Dean, M. Luster, A. E. Munson, and H. Amos, eds. Raven Press, New York.

Luster, M. I., Germolec, D. R., Burleson, G. R., Jameson, C. W., Ackermann, M. F., Lamm, K. R., and Hayes, H. T. 1987. Selective immunosuppression in mice of natural killer cell activity by ochratoxin A. Cancer Res. 47:2259-2263.

Manolova, Y., Manolov, G., Parvanova, L., Petkova-Bocharova, T., Castegnaro, M., and Chernozemsky, I. N. 1990. Induction of characteristic chromosomal aberrations, particularly X- trisomy, in cultured human lymphocytes treated by ochratoxin A, a mycotoxin implicated in Balkan endemic nephropathy. Mutat. Res. 231:143-149.

McKinley, E. R., Carlton, W. W., and Boon, G. D. 1982. Patulin mycotoxicosis in the rat: Toxicology, pathology and clinical pathology. Food Chem. Toxicol. 20:289-300.

Mekhancha-Dahel, C., Lafarge-Frayssinet, C., and Frayssinet, C. 1990. Immuno-suppressive effects of four trichothecene mycotoxins. Food Addit. Contam. 7:S94-S96.

Miller, J. D. 1992. Fungi as contaminants in indoor air. Atmos. Environ. 26A:2163-2172.

Miller, K., and Atkinson, H. A. C. 1986. The in vitro effects of trichothecenes on the immune system. Food Chem. Toxicol. 24:545-549.

Miller, K., and Atkinson, H. A. C. 1987. The in vitro effects of trichothecenes on the immune system. Arch. Toxicol. 11(Suppl.):321-324.

Minervini, F., Lucivero, G., Visconti, A., and Bottalico, C. 1992. Immunomodulatory effects of fusarochromanones TDP-1 and TDP-2. Nat. Toxins 1:15-18.

Mirocha, C. J., and Abbas, H. K. 1989. Chemistry, occurrence and toxicology of the hemorrhagic mycotoxin (wortmannin) produced by *Fusarium*. Pages 213-221 in: Mycotoxins and Phycotoxins 88. S. Natori, K. Hashimoto, and Y. Ueno, eds. Elsevier, New York.

Moore, C., Bilodeau, R., and Wiseman, B. 1990. Clinical aspects and consequences of mystery swine disease in nursery and grow-finish pigs. Pages 41-51 in: Proc. Mystery Swine Disease Committee Meeting, Denver, Colorado. Livestock Conservation Institute, Madison, WI.

Neldon-Ortiz, D. L., and Qureshi, M. A. 1991. Direct and microsomal activated aflatoxin

B₁ exposure and its effects on turkey peritoneal macrophage functions in vitro. Toxicol. Appl. Pharmacol. 109:432-442.

Neldon-Ortiz, D. L., and Qureshi, M. A. 1992. The effects of direct and microsomal activated aflatoxin B₁ on chicken peritoneal macrophages in vitro. Vet. Immunol. Immunopathol. 31:61-76.

Niyo, K. A., Richard, J. L., Niyo, Y., and Tiffany, L. H. 1988. Effects of T-2 mycotoxin ingestion on phagocytosis of *Aspergillus fumigatus* conidia by rabbit alveolar macrophages and on hematologic, serum biochemical and pathologic changes in rabbits. Am. J. Vet. Res. 49:2151-2160.

Otokawa, M. 1983. Immunological disorders. Pages 163-170 in: Trichothecenes: Chemical, Biological, and Toxicological Aspects. Y. Ueno, ed. Elsevier, New York.

Panangala, V. S., Giamgrone, J. J., Diener, U. L., Davis, N. D., Hoerr, F. J., Mitra, A., Schultz, R. D., and Wilt, G. R. 1986. Effects of aflatoxin on the growth performance and immune responses of weanling swine. Am. J. Vet. Res. 47:2062-2067.

Paucod, J.-C., Krivobok, S., and Vidal, D. 1990. Immunotoxicity testing of mycotoxins T-2 and patulin on BALB/c mice. Acta Microbiol. Hung. 37:331-339.

Pestka, J. J., and Bondy, G. S. 1990. Alteration of immune function following dietary mycotoxin exposure. Can. J. Physiol. Pharmacol. 68:1009-1016.

Pestka, J. J., and Dong, W. 1992. Serum IgE hyperelevation in B6C3F1 mice following pulsed dietary exposure to the trichothecene vomitoxin. FASEB (Fed. Am. Soc. Exp. Biol.) J. 6:A1875.

Pestka, J. J., and Forsell, J. H. 1988. Inhibition of human lymphocyte transformation by the macrocyclic trichothecenes roridin A and verrucarin A. Toxicol. Lett. 41:215-222.

Pestka, J. J., Tai, J-H., Witt, M. F., Dixon, D. E., and Forsell, J. H. 1987. Suppression of immune response in the B6C3F1 mouse after dietary exposure to the *Fusarium* mycotoxins deoxynivalenol (vomitoxin) and zearalenone. Food Chem. Toxicol. 25:297-304.

Pestka, J. J., Dong, W., Warner, R. L., Rasooly, L., Bondy, G. S., and Brooks, K. H. 1990a. Elevated membrane IgA⁺ and CD4⁺ (T helper) populations in murine Peyer's patch and splenic lymphocytes during dietary administration of the trichothecene vomitoxin (deoxynivalenol). Food Chem. Toxicol. 28:409-420.

Pestka, J. J., Warner, R. L., Dong, W., Rasooly, L., and Bondy, G. S. 1990b. Effects of dietary administration of the trichothecene vomitoxin (deoxynivalenol) on IgA and IgG secretion by Peyer's patch and splenic lymphocytes. Food Chem. Toxicol. 28:693-699.

Pier, A. C. 1981. Mycotoxins and animal health. Adv. Vet. Sci. Comp. Med. 25:185-243.

Pier, A. C. 1986. Immunomodulation in aflatoxicosis. Pages 143-148 in: Diagnosis of Mycotoxicosis. J. L. Richard and J. R. Thurston, eds. Martinus Nijhoff Publishers, Boston.

Pier, A. C., and McLoughlin, M. E. 1985. Mycotoxic suppression of immunity. Pages 507-519 in: Trichothecenes and Other Mycotoxins. J. Lacey, ed. John Wiley and Sons, Toronto.

Pier, A. C., Varman, M. J., Dahlgren, R. R., Belden, E. L., and Maki, L. R. 1986. Aflatoxin suppression of cell mediated immune response and interaction with T-2 toxin. Pages 423-434 in: Mycotoxins and Phycotoxins. P. S. Steyn and R. Vleggaar, eds. Elsevier, New York.

Pier, A. C., Belden, E. L., Ellis, J. A., Nelson, E. W., and Maki, L. R. 1989. Effects of cyclopiazonic acid and aflatoxin singly and in combination on selected clinical, pathological and immunological responses of guinea pigs. Mycopathologia 105:135-142.

Porcher, J.-M., Dahel, C., Lafarge-Frayssinet, C., Chu, F. S., and Frayssinet, C. 1988. Uptake and metabolism of T-2 toxin in relation to its cytotoxicity in lymphoid cells. Food Chem. Toxicol. 26:587-593.

Prior, M. G., and Sisodia, C. S. 1982. The effects of ochratoxin A on the immune response of Swiss mice. Can. J. Comp. Med. 46:91-96.

Qureshi, M. A., and Hagler, W. M. 1992. Effect of fumonisin-B_1 exposure on chicken macrophage function in vitro. Poult. Sci. 71:104-112.

Radovanovic, Z., Naunovic, T., and Velimirovic, D. 1984. Clustering of the upper urothelial tumours in a family. Oncology (Basel) 41:396-398.

Raisuddin, Singh, K. P., Zaidi, S. I. A., Saxena, A. K., and Ray, P. K. 1990. Effects of aflatoxin on lymphoid cells of weanling rat. J. Appl. Toxicol. 10:245-250.

Rasooly, L., and Pestka, J. J. 1992. Vomitoxin-induced dysregulation of serum IgA, IgM and IgG reactive with gut bacterial and self antigens. Food Chem. Toxicol. 30:499-504.

Rasooly, L., Abouzied, M., and Pestka, J. J. 1993. Polyreactive IgA antibodies in vomitoxin-induced IgA nephropathy. (Abstr.) Soc. Toxicol. Annu. Meet. p. 104.

Reddy, R. V., and Sharma, R. P. 1989. Effects of aflatoxin B_1 on murine lymphocytic functions. Toxicology 54:31-44.

Reddy, R. V., Taylor, M. J., and Sharma, R. P. 1987. Studies of immune function of CD-1 mice exposed to aflatoxin B_1. Toxicology 43:123-132.

Reddy, R. V., Taylor, M. J., and Sharma, R. P. 1988. Evaluation of citrinin toxicity on the immune functions of mice. J. Food Protect. 51:32-36.

Reubel, G. H., Gareis, M., and Amselgruber, W. M. 1987. Cytotoxicity evaluation of mycotoxins by an MTT bioassay. Mycotoxin Res. 3:85-96.

Richard, J. L., Thurston, J. R., Deyse, B. L., and Booth, G. D. 1975. Effect of ochratoxin and aflatoxin on serum proteins, complement activity, and antibody production to *Brucella abortus* in guinea pigs. Appl. Microbiol. 29:27-29.

Richard, J., Thurston, J. R., and Pier, A. C. 1978. Effects of mycotoxins on immunity. Pages 801-817 in: Toxins: Animal, Plant and Microbial. P. Rosenberg, ed. Pergamon Press, New York. Richard, J. L., Peden, W. M., Fichtner, R. E., and Cole, R. J. 1986. Effect of cyclopiazonic acid on delayed hypersensitivity to *Mycobacterium tuberculosis*, complement activity, serum enzymes, and bilirubin in guinea pigs. Mycopathologia 96:73-77.

Robbana-Barnat, S., Lafarge-Frayssinet, C., Cohen, H., Neish, G. A., and Frayssinet, C. 1988. Immunosuppressive properties of deoxynivalenol. Toxicology 48:155-166.

Rosenstein, Y., LaFarge-Frayssinet, C., Lespinats, G., Loisillier, F., Lafont, P., and Frayssinet, C. 1979. Immunosuppressive activity of *Fusarium* toxins. Effects on antibody synthesis and skin grafts of crude extracts, T-2 toxin and diacetoxyscirpenol. Immunology 36:111-117.

Salazar, S., Fromentin, H., and Mariat, F. 1980. Effects of diacetoxyscirpenol on experimental candidiasis of mice. C. R. Acad. Sci. Ser. 3, 290:877-878.

Scott, T. R., Rowland, S. M., Rodgers, R. S., and Bodine, A. B. 1991. Genetic selection for aflatoxin B_1 resistance influences chicken T-cell and thymocyte proliferation. Dev. Comp. Immunol. 15:383-391.

Shier, W. T., Abbas, H. K., and Mirocha, C. J. 1991. Toxicity of the mycotoxins fumonisins B_1 and B_2 and *Alternaria alternata* f. sp. *lycopersici* (AAL) in cultured mammalian cells. Mycopathologia 116:97-104.

Singh, J., Tiwari, R. P., Singh, G., Singh, S., and Vadehra, D. V. 1987. Biochemical and immunological effects of aflatoxin in rabbits. Toxicol. Lett. 35:225-230.

Sorenson, W. G., Simpson, J., and Castranova, V. 1985. Toxicity of the mycotoxin patulin for rat alveolar macrophages. Environ. Res. 38:407-416.

Sorenson, W. G., Gerberick, G. F., Lewis, D. M., and Castranova, V. 1986. Toxicity of mycotoxins for the rat pulmonary macrophage in vitro. Environ. Health Perspect. 66:45-53.

Sorenson, W. G., Frazer, D. G., Jarvis, B. B., Simpson, J., and Robinson, V. A. 1987. Trichothecene mycotoxins in aerosolized conidia of *Stachybotrys atra*. Appl. Environ. Microbiol. 53:1370-1375.

Speijers, G. J. A., Franken, M. A. M., and van Leewen, F. X. R. 1988. Subacute toxicity study of patulin in the rat: Effects on the kidney and the gastro-intestinal tract. Food Chem. Toxicol. 26:23-30.

Steyn, P. S. 1984. Ochratoxins and related dihydroisocoumarins. Pages 183-216 in: Mycotoxins—Production, Isolation, Separation and Purification. V. Betina, ed. Elsevier, New York.

Tai, J.-H., and Pestka, J. J. 1988a. Impaired murine resistance to *Salmonella typhimurium* following oral exposure to the trichothecene T-2 toxin. Food Chem. Toxicol. 26:691-698.

Tai, J.-H., and Pestka, J. J. 1988b. Synergistic interaction between the trichothecene T-2 toxin and *Salmonella typhimurium* lipopolysaccharide in C3H/HeN and C3H/HeJ mice. Toxicol. Lett. 44 191-200.

Tai, J.-H., and Pestka, J. J. 1990. T-2 toxin impairment of murine response to *Salmonella typhimurium*: A histopathologic assessment. Mycopathologia 109:149-155.

Taylor, M. J., Reddy, R. V., and Sharma, R. P. 1985. Immunotoxicity of repeated low level exposure to T-2 toxin, a trichothecene mycotoxin, in CD-1 mice. Mycotoxin Res. 1:57-64.

Taylor, M. J., Hughes, B. J., and Sharma, R. P. 1987. Dose and time related effects of T-2 toxin on mitogenic response of murine splenic cells in vitro. Int. J. Immunopharmacol. 9:107-113.

Taylor, M. J., Smart, R. A., and Sharma, R. P. 1989. Relationship of the hypothalamic-pituitary-adrenal axis with chemically induced immunomodulation. I. Stress-like response after exposure to T-2 toxin. Toxicology 56:179-195.

Taylor, M. J., Lafarge-Frayssinet, C., Luster, M. I., and Frayssinet, C. 1991. Increased endotoxin sensitivity following T-2 toxin treatment is associated with increased absorption of endotoxin. Toxicol. Appl. Pharmacol. 109:51-59.

Thompson, W. L., and Wannemacher, R. W. 1986. Structure-function relationships of 12,13-epoxytrichothecene mycotoxins in cell culture: Comparison to whole animal lethality. Toxicon 24:985-994.

Thurston, J. R., Richard, J. L., and Peden, W. M. 1986. Immunomodulation in mycotoxicoses other than aflatoxicosis. Pages 149-161 in: Diagnosis of Mycotoxicosis. J. L. Richard and J. R. Thurston, eds. Martius Nijhoff Publishers, Boston.

Tomar, R. S., Blakley, B. R., Schiefer, H. B., and DeCoteau, W. E. 1986. In vitro effects of 3-acetyl-deoxynivalenol on the immune response of human peripheral blood lymphocytes. Int. J. Immunopharmacol. 8:125-130.

Tomar, R. S., Blakley, B. R., and DeCoteau, W. E. 1987. Immunological responsiveness of mouse spleen cells after in vivo or in vitro exposure to 3-acetyl-deoxynivalenol. Food Chem. Toxicol. 25:393-398.

Tomar, R. S., Blakley, B. R., and DeCoteau, W. E. 1988. In vitro effects of T-2 toxin on the mitogen responsiveness and antibody-producing ability of human lymphocytes. Toxicol. Lett. 40:109-117.

Tryphonas, H., Iverson, F., So, Y., Nera, E. A., McGuire, P. F., O'Grady, L., Clayson, D. B., and Scott, P. M. 1986. Effects of deoxynivalenol (vomitoxin) on the humoral and cellular immunity of mice. Toxicol. Lett. 30:137-150.

Vidal, D. R. 1990. Proprietes immunosuppressives des mycotoxines du groupe des trichothecenes. Bull. Inst. Pasteur 88:159-192.

Vidal, D., and Mavet, S. 1989. In vitro and in vivo toxicity of T-2 toxin, a *Fusarium* mycotoxin, to mouse peritoneal macrophages. Infect. Immun. 57:2260-2264.

Virdi, J. S., Tiwari, R. P., Saxena, M., Khanna, V., Singh, G., Saini, S. S., and Vadehra, D. V. 1989. Effects of aflatoxin on the immune system of the chick. J. Appl. Toxicol. 9:271-275.

Voss, K. A., Norred, W. P., Hinton, D. M., Cole, R. J., and Dorner, J. W. 1990. Subchronic oral toxicity of cyclopiazonic acid (CPA) in male Sprague-Dawley rats. Mycopathologia 110:11-18.

Wiesinger, D., Gubler, H. U., Haefliger, W., and Hauser, D. 1974. Antiinflammatory activity of the new mould metabolite 11-desacetoxy-wortmannin and some of its derivatives. Experientia 30:135-136.

Yang, K. H., Kim, B. S., Munson, A. E., and Holsapple, M. P. 1986. Immunosuppression induced by chemicals requiring metabolic activation in mixed cultures of rat hepatocytes and murine splenocytes. Toxicol. Appl. Pharmacol. 83:420-429.

Chapter 9

Toxicology of Mycotoxins

D. B. Prelusky, B. A. Rotter, and R. G. Rotter

Numerous surveys have shown that while individual countries may encounter vastly different problems associated with mold and mycotoxin contamination, taken collectively these toxins are a worldwide concern uncontained by national borders. Toxicological diseases of animals (and man) associated with mold-infested feeds and foodstuffs have been recognized or suspected for centuries. However, the insidious nature of mycotoxin production and resulting disease states makes it difficult to determine the extent of the problem. Assigning a specific toxin as the certain cause of a usually poorly diagnosed mycotoxicosis has proven to be a difficult task. While considerable effort has been made to link field outbreaks to the suspected toxin, in most cases any means of corroboration is circumvented for a number of reasons, as outlined by both Ciegler (1975) and Osweiler (1990):

1) Mycotoxins often occur in very low concentrations and may be difficult to detect, or analysis does not always result in an accurate assessment of the toxic potential of the feed either because of weak analytical procedures or the presence of unidentified toxins.

2) By the time a mycotoxicosis is suspected, the food or feed has often been disposed of and is no longer available for analysis.

3) Host clinical signs and/or symptoms are often present as vague, subacute, or chronic conditions. For some mycotoxins, the only signs may be poor performance, unthriftiness, or an increased susceptibility to infectious diseases. While new mycotoxins are being constantly identified, their biological, medical, and veterinary medical impacts often have not yet been defined.

The first portion of this chapter was written by D. B. Prelusky. B. A. Rotter and R. G. Rotter contributed the last section, Interactive Effects of Mycotoxins and Other Fungal Metabolites.

4) Veterinarians and physicians are not trained sufficiently or familiar with symptomatology of mycotoxicoses even in acute cases. Mycotoxins occur sporadically, both seasonally and geographically; this makes it difficult to implement an organized diagnostic response.

5) The important toxicologic principle of "dose-response" has been difficult to apply to the diagnosis of mycotoxin-related disease. This is often due to the chronic nature of mycotoxicoses, their sporadic occurrence, possible interactions with other toxins or nutrients, and other management practices.

6) Interactions between individual mycotoxins have not been well studied. This is of particular importance because contaminated products usually contain mixtures of mycotoxins, many unidentified, which complicate defining clinical manifestations as the result of a single compound or the interaction of two or more toxins.

Whereas field outbreaks of a suspected mycotoxicosis rarely provide the opportunity to monitor and evaluate the situation properly, there have been many instances where clinical evaluation using pure toxins in a research setting provided sufficient evidence to establish solidly a relationship between toxin exposure and disease.

Robb (1990), in an attempt to define a disease as a true mycotoxicosis, established the following criteria:

1) The disease must be feed related.

2) No microorganisms of significant pathogenicity have been isolated from the case.

3) The disease is not contagious.

4) The disease is not transferable.

5) The disease is not infectious.

6) When feed is withdrawn, the animal shows signs of improvement. Additional criteria concerning the actual toxin were suggested by Osweiler (1990):

7) A known mycotoxin is identified from the feed or animals in amounts sufficient to cause a problem.

8) Mycotoxins isolated are known to produce a disease typical of the clinical signs presented.

9) Feeding the suspected ration (if available) to other healthy animals of the same species reproduces the disease.

Ergot

Ergotism is the general, nonspecific term associated with ergot toxicoses. St. Anthony's fire is an old common term for ergotism occurring in humans.

Etiology

Toxicity associated with consumption of the sclerotia (ergot bodies) produced by fungi is not a recently identified problem. Historically there is evidence to indicate that ergot poisonings have plagued man and animals alike for centuries. Fortunately, as a known human toxicosis, ergotism is no longer prevalent due to improvements in preventing contaminated grain products from entering the food chain. However, there always exists a considerable risk to domestic animals and poultry.

Found throughout Canada on a wide range of grains and pasture grasses, ergot bodies contain a large number of biologically active compounds, including a variety of ergotamine, ergopeptide pyrrolizidine, and lysergic acid alkaloids, collectively referred to as ergot mycotoxins. These sclerotia are commonly and exclusively produced by several species of the fungus *Claviceps,* although there is reported evidence to suggest that other genera of fungi are also capable of synthesizing ergot alkaloids (Lyons et al 1986, Bacon et al 1986, Yates et al 1985). Consequently, there may be a link to other feed-induced disease states with ergotlike signs that are not yet confirmed to be ergot caused, including tall fescue grass toxicity (fescue foot, summer syndrome), rye grass staggers, Bermuda grass tremors, and Dallis grass shakes (Thompson and Porter 1990, Robbins et al 1986, Lindley 1978). The role of grasses and their interaction with certain species of fungi is not fully understood in the etiology of the many ergotlike toxicity syndromes. A considerable amount of information concerning ergot alkaloids has been reported over the years because of their wide range of potent biological actions. However, it is apparent that not all the ergot compounds have been identified, nor have their toxicological effects been defined.

The severity of the problem in livestock and poultry depends on such factors as species, age, and general health status of the animal, period of exposure, types of ergot, and amount consumed. The toxic alkaloid content of ergot-contaminated grains and grasses can depend on the strain and stage of maturity of the fungus, type of host plant, growing condition, and geographic conditions. The total concentration and composition of the ergot toxins may be highly variable between individual sclerotia, and consequently throughout a contaminated field. Exposure to the large variety of ergot toxins can be manifested by a wide range of clinical signs.

There appear to be two forms of ergotism observed in animals exposed to ergot, gangrenous and convulsive. Cattle appear to be more sensitive to these toxins than mice, sheep, or swine (Oresnik et al 1980, Robbins et al 1986).

Cattle

Among the major clinical signs of ergot poisoning in animals are lameness and gangrene. Resulting from the toxins' peripheral vasoconstrictive activity, the disease progresses to vascular thrombosis, stasis of blood flow, and subsequent diminished blood supply to the feet, tail, ears, teats, and other extremities (Robbins et al 1986, Woods et al 1966, Solomons et al 1989, Holliman and Barnes 1990). A sufficiently severe toxicosis will eventually result in lameness in the hind limbs, possibly accompanied by local hypothermia and insensitivity. This is often followed by development of gangrenous skin of the feet and eventual separation of the hoof at the coronary band.

Occasionally cattle have been reported to develop convulsive ergotism, usually described as "staggers" or "tremors" (Robbins et al 1986). Clinical signs typically include evidence of hyperexcitability with mild shaking and incoordination, which develops into an enhanced level of agitation and fearfulness. Although not usually fatal unless acutely toxic, animals in this hyperactive condition may accidentally harm themselves.

Low-level ergot poisoning has been more difficult to diagnose under field conditions, particularly if "classical" symptoms are not evident and thus ergot is not the suspected agent. Reduction of milk production may occur prior to other signs of chronic, low-level ergot toxicosis. Other clinical signs recorded during ergot outbreaks in cattle have included reduced weight gain, emaciation, increased salivation, diarrhea, areas of cutaneous (skin) necrosis and epilation, erosion of oral mucosa, reproductive problems, agalactia, abortion, and elevated temperatures or heat stress (hyperthermia) that causes animals to increase water intake and urination and to seek shade or stand in water (Bacon et al 1986, Appleyard 1986, Hemken 1988, Schmidt and Marple 1988, Coppock et al 1989a, Fraser and Dorling 1983).

Clinical manifestations of fat necrosis have also been associated with ergot toxicity. This syndrome is exhibited as masses of hard fat located in adipose tissues, typically in the abdominal cavity (Bush et al 1979, Rumsey et al 1979, Stuedemann et al 1985). Larger deposits surround and constrict intestines and reproductive organs leading to digestive and/or calving problems. Symptoms of fat necrosis include weight loss, decreased feed intake, and bloat. Rumsey et al (1979) speculated that elevated body temperature caused by ergot triggers a lipolitic process and formation of these hard fat deposits.

Fescue foot is a severe toxicosis in cattle with clinical signs very similar to those of ergot-induced gangrene of extremities. Associated with consumption of *Acremonium coenaphialum* infested tall fescue forage grass, the relationship with ergot remains unconfirmed because of the absence of the ergot fungus *Claviceps*, and in many instances, no detection of known ergot alkaloids (Robbins et al 1986). However, increasing reports of various ergot alkaloids being identified in tall fescue strongly suggest that fescue toxicosis is ergot caused (Lyons et al 1986, Yates et al 1985, Solomons et al 1989, Schmidt and Marple 1988, Hemken 1990, Bacon et al 1986). Summer syndrome, which is also associated with tall fescue toxicity and characterized by reduced weight gain, rough hair coat, and increased respiration and temperature, may be a less severe manifestation of ergot toxicity (Lyons et al 1986, Bush et al 1979). Animals displaying signs of the summer syndrome have also developed fat necrosis (Stuedemann et al 1975).

Sheep and Goats

Naturally occurring ergotism in sheep and goats appears to be uncommon, although some instances have been reported. Lambs and goat kids seem to be more susceptible than older animals, as are ewes in the later stages of pregnancy (Greatorex and Mantle 1974). Clinical signs are less stereotyped than with cattle. In sheep, anorexia and dullness appear to be the most overt signs, and there is no evidence of either gangrenous lesions of the extremities or nervous disorders (Løken 1984, Cunningham et al 1944, Greatorex and Mantle 1973). Increased pulse rate and abortions have also been cited. The most obvious pathological findings were extensive inflammation and necrotic lesions throughout the gastrointestinal tract. Experimentally produced ergot toxicoses resulted in anorexia, hyperventilation, cold extremities, increased salivation, and occasionally necrosis of the tongue (Greatorex and Mantle 1973). Postmortem examination also showed extensive inflammation of the alimentary tract. In goats, gangrenous necrosis has been reported, but although they were severely lame, the animals did not lose hooves (Hibbs and Wolf 1982).

It is suspected that low-level, chronic exposure of sheep to ergots is more commonplace than reported, but because of their apparent greater tolerance compared with cattle, these occurrences are probably largely at subtoxic concentrations.

Horses

Ergot toxicosis in horses appears to be very rare. While no confirmed cases have been documented, there are several suspected instances of ergot poisoning through consumption of contaminated Bermuda grass hay (Greatorex and Mantle 1974). Although affecting mainly the central nervous system, clinical manifestations are not well defined. Incoordination, trembling, peculiar head movements, and licking motions have been observed. Routine blood and urine examinations, when performed, did not disclose abnormalities. Animals tended to recover quickly once the suspect feed was removed.

Swine

Pigs are considered to be more tolerant to the toxic effects of ergots than ruminants and horses (Whittemore et al 1976, Bakau et al 1988).

They are unlikely to show signs of peripheral gangrene, and in fact, may only be sensitive to the central nervous system disturbances caused by elevated ergot levels (Carlevaro 1956). However, while some reports indicate no appreciable adverse effects at naturally occurring ergot levels (Dignean et al 1986), others have noted reduced feed consumption and growth, agalactia, and abortion (Bakau et al 1988; Friend and MacIntyre 1969, 1970; Whittemore et al 1976, 1977).

Poultry

While the presence of ergots can have significant adverse effects on overall performance, poultry appear to tolerate higher levels of the toxin in feedstuffs than do ruminants, horses, and swine (Rotter et al 1985b). Older birds had an increased tolerance, and broilers appeared to be marginally more sensitive than Leghorns (Rotter et al 1985b, J. C. Young and Marquardt 1982, Swarbrick and Swarbrick 1968). Other fowl reported to be affected include ducks and bobwhite quail (Swarbrick and Swarbrick 1968, Manning et al 1990).

Signs of increasing ergotism are characterized by reduced feed consumption, depressed growth, incoordination, poor feathering, and vasoconstriction resulting in elevated blood pressure, restricted blood flow and subsequent necrosis of toes, beak, and occasionally skin (Rotter et al 1985a,b; J. C. Young and Marquardt 1982; O'Neil and Rae 1965; Bragg et al 1970; Vanderbrook and Vos 1940). Additionally, behavioral effects such as increasing nervousness can be observed as the toxicoses progress (Bragg et al 1970). At higher ergot concentrations in feed, increased mortality has been reported after several weeks (Rotter et al 1985b).

The prognosis for recovery from low to moderate toxicoses is good once the contaminated feed is removed. A tendency for birds to show a slight adaptation to the ergot as exposure time increases has also been reported (Rotter et al 1985b).

Tolerance Levels

Because of the numerous factors affecting exposure of animals at pasture, there is considerable difficulty in determining amounts of ergot mycotoxins ingested and relating this to the severity of toxicosis produced. In general, a maximum intake of dietary ergot sclerotia of 0.1% has been recommended for cattle, sheep, and horses, and up to 0.2% for swine. No-effect levels as high as 0.3–0.8% have been reported for poultry (Young 1979, Bragg et al 1970, O'Neil and Rae 1965, Dignean et al 1986). Since the average alkaloid content for Canadian ergot has been estimated to range from 0.24 to 0.31% (Young 1981, Dignean et al 1986), 0.1% in feeds would be 2.4–3.1 ppm total alkaloid content.

It has been reported that cows estimated to have consumed 240 or 173 mg of ergot alkaloids per day showed signs of lameness and tissue necrosis in 10 or 14 days, respectively, with subsequent declines in milk yield (Fraser and Dorling 1983). Rough estimates of the case history indicate that a 500-kg cow grazing contaminated meadow hay at a rate of 2.5% body weight per day would need to be exposed to 0.8% dietary ergot with an alkaloid content of 0.24%. This is equivalent to a field containing 19.2 ppm total alkaloid content, or eight times the recommended maximum level.

With sheep, Greatorex and Mantle (1974) reported that ergotamine administered at a level of 1 mg/kg produced clinical signs of ergot toxicosis after only one day, which progressed rapidly to death as early as five days if the dose was maintained. Løken (1984), feeding increasing levels of alkaloids to lambs, saw no effect after 27 days at 0.48 mg/kg of body weight (bw) per day, but toxicosis was manifested at daily doses of 1 mg/kg bw and higher.

With swine, it has been reported that contamination of diets with total ergot alkaloids at up to 4.5 ppm (0.2% ergot × 0.227% alkaloids) had no significant adverse effects on maintenance of pregnancy and on lactation, on piglet weights and growth, nor on formation of lesions (Dignean et al 1986). Friend and MacIntyre (1970), however, noted that marginally lower weight and feed performance began at levels as low as 0.5 ppm total alkaloids.

In growing chicks, a concentration of 3.1 ppm total ergot alkaloids produced statistically significant reduction in weight gains and feed efficiency, progressing to an 80% decline in weight gains by 24.6 ppm ergot (Rotter et al 1985b). At lower levels (<10 ppm) no signs of ergot poisoning other than decreased feed performance and lethargy were noted. However, above 12 ppm there was a significant and progressive increase in mortality, and only those birds that died showed lesions of peripheral necrosis. O'Neil and Rae (1965) indicated a tolerance level for chicks was 0.3% ergot (about 7.0 ppm) in the diet, with respect to egg production, feed consumption, and maintenance of body weight. Egg quality and hatchability were not affected at this level. Even at 9% ergot (about 200 ppm), mortality in mature hens was not affected.

Dicoumarol

The common term for disease produced by dicoumarol is (moldy) sweet clover poisoning.

Etiology

Dicoumarol poisoning has long been associated with the feeding of spoiled sweet clover hay or silage. Although it is not synthesized from simple substrates as are most mycotoxins, dicoumarol is nevertheless a product of fungal metabolsim in that, coumarol, a normal nontoxic constituent of sweet clover, is converted to the toxic component by certain fungal species infesting the crop (Scheel 1978). Outbreaks of severe sweet clover poisonings are rare in Canada. The disease is reported primarily where sweet clover is grown extensively as a forage crop. The clinical signs, while varying considerably depending on the species, are related to the anticoagulant action of dicoumarol. Although the exact mode of action of the toxin has not been confirmed completely, it is widely accepted to act by inhibiting vitamin K–epoxide reductase, which decreases the availability of vitamin K for participation in the final activation of vitamin K–dependent coagulation factors (prothrombin, factors VII, IX, and X). Consequently, this impedes the clotting mechanism of the animal's blood (Blakley 1985, Blood et al 1983, Pritchard et al 1983). Consumption of sufficient quantities of dicoumarol essentially causes the animal to bleed to death internally.

Although all animals are considered susceptible to dicoumarol poisoning, feeding practices usually limit the toxicoses to cattle, although affected sheep and horses have been reported. One estimate of cattle poisoning in Saskatchewan (1983) was 1.56 and 0.95 cases per 10,000 for beef and dairy cattle, respectively (Blakley 1985).

Cattle

Calves, yearlings, and pregnant cows near parturition seem to be more susceptible to the action of dicoumarol than older stock and lactating animals (Fraser and Nelson 1959, Goodman and Gilman 1990). Characteristic clinical manifestations include increased clotting and prothrombin times with accompanying internal and external hemorrhage, anemia, pallor of mucous membranes, weakness, and abortion (Blakley 1985, Radostits et al 1980). Any accidental trauma by animals in this toxic state may result in severe bleeding. The extent of bleeding depends on the amount of dicoumarol ingested. While the disease may not be obvious for several months if the exposure level is low, and animals may function normally, at higher toxin concentrations signs can be overtly evident within weeks (Alstad et al 1985, Blakley 1985).

With severe toxicoses, animals may suffer extensive hemorrhage into subcutaneous tissues, organs, body cavities, and joints, causing discomfort, pain, unsteadiness, weakness, pallor, and possibly lameness and recumbency. As toxicoses progress, other signs may include tachypnoea, bloat, tachycardia, hematomas, and possible death (Pritchard et al 1983, Radostits et al 1980, Cranwell 1983).

Sheep

While reports of sweet clover poisoning of sheep are few and unsubstantiated in Canada, cases of dicoumarol poisoning associated with the giant fennel weed (*Ferula communis* L.) have been confirmed in Mediterranean countries (Egyed et al 1981, Shlosberg and Egyed 1985). While it has been indicated that sheep are more resistant than cattle to the toxic effects of dicoumarol, field case studies seem to suggest that instances of poisoning in sheep are almost always fatal (Shlosberg and Egyed 1985). Clinical findings in sheep are similar to those reported in cattle, although there is a considerable dose-effect variation in sheep, whereas cattle show a more consistent response.

Horses

One case of sweet clover poisoning of a horse has been documented in Canada (McDonald 1980). Clinical findings showed a history of spontaneous epistaxis (nose bleeds) of 24 hr duration, increased prothrombin time, pallor of the mucous membranes, subcutaneous edema, and slight blood loss anemia. Although dicoumarol analysis was not obtained, the animal had free access to weathered sweet clover hay prior to the diagnosis.

Miscellaneous

Because of the potential presence of the rodenticide warfarin, a derivative of dicoumarol with similar anticoagulant activity, incidences of hemorrhagic diathesis should be properly investigated to confirm the causative agent and source. Several other anticoagulant rodenticides are also in the marketplace and should similarly be considered, including brodifacoum, diphacinone, chlorophacinone, and related compounds.

Trichothecenes

Alimentary toxic aleukia, fusariotoxicoses, and scabby grain or moldy corn toxicoses are syndromes caused by trichothecenes.

Etiology

Trichothecenes are distributed widely in cereal crops, to the extent of ubiquity in certain crops grown in specific regions. Fortunately, in North America, the trichothecenes found most prominently tend to be those of lesser toxicity, or at such low concentrations as not to present an immediate threat. In Canada and the United States the trichothecene most commonly encountered appears to be deoxynivalenol (DON, vomitoxin), and to lesser extents, nivalenol, T-2 toxin, HT-2 toxin, and, rarely, diacetoxyscirpenol (DAS). While all trichothecenes can be acutely lethal if ingested in sufficient quantities, the major problems faced by North American livestock producers with regard to contaminated feeds tend to be the subacute toxicoses arising from chronic, low-level consumption of these toxins, which leads to subtle, nonspecific effects associated with poor performance. Because of the large number of trichothecene-type compounds produced by the *Fusarium* fungi, it is not always possible to identify the causative toxins, even though the infecting fungal species is known. In fact, toxic responses above what chemical analyses of the feed would predict strongly suggest that other unidentified toxins are involved.

Although all animals can be affected following exposure to trichothecenes, the severity of toxicoses depends on the specific toxin(s), the degree and duration of exposure, and of course the species of animal involved. Swine and other monogastric animals (including humans) display the greatest sensitivity to these toxins. Chickens and turkeys appear to have a high tolerance, followed by ruminants with little evidence of naturally occurring trichothecene-induced toxicoses.

Besides characteristic, overt effects such as feed refusal and vomiting, trichothecenes are potent immunosuppressants and inhibitors of protein synthesis, which can predispose animals to other diseases and mask the underlying toxicoses (Pestka and Bondy 1990, Taylor et al 1989, Thurston et al 1986).

For all species suffering from chronic trichothecene-induced toxicoses, there is an excellent prognosis for recovery once the contaminated feed source is removed.

Swine

The first evidence that swine have been exposed to toxic levels of trichothecenes usually begins with decreased consumption of the contaminated feed and an accompanying slowed weight gain (Friend et al 1986a; Moore et al 1985; Chavez and Rheaume 1986; Weaver et al 1978a, 1981). As the toxicosis progresses, or contamination levels increase, clinical signs can include diarrhea, lethargy, blanched skin color, dermal irritation, hypothermia and shivering, intestinal hemorrhage, and eventual complete feed refusal. Although vomiting is uncommon under normal feeding conditions because swine regulate toxin ingestion by decreasing feed intake, as levels increase, acute exposure can result in signs of abdominal distress, teeth grinding, increased salivation, and emesis (Vesonder and Hesseltine 1980, Williams et al 1988, Young et al 1983). Weaver et al (1978a, 1981), reported oral lesions in swine following exposure to elevated levels of DAS or T-2 toxin. Death, although attributed to the more toxic trichothecenes (DAS, T-2), has rarely, if ever, been documented to have occurred under natural circumstances.

Because swine tend to regulate the amount of toxin ingested with contaminated feeds, extensive lesions are usually not evident in field cases (Patterson et al 1979; Harvey et al 1989a,b; Chavez and Rheaume 1986;

Friend et al 1986a). Occasionally the effects observed cannot be distinguished from those that would occur from reduced feed consumption (Young et al 1983). Tissue necrosis (lesions) of organs and along the gastrointestinal tract has been noted (Marpegan et al 1988). In contrast, it has also been reported that DON-contaminated diets tend to maintain integrity of the stomach mucosa (less ulceration) better than clean, control diets, which is probably a reflection of the animal's altered eating behavior (Friend et al 1986a). Some authors have reported abnormal blood chemistry parameters and blood cell dyscrasias, but in most instances these have resulted from acute studies following either intravenous or inhalation administration of toxins or exposing animals to very high levels of contaminated feed (Weaver et al 1978a; Harvey et al 1990; Pang et al 1986, 1987a,b,c 1988; Lorenzana et al 1985; Beasley et al 1986; Coppock et al 1989a,b).

Reproductive problems do not appear to be a problem for swine with ingestion of trichothecenes at levels typically found in grains and feedstuffs (Friend et al 1983, 1986b), although nutritional problems could arise if feeding behavior were significantly altered. Experimentally induced T-2 toxicosis, however, results in infertility, small litter size, small piglets, and abortion (Weaver et al 1978b).

Poultry

Chickens and turkeys appear to be relatively tolerant of trichothecenes. Compared to swine, relatively high toxin levels in feed are needed to elicit a response. Chronic feeding studies involving T-2 toxin or DAS reveal reduced feed intake, slight reduction in weight gain, oral lesions, necrosis of certain tissues (lymphoid, hematopoietic, alimentary mucosal), possible neural disorders (abnormal wing positioning, hysteroid seizures, impaired righting reflex), and abnormal feathering (Burditt et al 1983, Hoerr et al 1982, Dziuk et al 1979, Wyatt et al 1973, Ademoyero and Hamilton 1989). Turkeys appear to be more susceptible to dietary T-2 toxin than chickens (Richard et al 1978). Similar studies with DON, however, have mostly shown that, except for a transitory decrease in hemoglobin levels, or a very slight effect on egg quality, there was no significant evidence that this particular toxin affects poultry performance (Kubena and Harvey 1988; Lun et al 1986; Kubena et al 1985, 1987a,b, 1989a; Hamilton et al 1983, 1985a,b; Moran et al 1987). Huff et al (1986), did report slowed weight gain, anemia, and various altered blood chemistry parameters in broilers fed DON-contaminated feed and concluded that rapidly growing broilers are more sensitive than layers to the feed refusal syndrome induced by these toxins.

In one study in which broiler chickens were intubated with a large single dose of DON, acute toxicosis was characterized by irritation of the upper gastrointestinal tract, hemorrhaging throughout the body, neurologic disturbances, kidney dysfunction, and an undefined inflammatory response (Huff et al 1981). For the birds that survived, recovery was rapid, and no clinical signs or lesions were noted seven days after dosing.

While trichothecene levels in typical contaminated grains are too low to produce observable effects in poultry, numerous studies have indicated that mixtures of toxins may potentially interact to cause increased toxicity. (Kubena et al 1989a,b; Branton et al 1989; Mannion and Blaney 1988; Huff et al 1986, 1988a,b,c; Manley et al 1988). Combinations involving DON, T-2 toxin, aflatoxin, and zearalenone have resulted in effects greater than the individual mycotoxins.

Cattle

Ruminants appear to be relatively tolerant of the adverse affects of trichothecenes. While all the factors which afford these animals protection are not known, much is due to the ability of the rumen microflora to detoxify these toxins. Nevertheless, there is circumstantial evidence to indicate that toxicoses have occurred under field conditions, particularly when the more toxic compounds (DAS, T-2) are involved. However, these have been difficult or impossible to reproduce experimentally. Investigations of several field outbreaks have concluded that trichothecenes are a probable causative agent in a hemorrhagic disease observed in cattle where clinical and pathological findings include reduced feed intake, decreased milk yield, hair loss, delayed clotting and prothrombin times, bloody diarrhea, and multiple petechial hemorrhages of the mucous membranes (Hsu et al 1972, Petrie et al 1977). However, many of the clinical signs produced experimentally, usually involving life-threatening doses of toxin (Gentry et al 1984, Coppock et al 1989b, Mann et al 1983, Rukhlyada 1989), have not been reproducible in other studies, particularly at the lower toxin levels more likely to be found in contaminated feedstuffs (Matthews et al 1977, Osweiler et al 1981, Patterson et al 1979). Because hemorrhagic problems have occurred almost exclusively in field situations, it is very possible the syndrome is caused either by a mixture of trichothecene toxins resulting in a synergistic interaction, or by other, unidentified compounds. Other findings reported at different exposure levels have included various blood dyscrasias and suppression of immune function, organ weight changes of thymus and adrenals, intestinal necrosis, ulceration around mouth and nostrils, and dermatitis (Kosuri et al 1970, Gabal et al 1986, Pier et al 1976).

Tolerance Levels

Considerable effort has been made to establish tolerance levels in swine, one of the species more sensitive to the effects of trichothecenes. Using diets containing added pure DON, adverse effects (decreased feed consumption, reduced weight gain) were initially observed at 5 ppm (Trenholm et al 1984); however, it is evident that additional toxins present in naturally contaminated grains can substantially reduce the first effect level measured as a function of DON concentration. Diets containing DON as low as 1–2 ppm have been reported to depress feed consumption significantly, with total feed refusal observed at 12 ppm, and vomiting at 20–23 ppm (Friend et al 1982 1986a, Marpegan et al 1988, Abbas et al 1986, Young et al 1983). There do not appear to be any reproductive problems directly associated with DON, even at higher concentrations, although very often there is also present sufficient zearalenone to cause an estrogenic syndrome in swine (Friend et al 1986a,b; Williams et al 1988; Moore et al 1985).

There are few reports involving effects of dietary T-2 toxin on swine. Weaver et al (1978a) estimated the no effect level to be less than 1 ppm in feed. At 1–8 ppm there appears to be a moderate, although significant effect on body weight and feed consumption. And, while young swine will tolerate 10–12 ppm, studies have shown that, at these higher levels, this toxin produces significant problems in weight gain, blood chemistry, and fertility (Harvey et al 1990, Weaver et al 1978b). Total feed refusal was observed at 16 ppm.

With purified DAS, the no-effect level in swine feed was less than 2 ppm (Weaver et al 1981). At 2–9 ppm DAS in the ration, oral lesions, as well as intestinal necrosis and subsequent hemorrhaging were observed. Decreased feeding was also evident, presumably as a result of the combined effects of central nervous system mediated reductions in appetite and the oral and intestinal lesions. Complete feed refusal occurred at 10 ppm of DAS.

Poultry appear to be capable of tolerating a relatively high concentration level of DON in their diet, and to a lesser extent T-2 toxin and DAS. At DON levels typically found in contaminated feedstuffs (0.35–8.0 ppm), there have been no indications of any problems (Hamilton et al 1983, 1985a,b, 1986); in fact, some birds appear to prefer a diet containing low levels of DON (Hamilton et al 1986). Concentrations of DON up to 82.8 ppm have been fed to Leghorn hens for 27 days with no effect on performance and no production of lesions (Lun et al 1986), although several other studies have reported very slight lesions, and reductions in egg quality do occur with time at 18 ppm (Kubena and Harvey 1988; Kubena et al 1985, 1987a,b) and above (Moran et al 1987). Fusarenon-X, a metabolite closely related to DON, was fed to broiler chicks at up to 8 ppm for a six-week period and without significant effects on performance or blood parameters, and no lesions were noted (Sato et al 1981).

DAS and T-2 toxin have produced dose-related oral lesions in broiler chickens at levels as low as 1 ppm of feed (Wyatt et al 1973, Hoerr et al 1982); more significant effects on performance occurred at 4 ppm and higher, including depressed feed intake and retarded growth, various blood chemistry changes, and neurotoxicity (Burditt et al 1983, Wyatt et al 1973).

Oral lesions were observed in turkey poults at 5 ppm T-2 and reduced weight gains at 10 ppm (Richard et al 1978, Dziuk et al 1979). A direct comparison feeding trial showed poults to be marginally more sensitive to the toxin than chicks (Richard et al 1978). Although one study did show 3 ppm of T-2 in naturally contaminated feed to be lethal to geese (Palyusik and Koplik-Kovács 1975), trichothecenes generally do not result in increased mortality, requiring levels of several hundred parts per million to result in death (Hoerr et al 1982). Similarly, in a toxicosis outbreak attributed to T-2 toxin that affected domestic ducks, geese, horses, and swine, mortalities occurred only in the geese, suggesting a greater sensitivity than the other species (Greenway and Puls 1976).

Ruminants are relatively insensitive to trichothecene-contaminated diets. Feeding dairy cows DON at 6.4 ppm for six weeks (Trenholm et al 1985) or at 66 ppm for five days (Côté et al 1986) produced no evidence of illness that could be attributed to the toxin, including feed consumption, weight gain, or milk production. One study showed a decrease in milk production in animals given a ration containing as little as 0.5 ppm of DON (Whitlow and Hagler 1987). The toxin source was from naturally contaminated grain and because of its unusual toxicity, it was probable that other, unidentified toxins were also contributing to the effect.

Harvey et al (1986) reported no problems in lambs fed a diet containing 15.6 ppm DON for four weeks. Prelusky et al (1987) found that of two one-year-old ewes offered a diet containing 880 ppm of DON for three days, one consumed it completely and the second refused about two-thirds of the total.

An initial response of decreased feed consumption by calves occurred in experiments with purified T-2 toxin, but there were no other obvious clinical manifestations at a total dietary level of 10 to 20 ppm (equivalent to 0.2–0.6 mg/kg bw of T-2 toxin per day) (Patterson et al 1979, Mann et al 1983, Osweiler et al 1981), far above levels typically found in contaminated feedstuffs. Various changes in hematology or clinical chemistry were associated with the higher level (Osweiler et al 1981, Rukhlyada 1989). It appears that calves are slightly more tolerant of DAS than T-2 toxin (Patterson et al 1979). Levels as low as 1.2 ppm of T-2 toxin total dietary concentration in naturally contaminated feed sources (feed composed of 60% 2.0 ppm T-2 toxin in moldy corn) consumed over several months were associated with lethal toxicoses in dairy cattle, although the authors admitted the toxin levels may have been much higher than reported (Hsu et al 1972). Severe health problems have been reported in field situations with undefined levels of T-2 toxin (Petrie et al 1977).

Zearalenone

F-2 toxicosis and hyperestrogenism are two diseases caused by zearalenone.

Etiology

Analytical surveys continually show the presence of low concentrations of zearalenone (ZEN) in certain grains grown in southern Canada and the upper-midwestern United States. ZEN and related metabolites possess strong estrogenic activity. Consequently, when they are fed to domestic animals in sufficient amounts, they can result in severe reproductive and infertility problems. Clinical reports of hyperestrogenism in swine were published as far back as the 1920s (Buxton 1927, McNutt et al 1928).

Because of its anabolic properties, zeranol (Ralgro), a synthetically produced chemical that very closely resembles zearalenone, has been used commercially as a growth promoter in beef cattle and sheep. Characteristics of hyperestrogenism can be exhibited in treated animals (Huston et al 1980, Perry et al 1970, Baldwin et al 1983).

Swine

Swine appear to be the most sensitive of the domestic animals and consequently the species most often reported to show problems associated with ZEN. Enlargement or swelling and reddening of the vulva in gilts and sows, generally termed vulvovaginitis, seems to be the most prevalent abnormality reported. Vaginal and rectal prolapses may also occur (L. G. Young and King 1986a; Sydenham et al 1988; Young et al 1981, 1982; Blaney et al 1984; Farnworth and Trenholm 1981; Bristol and Djurickovic 1971). Other signs may include swelling of the mammary glands and atrophy of the ovaries. In young males, the toxin can cause feminization—which includes swelling of the prepuce, testicular atrophy, and enlargement of the mammary glands—but does not appear to affect reproductive potential when they mature (Ruhr 1979). In boars, reduced libido and a marginal reduction in sperm quality have been reported, but there does not seem to be any effect on overall conception rate (Bristol and Djurickovic 1971, Berger et al 1981, L. G. Young and King 1986b).

In sows, interference with reproduction is the major problem. Low levels of ZEN have been reported to cause delayed first estrus, infertility characterized by continuous estrus, pseudopregnancy, and ovarian abnormalities. Other disorders include reduced litter size, loss of pregnancy, poor postweaning rebreeding performance, agalactia, and possibly increased fetal mortality, although some studies have disputed this last effect (Edwards et al 1987a,b; Green et al 1990; Young et al 1990; Friend et al 1990; Glavits et al 1983; Long and Diekman 1986; Gardner et al 1963). Few biochemical or structural alterations associated with chronic ZEN ingestion have been observed, and these are essentially limited to elevated serum progesterone levels, decreased prolactin and luteinizing hormone concentrations, and various minor histological lesions in the reproductive organs and adrenal and thyroid glands (Kurtz et al 1969).

Cattle and Ruminants

Field cases of zearalenone-induced problems in cattle and sheep have been well documented. Clinical manifestations are typically associated with a hyperestrogenic syndrome, including restlessness, udder enlargement, mucoid vaginal discharge, diarrhea, disrupted behavior during estrus, continuous estrus, infertility, vaginitis, decreased milk yield, and abortion. In prepubertal heifers, ZEN intoxication causes precocious mammary development and sterility (Bloomquist et al 1982, Coppock et al 1990, Kallela and Ettala 1984, Khamis et al 1986, Roine et al 1971, Schuh and Baumgartner 1988). Neither routine blood chemistries and hematological parameters nor gross and histologic appearance of tissue, however, were notably altered by the toxin, even at highly elevated concentrations (Weaver et al 1986a,b).

In sheep, problems with reproductive performance were manifested as depressed ovulation rates, increased duration of estrus, and lower lambing percentages (Jagusch et al 1986, Smith et al 1990).

Poultry

Except at extremely high levels of contamination, poultry do not appear to be significantly affected by ingestion of ZEN, although there is some indication that turkey poults are slightly more sensitive than layer and broiler chicks (Chi et al 1980a,b; Allen et al 1981; Lee et al 1985; Mirocha and Christensen 1974). At typical dietary levels, no changes in feed consumption, weight gain, egg production or quality, blood chemistry and hematology parameters, gross and histologic appearance of tissues, or behavior have been reported. With elevated amounts, or experimentally induced toxicoses in male turkeys, clinical signs still remain largely non-specific but include reduced feed efficiency, some organ weight changes, reduced fertility, and behavior changes. Instances where naturally contaminated diets resulted in increased mortality of birds have been reported, but it is unlikely that ZEN, although present, was the causative agent (Bock et al 1986, Meronuck et al 1970).

Tolerance Levels

Because of the considerable sensitivity of swine to zearalenone, it is difficult to confirm a no-effect level in this species. Several authors have reported definite dose-related reproductive dysfunctions in gilts at levels as low as 1–3 ppm and instances of vulvovaginitis at even lower levels (Friend et al 1990, L. G. Young and King 1986a, Young et al 1982). Various studies indicate that noticeable effects on reproductive efficiency in young gilts would probably begin appearing above 1 ppm dietary ZEN, with problems increasing rapidly as levels increased. Serious breeding problems have been observed at 5–6 ppm, although, once pregnant, levels up 15 ppm do not appear to affect litter development (Edwards et al 1987a,b; Long and

Diekman 1986, L. G. Young and King 1986a; Young et al 1982, 1990). Furthermore, Green et al (1990) indicated removal of ZEN-contaminated feed two weeks before insemination restored normal reproduction efficiency of gilts.

In boars, although 9 ppm of ZEN affects sperm quality slightly, it has no overall effect on impregnation (L. G. Young and King 1986a). Other studies have shown levels up to 200 ppm for eight weeks did not adversely diminish the reproductive potential of mature boars (Ruhr 1979).

In cattle, there have been only a few documented cases of infertility attributed to ZEN ingestion. Again problems associated with naturally contaminated feedstuffs were much more serious than when the purified toxin was used. Weaver et al (1986a) fed pure ZEN, equivalent to 50 ppm in total ration, to dairy heifers over three estrus cycles and observed no effect on conception rate or other parameters measured. However, Mirocha et al (1968) reported fertility disturbances in a herd of 150 dairy cows exposed to just 14 ppm of naturally produced ZEN in hay, and Roine et al (1971) noted estrus problems in 21 cows consuming moldy hay containing no measurable ZEN, but only the fungus (*Fusarium graminearum*) capable of producing the estrogenic toxin. In Finland, a series of early abortions was attributed to naturally contaminated hay containing 10 ppm of ZEN consumed over an unknown length of time; the problem ceased once the toxic feed was removed (Kallela and Ettala 1984).

It has been suggested that zearalenone by itself is not an important factor in dairy cow productivity (Weaver et al 1986a,b), but levels as low as 1.5 ppm of ZEN in combination with other known or unidentified toxins (i.e., *Fusarium* trichothecenes) have been reported to affect fertility and productivity seriously (Coppock et al 1990, Schuh and Baumgartner 1988, Bloomquist et al 1982).

Sheep may be moderately more sensitive than cattle to the effects of ZEN. Smith et al (1990) reported that ewes administered 3 mg of pure ZEN a day (equivalent to 2 ppm at 1.5 kg daily feed intake) before mating experienced depressed ovulation rates and lower lambing percentages. When exposure of ewes to ZEN at up to 24 mg/day was begun five days after mating, for a 10-day period, however, there was no effect on pregnancy rates or embryonic losses.

Poultry are very resistant to the toxicity of zearalenone. Several reports of broiler chickens and turkey poults fed up to 800 ppm of pure ZEN noted only very minor effects, and these occurred at the higher exposure levels. Included were decreased numbers of leukocytes, hypertrophy of some oviducts, decreased male broiler comb and testes weight, and increased development of dewlaps and carbuncles in male turkeys (Allen et al 1981, Chi et al 1980a). Similarly, Chi et al (1980b) orally administered to female White Leghorn chickens up to 800 mg/kg bw of ZEN (equivalent to several thousand parts per million of dietary ZEN) for seven consecutive days and noted, only at the upper dose levels, decreased weight gain and some organ weight changes. Birds receiving a single oral dose of 15 g/kg bw all survived with no noticeable symptoms.

Fumonisins

Equine leukoencephalomalacia and porcine pulmonary edema syndrome are toxicoses caused by fumonisins.

Etiology

Fumonisins were identified as the causative agent in several animal syndromes within the past five years. These problems have long been related to contamination with *Fusarium moniliforme*, the major fungus shown to produce these toxins, but the chemical nature of the responsible mycotoxin remained unknown until recently (Wilson and Maronpot 1971, Marasas et al 1976, Kriek et al 1981). Although the extent of fumonisin-induced problems is not known, the mold is one of the most prevalent fungi associated with certain grains such as corn (Marasas et al 1984). Consequently fumonisins are emerging as a highly visible animal and human health safety concern (Kellerman et al 1990, Wilson et al 1990, Norred et al 1989, Marasas et al 1988b).

Horses

Equine leukoencephalomalacia (ELEM) is a disease that tends to occur in sporadic, seasonal, epidemic-like outbreaks and is characterized by extensive damage to brain tissue. Apathy and decreased feed consumption may be the initial signs, which are followed quickly by various neurologic signs including ataxia, facial paralysis, abnormal head movement, aimless walking or circling, unilateral blindness, marked stupor, and increased excitability. As neurologic dysfunction progresses, lameness and recumbency may occur, followed by possible seizures and death. Depending on the amount of toxic material consumed, death can occur as quickly as several hours from the first onset of clinical signs (Wilson et al 1991, Kellerman et al 1990, Haliburton and Buck 1986, Marasas et al 1988b, Ivanov et al 1959). Overall mortality in horses exhibiting ELEM is generally low (less than 25%), although outbreaks in small herds may result in larger percentages. Animals that survive usually do so with some degree of permanent neurologic dysfunction.

The lesions of ELEM are well defined, consisting of liquefactive necrosis in one or both cerebral hemispheres. Depending on the severity of toxicoses, tissue damage ranges from subtle, microscopic lesions, to enlarged, irregular necrotic cavities where white matter of the brain has essentially disintegrated

(Marasas et al 1988b, Haliburton and Buck 1986, Schwarte et al 1937, Kellerman et al 1990, Brownie and Cullin 1987, Masri 1988). Edema, mineralization, minor hemorrhaging, mild to moderate fibrosis, and perivascular mononuclear cell infiltration may be noted in the surrounding malacic area. Other abnormalities reported have been elevated levels of aspartate aminotransferase, gamma-glutamyl-transferase, and blood ammonia. In some instances, the cerebrospinal fluid was characterized by increased numbers of leukocytes and elevated protein concentrations.

A hepatoxic syndrome has also been associated with fumonisin toxicosis in horses but is not well defined. The liver atrophies, becomes firm, and may acquire a yellowish discoloration. Jaundice and small visceral hemorrhages have also been reported. Histopathologic findings include centrilobular necrosis and fibrosis, cellular infiltration of neutrophils, bile duct proliferation, fatty degeneration, and inflammation. Except for mild swelling of the kidneys, no damage to other organs has been observed (Haliburton and Buck 1986, Kellerman et al 1990, Marasas et al 1988b, Uhlinger 1991).

Swine

While there is some evidence to suggest that outbreaks of porcine pulmonary edema (PPE) are caused by *F. moniliforme*, and presumably by the fumonisin mycotoxins it produces, available information is very limited. Studies have been made on swine fed feed contaminated with fumonisin-B_1 (FB_1) or extracts of *F. moniliforme* cultures containing high levels of the toxin (Colvin and Harrison 1992, Haschek et al 1992, Kriek et al 1981, Harrison et al 1990). At lower levels of fumonisin ingestion, clinical signs appear to be subtle and not well defined, although the liver appears to be a constant target organ in swine. At high-level exposures consistent with those sometimes encountered in field cases, animals showed rapid death from massive pulmonary edema or hydrothorax. Other clinical findings included elevated serum cholesterol levels and liver enzymes and abnormalities in membrane lipid turnover. Lesions included intracellular membrane degeneration and plasma membrane changes. Pancreatic lesions have been reported in cases of suspected FB_1 toxicoses. Most other organs do not appear to be damaged in swine.

Other

F. moniliforme contamination of grainstuffs has been associated with several other disease states in other species. Experimentally induced toxicosis in sheep is manifested by severe nephrosis and hepatosis (Kriek et al 1981). High concentrations of fumonisins have been shown to correlate with the increased incidence of human esophageal cancer in China and South Africa (Yang 1980, Marasas et al 1988a). However, while there is no direct indication that those toxins are the causative agents of this form of human cancer, there is evidence to implicate fumonisins as hepatocarcinogenic in rats (Gelderblom et al 1988, Wilson et al 1985, Voss et al 1990). The mechanism for promoting liver cancer has been speculated to be the potent and specific inhibition of sphingosine biosynthesis in rat hepatocytes (Norred et al 1992). In rats it has also been demonstrated that consumption of *F. moniliforme* contaminated diets can affect brain concentrations of various neurotransmitters (Porter et al 1990).

Tolerance Levels

Because fumonisins have only recently been identified, there is limited information detailing dose-effect relationships either from field cases or in the laboratory. Studies that have been carried out used either crude extracts of *F. moniliforme* cultures or naturally contaminated grains; consequently, the likelihood of other toxins being present in addition to those identified is very real.

Plattner et al (1990) reported on two field outbreaks in which several horses died with typical signs and/or lesions of ELEM after consuming contaminated feed. In the first case, four out of four animals died rapidly at 150 ppm FB_1 plus 17 ppm FB_2, and in the second incident, six out of 14 horses died after ingesting 20 ppm FB_1 plus traces of FB_2 in corn for approximately one month. In this second outbreak, neurologic dysfunctions were more evident. Wilson et al (1990) and Kellerman et al (1990) concluded that the ingestion of FB_1 at 0.6–2.1 or 1.25–4.0 mg/kg bw per day, respectively, would probably induce ELEM in horses and subsequently death. This is approximately equivalent to 35–125 ppm FB_1 in the diet, although the contribution of other fumonisins has not been taken into account.

There is a strong indication that toxicological problems in horses can develop at levels much lower than 20 ppm FB_1. This was supported by Thiel et al (1991), who reported that 14 different feed samples associated with confirmed cases of ELEM contained 1.3–27.0 ppm FB_1 and 0.1–12.6 ppm FB_2, and Ross et al (1991) who found FB_1 levels ranged from less than 1 ppm to 126 ppm in 98 samples associated with 44 cases of ELEM. The extent of the toxicoses associated with these toxin levels was not provided. In comparison, a study in which ponies were fed naturally contaminated corn screenings reported at 8 ppm FB_1 only minor nonspecific lesions in liver, kidney and brain stem; at 15–22 ppm, mild behavior changes, with one out of four animals dying from ELEM; and at 44 ppm and above, severe ELEM-related problems and mortality (Wilson et al 1992).

No low-level feeding studies have been described for swine. Ross et al (1991) reported feeds associated with PPE ranged from less than 1 ppm to more than 300 ppm FB_1, but due to problems of sample representativeness and lack of information concerning other possible toxins, the dose-response relationship for fumonisins could not be determined. Animals intubated with FB_1 at 4 mg/kg bw (about 50 ppm in diet) developed evidence of liver dysfunction by four weeks. At higher levels, severe liver disease and death occurred within seven to eight days, and at 16 mg/kg, pigs died within three to five days from massive pulmonary edema (Colvin and Harrison 1992). A similar study which included feeding >200 ppm total $FB_1 + FB_2$ in naturally contaminated corn screenings induced pulmonary edema, pancreatic lesions, and liver injury, although not all animals died (Haschek et al 1992). It has been suggested that pulmonary edema is reversible if the contaminated feed is removed early enough.

With rats, primary hepatocellular carcinomas were associated with long-term consumption of high levels of FB_1-contaminated diets (Gelderblom et al 1991). At 50 ppm semipurified (>90%) FB_1, a majority of rats developed the hepatocarcinoma after 18 months of exposure, with several affected animals also having metastases to the heart, lung, or kidney.

Ochratoxin

Etiology

The ochratoxins consist of several chemically related, highly toxic metabolites produced by molds belonging to the genera *Aspergillus* and *Penicillium*. These compounds have been found as contaminants of grain in North America and most European countries. The extent of available evidence strongly indicates that both acute and chronic forms of ochratoxin toxicoses occur in domestic livestock and poultry. A potent nephrotoxin and teratogen, ochratoxin A (OTA) has been implicated as the causative agent responsible for human endemic nephropathy prevalent in areas of the Balkan countries.

OTA typically coexists with low levels of citrinin, another nephrotoxic mycotoxin. Although citrinin is comparably less toxic than OTA and its toxicosis not well defined, its principal effects are similar to that of OTA-induced nephropathy in susceptible species (Krough 1978).

Swine

Swine are particularly susceptible to the effects of OTA, with the kidney as the target organ. OTA has been linked to mycotoxic porcine nephropathy, although other factors (i.e., citrinin) may also be involved as causal determinants (Krough et al 1979, Elling et al 1985). The first indication of a low-level toxicosis may not be evident until slaughter and inspection. Feeding diets with low concentrations of OTA, while sufficient to induce preliminary kidney damage, typically may not produce any overt symptoms, because clinical signs can be very subtle (Lippold et al 1987, Mortensen et al 1983). Visual signs of kidney damage are swollen and discolored appearance, and the initial histological abnormalities may include periglomerular fibrosis, tubular degeneration and atrophy, and interstitial fibrosis (Golinski et al 1984, Cook et al 1986, Elling 1983, Elling et al 1985, Tapia and Seawright 1985). Changes in biochemical and hematologic values have also been reported but are not well defined (Harvey et al 1989b, Lippold et al 1987, Tapia and Seawright 1985). As the nephropathy progresses, the kidney can enlarge to several times the normal size and show pronounced gross changes in color and texture, icterus, and advanced cellular damage. At higher toxin levels animals may show decreased performance and depressed weight gains (Huff et al 1988b, Cook et al 1986). Damage to other organs such as the liver and urinary bladder can occur. Excess water consumption, diarrhea, increased frequency of urination, and dehydration are signs of advanced renal damage, although complete renal failure probably will not occur even after chronic exposure to OTA (Krogh et al 1979).

OTA has been shown in vitro to inhibit porcine immune response (Holmberg et al 1988), but no correlation between contaminated feed and signs of infectious disease has been observed (Holmberg et al 1990). Ochratoxicosis in swine has not been associated with reproductive problems (Shreeve et al 1977, Mortensen et al 1983).

Poultry

OTA has been demonstrated to have numerous deleterious effects on domestic fowl, which can result in significant production losses for layer, broiler, and turkey operations. Overt signs at low toxin levels include weakness, possible anemia, decreased feed consumption, decreased egg production, and reduced growth rate. Higher OTA concentrations can severely increase bird mortality (Gibson et al 1989, 1990; Sreemannarayana et al 1989; Harvey et al 1987a; Huff et al 1988a,b,c; Burditt et al 1984; Jayakumar et al 1988; Hamilton et al 1982; Shirley and Tohala 1983). Turkeys may be more sensitive to OTA-induced feed refusal than chickens (Burditt et al 1984). Other clinical manifestations tend to be more subtle. There is strong evidence to indicate that OTA can significantly alter avian renal function. Chronic consumption of OTA has been associated with spontaneously occurring avian nephropathy (Elling et al 1975, Hamilton et al 1982). Pathophysiological changes include decreased urine concentration, decreased glomerular filtration rate, impairment of proximal tubular function, and degeneration and ultrastructural alterations in renal cellular integrity (Huff et al 1975; Krogh et al 1976; Svendsen and Skadhouge 1976; Glahn et al 1988, 1989; Dwivedi and Burns 1984a). Swelling and color changes of the kidneys have been reported as one of the more consistent lesions of ochratoxicosis (Huff et al 1975). Renal dysfunction can be manifested as an increase in urine flow (diuresis) or an increase in intestinal water loss (diarrhea).

OTA also affects other organs. Increases in the relative weights of the liver, spleen, pancreas, proventriculus, gizzard, heart, and testes are reported (Gibson et al 1989, Jayakumar et al 1988, Sreemannarayana et al 1989, Huff et al 1988b). Hemorrhages in the kidneys and liver as well as digestive tract have been noted (Chu and Chang 1971). Numerous biochemical and hematological alterations may occur, particularly elevated serum uric acid and creatinine levels as an indication of nephrotoxicity (Huff et al 1988b, Bailey et al 1990, Niemiec et al 1989, Harvey et al 1987b, Gibson et al 1989, Sreemannarayana et al 1989). Suppression of protein synthesis and the immune function have also been reported (Dwivedi and Burns 1984b, Bailey et al 1989, Singh et al 1990). There is also some suggestion that skeletal growth and strength may be affected in the form of generalized skeletal osteopenia (Duff et al 1987, Huff et al 1980). With regard to the underlying mechanism of skeleted toxicity, it remains unclear whether the toxin causes certain vitamin imbalances or if other OTA syndromes of intestinal malabsorption and nephrotoxic changes are influencing mineral homeostasis.

Ruminants

The potential toxic effects of ochratoxins in cattle and other ruminants have not been well characterized. Although experimentally induced toxicosis has been reported, documented instances of ochratoxicoses occurring in the field are rare. Kiessling et al (1984) reported that OTA is efficiently hydrolyzed enzymatically by rumen microflora to nontoxic metabolites. Consequently ruminants are well protected from levels of OTA that are typically found in contaminated feeds (Patterson et al 1981).

Experimentally induced toxicosis in ruminants using high doses of OTA will produce clinical signs that potentially include depression, reduced feed consumption, decreased weight gain, diarrhea, decreased milk production, kidney and liver damage, dehydration associated with renal dysfunction, altered clinical chemistry, and possible death at very elevated doses (Pier et al 1976, Lloyd 1980, Chu 1984). Exposing pregnant animals to high doses of OTA does not appear to affect the pregnancy (Still 1973, Munro et al 1973). Preruminant animals appear to be more sensitive to the toxin than older ruminant calves (Sreemannarayana et al 1988).

Other Toxic Effects

Besides being a major factor in nephropathy in various species, laboratory studies have demonstrated OTA to be a possible cause of other serious problems. Several authors have demonstrated the carcinogenicity of OTA via tumor formation in the livers and kidneys of mice fed the toxin at high doses for chronic periods (Bendele et al 1985, Kanizawa and Suzuki 1978, Kanizawa 1984). The capability of low doses of OTA to damage

DNA has further supported the carcinogenic potential of the toxin (Kane et al 1986).

Other toxic effects of ochratoxins include immune suppression (Prior and Sisodia 1982, Hong et al 1988, Campbell et al 1983, Germolec et al 1987), teratogenicity (Hood et al 1976, Mayura et al 1982, Hayes et al 1974, Brown et al 1976), and hemorrhages (Galtier et al 1976, Huff et al 1983, Doerr et al 1981).

Tolerance Levels

While swine are relatively sensitive to the presence of ochratoxins in their diets, toxic effects tend to be manifested more as cellular and organ damage rather than overt clinical signs. In chronic feeding studies, animals exposed to as low as 0.2 ppm developed nephropathy after several months (Krogh et al 1974). The higher the toxin level, the more quickly lesions developed. At 1.0–1.4 ppm, swelling and discoloration of the kidneys sometimes appeared as soon as two weeks after the onset of exposure (Elling 1983, Elling et al 1985). Blood chemistry and initial signs of renal dysfunction (increased urination, increased water intake) may also be apparent. Reduced feed consumption and decreased weight gains may become evident with diets containing greater than 2 ppm of OTA (Tapia and Seawright 1985, Lippold et al 1987, Huff et al 1980, Harvey et al 1989b). Death of swine from OTA-contaminated diets is unlikely, even after chronic exposure, unless unusually high concentrations (25–50 ppm) are involved (Szczech et al 1973). To the authors' knowledge, the reversibility of nephropathy in swine if the toxin is removed has not been evaluated.

Poultry appear to have a sensitivity to the effects of OTA similar to that of swine; however, this is difficult to confirm because few low-level (<1 ppm) feeding trials have been published. Micco et al (1988) found no clinical signs or lesions in chickens fed 50 ppb of OTA for up to 169 days. Prior et al (1980) noted reduced feed intake at 0.5 ppm, yet at 1 ppm of OTA Reichmann et al (1982) observed no adverse effect on growth, feed consumption, pathology, or blood parameters. Others have reported that although they have also seen no effects on growth at 1 ppm, changes in clinical chemistry values were beginning to show (Huff et al 1988b, Sreemannarayana et al 1989).

It appears that in feeding trials involving pure OTA, 2 ppm in the diet is where typical signs of ochratoxicosis become evident, including decreased weight gains, reduced feed consumption, decreased egg production, and increased water intake, diarrhea, or excess urination characteristic of renal dysfunction (Prior and Sisodia 1978, Gibson et al 1989, Dwivedi and Burns 1984a). At 2 ppm OTA, definite changes in blood biochemistry were also manifested (Dwivedi and Burns 1984b, Singh et al 1990, Bailey et al 1989, Huff et al 1988b). In comparison, during naturally occurring outbreaks, broilers showed decreased growth and layers reduced production with OTA levels as low as 0.3 ppm, suggesting a toxic contribution by other mycotoxins that may have been present (Hamilton et al 1982, Krogh et al 1976).

As toxin levels increase, so do associated problems. At 4 ppm of OTA in total diet, Gibson et al (1989, 1990) noted an increase in mortality (up to 35%), and Duff et al (1987) reported skeletal abnormalities in broiler chicks and turkey poults. Above 4 ppm, damage to other organs, delayed sexual maturity, and increasing mortality have been observed (Choudhury et al 1971, Sreemannarayana et al 1989).

Because ruminants are relatively tolerant of the toxic effects of ochratoxins, extremely high levels are necessary to elicit a response. It has been estimated that the lethal dose in calves is at least 10 times that of swine (Ribelin 1978). Administered orally at 0.16–2.5 mg/kg bw (about 10–160 ppm in the diet) for five days, OTA had no effect on milk production nor on pregnancy in cows (Ribelin 1978, Still 1973). Only at much higher levels (single oral dose of 13 mg/kg, about 856 ppm in feed) could problems be induced, and these included transitory anorexia, diarrhea, and decreased milk yield.

It appears that most of the tolerance afforded ruminants can be attributed to the metabolic and detoxifying activity of the rumen microflora. Consequently, preruminant calves are considerably more sensitive to the toxins' effects. Studies have demonstrated that whereas an oral dose as low as 1 mg OTA/kg was fatal to preruminant calves, calves with a functional rumen survived 2 mg/kg without any notable ill effects (Sreemannarayana et al 1988, Pier et al 1976).

Ovine Ill-Thrift

The disease of ill-thrift in sheep is poorly characterized. Essentially, it has been used as an ill-defined diagnosis for poor performance (growth rates) in the presence of an abundant feed source. It appears to affect primarily young sheep of either sex and occurs typically during the summer and autumn seasons (Hartley and Grant 1961). It can be a complex problem attributable to many causes (i.e., trace element imbalances, parasitism, malabsorption or other distinctive syndromes, and pathological or clinical abnormalities) most of which can be eliminated by supplemental feeding, vaccination, and proper management.

Occasionally, the syndrome appears in the absence of any notable source, for example, periodically and in very localized areas of Nova Scotia. Besides a reduced growth rate, the only other reported signs have been a multifold decrease in the number of viable rumen bacteria (Brewer et al 1971). While there appears to be no confirmed cause, Brewer and others (1971, 1972b, 1982; Brewer and Taylor 1980) reported they initially observed a significant increase in the amounts of fungi present in the pastures where the problem was occurring, and they subsequently identified several toxic secondary metabolites as possible causative agents. Three mycotoxins, characterized as isocyanide-containing compounds, seem to be produced by the fungus *Trichoderma hamatum*, which is commonly isolated in soils of permanent pastures. Brewer and co-workers further speculated that ingestion of these toxins impedes function of rumen bacteria, which consequently affects nutrition and growth. It was observed that when one metabolite (isocyanide III; 3-[3-isocyanacyclo-pent-2-enylidene-]propionic acid) was consumed by sheep (2.5 mg/kg per day), no overt toxicologic effects were noted, except for a decrease in growth rate and an associated reduction in effective digestion of feed (Brewer and Taylor 1981).

Other compounds such as chetonin, produced by *Chaetomium* species, have also been identified and shown to induce weight loss at sublethal oral doses (Brewer et al 1972a). It has been suggested that ill-thrift may be caused by a mixture of toxins in pasture, which individually would prove difficult to incriminate as the primary cause of the syndrome.

Interactive Effects of Mycotoxins and Other Fungal Metabolites

Infection of agricultural products by various species of fungi can occur in the field or during storage, and the contaminating mold(s) can produce a diverse variety of secondary metabolites, including mycotoxins. The complex processes involved in toxin production by a fungus make it difficult to predict which toxin(s) will preferentially be produced and at what concentration(s). If one mycotoxin occurs in a contaminated commodity, there is a distinct probability that other mycotoxins will also be present, produced by a single or several contaminating species of mold. In addition, since animal feeds are usually composed of several grain types and sources, each containing different fungal metabolites, there is a great potential for a variety of these metabolites to occur in a single feed mix. Coppock et al (1989a) described a case in which dairy calves were exposed to ergot (*Claviceps purpurea*) alkaloids and deoxynivalenol, and Barnikol et al (1985) reported illness in newborn piglets related to the presence of ergot and T-2 toxin in a sow's diet. While concentrations of individual fungal metabolites in naturally contaminated grain usually tend to be low, there is a potential for interaction between metabolites resulting in net effects different from what would be expected from addition of the effects of each metabolite alone. Unfortunately, much of the research concerning the effects of mycotoxins and other fungal metabolites on animal performance has involved either individual pure metabolites, usually mycotoxins, or chemically uncharacterized contaminated grain. Very little is known about interactions between different fungal metabolites in the same feedstuff or diet mixture. There are many fungal metabolites that either have not yet been identified or for which the mode of action and/or effects are not yet understood. Discussion of ergot alkaloids and other fungal metabolites, with regard to interactive effects, is a special situation. The alkaloid composition and total content is highly variable among sclerotia even within the same ergot body, field, and geographic region (Young 1981). An examination of inter-actions between ergot alkaloids and other metabolites (mycotoxins) even today is not possible due to a lack of understanding of interactions between the alkaloids themselves. As such, the case of ergot will not be discussed in this section.

Major obstacles to examining interactions between fungal metabolites are the isolation, production, and characterization of large quantities and the assessment of their individual toxicity prior to any interaction study. The interactions are complex and, in animal systems, they may only be detected as altered clinical signs. While the interactive effects of certain metabolite combinations may result in a net effect that is less than expected on the basis of their individual effects, it is the additive and synergistic (greater than additive) effects which are of concern. Metabolite and myco-toxin interactions in the field are difficult to diagnose, and each interaction has to be completely characterized to be quickly recognized under production conditions. A great deal of direct, and even more indirect, evidence suggests that mycotoxins, and possibly other metabolites, interact to various degrees.

In eastern Canada, the toxicity of *Fusarium* species cannot be explained only by the presence of two mycotoxins commonly found in mold-contaminated grain, DON and ZEN. Feeding studies with swine comparing diets containing naturally occurring with pure DON indicated that DON was not the sole cause of adverse effects (Trenholm et al 1984, Foster et al 1986). Major differences in pig performance were seen both for different sources of contaminated grain and methods of storage. In addition, there was no correlation between the dietary DON concentration and either feed consumption or weight gain. In most cases of natural exposure, as opposed to experimental, the exact chemical composition of the contaminated feed, including the mold itself, is not completely known, and so its effects cannot be duplicated in feeding studies. This further supports the hypothesis that there are either undetected toxins and/or possible interactions between several toxins (Foster et al 1986). Interactions between DON and other metabolites could account for some of the differences seen between pure and naturally occurring DON in field outbreaks involving severe toxicosis at <2 mg/kg of DON in the diet (Foster et al 1986, Côté et al 1984). Similarly, R. G. Rotter et al (1989) observed a concentration-dependent depression in feed intake and reduction in weight gain in Leghorn chicks fed a heavily mold-contaminated source of barley that contained no detect-able toxins. Although they postulated the effects were directly related to the presence of the mold, the likelihood that there were unidentified, interacting metabolites is also possible. It is probable that most reported cases of mycotoxicosis are the results of the action of several toxins, although a single highly toxic material could be the primary toxicant.

R. G. Rotter et al (1992) examined potential interactions between pure DON and several other *Fusarium* metabolites (3-acetyl deoxynivalinol [ADON], 15-ADON, sambucinol, dihydroxycalonectrin, and culmorin) on the performance of growing pigs. The only significant effects were seen in feed consumption, weight gain, and feed efficiency due to the presence of DON in the diet. Small, but not significant differences ($P > 0.05$) were seen in growth and feed consumption for sambucinol and culmorin, with and without DON. The data indicated that these specific fungal metabolites tested in combination with DON did not interact with DON in growing pigs to any marked degree.

Coffey et al (1990) examined the effects of aflatoxin B_1, ZEN, and T-2 toxin on young pigs. Aflatoxin B_1, alone and with T-2 toxin, reduced weight gain in female pigs, but 1 mg/kg of ZEN in the diet combined with aflatoxin B_1 prevented growth depression caused by aflatoxin B_1. The effects of dietary aflatoxin and DON were evaluated in growing crossbred barrows (Harvey et al 1989a). A combination of DON and aflatoxin in the diet did not result in effects different from aflatoxin alone, suggesting that these two toxins did not interact.

In general, research on mycotoxin interactions in swine is limited because large quantities of toxins are required to conduct proper experiments. Swine appear to be very sensitive to DON and ZEN, whereas ruminants and poultry are more tolerant (Trenholm et al 1984). The relative tolerance of poultry to DON is due to a combination of poor absorption of the toxin after oral administration, extensive metabolism (detoxification), and DON's rapid elimination from the body (Prelusky et al 1986). As a result, poultry studies with DON require relatively high dietary concentrations of the toxin. Huff et al (1986), examining the effects of 16 mg/kg of DON and 2.5 mg/kg of aflatoxin in the diet, concluded that the net effect of these toxins combined was additive. In another study, Kubena et al (1989a) showed that DON and T-2 toxin together (16 and 4 mg/kg of toxin in the diet, respectively) caused an increase in the severity of oral lesions and marked reductions in body weight gains (one to 21 days) and mean corpuscular volume and serum chemistry compared with that shown for each toxin alone.

Combinations of aflatoxin and T-2 toxin (2.5 and 4.0 mg/kg of toxin in the diet, respectively; Huff et al 1988a) and OTA and T-2 toxin (2 and 4 mg/kg of toxin in the diet, respectively; Kubena et al 1989b) were also examined in broiler chicks. Aflatoxin and T-2 together were synergistic in toxicity ($P < 0.05$), seen as decreased body weights, increased relative organ weights (organ g/100 g body weight) of the kidney, gizzard, and heart, and decreased mean corpuscular volume and serum potassium (Huff et al 1988a). It is possible that aflatoxin primarily affects protein synthesis during transcription, whereas T-2 toxin has its effect during translation, partially accounting for their synergistic interaction. Possible signs of synergistic toxicity should be monitored because of occurrence of these mycotoxins and the severity of their interactive toxicity. This poses a potential threat to the poultry industry. A combination of OTA and T-2 toxin may also be more toxic for some parameters (elevated serum triglyceride levels, decreased gamma-glutamyl-transferase and calcium levels) than the individual mycotoxins. Huff et al (1988c) reported that the effects of aflatoxin, a hepatotoxin, and OTA, a nephrotoxin, were synergistic when simultaneously fed to broiler chickens. The target organ was kidney, presumably due to a lower regenerative capability than that of liver, and an increase in liver lipid (diagnostic lesion of aflatoxicosis) was absent when OTA was present (Huff and Doerr 1981). These opposite effects indicate the complexity of the problem in dealing with metabolite interactions. In contrast, Doerr et al (1982) reported that OTA in combination with citrinin was antagonistic when fed to broiler chickens. Giroir et al (1991) examined the influence of kojic acid and aflatoxin in male broilers and found that the combination of these two mycotoxins resulted in an antagonistic effect through an increase in a mean corpuscular hemoglobin.

Several mycotoxin combinations have been tested in poultry and swine, from which aflatoxin plus OTA and aflatoxin plus T-2 toxin appear to be the most toxic (Huff et al 1988c).

Various other bioassay systems, including yeast, caterpillars, chick embryo assays, and mice have been used to examine mycotoxin interactions. Koshinsky and Khachatourians (1992) measured the effects of HT-2 toxin, roridin A, and T-2 toxin individually or as binary mixtures on growth of the yeast *Kluyveromyces marxianus*. The interaction was affected by the ratio of individual toxins and the percent inhibition of yeast growth; it changed from antagonistic at a low percentage of growth inhibition to synergistic at a high percentage (T-2 toxin and roridin A or T-2 toxin and HT-2 toxin). Dowd et al (1989) fed various *F. graminearum* metabolites to caterpillars and concluded that DON (25 mg/kg of diet) was the most toxic compound tested. Other metabolites in combination with DON, such as culmorin (10 mg/kg of diet), dihydroxycalonectrin (1 mg/kg of diet) or sambucinol (1 mg/kg of diet), promoted strong synergistic effects. Dowd (1988) investigated the effect of coapplying the fungal metabolites kojic acid and fusaric acid with selected mycotoxins, allochemicals, and insecticides. Kojic acid enhanced the toxicity of aflatoxin B_1, while fusaric acid enhanced the toxicity of some trichothecenes to two insect species.

Selected mycotoxins have also been tested with chick embryo. Vesela et al (1983) examined the embryotoxic potential of pure OTA and citrinin, while B. A. Rotter et al (1991) examined the response of DON, 15-ADON, and HT-2 toxin in a dose-controlled experiment. The combined effects of ochratoxin and citrinin and between any of the two trichothecenes were additive. Examination of interactions between T-2 toxin and DON in mice showed a possible interaction ($P = 0.064$) between the two toxins after seven days of exposure. A further depression in weight gain was observed only at a lower T-2 toxin concentration (2 mg/kg of diet) when DON concentration increased from 4 to 8 mg/kg of diet (B. A. Rotter et al 1992).

Grains or feeds thought to cause animal disorders may often contain more than one mycotoxin as well as other components that enhance the toxicity properties of toxins. The toxicity of fungal metabolite–contaminated feed cannot be estimated by determining the concentration of a single toxin. Analysis for various mycotoxins and a better knowledge of their potential interactions in various animal species are needed. Information on this subject is very limited, mainly because of the complexity of fungal metabolism and interactions between fungal metabolites and other factors, as well as the difficulty in recreating field cases in the laboratory. In vivo studies for mycotoxin interactions should involve not only performance changes but also changes at the biochemical level so they can be used in the diagnosis of multiple contaminations.

Literature Cited

Abbas, H. K, Mirocha, C. J., and Tuite, J. 1986. Natural occurrence of deoxynivalenol 15-acetyl-deoxynivalenol, and zearalenone in refused corn stored since 1972. Appl. Environ. Microbiol. 51:841-843.

Ademoyero, A. A., and Hamilton, P. B. 1989. Influence of degree of acetylation of scirpenol mycotoxins on feed refusal by chickens. Poult. Sci. 68:854-856.

Allen, N. K., Mirocha, C. J., Weaver, G., Aakhus-Allen, S., and Bates, F. 1981. Effects of dietary zearalenone on finishing broiler chickens and young turkey poults. Poult. Sci. 60:124-131.

Alstad, A. D., Casper, H. H., and Johnson, L. J. 1985. Vitamin K treatment of sweet clover poisoning in calves. J. Am. Vet. Med. Assoc. 187:729-731.

Appleyard, W. T. 1986. Outbreak of bovine abortion attributed to ergot poisoning. Vet. Rec. 118:48-49.

Bacon, C. W., Lyons, P. C., Porter, J. K., and Robbins, J. D. 1986. Ergot toxicity from endophyte-infected grasses: A review. Agron. J. 78:106-116.

Bailey, C. A., Gibson, R. M., Kubena, L. F., Huff, W. E., and Harvey, R. B. 1989. Ochratoxin A and dietary protein. 2. Effects on hematology and various clinical chemistry measurements. Poult. Sci. 68:1664-1671.

Bailey, C. A., Gibson, R. M., Kubena, L. F., Huff, W. E., and Harvey, R. B. 1990. Impact of L-phenylalanine supplementation on the performance of three-week-old broilers fed diets containing ochratoxin A. 2. Effects on hematology and clinical chemistry. Poult. Sci. 69:420-425.

Bakau, B. J. K., Peacock, A. J., Love, R. J., and Bryden, W. L. 1988. Toxicity of ergots of *Claviceps purpurea* in growing pigs subjected to different ambient temperatures. (Abstr.) Proc. Nutr. Soc. Aust. Annu. Conf. 13:136.

Baldwin, R. S., Williams, R. D., and Terry, M. K. 1983. Zeranol: A review of the metabolism, toxicology and analytical methods for detection of tissue residues. Regul. Toxicol. Pharmacol. 3:9-25.

Barnikol, H., Thalmann, A., and Wengert, D. 1985. Hautschaden bei neugeborenen Ferkeln in Zusammenhang mit einem Fusarientoxin (T-2-toxin) und Mutterkown. Tierartzl. Umsch. 40:658-666.

Beasley, V. R., Swanson, S. P., Corley, R. A., Buck, W. B., Koritz, G. D., and Burmeister, H. R. 1986. Pharmacokinetics of the trichothecene mycotoxin T-2 toxin, in swine and cattle. Toxicon 24:13-23.

Bendele, S. A., Carlton, W. W., Krogh, P., and Lillehoj, E. B. 1985. Ochratoxin A carcinogenesis in the (C57BL/6YXC3HF)1 mouse. J. Nat. Cancer Inst. 75:733-739.

Berger, T., Esbenshade, K. L., Diekman, M. A., Hoagland, T., and Tuite, J. 1981. Influence of prepubertal consumption of zearalenone on sexual development of boars. J. Anim. Sci. 53:1559-1564.

Blakley, B. R. 1985. Moldy sweet clover (dicoumarol) poisoning in Saskatchewan cattle. Can. Vet. J. 26:357-360.

Blaney, B. J., Bloomfield, R. C., and Moore, C. J. 1984. Zearalenone intoxication of pigs. Aust. Vet. J. 61:24-27.

Blood, D. C., Radostits, O. M., and Henderson, J. A., eds. 1983. Moldy sweetclover poisoning. Pages 1170-1172 in: Veterinary Medicine, 6th ed. Bailliere and Tindall, London.

Bloomquist, C., Davidson, J. N., and Pearson, E. G. 1982. Zearalenone toxicosis in prepubertal dairy heifers. J. Am. Vet. Med. Assoc. 180:164-165.

Bock, R. R., Shore, L. S., Samberg, Y., and Perl, S. 1986. Death in broiler breeders due to salpingitis: Possible role of zearalenone. Avian Pathol. 15:495-502.

Bragg, D. B. Salem, H. A., and Devlin, T. J. 1970. Effect of dietary triticale ergot on the performance and survival of broiler chicks. Can. J. Anim. Sci. 50:259-264.

Branton, S. L., Deaton, J. W., Hagler, W. M. Jr., Maslin, W. R., and Hardin, J. M. 1989. Decreased egg production in commercial laying hens fed zearalenone- and deoxynivalenol-contaminated grain sorghum. Avian Dis. 33:804-808.

Brewer, D., and Taylor, A. 1980. Ovine ill-thrift in Nova Scotia. 6. Quantitative description of the fungal flora of soils of permanent pasture. Proc. Nova Scotia Inst. Sci. 30:101-133.

Brewer, D., and Taylor, A. 1981. Ovine ill-thrift in Nova Scotia. 7. *Trichoderma hamatum* isolated from pasture soil. Mycopathologia 76:167-173.

Brewer, D., Calder, F. W., MacIntyre, T. M., and Taylor, A. 1971. Ovine ill-thrift in Nova Scotia. 1. The possible regulation of the rumen flora in sheep by the fungal flora of permanent pasture. J. Agric. Sci. 76:465-477.

Brewer, D., Duncan, J. M., Jerram, W. A., Leach, C. K., Safe, S., Taylor, A, Vining, L. C., Archibald, R. McG., Stevenson, R. G., Mirocha, C. J., and Christensen, C. M. 1972a. Ovine ill-thrift in Nova Scotia. 5. The production and toxicology of chetomin, a metabolite of *Chaetomium* spp. Can. J. Microbiol. 18:1129-1137.

Brewer, D., Taylor, A., and Hoehn, M. M. 1972b. Ovine ill-thrift in Nova Scotia. II. The production of antibiotics by fungi isolated from forest and marshland soil. J. Agric. Sci. 78:257-264.

Brewer, D., Feicht, A., Taylor, A., Keeping, J. W., Taha, A. A., and Thaller, V. 1982. Ovine ill-thrift in Nova Scotia. 9. Production of experimental quantities of isocyanide metabolites of *Trichoderma hamatum*. Can. J. Microbiol. 2:1252-1260.

Bristol, F. M., and Djurickovic, S. 1971. Hyperestrogenism in female swine as the result of feeding mouldy corn. Can. Vet. J. 12:132-135.

Brown, M. H., Szczech, G. M., and Purmalis, B. P. 1976. Teratogenic and toxic effects of ochratoxin A in rats. Toxicol. Appl. Pharmacol. 37:331-338.

Brownie, C. F., and Cullen, J. 1987. Characterization of an experimentally induced equine leukoencephalomalacia (ELEM) in ponies (*Equus caballus*): Preliminary report. Vet. Human Toxicol. 29:34-38.

Burditt, S. J., Winston, M., Hagler, W. M., Jr., and Hamilton, P. B. 1983. Survey of molds and mycotoxins for their ability to cause feed refusal in chickens. Poult. Sci. 62:2187-2191.

Burditt, S. J., Hagler, W. M., Jr., and Hamilton, P. B. 1984. Feed refusal during ochratoxicosis in turkeys. Poult. Sci. 63:2172-2174.

Bush, L., Boling, J., and Yates, S. 1979. Animal disorders. Pages 247-292 in: Tall Fescue. R. C. Buckner and L. P. Bush, eds. Am. Soc. Agron. Monogr. 20.

Buxton, E. A. 1927. Mycotic vaginitis in swine. Vet. Med. 22:451-452.

Campbell, M. L., May, J. D., Huff, W. E., and Doerr, J. A. 1983. Evaluation of immunity of young broiler chickens during simultaneous aflatoxicosis and ochratoxicosis. Poult. Sci. 62:2138-2144.

Carlevaro, G. 1956. L'ergotismo in patologia comparata. Zooprofilassi 11:587-595.

Chavez, E. R., and Rheaume, J. A. 1986. The significance of the reduced feed consumption observed in growing pigs fed vomitoxin-contaminated diets. Can. J. Anim. Sci. 66:277-287.

Chi, M. S., Mirocha, C. J., Kurtz, H. J., Weaver, G. A., Bates, F., Robison, T., and Shimoda, W. 1980a. Effect of dietary zearalenone on growing broiler chicks. Poult. Sci. 59:531-536.

Chi, M. S., Mirocha, C. J., Weaver, G. A., and Kurtz, H. J. 1980b. Effect of zearalenone on female White Leghorn chickens. Appl. Environ. Microbiol. 39:1026-1030.

Choudhury, H., Carlson, C. W., and Semeniuk, G. 1971. A study of ochratoxin toxicity in hens. Poult. Sci. 50:1855-1859.

Chu, F. S. 1984. Studies on Ochratoxins. CRC Crit. Rev. Toxicol. 2:499-524.

Chu, F. S., and Chang, C. C. 1971. Sensitivity of chicks to ochratoxin. J. Assoc. Off. Anal. Chem. 54:1032-1034.

Ciegler, A. 1975. Mycotoxins: Occurrence, chemistry, biological activity. Lloydia 38:21-35.

Coffey, M. T., Hagler, W. M., Jr., Jones, E. E., and Cullen, J. M. 1990. Interactive effects of multiple mycotoxin contamination of swine diets. Pages 117-128 in: Biodeterioration Research 3. C. Llewellyn and C. E. O'Rear, eds. Plenum Books, New York.

Colvin, B. M., and Harrison, L. R. 1992. Fumonisin-induced pulmonary edema/hydrothorax in swine. Mycopathologia 117:79-82.

Cook, W. O., Osweiler, G. D., Anderson, T. D., and Richard, J. L. 1986. Ochratoxicosis in Iowa swine. J. Am. Vet. Med. Assoc. 188:1399-1402.

Coppock, R. W., Mostrom, M. S., Simon, J., McKenna, D. J., Jacobsen, B., and Szlachta, H. L. 1989a. Cutaneous ergotism in a herd of dairy calves. J. Am. Vet. Med. Assoc. 194:549-551.

Coppock, R. W., Hoffmann, W. E., Gelberg, H. B., Bass, D., and Buck, W. B. 1989b. Hematologic changes induced by intravenous administration of diacetoxyscirpenol in pigs, dogs, and calves. Am. J. Vet. Res. 50:411-415.

Coppock, R. W., Mostrom, M. S., Sparling, C. G., Jacobsen, B., and Ross, J. C. 1990. Apparent zearalenone intoxication in a dairy herd from feeding spoiled acid-treated corn. Vet. Human Toxicol. 32:246-248.

Côté, L. M., Reynolds, J. D., Vesonder, R. F., Buck, W. B., Swanson, S. P., Coffey,

R. T., and Brown, D. C. 1984. Survey of vomitoxin-contaminated feed grains in midwestern United States, and associated health problems in swine. J. Vet. Med. Assoc. 184:189-192.

Côté, L. M., Dahlem, A. M., Yoshizawa, I., Swanson, S. P., and Buck, W. B. 1986. Excretion of deoxynivalenol and its metabolite in milk, urine, and feces of lactating dairy cows. J. Dairy Sci. 69:2416-2423.

Cranwell, M. P. 1983. Acute fatal haemorrhagic syndrome in dairy cows. Vet. Rec. 112:486.

Cunningham, I. J., Swan, J. B., and Hopkirk, C. E. M. 1944. The symptoms of ergot poisoning in sheep. N.Z. Sci. Technol. 26a:121-124.

Dignean, M. A., Schiefer, H. B., and Blair, R. 1986. Effects of feeding ergot-contaminated grain to pregnant and nursing sows. J. Vet. Med. Assoc. 33:757-766.

Doerr, J. A., Huff, W. E., Hamilton, P. B., and Lillehoj, E. B. 1981. Severe coagulopathy in young chickens produced by ochratoxin A. Toxicol. Appl. Pharmacol. 59:157-163.

Doerr, J. A. Campbell, M. L., and Huff, W. E. 1982. Interaction between dietary citrinin and ochratoxin A in broiler chickens. Poult. Sci. 61:1453.

Dowd, P. F. 1988. Toxicological and biochemical interactions of the fungal metabolities fusaric acid and kojic acid with xenobiotics in *Heliothis zea* (F.), and *Spodoptera frugiperda* (J. E. Smith). Pestic. Biochem. Physiol. 32: 123-134.

Dowd, P. F., Miller, J. D., and Greenhalgh, R. 1989. Toxicity and some interactions of some *Fusarium graminearum* metabolites to caterpillars. Mycologia 81:646-650.

Duff, S. R. I., Burns, R. B., and Dwivedi, P. 1987. Skeletal changes in broiler chicks and turkey poults fed diets containing ochratoxin A. Res. Vet. Sci. 43:301-307.

Dwivedi, P., and Burns, R. B. 1984a. Pathology of ochratoxicosis A in young broiler chicks. Res. Vet. Sci. 36:92-103.

Dwivedi, P., and Burns, R. B. 1984b. Effect of ochratoxin A on immunoglobulins in broiler chicks. Res. Vet. Sci. 36:117-121.

Dziuk, H. E., Nelson, G. H., Duke, G. E., Maheswaran, S. K., Chi, M. S., and Mirocha, C. J. 1979. Effects of T-2 toxin in the diet of turkey poults. Proc. Annu. Meet. Am. Assoc. Vet. Lab. Diagn. 22:215-222.

Edwards, S., Cantley, T. C. Rottinghaus, G. E., Osweiler, G. D., and Day, B. N. 1987a. The effects of zearalenone on reproduction in swine. 1. The relationship between ingested zearalenone dose and anestrus in non-pregnant, sexually mature gilts. Theriogenology 28:43-49.

Edwards, S., Cantley, T. C., and Day, B. N. 1987b. The effects of zearalenone on reproduction in swine. II. The effect on puberty attainment and postweaning rebreeding performance. Theriogenology 28:51-58.

Egyed, M. N., Shlosberg, A., Handji, V., and Rapaport, E. 1981. Changes in prothrombin time in subclinical experimental *Ferula communis* poisoning in sheep. Refu. Vet. 38:128-135.

Elling, F. 1983. Feeding experiments with ochratoxin A-contaminated barley to bacon pigs. IV. Renal lesions. Acta Agric. Scand. 33:153-159.

Elling, F., Hald, B., Jacobsen, C., Krogh, P. 1975. Spontaneous toxic nephropathy in poultry associated with ochratoxin A. Acta Pathol. Microbiol. Scand. 83:739-741.

Elling, F., Nielsen, J. P., Lillehoj, E. B., Thomassen, M. S., and Stormer, F. C. 1985. Ochratoxin A-induced porcine nephropathy: Enzyme and ultrastructure changes after short-term exposure. Toxicon 23:247-254.

Farnworth, E. R., and Trenholm, H. L. 1981. The effect of acute administration of the mycotoxin zearalenone to female pigs. J. Environ. Sci. B16:239-252.

Foster, B. C., Trenholm, H. L., Friend, D. W., Thompson, B. K., and Hartin, K. E. 1986. Evaluation of different sources of deoxynivalenol (vomitoxin) fed to swine. Can. J. Anim. Sci. 66:1149-1154.

Fraser, C. M., and Nelson, J. 1959. Sweet clover poisoning in newborn calves. J. Am. Vet. Med. Assoc. 135:283-286.

Fraser, D. M., and Dorling, P. R. 1983. Suspected ergotism in two heifers. Aust. Vet. J. 60:303-305.

Friend, D. W., and MacIntyre, T. M. 1969. The digestibility of rye and its value in pelleted rations for pigs. Can. J. Anim. Sci. 49:375-381.

Friend, D. W., and MacIntyre, T. M. 1970. Effect of rye ergot on growth and N-retention in growing pigs. Can. J. Comp. Med. 34:198-202.

Friend, D. W., Trenholm, H. L., Elliot, J. I., Thompson, B. K., and Hartin, K. E. 1982. Effect of feeding vomitoxin-contaminated heat to pigs. Can. J. Anim. Sci. 62:1211-1222.

Friend, D. W., Trenholm, H. L., Fiser, P. S., Thompson, B. K., and Hartin, K. E. 1983. Effect on dam performance and fetal development of deoxynivalenol-(vomitoxin) contaminated heat in the diet of pregnant gilts. Can. J. Anim. Sci. 63:689-698.

Friend, D. W., Trenholm, H. L., Thompson, B. K., Fiser, P. S., and Hartin, K. E. 1986a. Effect of feeding diets containing deoxynivalenol (vomitoxin)-contaminated heat or corn on the feed consumption, weight gain, organ weight and sexual development of male and female pigs. Can. J. Anim. Sci. 66:765-775.

Friend, D. W., Thompson, B. K., Trenholm, H. L., Hartin, K. E., and Prelusky, D. B. 1986b. Effects of feeding deoxynivalenol (DON)-contaminated wheat diets to pregnant and lactating gilts and on their progeny. Can. J. Anim. Sci. 66:229-236.

Friend, D. W., Trenholm, H. L., Thompson, B. K., Hartin, K. E., Fiser, P. S., Asem, E. K., and Tsang, B. K. 1990. The reproductive efficiency of gilts fed very low levels of zearalenone. Can. J. Anim. Sci. 70:635-645.

Gabal, M. A., Awad, Y. L., Morcos, M. B., Barakat, A. M., and Malik, G. 1986. Fusariotoxicoses of farm animals and mycotoxic leukoencephalomalacia of the equine associated with the finding of trichothecenes in feedstuffs. Vet. Human Toxicol. 28:207-212.

Galtier, P., More, J., and Alvinerie, M. 1976. Acute and short-term toxicity of ochratoxin A in 10-day old chicks. Food Cosmet. Toxicol. 14:129-131.

Gardner, M. L., First, N. L., and Casida, L. E. 1963. Effect of exogenous estrogens on corpus luteum maintenance in gilts. J. Anim. Sci. 22:132-134.

Gelderblom, W. C. A., Jaskiewicz, K., Marasas, W. F. O., Thiel, P. G., Horak, R. M., Vleggaar, R., and Kriek, N. P. J. 1988. Fumonisins: Novel mycotoxins with cancer-promoting activity produced by *Fusarium moniliforme*. Appl. Environ. Microbiol. 54:1806-1811.

Gelderblom, W. C. A., Krick, N. P. J., Marasas, W. F. O., and Thiel, P. G. 1991. Toxicity and carcinogenicity of the *Fusarium moniliform* metabolite, fumonisin B_1, in rats. Carcinogenesis 12:1247-1251.

Gentry, P. A., Ross, M. L., and Chan, P. K.-C. 1984. Effect of T-2 toxin on bovine hematological and serum enzyme parameters. Vet. Human Toxicol. 26:24-28.

Germolec, D. R., Burleson, G. R., Jameson, C. W., Ackermann, M. F., Lamm, K. R., Hayes, H. T., and Luster, M. I. 1987. Depression of natural killer cell activity by ochratoxin A. Cancer Res. 47:2259-2263.

Gibson, R. M., Bailey, C. A., Kubena, L. F., Huff, W. E., and Harvey, R. B. 1989. Ochratoxin A and dietary protein. 1. Effects on body weight, feed conversion, relative organ weight, and mortality in three-week-old broilers. Poult. Sci. 68:1658-1663.

Gibson, R. M., Bailey, C. A., Kubena, L. F., Huff, W. E., and Harvey, R. B. 1990. Impact of L-phenylalanine supplementation on the performance of three-week-old broilers fed diets containing ochratoxin A. 1. Effects on body weight, feed conversion, relative organ weight, and mortality. Poult. Sci. 69:414-419.

Giroir, L. E., Huff, W. E., Kubena, L. F., Harvey, R. B., Elissalde, M. H., Witzel, D. A., Yersin, A. G., and Ivie, G. W. 1991. The individual and combined toxicity of kojic acid and aflatoxin in broiler chickens. Poult. Sci. 70:1351-1356.

Glahn, R. P., Wideman, R. F., Jr., Evangelisti, J. W., and Huff, W. E. 1988. Effects of ochratoxin A alone and in combination with citrinin on kidney function of single comb white Leghorn pullets. Poult. Sci. 67:1034-1042.

Glahn, R. P., Shapiro, R. S., Vena, V. E., Wideman, R. F., Jr., and Huff, W. E. 1989. Effects of chronic ochratoxin a and citrinin toxicosis on kidney function of single comb white Leghorn pullets. Poult. Sci. 68:1205-1212.

Glavits, R. Sandor, G. S., Vanyi, A., and Gajdacs, G. 1983. Reproductive disorders caused by trichothecene mycotoxins in a large-scale pig herd. Acta Vet. Hung. 31:173-180.

Golinski, P., Hult, K., Grabarkiewicz-Szczesna, J., Chelkowski, J. Kneblewski, P., and Szebiotko, K. 1984. Mycotoxic porcine nephropathy and spontaneous occurrence of ochratoxin A residues in kidneys and blood of Polish swine. Appl. Environ. Microbiol. 47:1210-1212.

Goodman, L. S., and Gilman, A., eds. 1990. Anticoagulant, thrombolytic and antiplatelet drugs. Pages 1321-1322 in: The Pharmacological Basis of Therapeutics. 8th ed. Pergamon Press, New York.

Greatorex, J. C., and Mantle, P. G. 1973. Experimental ergotism in sheep. Res. Vet. Sci. 15:337-346.

Greatorex, J. C., and Mantle, P. G. 1974. Effect of rye ergot on the pregnant sheep. J. Reprod. Fertil. 37:33-41.

Green, M. L., Diekman, M. A., Malayer, J. R., Scheidt, A. B., and Long, G. G. 1990. Effect of prepubertal consumption of zearalenone on puberty and subsequent reproduction of gilts. J. Anim. Sci. 68:171-178.

Greenway, J. A., and Puls, R. 1976. Fusariotoxicosis from barley in British Columbia. I. Natural occurrence and diagnosis. Can. J. Comp. Med. 40:12-15.

Haliburton, J. C., and Buck, W. B. 1986. Equine leucoencephalomalacia: An historical review. Pages 75-79 in: Diagnosis of Mycotoxicoses: Current Topics in Veterinarian Medicine and Animal Science. J. L. Richard and J. R. Thurston, eds. Martinus Nijhoff Publishers, Dordrecht.

Hamilton, P. B., Huff, W. E., Harris, J. R., Wyatt, R. D. 1982. Natural occurrences of ochratoxicosis in poultry. Poult. Sci. 61:1832-1841.

Hamilton, R. M. G., Trenholm, H. L., and Thompson, B. K. 1983. Effects of feeding chicks and laying hens diets that contain vomitoxin (deoxynivalenol)-contaminated wheat. Proc. Annu. Conf. Nutr. Feed Manufact. 19:48-55.

Hamilton, R. M. G., Trenholm, H. L., Thompson, B. K., and Greenhalgh, R. 1985a. The tolerance of white Leghorn and broiler chicks, and turkey poults to diets that contained deoxynivalenol (vomitoxin)-contaminated wheat. Poult. Sci. 64:273-286.

Hamilton, R. M. G., Thompson, B. K., Trenholm, H. L., Fiser, P. S., and Greenhalgh, R. 1985b. Effects of feeding white Leghorn hens diets that contain deoxynivalenol (vomitoxin)-contaminated wheat. Poult. Sci. 64:1840-1852.

Hamilton, R. M. G., Thompson, B. K., and Trenholm, H. L. 1986. The effects of deoxynivalenol (vomitoxin) on dietary preference of white Leghorn hens. Poult. Sci. 65:288-293.

Harrison, L. R., Colvin, B. M., Greene, J. T., Newman, L. E., and Cole, J. R. Jr. 1990. Pulmonary edema and hydrothorax in swine produced by fumonisin B$_1$, a toxic metabolite of *Fusarium moniliforme*. J. Vet. Diagn. Invest. 2:217-221.

Hartley, W. J., and Grant, A. B. 1961. A review of selenium responsive diseases in New Zealand livestock. (Abstr.) Proc. Fed. Am. Soc. Exp. Biol. 20:679.

Harvey, R. B., Kubena, L. F., Corrier, D. E., Witzel, D. A., Phillips, T. D., and Heidelbaugh, N. D. 1986. Effects of deoxynivalenol in a wheat ration fed to growing lambs. Am. J. Vet. Res. 47:1630-1632.

Harvey, R. B., Kubena, L. F., Lawhorn, D. B., Fletcher, O. J., and Phillips, T. D. 1987a. Feed refusal in swine fed ochratoxin-contaminated grain sorghum: Evaluation of toxicity in chicks. J. Am. Vet. Med. Assoc. 190:673-675.

Harvey, R. B., Kubena, L. F., Naqi, S. A., Gyimah, J. E., Corrier, D. E., Panigrahy, B., and Phillips, T. D. 1987b. Immunologic effects of low levels of ochratoxin A in ovo: Utilization of a chicken embryo model. Avian Dis. 31:787-791.

Harvey, R. B., Kubena, L. F., Huff, W. E., Corrier, D. E., Clark, D. E., and Phillips, T. D. 1989a. Effects of aflatoxin, deoxynivalenol, and their combinations in the diets of growing pigs. Am. J. Vet. Res. 50:602-607.

Harvey, R. B., Huff, W. E., Kubena, L. F., and Phillips, T. D. 1989b. Evaluation of diets contaminated with aflatoxin and ochratoxin fed to growing pigs. Am. J. Vet. Res. 50:1400-1405.

Harvey, R. G., Kubena, L. F., Huff, W. E., Corrier, D. E., Rottinghaus, G. E., and Phillips, T. D. 1990. Effects of treatment of growing swine with aflatoxin and T-2 toxin. Am. J. Vet. Res. 51:1688-1693.

Haschek, W. M., Motelin, G., Ness, D. K., Harlin, K. S., and Hall, W. F., Vesonder, R. F., Peterson, R. E., and Beasley, V. R. 1992. Characterization of fumonisin toxicity in orally and intravenously dosed swine. Mycopathologia 117:83-96.

Hayes, W. A., Hood, R. D., and Humphrey, L. L. 1974. Teratogenic effects of ochratoxin A in mice. Teratology 9:93-97.

Hemken, R. W. 1988. Recent advances in the knowledge of the fescue endophyte— dairy cattle. J. Anim. Sci. 66(Suppl. 1):60.

Hemken, R. W. 1990. Toxic alkaloids associated with endophytic fungi as a causative agent in tall fescue toxicosis. (Abstr.) Am. Chem. Soc. Proc. Meet. 200(1-2):38.

Hibbs, C. M., and Wolf, N. 1982. Ergot toxicosis in young goats. Mod. Vet. Pract. 63:126-128.

Hoerr, F. J., Carlton, W. W., Yagen, B., and Joffe, A. Z. 1982. Mycotoxicosis caused by either T-2 toxin or diacetoxyscirpenol in the diet of broiler chickens. Fundam. Appl. Toxicol. 2:121-124.

Holliman, A., and Barnes, J. 1990. Ergotism in young cattle. Vet. Rec. 127:388.

Holmberg, T., Thuvander, A., Hult, K. 1988. Ochratoxin A as a suppressor of mitogen-induced blastogenesis of porcine blood lymphocytes. Acta Vet. Scand. 29:219-223.

Holmberg, T., Breitholtz, A., Bengtsson, A., and Hult, K. 1990. Ochratoxin A in swine blood in relation to moisture content in feeding barley at harvest. Acta Agric. Scand. 40:201-204.

Hong, H. H. L., Jameson, C. W., and Boorman, G. A. 1988. Residual hematopoietic effect in mice exposed to ochratoxin A prior to irradiation. Toxicology 53:57-67.

Hood, R. D., Naughton, M. J., and Hayes, A. W. 1976. Prenatal effects of ochratoxin A in hamsters. Teratology 13:11-14.

Hsu, I.-C., Smalley, E. B., Strong, F. M., and Ribelin, W. E. 1972. Identification of T-2 toxin in moldy corn associated with a lethal toxicosis in dairy cattle. Appl. Microbiol. 24:685-690.

Huff, W. E., and Doerr, J. A. 1981. Synergism between aflatoxin and ochratoxin A in broiler chickens. Poult. Sci. 60:550-555.

Huff, W. E., Wyatt, R. D.,and Hamilton, P. B. 1975. Nephrotoxicity of dietary ochratoxin A in broiler chickens. Appl. Microbiol. 30:48-51.

Huff, W. E., Doerr, J. A., Hamilton, P. B., Hamann, D. D., Peterson, R. E., and Ciegler, A. 1980. Evaluation of bone strength during aflatoxicosis and ochratoxicosis. Appl. Environ. Microbiol. 40:102-107.

Huff, W. E., Doerr, J. A., Hamilton, P. B., and Vesonder, R. F. 1981. Acute toxicity of vomitoxin (deoxynivalenol) in broiler chickens. Poult. Sci. 60:1412-1414.

Huff, W. E., Doerr, J. A., Wabeck, C. J., Chaloupka, G. W., May, J. D., Merkley, J. W. 1983. Individual and combined effects of aflatoxin and ochratoxin A on bruising in broiler chickens. Poult. Sci. 62:1764-1771.

Huff, W. E., Kubena, L. F., Harvey, R. B., Hagler, W. M., Jr., Swanson, S. P., Phillips, T. D., and Creger, C. R. 1986. Individual and combined effects of aflatoxin and deoxynivalenol (DON), vomitoxin. in broiler chickens. Poult. Sci. 65:1291-1298.

Huff, W. E., Harvey, R. B., Kubena, L. F., and Rottinghaus, G. E. 1988a. Toxic synergism between aflatoxin and T-2 toxin in broiler chickens. Poult. Sci. 67:1418-1423.

Huff, W. E., Kubena, L. F., and Harvey, R. B. 1988b. Progression of ochratoxicosis in broiler chickens. Poult. Sci. 67:1139-1146.

Huff, W. E., Kubena, L. F., Harvey, R. B., and Doerr, J. A. 1988c. Mycotoxin interactions in poultry and swine. J. Anim. Sci. 66:2351-2355.

Huston, J. E., Davis, D. I., Menzies, C. S., and Kramer, D. C. 1980. Effects of zeranol on growth and reproduction in heifers. Southwest Vet. 33:209-212.

Ivanov, X., Yuan, C. K., Wang, C. H., and Fang, S. C. 1959. Toxic encephalomalacia (mouldy corn poisoning) of solipedous domestic animals in China. Acta Vet. Zootech. Sin. 4:30-37

Jagusch, K. T., Gray, M. H., Maclean, K. S., Towers, N. R., di Menna, M. E., and McMillan, W. H. 1986. The cause of reproductive loss in Gisborne-East Coast ewe flocks. Proc. N.Z. Soc. Anim. Prod. 46:251-253.

Jayakumar, P. M., Valsala, K. V., and Rajan, A. 1988. Testicular pathology in experimental ochratoxicosis in ducks. Indian J. Vet. Pathol. 12:37-41.

Kallela, K., and Ettala, E. 1984. The oestrogenic *Fusarium* toxin (zearalenone) in hay as a cause of early abortions in the cow. Nord. Veterinaermed. 36:305-309.

Kane, A., Creppy, E. E., Roth, A., Roschenthaler, R., and Dirheimer, G. 1986. Distribution of the [³H]-label from low doses of radioactive ochratoxin A ingested by rats, and evidence for DNA single-strand breaks caused in liver and kidneys. Arch. Toxicol. 38:219-224.

Kanizawa, M. 1984. Synergistic effect of citrinin on hepatorenal carcinogenesis of ochratoxin A in mice. Pages 245-254 in: Toxigenic Fungi, Their Toxins and Health Hazard. Developments in Feed Science no. 7. H. Kurata, Y. Ueno, eds. Elsevier Science Publishers New York.

Kanizawa, M., and Suzuki, S. 1978. Induction of renal and hepatic tumours in mice by ochratoxin A, a mycotoxin. Gann 69:599-600.

Kellerman, T. S., Marasas, W. F. O., Thiel, P. G., Gelderblom, W. C. A., Cawood, M., and Coetzer, J. A. W. 1990. Leukoencephalomalacia in two horses induced by oral dosing of fumonisin B_1. Onderspoort J. Vet. Res. 57:269-275.

Khamis, Y., Hammad, H. A., and Hemeida, N. A. 1986. Mycotoxicosis with oestrogenic effect in cattle. Zuchthygiene 21:233-236.

Kiessling, K. H., Pettersson, H., Sandholm, K., and Olsen, M. 1984. Metabolism of aflatoxin, ochratoxin, zearalenone, and three trichothecenes by intact rumen fluid, rumen protozoa and rumen bacteria. Appl. Environ. Microbiol. 47:1070-1073.

Koshinsky, H. A., and Khachatourians, G. G. 1992. Trichothecene synergism, additivity, and antagonism: The significance of the maximally quiescent ratio. Nat. Toxins 1:38-47.

Kosuri, N. R., Grove, M. D., Yates, J. G., Tallent, W. H., Ellis, J. J., Wolff, I. A., and Nicholas, R. E. 1970. Response of cattle to mycotoxins of *Fusarium tricinctum* isolated from corn and fescue. J. Am. Vet. Med. Assoc. 157:938-940.

Kriek, N. P. J., Kellerman, T. S., and Marasas, W. F. O. 1981. A comparative study of the toxicity of *Fusarium verticilliodes* (=*F. moniliforme*) to horses, primates, pigs, sheep and rats. Onderspoort J. Vet. Res. 48:129-131.

Krogh, P. 1978. Causal associations of mycotoxic nephropathy. Acta Pathol. Microbiol. Scand. 81:689-695.

Krogh, P., Axelson, N. H., Elling, F., Gyrd-Hansen, N., Hald, B., Hyldgaard-Jansen, J., Larsen, A. E., Madsen, A., Mortensen, H. P. Moller, T., Petersen, O. K., Ravnskov, U., Rostgaard, M., and Aalund, O. 1974. Experimental porcine nephropathy: Changes of renal function and structure induced by ochratoxin A-contaminated feed. Acta Pathol. Microbiol. Scand. A246:21.

Krogh, P., Elling, F., Hald, B., Jylling, B., Petersen, V. E., Skadhauge, E., and Svendsen, C. K. 1976. Experimental avian nephropathy. Acta Pathol. Microbiol. Scand. 84:215-221.

Krogh, P., Elling, F., Friis, C., Hald, B., Larsen, A. E., Lillehoj, E. B., Madsen, A., Mortensen, H. P., Rasmussen, F., and Ravnskov, U. 1979. Porcine nephropathy induced by long-term ingestion of ochratoxin A. Vet. Pathol. 16:466-475.

Kubena, L. F., and Harvey, R. B. 1988. Response of growing Leghorn chicks to deoxynivalenol-contaminated wheat. Poult. Sci. 67:1778-1780.

Kubena, L. F., Swanson, S. P., Harvey, R. B., Fletcher, O. J., Rowe, L. D., and Phillips, T. D. 1985. Effects of feeding deoxynivalenol (vomitoxin)-contaminated wheat to growing chicks. Poult. Sci. 64:1649-1655.

Kubena, L. F., Harvey, R. B., Corrier, D. E., Huff, W. E., and Phillips, T. D. 1987a. Effects of feeding deoxynivalenol (DON, vomitoxin)-contaminated wheat to female white Leghorn chickens from day old through egg production. Poult. Sci. 66:1612-1618.

Kubena, L. F., Harvey, R. B., Phillips, T. D., Holman, G. M., and Creger, C. R. 1987b. Effects of feeding mature white Leghorn hens diets that contain deoxynivalenol (vomitoxin). Poult. Sci. 66:55-58.

Kubena, L. F., Huff, W. E. Harvey, R. B., Phillips, T. D., and Rottinghaus, G. E. 1989a. Individual and combined toxicity of deoxynivalenol and T-2 toxin in broiler chicks. Poult. Sci. 68:622-626.

Kubena, L. F., Harvey, R. B., Huff, W. E., Corrier, D. E., Phillips, T. D., and Rottinghaus, G. E. 1989b. Influence of ochratoxin A and T-2 toxin singly and in combination on broiler chickens. Poult. Sci. 68:867-872.

Kurtz, H. J., Nairn, M. E., Nelson, G. H., Christensen, C. M., and Mirocha, C. J. 1969. Histologic changes in the genital tracts of swine fed estrogenic mycotoxins. Am. J. Vet. Res. 30:551-556.

Lee, Y.-W., Mirocha, C. J., Schroeder, D. J., and Hamre, M. L. 1985. The effect of a purified water-soluble fraction of a *Fusarium roseum* "*Graminearum*" culture on reproduction of white Leghorn females. Poult. Sci. 64:1077-1082.

Lindley, W. M. 1978. Ergot toxicosis. Mod. Vet. Pract. 59:463-464.

Lippold, C. C., Stothers, S. C., Frohlich, A. A., and Marquardt, R. R. 1987. Effect of ochratoxin-A (OA) contaminated barley on young growing pigs. (Abstr.) Can.

J. Anim. Sci. 67:1205.

Lloyd, W. E. 1980. Citrinin and ochratoxin toxicosis in cattle in the United States. Pages 436-439 in: Proc. Int. Symp. Vet. Lab. Diagn. 2nd, vol. 3.

Løken, T. 1984. Ergot from meadow grass in Norway—Chemical composition and toxicological effects in sheep. Nord. Veterinaermed. 36:259-265.

Long, G. G., and Diekman, M. A. 1986. Characterization of effects of zearalenone in swine during early pregnancy. Am. J. Vet. Res. 47:184-187.

Lorenzana, R. M., Beasley, V. R., Buck, W. B., and Ghent, A. W. 1985. Experimental T-2 toxicosis in swine. II. Effect of intravascular T-2 toxin on serum enzymes and biochemistry, blood coagulation, and hematology. Fundam. Appl. Toxicol. 5:893-901.

Lun, A. K., Young, L. G., Moran, E. T., Jr., Hunter, D. B., and Rodriguez, J. P. 1986. Effects of feeding hens a high level of vomitoxin-contaminated corn on performance and tissue residues. Poult. Sci. 65:1095-1099.

Lyons, P. C., Plattner, R. D., and Bacon, C. W. 1986. Occurrence of peptide and clavine ergot alkaloids in tall fescue grass. Science 232:487-489.

Manley, R. W., Hulet, R. M., Meldrum, J. B., and Larsen, C. T. 1988. Turkey poult tolerance to diets containing deoxynivalenol (vomitoxin) and salinomycin. Poult. Sci. 67:149-152.

Mann, D. D., Buening, G. M., Hook, B., and Osweiler, G. D. 1983. Effects of T-2 mycotoxin on bovine serum proteins. Am. J. Vet. Res. 44:1757-1759.

Manning, J. G., Hargis, B. M., Moore, R. W., Edwards, J. F., Scanlan, C. M., and Creger, C. R. 1990. Recent investigations of an ergot-like feed associated toxicosis in commercial bobwhite quail. Poult. Sci. 69(Suppl. 1):87.

Mannion, P. F., and Blaney, B. J. 1988. Responses of meat chickens offered 4-deoxynivalenol- and zearalenone-containing wheat, naturally infected with *Fusarium graminearum*. Aust. J. Agric. Res. 39:533-540.

Marasas, W. F. O., Kellerman, T. S., Pienaar, J. G., and Naude, T. W. 1976. Leukoencephalomalacia: A mycotoxicosis of equidae caused by *Fusarium moniliforme* Sheldon. Onderspoort J. Vet. Res. 43:113-122.

Marasas, W. F. O., Nelson, P. E., and Toussoun, T. A., eds. 1984. Section Liseola. Pages 216-252 in: Toxigenic *Fusarium* Species: Identity and Mycotoxicology. The Pennsylvania State University Press, University Park.

Marasas, W. F. O., Jaskiewicz, K., Venter, F. S., and Van Schalkwyk, D. J. 1988a. *Fusarium moniliforme* contamination of maize in oesophageal cancer areas in Transkei. S. Afr. Med. J. 74:110-114.

Marasas, W. F. O., Kellerman, T. S., Gelderblom, W. C. A., Coetzer, J. A. W., Thiel, P. G., and Van der Lugt, J. J. 1988b. Leukoencephalomalacia in a horse induced by fumonisin B_1 isolated from *Fusarium moniliforme*. Onderspoort J. Vet. Res. 55:197-203.

Marpegan, M. R., Perfumo, C. J., Godoy, H. M., Sala de Miguel, M., Diaz, E., and Risso, M. A. 1988. Feed refusal of pigs caused by *Fusarium* mycotoxins in Argentina. J. Vet. Med. Assoc. 35:610-616.

Masri, M. 1988. Review of mycotoxic equine encephalomalacia. Proc. Annu. Vet. Med. Forum 6:124-126.

Matthews, J. G., Patterson, D. S. P., Roberts, B. A., and Shreeve, B. J. 1977. T-2 toxin and haemorrhagic syndromes of cattle. Vet. Rec. 102:391.

Mayura, K., Reddy, R. V., Hayes, A. W., and Berndt, W. O. 1982. Embryocidal, fetotoxic and teratogenic effects of ochratoxin A in rats. Teratology 25:175-185

McDonald, G. K. 1980. Moldy sweetclover poisoning in a horse. Can. Vet. J. 21:250-251.

McNutt, S. H., Purwin, P., and Murray, C. 1928. Vulvovaginitis in swine, preliminary report. J. Am. Vet. Med. Assoc. 73:484-492.

Meronuck, R. A., Garren, K. H., Christensen, C. M., Nelson, G. H., and Bates, F. 1970. Effects on turkey poults and chicks of rations containing corn invaded by *Penicillium* and *Fusarium* species. Am. J. Vet. Res. 31:551-555.

Micco, C., Miraglia, M., Benelli, L., Onori, R., Ioppolo, A., and Mantovani, A. L. 1988. Long term administration of low doses of mycotoxins in poultry. 2. Residues of ochratoxin A and aflatoxins in broilers and laying hens after combined administration of ochratoxin A and aflatoxin B1. Food Addit. Contam. 5:309-314.

Mirocha, C. J., and Christensen, C. M. 1974. Oestrogenic mycotoxins synthesized by Fusarium. Pages 129-149 in: Mycotoxins. J. F. H. Purchase, ed. Elsevier, Amsterdam.

Mirocha, C. J., Harrison, J., Nichols, A. A., and McClintock, M. 1968. Detection of fungal estrogen (F-2) in hay associated with infertility in dairy cattle. Appl. Microbiol. 16:797-798.

Moore, C. J., Blaney, B. J., Spencer, R. A., and Dodman, R. L. 1985. Rejection by pigs of mouldy grain containing deoxynivalenol. Aust. Vet. J. 62:60-62.

Moran, E. T., Jr., Ferket, P. R., and Lun, A. K. 1987. Impact of high dietary vomitoxin on yolk yield and embryonic mortality. Poult. Sci. 66:977-982.

Mortensen, H. P., Hald, B., Larsen, A. E., and Madsen, A. 1983. Ochratoxin A contaminated barley for sows and piglets. Pig performance and residues in milk and pigs. Acta Agric. Scand. 33:349-352.

Munro, I. C., Scott, P. M., Moodie, C. A., and Willes, R. F. 1973. Ochratoxin A— Occurrence and toxicity. J. Am. Vet. Med. Assoc. 163:1269-1273.

Niemiec, J., Brodacki, A., and Scholtyssek, S. 1989. Changes in blood serum proteins of broilers and layers as a result of influence of ochratoxin A in the feed. Archiv. Gefluegelkd. 53:108-111.

Norred, W. P., Plattner, R. D., Voss, K. A., Bacon, C. W., and Porter, J. K. 1989. Natural occurrence of fumonisins in corn associated with equine leukoencephalomalacia (ELEM). (Abstr.) Toxicologist 9:258.

Norred, W. P., Wang, E., Yoo, H. W., Riley, R. T., and Merrill, A. H. 1992. In vitro toxicology of fumonisins and the mechanistic implications. Mycopathologia 117:73-78.

O'Neil, J. B., and Rae, W. J. 1965. Ergot tolerance in chicks and hens. (Abstr.) Poult. Sci. 44:1404.

Oresnik, A., Salobir, K., and Lavric, A. 1980. A comparison of the effects of natural ergot alkaloids on laboratory mice and dairy cows. Proc. Int. Congr. Cattle Dis. 11:1378-1380.

Osweiler, G. D. 1990. Mycotoxins and livestock: What role do fungal toxins play in illness and production losses? Vet. Med. 85:89-94.

Osweiler, G. D., Hooks, B. S., Mann, D. D., Buening, G. M., and Ruttinghaus, G. E. 1981. Effects of T-2 toxin in cattle. Proc. U.S. Anim. Health Assoc. 85:214-231.

Palyusik, M., and Koplik-Kovács, E. 1975. Effect on laying geese of feeds containing the fusariotoxins T2 and F2. Acta Vet. Acad. Sci. Hung. 25:363-368.

Pang, V. F., Adams, J. H., Beasley, V. R., Buck, W. B., and Haschek, W. M. 1986. Myocardial and pancreatic lesions induced by T-2 toxin, a trichothecene mycotoxin, in swine. Vet. Pathol. 23:310-319.

Pang, V. F., Lorenzana, R. M., Beasley, V. R., Buck, W. B., and Haschek, W. M. 1987a. Experimental T-2 toxicosis in swine. III. Morphologic changes following intravascular administration of T-2 toxin. Fundam. Appl. Toxicol. 8:298-309.

Pang, V. F., Swanson, S. P., Beasley, V. R., Buck, W. B., and Haschek, W. M. 1987b. The toxicity of T-2 toxin in swine following topical application. I. Clinical signs, pathology, and residue concentrations. Fundam. Appl. Toxicol. 9:41-49.

Pang, V. F., Felsburg, P. J., Beasley, V. R., Buck, W. B., and Haschek, W. M. 1987c. The toxicity of T-2 toxin in swine following topical application. II. Effects on hematology, serum biochemistry, and immune response. Fundam. Appl. Toxicol. 9:50-59.

Pang, V. F., Lambert, R. J., Felsburg, P. J., Beasley, V. R., Buck, W. B., and Haschek, W. M. 1988. Experimental T-2 toxicosis in swine following inhalation exposure: Clinical signs and effects of hematology, serum biochemistry, and immune response. Fundam. Appl. Toxicol. 11:100-109.

Patterson, D. S. P., Matthews, J. G., Shreeve, B. J., Roberts, B. A., McDonald, S. M., and Hayes, A. W. 1979. The failure of trichothecene mycotoxins and whole cultures of *Fusarium tricinctum* to cause experimental haemorrhagic syndromes in calves and pigs. Vet. Rec. 105:252-255.

Patterson, D. S. P., Shreeve, B. J., Roberta, B. A., Berrett, S., Brush, P. J., and Glancy, E. M. 1981. Effect on calves of barley naturally contaminated with ochratoxin A and groundnut meal contaminated with low concentrations of aflatoxin B_1. Res. Vet. Sci. 31:213-218.

Perry, T. W., Stob, M., Huber, D. M., and Petterson, R. C. 1970. Effect of subcutaneous implantation of resorcylic acid lactone on performance of growing and finishing beef cattle. J. Anim. Sci. 31:789-793.

Pestka, J. J., and Bondy, G. S. 1990. Alteration of immune function following dietary mycotoxin exposure. Can. J. Physiol. Pharmacol. 68:1009-1016.

Petrie, L., Robb, J., and Stewart, A. F. 1977. The identification of T-2 toxin and its association with a haemorrhagic syndrome in cattle. Vet. Rec. 101:326.

Pier, A. C., Cysewski, S. J., Richards, J. L., Baetz, A. L., and Mitchell, L. 1976. Experimental mycotoxicoses in calves with aflatoxin, ochratoxin, rubratoxin, and T-2 toxin. Proc. U.S. Anim. Health Assoc. 80:130-148.

Plattner, R. D., Norred, W. P., Bacon, C. W., Voss, K. A., Peterson, R., Shackelford, D. D., and Weisleder, D. 1990. A method of detection of fumonisins in corn samples associated with field cases of equine leukoencephalomalacia. Mycologia 82:698-702.

Porter, J. K., Voss, K. A., Bacon, C. W., and Norred, W. P. 1990. Effect of *Fusarium moniliforme* and corn associated with equine leukoencephalomalacia on rat neurotransmitters and metabolites. Proc. Soc. Exp. Med. 194:265-269.

Prelusky, D. B., Hamilton, R. M. G., Trenholm, H. L., and Miller, J. D. 1986. Tissue distribution and excretion of radioactivity following administration of ^{14}C-labelled deoxynivalenol to white Leghorn hens. Fundam. Appl. Toxicol. 7:635-645.

Prelusky, D. B., Veira, D. M., Trenholm, H. L., and Foster, B. C. 1987. Metabolic fate and elimination in milk, urine and bile of deoxynivalenol following administration to lactating sheep. J. Environ. Sci. Health B22:125-148.

Prior, M. G., and Sisodia, C. S. 1978. Ochratoxicosis in white Leghorn hens. Poult. Sci. 57:619-623.

Prior, M. G., and Sisodia, C. S. 1982. The effects of ochratoxin A on the immune response of Swiss mice. Can. J. Comp. Med. 46:91-96.

Prior, M. G., O'Neil, J. B., and Sisodia, C. S. 1980. Effects of ochratoxin A on growth response and residues in broilers. Poult. Sci. 59:1254-1257.

Pritchard, D. G., Markson, L. M., Brush, P. J., Sawtell, J. A. A., and Bloxham, P. A. 1983. Haemmorrhagic syndrome of cattle associated with the feeding of sweet vernal (*Anthoxanthum oderatum*) hay containing dicoumarol. Vet. Rec. 113:78-84.

Radostits, O. M., Searcy, G. P., and Mitchall, K. G. 1980. Moldy sweetclover poisoning in cattle. Can. Vet. J. 21:155-158.

Reichmann, K. G., Blaney, B. J., Connor, J. K., Runge, B. M. 1982. The significance of aflatoxin and ochratoxin in the diet of Australian chickens. Aust. Vet. J. 58:211-212.

Ribelin, W. E. 1978. Ochratoxicosis in cattle. Pages 28-36 in: Mycotoxic Fungi, Mycotoxins, Mycotoxicoses. Vol. 2, Mycotoxicoses of Domestic and Laboratory Animals, Poultry and Aquatic Invertebrates and Vertebrates. T. D. Wyllie, L. G. Morehouse, eds. Marcel Dekker, New York.

Richard, J. L., Cysewski, S. J., Pier, A. C., and Booth, G. D. 1978. Comparison of effects of dietary T-2 toxin on growth, immunogenic organs, antibody formation and pathologic changes in turkeys and chickens. Am. J. Vet. Res. 39:1674-1679.

Robb, J. 1990. Effects of mycotoxins on animal performance. Pages 61-76 in: Recent Advances in Animal Nutrition. W. Haresign and D. J. A. Cole, eds. Butterworths, Toronto.

Robbins, J. D., Porter, J. K., and Bacon, C. W. 1986. Occurrence and clinical manifestations of ergot and fescue toxicoses. Pages 61-74 in: Diagnosis of Mycotoxicoses. J. L. Richard and J. R. Thurston, eds. Martinus Nijhoff Publishers, Boston.

Roine, K., Korpinen, E.-L., and Kallela, K. 1971. Mycotoxicosis as a probable cause of infertility in dairy cows. Nord. Veterinaermed. 23:623-633.

Ross, P. F., Rice, L. G., Plattner, R. D., Osweiler, G. D., Wilson, T. M., Owens, D. L., Nelson, H. A., and Richard, J. L. 1991. Concentrations of fumonisin B_1 in feeds associated with animal health problems. Mycopathologia 114:129-135.

Rotter, B. A., Thompson, B. K., Prelusky, D. B., and Trenholm, H. L. 1991. Evaluation of potential interactions involving trichothecene mycotoxins using the chick embryotoxicity bioassay. Arch. Environ. Contam. Toxicol. 21:621-624.

Rotter, B. A., Rotter, R. G., Thompson, B. K., and Trenholm, H. L. 1992. Investigations in the use of mice exposed to mycotoxins as a model for growing pigs. J. Environ. Health 37:295-305.

Rotter, R. G., Marquardt, R. R., and Young, J. C. 1985a. Effect of ergot from different sources and of fractionated ergot on the performance of growing chicks. Can. J. Anim. Sci. 65:953-961.

Rotter, R. G., Marquardt, R. R., and Crow, G. H. 1985b. A comparison of the effect of increasing dietary concentrations of wheat ergot on the performance of Leghorn and broiler chicks. Can. J. Anim. Sci. 65:963-974.

Rotter, R. G., Frohlich, A. A., Marquardt, R. R., and Abramson, D. 1989. Comparison of the effects of toxin-free and toxin-containing mold-contaminated barley on chick performance. Can. J. Anim. Sci. 69:247-259.

Rotter, R. G., Thompson, B. K., Trenholm, H. L., Prelusky, D. B., Hartin, K. E., and Miller, J. D. 1992. A preliminary examination of potential interactions between deoxynivalenol (DON), and other selected *Fusarium* metabolites in growing pigs. Can. J. Anim. Sci. 72: 107-116.

Ruhr, L. P. 1979. The effect of the mycotoxin zearalenone on fertility in the boar. Ph.D. thesis. University of Missouri, Columbia.

Rukhlyada, V. V. 1989. Experimental poisoning of cattle with T-2 mycotoxin. Veterinariya (Mosc.) 2:52-54.

Rumsey, T. S., Stuedemann, J. A., Wilkinson, S. R., and Williams, D. J. 1979. Chemical composition of necrotic fat lesions in beef cows grazing fertilized "Kentucky 31″ tall fescue. J. Anim. Sci. 48:673-682.

Sato, S., Ohya, T., Homma, S., and Tatsuno, T. 1981. Effects of fusarenon-X on serological responses of chicks inoculated with Newcastle disease vaccine. Proc. Jpn. Assoc. Mycotoxicol. 13:43-45.

Scheel, L. D. 1978. The toxicity of sweet clover and coumarin anticoagulants. Pages

121-142 in: Mycotoxic Fungi, Mycotoxins, Mycotoxicoses, An Encyclopedic Handbook. T. D. Wyllie and L. G. Morehouse, eds. Marcel Dekker, New York.

Schmidt, S. P., and Marple, D. N. 1988. Recent advances in knowledge of fescue endophyte—Beef cattle. J. Anim. Sci. 66(Suppl 1):60.

Schuh, M., and Baumgartner, W. 1988. Microbiological and mycotoxicological contaminated feedstuffs as disease causing agents in cattle. Wien. Tierartzl. Monatsschr. 75:329-332.

Schwarte, L. H., Biester, H. E and Murray, C. 1937. A disease of horses caused by feeding moldy corn. J. Am. Vet. Med. Assoc. 90:76-85.

Shirley, H. V., and Tohala, S. H. 1983. Ochratoxicosis in laying hens. Univ. Tenn. Agric. Res. Stn. Annu. Sci. Prog. Rep. 1982. 83-08:52-53.

Shlosberg, A., and Egyed, M. N. 1985. Experimental *Ferula communis* (giant fennel) toxicosis in sheep. Zentralbl. Veterinaermed. Reihe A. 32:778-784.

Shreeve, B. J., Patterson, D. S., Pepin, G. A., Roberts, B. A., and Wrathall, A. E. 1977. Effects of feeding ochratoxin to pigs during early pregnancy. Br. Vet. J. 133:412-417.

Singh, G. S. P., Chauhan, H. V. S., Jha, G. J., and Singh, K. K. 1990. Immunosuppression due to chronic ochratoxicosis in broiler chicks. J. Comp. Pathol. 103:399-410.

Smith, J. F., di Menna, M. E., and McGowan, L. T. 1990. Reproductive performance of Coopworth ewes following oral doses of zearalenone before and after mating. J. Reprod. Fertil. 89:99-106.

Solomons, R. N., Oliver, J. W., and Linnabary, R. D. 1989. Reactivity of dorsal pedal vein of cattle to selected alkaloids associated with *Acremonium coenophialum*-infected fescue grass. Am. J. Vet. Res. 50:235-238.

Sreemannarayana, O., Frohlich, A. A., Vitti, T. G., Marquardt, R. R., and Abramson, D. 1988. Studies of the tolerance and disposition of ochratoxin A in young calves. J. Anim. Sci. 66:1703-1711.

Sreemannarayana, O., Marquardt, R. R., Frohlich, A. A., Abramson, D., and Phillips, G. D. 1989. Organ weights, liver constituents, and serum components in growing chicks fed ochratoxin A. Arch. Environ. Contam. Toxicol. 18:404-410.

Still, P. E. 1973. Mycotoxins as possible causes of abortion in dairy cattle. Ph.D. dissertation. University of Wisconsin, Madison.

Stuedemann, J. A., Rumsey, T. S., Bond, J., Wilkinson, S. R., Bush, L. P., Williams, D. J., and Caudly, A. B. 1985. Association of blood cholesterol with occurrence of fat necrosis in cows and tall fescue summer toxicosis in steers. Am. J. Vet. Res. 46:1990-1995.

Stuedemann, J. A., Wilkinson, S. R., Williams, D. J., Ciordia, H., Ernst, J. W., Jackson, W. A., and Jones, J. B. Jr. 1975. Long-term broiler litter fertilization of tall fescue pastures and health and performance of beef cows. (Abstr.) Proc. Int. Symp. Livestock Wastes 3:264.

Svendsen, C., and Skadhauge, E. 1976. Renal function in hens fed graded dietary levels of ochratoxin A. Acta Pharmacol. Toxicol. 38:186-194.

Swarbrick, O., and Swarbrick, J. T. 1968. Suspected ergotism in ducks. Vet. Rec. 82:76-77.

Sydenham, E. W., Thiel, P. G., and Marasas, W. F. O. 1988. Occurrence and chemical determination of zearalenone and alternariol monomethyl ether in sorghum-based mixed feeds associated with an outbreak of suspected hyperestrogenism in swine. J. Agric. Food Chem. 36:621-625.

Szczech, G. M., Carlton, W. W., Tuite, J., and Caldwell, R. 1973. Ochratoxin A toxicosis in swine. Vet. Pathol. 10:347-364.

Tapia, M. O., and Seawright, A. A. 1985. Experimental combined aflatoxin B_1 and ochratoxin A intoxication in pigs. Aust. Vet. J.62:33-37.

Taylor, M. J., Pang, V. F., and Beasley, V. R. 1989. The immunotoxicity of trichothecene mycotoxins. Pages 2-37 in: Trichothecene Mycotoxicosis: Pathophysiologic Effects. Vol. 2. V. R. Beasley, ed. CRC Press, Boca Raton, FL.

Thiel, P. G., Shephard, G. S., Sydenham, E. W., Marasas, W. F. O., Nelson, P. E., and Wilson, T. M. 1991. Levels of fumonisins B_1 and B_2 in feeds associated with confirmed cases of equine leukoencephalomalacia. J. Agric. Food Chem. 39:109-111.

Thompson, F. N., and Porter, J. K. 1990. Tall fescue toxicosis in cattle: Could there be a public health problem here? Vet. Human Toxicol. 32:51-56.

Thurston, J. R., Richard, J. L., and Peden, W. M. 1986. Importance of mycotoxins as suppressors of immunity. Pages 53-58 in: Proc. Int. Conf. Mycoses 6th. Pan American Health Organization. WHO Sci. Publ. 479.

Trenholm, H. L., Hamilton, R. M. G., Friend, D. W., Thompson, B. K., and Hartin, K. E. 1984. Feeding trials with vomitoxin (deoxynivalenol)-contaminated wheat: Effects on swine, poultry and dairy cattle. J. Am. Vet. Med. Assoc. 185:527-531.

Trenholm, H. L., Thompson, B. K., Hartin, K. E., Greenhalgh, R., and McAllister, A. J. 1985. Ingestion of vomitoxin (deoxynivalenol)-contaminated wheat by nonlactating dairy cows. J. Dairy Sci. 68:1000-1005.

Uhlinger, C. 1991. Clinical and epidemiologic features of an epizootic of equine leukoencephalomalacia. J. Am. Vet. Med. Assoc. 198:126-128.

Vanderbrook, M. J., and Vos, B. J., Jr. 1940. Pharmacodynamics of the domestic fowl with respect to ergonovine and ergotamine. Q. J. Exp. Physiol. 30:173-185.

Vesela, D., Vesely, D., and Jelinek, R. 1983. Toxic effects of ochratoxin A and citrinin, alone and in combination, on chicken embryos. Appl. Environ. Microbiol. 45:91-93.

Vesonder, R. F., and Hesseltine, C. W. 1980. Vomitoxin: Natural occurrence on cereal grains and significance as a refusal and emetic factor to swine. Process Biochem. 16:12-15.

Voss, K. A., Plattner, R. D., Bacon, C. W., and Norred, W. P. 1990. Comparative studies of hepatotoxicity and fumonisin B_1 and B_2 content of water and chloroform/methanol extracts of *Fusarium moniliforme* strain MRC 826 culture material. Mycopathologia 112:81-92.

Weaver, G. A., Kurtz, H. J., Bates, F. Y., Chi, M. S., Mirocha, C. J., Behrens, J. C., and Robison, T. S. 1978a. Acute and chronic toxicity of T-2 mycotoxin in swine. Vet. Rec. 103:531-535.

Weaver, G. A., Kurtz, H. J., Mirocha, C. J., Bates, F. Y., Behrens, J. C., and Robison, T. S. 1978b. Effect of T-2 toxin on porcine reproduction. Can. Vet. J. 19:310-314.

Weaver, G. A., Kurtz, H. J., Bates, F. Y., Mirocha, C. J., Behrens, J. C., and Hagler, W. M. 1981. Diacetoxyscirpenol toxicity in pigs. Res. Vet. Sci. 31:131-135.

Weaver, G. A., Kurtz, H. J., Behrens, J. C., Robison, T. S., Seguin, B. E., Bates, F. Y., and Mirocha, C. J. 1986a. Effect of zearalenone on the fertility of virgin dairy heifers. Am. J. Vet. Res. 47:1395-1397.

Weaver, G. A., Kurtz, H. J., Behrens, J. C., Robison, T. S., Seguin, B. E., Bates, F. Y., and Mirocha, C. J. 1986b. Effect of zearalenone on dairy cows. Am. J. Vet. Res. 47:1826-1828.

Whitlow, L. W., and Hagler, W. M., Jr. 1987. The association of productivity losses in dairy cows with deoxynivalenol. Pages E3-13 in: Symp. Recent Dev. Study Mycotoxins. Kaiser Aluminum and Chemical Corp., Oakland, CA.

Whittemore, C. T., Macer, R. C. F., Miller, J. K., and Mantle, P. G. 1976. Some consequences of the ingestion by young and growing pigs of feed contaminated with ergot. Res. Vet. Sci. 20:61-69.

Whittemore, C. T., Miller, J. K., and Mantle, P. G. 1977. Further studies concerning the toxicity of ingested ergot sclerotia to young and growing pigs. Res. Vet. Sci. 22:146-150.

Williams, K. C., Blaney, B. J., and Magee, M. H. 1988. Responses of pigs fed wheat naturally infected with *Fusarium graminearum* and containing the mycotoxins 4-deoxynivalenol and zearalenone. Aust. J. Agric. Res. 39:1095-1105.

Wilson, B. J., and Maronpot, R. R. 1971. Causative fungus agent of leukoencephalomalacia in equine animals. Vet. Rec. 88:484-486.

Wilson, T. M., Nelson, P. E., and Knepp, C. R. 1985. Hepatic neoplastic nodules, adenofibrosis and cholangiocarcinomas in male Fisher 344 rats fed corn naturally contaminated with *Fusarium moniliforme*. Carcinogenesis 6:1155-1160.

Wilson, T. M., Ross, P. F., Owens, D. L., Rice, L. G., Plattner, R. D., Reggiardo, C., Noon, T. H., and Pickrell, J. W. 1990. Fumonisin B_1 levels associated with an epizootic of equine leukoencephalomalacia. J. Vet. Diagn. Invest. 2:213-216.

Wilson, T. M., Ross, P. F., and Nelson, P. E. 1991. Fumonisin mycotoxins and equine leukoencephalomalacia. J. Am. Vet. Med. Assoc. 198:1104.

Wilson, T. M., Ross, P. F., Owens, D. L., Rice, L. G., Green, S. A., Jenkins, S. J., and Nelson, H. A. 1992. Experimental reproduction of ELEM: A study to determine the minimum toxic dose in ponies. Mycopathologia 17:115-120.

Woods, A. J., Bradley-Jones, J., Mantle, P. G. 1966. An outbreak of gangrenous ergotism in cattle. Vet. Rec. 78:742-749.

Wyatt, R. D., Hamilton, P. B., and Burmeister, H. R. 1973. The effects of T-2 toxin in broiler chickens. Poult. Sci. 52:1853-1859.

Yang, C. S. 1980. Research on esophageal cancer in China: A review. Cancer Res. 40:2633-2644.

Yates, S. G., Plattner, R. D., and Garner, G. B. 1985. Detection of ergopeptine alkaloids in endophyte infected, toxic Ky-31 tall fescue by mass spectrometry/mass spectrometry. J. Agric. Food Chem. 33:719-722.

Young, J. C. 1979. Ergot contamination of feedstuffs. Feedstuffs 51:23-33.

Young, J. C. 1981. Variability in the content and composition of alkaloids found in Canadian ergot. I. Rye. J. Environ. Sci. Health 16B:83-111.

Young, J. C., and Chen, Z.-J. 1982. Variability in the content and composition of alkaloids found in Canadian ergot. III. Triticale and barley. J. Environ. Sci. Health 17B:93-107.

Young, J. C., and Marquardt, R. R. 1982. Effects of ergotamine tartrate on growing chickens. Can. J. Anim. Sci. 62:1181-1191.

Young, L. G., and King, G. J. 1986a. Low concentrations of zearalenone in diets of mature gilts. J. Anim. Sci. 63:1191-1196.

Young, L. G., and King, G. J. 1986b. Low concentrations of zearalenone in diets of boars for a prolonged period of time. J. Anim. Sci. 63:1197-1200.

Young, L. G., Vesonder, R. F., Funnell, H. S., Simons, I., and Wilcock, B. 1981. Moldy corn in diets of swine. J. Anim. Sci. 52:1312-1318.

Young, L. G., King, G. J., McGirr, L., and Sutton, J. C. 1982. Moldy corn in diets of gestating and lactating swine. J. Anim. Sci. 54:976-982.

Young, L. G., McGirr, L., Valli, V. E., Lumsden, J. H., and Lun, A. 1983. Vomitoxin in corn fed to young pigs. J. Anim. Sci. 57:655-664.

Young, L. G., Ping, H., and King. G. J. 1990. Effects of feeding zearalenone to sows on rebreeding and pregnancy. J. Anim. Sci. 68:15-20.

Chapter 10

Residues in Food Products
of Animal Origin

D. B. Prelusky

The concern about chemical and toxic residues in the food chain, whether they occur naturally or are introduced by man, is not a new issue. There is a great diversity of potentially toxic chemical substances that may be introduced into foods at various times of production. Even though the extent of risk posed by these residues has been a matter of much public debate, it has become overwhelmingly evident that provision for a safe food supply should be a fundamental component of any modern society. It is important to determine the nature and extent of these residues and to evaluate their potential as health hazards to the general public.

It would be misleading to state that mycotoxins in the food chain are not of concern, although the risk to consumers in a modern society can be kept to a minimum. Known, commonly occurring toxins, particularly those that present a considerable potential health threat, are monitored constantly in foodstuffs; others have been studied and found not to be transmitted into animal-derived food products such as eggs, meat, or milk. However, an obvious lack of information about many other toxins does not allow for complacency. An integrated understanding of the toxicology and biological fate of toxins once ingested is necessary to make rational evaluations of the risk of consuming products from animals exposed to mycotoxins.

Trichothecenes

Poultry

While poultry are capable of tolerating relatively high concentrations of trichothecenes in their diet, there is sufficient information to indicate that very trace amounts can be transmitted into meat and eggs. Once the contaminated feed source is removed, however, it appears that residues quickly decline to negligible levels.

Various authors have reported that the following deoxynivalenol (DON) feeding trials in chickens resulted in no detectable residues in tissues (4 ppm for 28 days, 9 or 18 ppm for 35 days, 83 ppm for 27 days) or in eggs (5 ppm for 190 days, 18 ppm for 28 days, 83 ppm for 27 days) (El-Banna et al 1983; Kubena et al 1985, 1987; Lun et al 1986). Detection limits for DON residues in the above studies were 10 ng/g of sample. In studies using the radiolabeled toxin, others have demonstrated that low residual levels can be found following ingestion of DON, although the chemical nature of the contaminant isn't always known. Following a single oral administration of ^{14}C-DON (2.2 mg) to layers, up to 0.087% of the dose was detected in the first egg laid after dosing, which was equivalent to about 32 ppb (Prelusky et al 1987a). At 20 ppm of ^{14}C-toxin in feed consumed for six days (2.2 mg of DON per day), radioactivity levels increased with each subsequent egg laid through the last day's exposure to the toxin; maximum levels measured were equivalent to 148 and 38 ppb of DON (total metabolites) in the yolk and egg white, respectively (total egg = 70 ppb). This amounted to approximately 0.19% of the daily consumed dose (Prelusky et al 1987a). The maximum tissue level detected during the six-day trial was in the kidney and equivalent to about 60 ppb (Prelusky et al 1986). Various other tissues measured 10–42 ppb. Six days after removal of the adulterated feed, minute levels (<13 ppb) could still be detected in most tissues. During a longer-term study in which layers were provided with a diet containing 5.5 ppm of ^{14}C-DON for a 65-day period, maximum radioactivity in eggs plateaued early (day 8), and measured only about 28 ppb in DON-equivalents (Prelusky et al 1989).

The transmission of radioactivity into eggs following intubation of layers with ^{3}H-labeled T-2 toxin has been studied (Chi et al 1978a). In single-dosed birds, maximum residues in eggs occurred 24 hr after dosing and accounted for approximately 0.17% of the administered dose. In multidosed birds (equivalent to 1.6 ppm of dietary T-2 for eight days), total residues appeared to plateau after the fourth day and measure about 19 ppb (0.9 μg/50 g egg).

Chi et al (1978b) also were able to detect tissue residues up to 60 ppb of T-2, on the basis of radioactivity, following intubation of broiler chickens with 0.5 mg of ^{3}H-T-2 toxin/kg body weight (about 6–7 ppm of dietary T-2). In comparison, Hofmann (1980) was unable to measure T-2 residues in edible tissues of broilers fed ad libitum a diet containing 15 ppm of T-2 toxin (the detection limit was 15 ppb); however, there were no analyses for possible metabolites. Several studies have reported T-2 toxin undergoes extensive metabolism in the chicken (Yoshizawa et al 1980, Visconti and Mirocha 1985).

The transmission of trichothecenes in milk has also been investigated, although relatively high toxin concentrations need to be ingested to produce detectable residues. Prelusky et al (1984) found only trace levels (<4 ppb) of DON and/or metabolites in milk following a single oral dose of 0.92 g (about 1.8 mg/kg) of the toxin. Longer-term feeding studies appear to result in higher concentrations of residues. Côté et al (1986) fed 66 ppm of DON to dairy cows for five days (about 300 mg/day) and found up to 30 ppb in milk of the metabolite DOM-1; the parent toxin itself was not detected. Prelusky et al (1987b) fed 880 ppm of DON ad libitum to lactating ewes for three days and measured up to 220 ppb of total residues, the greater proportion being in the conjugated form of DOM-1. The extent of DON residues in tissues of dairy cows has not been determined.

T-2 toxin appears to be transmitted into milk marginally more than DON, possibly due to its more lipophilic nature. A dairy cow dosed orally with radiolabeled T-2 toxin over a several-day period at roughly 36 ppm in the diet yielded 0.2% of the administered radioactivity in the milk, with the maximum residual concentration equivalent to 37 ppb (Yoshizawa et al 1981). In a similar study using unlabeled toxin, a cow intubated with 182 mg of T-2 for 15 days (about 50 ppm in the diet), yielded T-2 levels in milk ranging from 10 to 160 ppb during the study; other possible metabolites were not quantitated (Robison et al 1979a). It has been demonstrated that T-2 toxin undergoes extensive and very rapid metabolism in the cow (Yoshizawa et al 1981, Beasley et al 1986, Chatterjee et al 1986). In addition, these latter metabolism studies also reported no substantial accumulation of any residues in organs or tissues, even though animals had received relatively high oral doses of the toxin (0.5–3.6 mg/kg).

Several studies have demonstrated that residues probably do not accumulate in swine exposed to low to medium levels of DON in the diet. Pollmann et al (1985) detected maximum residues of just 23 ppb in tissues of starter pigs fed 1.2–3.6 ppm of DON-contaminated diet for three weeks. Côté et al (1985) fed several contaminated diets (0.7–5.8 ppm of DON) to piglets for a five-week period and found only marginally detectable residual amounts in the swine that consumed the higher toxin level. Prelusky and Trenholm (1992) found that in swine consuming 6.0–7.6 ppm of DON for periods of three to seven weeks, the greater proportion of tissue samples (87%) contained less than 10 ppb of DON residue, although a couple of samples did measure above 50 ppb. It is evident that DON is rapidly eliminated by the pig, and removal of the contaminated feed source should result in a rapid decrease in any residual levels (Prelusky et al 1988, Prelusky and Trenholm 1991, Coppock et al 1985, Pollmann et al 1985).

No feeding trials to determine T-2 toxin residues in swine tissues have been carried out to date. While some information on the distribution of the toxin is available following single oral and intravenous doses, it is sufficient to indicate that T-2 toxin is extensively metabolized and rapidly eliminated from swine and would probably not accumulate in tissue if consumed at levels typically encountered in contaminated feeds (Beasley et al 1986, Robison et al 1979b). Additionally, in one study where a sow was provided with 12 ppm of T-2 toxin in the feed for 220 days, 76 ppb of T-2 toxin was detected in the sole milk sample taken on day 190 (Robison et al 1979a).

Similar to T-2 toxin, diacetoxyscirpenol (DAS) is rapidly and extensively metabolized in the pig (Coppock et al 1987, Bauer et al 1985). Consequently, while it is unlikely that accumulation of the parent toxin would occur in naturally exposed animals, the toxicity of DAS should warrant further study on the toxicity and tissue distribution of its metabolites.

Ochratoxin

Because ochratoxin A is an extremely toxic compound that elicits both nephrotoxicity as well as hepatotoxicity, considerable effort has been made to monitor residues in food production animals. Because of the nature of toxin production (storage mycotoxin), it has been noted that residue occurrence and concentration are probably higher in the spring than directly after harvest (Szebiotko et al 1981).

Swine

Ochratoxin tends to be a mycotoxin found predominantly in northern climates. Much of the information concerning ochratoxin in swine comes from Europe, particularly Denmark, where maximum tissue residue levels are legislated. Kidneys are examined during slaughterhouse processing, and those suspected of nephropathy are analyzed for ochratoxin A (OTA). Some relationships between kidney and meat levels as well as blood and kidney levels have been established. Attempts are now being made to use blood residue levels as an indication of exposure, not only as a screen to prevent contaminated meat from entering the food chain, but before slaughter to allow time for the toxin to be cleared from the bodies of positive animals and thereby prevent condemnation of carcasses. It has been reported that a period of four weeks on an OTA-free diet will essentially eliminate residues from the kidneys and/or carcass (Krogh et al 1976a).

Based on a 25 ppb OTA maximum level in kidney, Denmark during 1980–1983 had a condemnation rate ranging from 1.6 to 14.6 carcasses per 100,000 slaughterings (Büchmann and Hald 1985). Between 10 and 29% of suspected kidneys tested positive above the 25 ppb limit. Other published surveys of ochratoxin residue occurrences (1975–1987) are listed in Table 10.1.

Various feeding trials to establish residual levels in pork have been carried out. Krogh et al (1974) exposed swine for three to four months to feed containing 0.2, 1.0, and 4.0 ppm of OTA. Highest residual levels were found in the kidney (50 ppb) in animals fed the highest dietary OTA; lesser amounts were measured in the liver, muscle, and fat. Furthermore, a good correlation was shown to exist between diet contamination level and residue level in the four tissues analyzed. However, Mortensen et al (1983b), indicated that exposure time, as well as toxin level, determined residual levels.

Feeding creep feed containing 0.5–1 ppm of OTA to piglets from two to eight weeks of age resulted in kidney and liver OTA levels of 16–23 and 5–11 ppb, respectively (Mortensen et al 1983a). In another study, older animals were fed a diet containing 1.38 ppm of OTA for periods of two to eight weeks, which resulted in kidney residue levels of 45–71 ppb (Elling 1983). Moreover, these high levels were found early in the study, yet a discernable nephropathy was not evident until four weeks. Consequently, despite inspections, OTA-contaminated pork could still enter the food chain. However, because OTA has been found in blood at a much higher rate than the frequency that nephropathy has been identified (Bauer et al 1984, Hult et al 1980), analysis of blood may be a better indicator of low level, or early exposure to the toxin. Marquardt et al (1988) analyzed 1,200 swine blood samples collected from slaughter houses in western Canada during 1986 and found 3.6–4.2% incidence of OTA greater than 20 ppb, suggesting a previously undiagnosed low level exposure. Mortensen et al (1983b) reported a blood-kidney ratio for OTA residues of approximately 10:1 and that as long as the content of OTA in the blood was less than 200 ppb (serum), the risk of nephropathy was relatively minor.

Poultry

Unlike their response to many other types of mycotoxins (i.e., trichothecenes, zearalenone), poultry are sensitive to low levels of ochratoxins, and consequently large amounts of the toxin are not usually consumed. Still, even at low levels of contamination, residues in tissues have been reported, and it appears these levels rise slowly the longer birds ingest the contaminated feed. Under natural conditions, residues up to 29 ppb

TABLE 10.1
Surveys of Ochratoxin Residues

Country	No. of Kidneys Analyzed	Percent of Pigs Slaughtered	Percent of Kidneys Positive	Ochratoxin A Concentration (ppb)	References
Poland	113	0.05	9	2–23	Golinski et al 1984
Sweden	129	0.07	25	2–104	Rutqvist et al 1977
Belgium	385	0.09	17	0.2–12	Rousseau and Van Peteghem 1989
Denmark	60	0.07	35	2–68	Krogh 1977
Hungary	122	0.02	39	2–100	Sandor et al 1982

in muscle have been found at slaughter, although the degree of exposure was unknown (Elling et al 1975). Liver and kidney typically contain the highest residues, but not always. Muscular tissue has also tested positive. Micco et al (1987) measured trace amounts (1–11 ppb) in tissues of broilers and layers fed 50 ppb of OTA for up to 169 days; broilers tended to have the higher levels.

In hens fed 1 ppm of OTA for five weeks, levels in the liver and kidney were just 1.5–2.5 ppb and 3–10 ppb, respectively (Reichmann et al 1982). In comparison, with broilers also fed 1 ppm of OTA for eight weeks, Prior et al (1980) was unable to detect residues in tissues, but at 2 ppm in the diet found 24 and 41 ppb in liver and kidney, respectively. Hens fed a similar toxin level (1 ppm) for almost one year had kidney, liver, and muscle residues up to 50 ppb (Krogh et al 1976b). No OTA was found in the eggs. At 4 ppm in the diet, kidneys contained 31 ppb of OTA (Prior and Sisodia 1978); at 5.2 ppm for one month, liver measured 9–18 ppb, and trace levels (1.6–4.0 ppb) were found in yolk (Bauer et al 1988); and at 10 ppm, kidney contained 1–6 ppb OTA, and eggs measured 0.7–1.3 ppb residues (Juszkiewicz et al 1982).

Several of these studies reported that OTA residues in tissues decrease rapidly following removal of the contaminated feed source. The time required to produce nondetectable or trace levels varied from 24 hr to several weeks, depending on extent of exposure and tissue analyzed (Micco et al 1987, 1988; Prior et al 1980).

Ruminants

Ochratoxin residues are generally not found to any significant extent in tissues of ruminants, or in milk obtained from them. The capability of rumen microflora to extensively metabolize OTA greatly impedes absorption of the toxin in its original form (Kiessling et al 1984, Hult et al 1976). To six calves fed 0.39–0.54 ppm dietary OTA for 87 days (total amount consumed was 155 mg OTA) only very trace residues (0–5 ppb) of the parent toxin were detected in the kidneys, along with 5–10 ppb of the nontoxic metabolite ochratoxin α (Patterson et al 1981). No residues were measured in liver, plasma, or urine. In a study where two dairy cows were fed 0.32–1.12 ppm of OTA in the ration for 11 weeks, a residue of only 5 ppb was found in the kidney of one animal, and none was detected in any other tissues or in the milk. The metabolite ochratoxin α could not be found in any tissues (Shreeve et al 1979). Ribelin et al (1978) calculated that it would require ingestion of at least 1.66 mg of OTA/kg body weight for four days before any detectable residues were carried over in the milk. This would be equivalent to a dietary intake of approximately 60 ppm of OTA.

Zearalenone

Because of the potent estrogenic activity of zearalenone (ZEN), numerous studies have been carried out to determine the extent this toxin and its metabolites are transmitted into animal-derived products. The problem is possibly increased further by the use of a synthetic derivative of ZEN, zeranol (Ralgro), which is approved as a growth-promoting agent only in beef cattle and sheep, but it could unethically be used in other livestock as well (Sundlof and Strickland 1986). ZEN's potential as a human health risk, particularly to children, has been carefully evaluated (Kuiper-Goodman et al 1987). Based on available information to determine Canadian exposure levels, no adverse health effects are anticipated, although further study is always warranted.

Ruminants

While ZEN and its metabolites are capable of being transmitted to tissues and milk of ruminants at elevated feed levels, there is limited evidence of carryover at levels more typical of a natural exposure. Shreeve et al (1979) fed 0.38–1.93 ppm of ZEN to dairy cows for 11 weeks prior to slaughter and were unable to detect the toxin in muscle, liver, kidney, blood, or milk. At 544 mg/day (about 40 ppm of dietary ZEN) for 21 days, Prelusky et al (1990) found less than 6 ppb of total residues in the milk. Administering up to 6.0 g of ZEN during one day, maximum total residues amounted to only 16 ppb; total recovery in milk was less than 0.01% of the dose. Similarly, a single 5.0-g dose of ZEN yielded only "traces" (<1 ppb) of total ZEN and metabolites in cow's milk (Hagler et al 1980). In comparison, Mirocha et al (1981) fed 25 ppm of ZEN to a single cow for eight days (about 200 mg of ZEN/day) and found up to 1,360 ppb of total metabolites in milk. Total recovery toxin consumed in milk during this study was estimated at about 0.7%, considerably higher than most other studies. The capacity of rumen protozoa to metabolize ZEN once ingested could account for the wide discrepancies in transmissions of residues to cow's milk, and rumen microflora would be affected by the animals' diet.

In one study performed with sheep, in which a lactating ewe received 1.8 g of ZEN, only very trace residue levels (1–2 ppb) were found in the milk, but apparently this was sufficient to cause signs of hyperestrogenisis in a nursing lamb within 10 days of dosing (Hagler et al 1980).

In a study with zeranol (Ralgro), where cattle were implanted with the growth promoter, clearance from the body was rapid enough that no accumulation of the drug occurred in any edible tissues (Sharp and Dyer 1972). In another study where cattle were implanted with doses of up to 168 mg, very trace levels could be measured at slaughter five days later, but there were less than 2 ppb of total residues in muscle and kidney and only slightly higher amounts (8 ppb total) in liver (Chichila et al 1988). The U.S. Food and Drug Administration approved dose of 36 mg of zeranol resulted in total tissue levels of only 1–2 ppb. In 1989, the U.S. FDA reduced the clearance time for zeranol ear implants for beef cattle from 65 days to none (Food Chemical News 1989). Residue limits for uncooked edible tissues of cattle were also increased from nondetectable to 150 ppb in muscle, 300 ppb in liver, 450 ppb in kidney, and 600 ppb in fat. Regulations, however, still require no measurable zeranol residues in sheep.

Poultry

There is evidence for the possible transmission of zearalenone residues into poultry products, but the limited available information is based on studies with unnaturally elevated amounts of toxin. There are no current data on residues resulting from natural low-level exposure.

Mirocha et al (1982) fed broilers a diet with 100 ppm of ZEN for eight days (about 10 mg/day) prior to intubation with 5.0 mg of radiolabeled ZEN per kilogram of body weight and detected relatively high residues in most tissues, particularly the liver and gizzard, for several days after dosing. Male turkey poults were fed a ration containing 800 ppm of ZEN for a two-week period; total residues measured, on average: lung, 258 ppb; heart, 295 ppb; kidney, 599 ppb; and liver, 2,991 ppb (Olsen et al 1986).

Experimental transmission of ZEN into eggs has also been reported. A single dose of 10 mg/kg body weight of radiolabeled ZEN (an approximate daily dietary level 100 ppm) administered to laying hens resulted in residues of up to 2,000 ppb in eggs within 72 hr after dosing, the greater proportion being found in the yolk, as compared to white (albumen). Approximately 1% of the administered toxin was transmitted to eggs during the initial 72 hr after dosing (Dailey et al 1980). It was speculated that because of the lipophilic nature of the residues, a prolonged exposure period might result in accumulation in egg yolk. No further work has been carried out.

Swine

Although pigs are extremely sensitive to low levels of ZEN, higher levels do not usually cause feed refusal. Because of the readily noticeable signs (red swollen vulva, vaginitis) associated with ZEN toxicoses, exposure to the toxin becomes apparent quickly, allowing for removal of contaminated feed.

Few studies have been carried out to determine the extent of residues in swine tissues. At 40 ppm of ZEN fed to pigs for four weeks, maximum total residues in liver reached 438 ppb (James and Smith 1982). Ványi et al (1983) fed dietary ZEN at 40 ppm to sows for nine to 14 days and found maximum residues in milk of 120 ppb, mostly as metabolites of the parent toxin. Residues dropped rapidly to below detection levels once the contaminated ration was removed. In a similar study, also at 40 ppm for nine days, peak levels of metabolites in milk measured a higher 575–790 ppb (Palyusik et al 1980). Indications were that at least five days following removal of toxin exposure were required to eliminate residues from milk. Similarly, in a study where a pig ingested 5 ppm of ZEN-contaminated feed for four days, four to five days were necessary to clear any detectable metabolites from the body (plasma, urine) (Olsen et al 1985).

Other Mycotoxins

Little or no investigations have been made on the following toxins and transmission of their residues to animal-derived products. The risk of toxicity from these compounds in such products, however, is perceived to be very slight.

DICOUMAROL

Dicoumarol appears to be well absorbed following ingestion by ruminants. Levels of 1.2–5.2 ppm in serum were detected in calves that died following consumption of a naturally contaminated diet containing 15 ppm of the toxin for 20–46 days (Pritchard et al 1983). Toxin levels were also determined in clinical cases of cattle suffering from sweet clover poisoning (Blakley 1985). Average levels reported for various tissues analyzed were muscle, 5.9 ppm; liver, 5.4 ppm; kidney, 3.2 ppm; neonatal liver, 2.8 ppm; plasma, 1.3 ppm; serum, 0.47 ppm.

ERGOT ALKALOIDS

Ergot alkaloids appear not to be transmitted into tissues and milk following ingestion by livestock and poultry. An early study reported that cows given sufficient contaminated feed to cause typical ergotism did not have detectable residues in either meat or milk (Cunningham et al 1944). Analyses for alkaloids were carried out both chemically and biologically. Similarly, Young and Marquardt (1981) showed that ergotamine did not accumulate in tissues of growing broiler or layer chicks, and only when extremely high toxin levels (90–810 ppm) were fed for 10 days could very trace amounts (<5 ppb) be found. With pigs fed a diet heavily contaminated with ergot (>100 ppm) for periods of up to several weeks, there was no evidence of alkaloid accumulation in the tissues (Whittemore et al 1976).

FUMONISINS

Fumonisins have recently been identified as a potential health concern to domestic animals and humans alike (Ross et al 1991, Gelderblom et al

1988, Thiel et al 1992, Marasas et al 1988). The actual extent of the problem and degree of risk have yet to be determined. There are few studies on the problem of fumonisin and fumonisin metabolite residues in animal products. Short-term studies involving oral and intravenous dosing of ^{14}C-labeled fumonisin (2 mg/kg body weight) in laying hens resulted in no detectable radioactivity in eggs 24 h after dosing. Only trace amounts were detected in liver, kidney and cecum (Vudathala et al 1993). Similar studies in swine at about 0.5 mg/kg body weight revealed that fumonisin was poorly absorbed. The fraction that was absorbed was concentrated in the liver (Prelusky et al, *in press*). Additionally, 4 h after intraperitoneal administration of ^{14}C-labeled fumonisin to rats, Shephard et al (1992) found 1% of the dose remained in the liver, with trace amounts in kidney. All other radioactivity had been eliminated by this time via feces and urine.

Literature Cited

Bauer, J., Gareis, M., and Gedek, B. 1984. Zum nachweis und vorkommen von ochratoxin A bei schlachtschweinen. Berl. Muench. Tierarztl. Wochenschr. 97:279-283.

Bauer, J., Bollwahn, W., Gareis, M., Gedek, B., and Heinritzi, K. 1985. Kinetic profiles of diacetoxyscirpenol and two of its metabolites in blood serum of pigs. Appl. Environ. Microbiol. 49:842-845.

Bauer, J., Niemiec, J., and Scholtyssek, S. 1988. Ochratoxin A im legehennen futter. 2. Mitteilung: Rückstande in serum, leber, und ei. Arch. Gefluegelkd. 52:71-75.

Beasley, V. R., Swanson, S. P., Corley, R. A., Buck, W. B., Koritz, G. D., and Burmeister, H. R. 1986. Pharmacokinetics of the trichothecene mycotoxin, T-2 toxin, in swine and cattle. Toxicon 24:13-23.

Blakley, B. R. 1985. Moldy sweet clover (dicoumarol) poisoning in Saskatchewan cattle. Can. Vet. J. 26:357-360.

Büchmann, N. B., and Hald, B. 1985. Analysis, occurrence and control of ochratoxin A residues in Danish pig kidneys. Food Addit. Contam. 2:193-199.

Chatterjee, K., Pawlosky, R. J., Treeful, L., and Mirocha, C. J. 1986. Kinetic study of T-2 toxin metabolites in a cow. J. Food Saf. 8:25-34.

Chi, M. S., Robison, T. S., Mirocha, C. J. Behrens, J. C., and Shimoda, W. 1978a. Transmission of radioactivity into eggs from laying hens (*Gallus domesticus*) administered tritium labeled T-2 toxin. Poult. Sci. 57:1234-1238.

Chi, M. S., Robison, T. S., Mirocha, C. J., Swanson, S. P., and Shimoda, W. 1978b. Excretion and tissue distribution of radioactivity from tritium labeled T-2 toxin in chicks. Toxicol. Appl. Pharmacol. 45:391-402.

Chichila, T. M. P., Silvestre, D., Covey, T. R., and Henion, J. D. 1988. Distribution of zeranol in bovine tissues determined by selected ion monitoring capillary gas chromatography/mass spectrometry. J. Analyt. Toxicol. 12:310-318.

Coppock, R. W., Swanson, S. P., Gelberg, H. B., Koritz, G. D., Hoffmann, W. E., Buck, W. B., and Vesonder, B. S. 1985. Preliminary study of the pharmacokinetics and toxicopathy of deoxynivalenol (vomitoxin) in swine. Am. J. Vet. Res. 46:169-174.

Coppock, R. W., Swanson, S. P., Gelberg, H. B., Koritz, G. D., Buck, W. B., and Hoffmann, W. E. 1987. Pharmacokinetics of diacetoxyscirpenol in cattle and swine: Effects of halothane. Am. J. Vet. Res. 48:691-695.

Côté, L. M., Beasley, V. R., Bratich, P. M., Swanson, S. P., Shivaprassad, H. L., and Buck, W. B. 1985. Sex related reduced weight gains in growing swine fed diets containing deoxynivalenol. J. Anim. Sci. 61:942-950.

Côté, L. M., Dahlem, A. M., Yoshizawa, T., Swanson, S. P., and Buck, W. B. 1986. Excretion of deoxynivalenol and its metabolite in milk, urine, and feces of lactating dairy cows. J. Dairy Sci. 69:2416-2423.

Cunningham, I. J., McIntosh, I. G., and Swan, J. B. 1944. The non-toxicity of milk and meat from ergotized cattle. N.Z. J. Sci. Technol. 26:125-136.

Dailey, R. E., Reese, R. E., and Brouwer, E. A. 1980. Metabolism of ^{14}C-zearalenone in laying hens. J. Agric. Food Chem. 28:286-291.

El-Banna, A. A., Hamilton, R. M. G., Scott, P. M., Trenholm, H. L. 1983. Non-transmission of deoxynivalenol (vomitoxin) to eggs and meat in chickens fed deoxynivalenol-contaminated diets. J. Agric. Food Chem. 31:1381-1384.

Elling, F. 1983. Feeding experiments with ochratoxin A-contaminated barley to bacon pigs. IV. Renal lesions. Acta Agric. Scand. 33:153-159.

Elling, F., Hald, B., Jacobsen, C., and Krogh, P. 1975. Spontaneous toxic nephropathy in poultry associated with ochratoxin A. Acta Pathol. Microbiol. Scand. Sect. A Pathol. 83:739-741.

Food Chemical News. 1989. Withholding period for zeranol in cattle reduced to zero. Aug. 7, 1989, p. 44.

Gelderblom, W. C. A., Jaskiewicz, K., Marasas, W. F. O., Thiel, P. G., Horak, R. M., Vleggaar, R., and Kriek, N. P. J. 1988. Fumonisins–Novel mycotoxins with cancer-promoting activity produced by *Fusarium moniliforme*. Appl. Environ. Microbiol. 54:1806-1811.

Golinski, P., Hult, K., Grabarkiewicz-Szczesna, J., Chelkowski, J., Kneblewski, P., and Szebiotko, K. 1984. Mycotoxic porcine nephropathy and spontaneous occurrence of ochratoxin A residues in kidneys and blood of Polish swine. Appl. Environ. Microbiol. 47:1210-1212.

Hagler, W. M., Danko, G., Horvath, L., Palyusik, M., and Mirocha, C. J. 1980. Transmission of zearalenone and its metabolite into ruminant milk. Acta Vet. Acad. Sci. Hung. Tomus 28:209-216.

Hofmann, G. 1980. Investigation into the carry-over of T-2 toxin in chicken. Fleischwirtschaft 60:1908-1910.

Hult, K., Hokby, E., Gatenbeck, S., and Rutqvist, L. 1980. Ochratoxin A in blood from slaughter pigs in Sweden: Use in evaluation of toxin content of consumed feed. Appl. Environ. Microbiol. 39:828-830.

Hult, K., Teiling, A., and Gatenbeck, S. 1976. Degradation of ochratoxin A by a ruminant. Appl. Environ. Microbiol. 32:443-444.

James, L. J., and Smith, T. K. 1982. Effect of dietary alfalfa on zearalenone toxicity and metabolism in rats and swine. J. Anim. Sci. 55:110-118.

Juszsewicz, T., Piskorska-Pliszczynska, J., and Wisniewska, H. 1982. Ochratoxin A in laying hens: Tissue disposition and passage into eggs. Proc. IUPAC Int. Symp. Mycotoxins Phycotoxins 5:122-125.

Kiessling, K.-H., Pettersson, H., Sandholm, K., and Olsen, M. 1984. Metabolism of aflatoxin, ochratoxin, zearalenone, and three trichothecenes by intact rumen fluid, rumen protozoa, and rumen bacteria. Appl. Environ. Micro. 47:1070-1073.

Krogh, P. 1977. Ochratoxin A residues in tissues of slaughter pigs with nephropathy.

Nord. Veterinaermed. 29:402-405.

Krogh, P., Axelsen, N. H., Elling, F., Gyrd-Hansen, N., Hald, B., Hyldgaard-Jensen, J., Larsen, A. E., Madsen, A., Mortensen, H. P., Møller, T., Petersen, O. K., Ravnskov, U., Rostgaard, M., and Aalund, O. 1974. Experimental porcine nephropathy. Acta Pathol. Microbiol. Scand. Sect. A Pathol. 246(Suppl.):1-21.

Krogh, P., Elling, F., Hald, B., Larsen, A. E., Lillehoj, E. B., Madsen, A., and Mortensen, H. P. 1976a. Time dependent disappearance of ochratoxin A residues in tissues of bacon pigs. Toxicology 6:235-242.

Krogh, P. Elling, F., Hald, B., Jylling, B., Petersen, V. E., Skadhauge, E., and Svendsen, C. K. 1976b. Experimental avian nephropathy. Acta Pathol. Microbiol. Scand. Sect. A Pathol. 84:215-221.

Kubena, L. F., Swanson, S. P., Harvey, R. B., Fletcher, O. J., Rowe, L. D., and Phillips, T. D. 1985. Effects of feeding deoxynivalenol (vomitoxin)-contamined wheat to growing chicks. Poult. Sci. 64:1649-1655.

Kubena, L. F., Harvey, R. B., Corrier, D. E., Huff, W. E., and Phillips, T. D. 1987. Effects of feeding deoxynivalenol (DON, vomitoxin)-contaminated wheat to female White Leghorn chickens from day old through egg production. Poult. Sci. 66:1612-1618.

Kuiper-Goodman, T., Scott, P. M., and Watanabe, H. 1987. Risk assessment of the mycotoxin zearalenone. Regul. Toxicol. Pharmacol. 7:253-306.

Lun, A. K., Young, L. G., Moran, E. T. Jr., Hunter, D. B., and Rodriguez, J. P. 1986. Effects of feeding hens a high level of vomitoxin-contaminated corn on performance and tissue residues. Poult. Sci. 65:1095-1099.

Marasas, W. F. O., Jaskiewicz, K., Venter, F. S., and Van Schalkwyk, D. J. 1988. *Fusarium moniliforme* contamination of maize in oesophageal cancer areas in Transkei. S. Afr. Med. J. 74:110-114.

Marquardt, R. R., Frohlich, A. A., Sreemannarayana, O., Abramson, D., and Bernatsky, A. 1988. Ochratoxin A in blood from slaughter pigs in western Canada. Can. J. Vet. Res. 52:186-190.

Micco, C., Miraglia, M., Onori, R., Ioppolo, A., and Mantovani, A. 1987. Long-term administration of low doses of mycotoxins in poultry. 1. Residues of ochratoxin A in broilers and laying hens. Poult. Sci. 66:47-50.

Micco, C., Miraglia, M., Benelli, L., Onori, R., Ioppolo, A., and Mantovani, A. 1988. Long term administration of low doses of mycotoxins in poultry. 2. Residues of ochratoxin A and aflatoxins in broilers and laying hens after combined administration of ochratoxin A and aflatoxin B_1. Food Addit. Contam. 5:309-314.

Mirocha, C. J., Pathre, S. V., and Robison, T. S. 1981. Comparative metabolism of zearalenone and transmission into bovine milk. Food Cosmet. Toxicol. 19:25-30.

Mirocha, C. J., Robison, T. S., Pawlosky, R. J., and Allen, N. K. 1982. Distribution and residue determination of ^{3}H-zearalenone in broilers. Toxicol. Appl. Pharmacol. 66:77-87.

Moran, E. T., Jr., Ferket, P. R., and Lun, A. K. 1987. Impact of high dietary vomitoxin on yolk yield and embryonic mortality. Poult. Sci. 66:977-982.

Mortensen, H. P., Hald, B., Larsen, A. E., and Madsen, A. 1983a. Ochratoxin A contaminated barley for sows and piglets. Pig performance and residues in milk and pigs. Acta Agric. Scand. 33:349-352.

Mortensen, H. P., Hald, B., and Madsen, A. 1983b. Feeding experiments with ochratoxin A contaminated barley for bacon pigs. 5. Ochratoxin A in pig blood. Acta Agric. Scand. 33:235-239.

Olsen, M., Malmlof, K., Pettersson, H., Sandholm, K., and Kiessling, K.-H. 1985. Plasma and urinary levels of zearalenone and α-zearalenol in a prepubertal gilt fed zearalenone. Acta Pharmacol. Toxicol. 56:239-243.

Olsen, M., Mirocha, C. J., Abbas, H. K., and Johansson, B. 1986. Metabolism of high concentrations of dietary zearalenone by young male turkey poults. Poult. Sci. 65:1905-1910.

Palyusik, M., Harrach, B., Mirocha, C. J., and Pathre, S. V. 1980. Transmission of zearalenone and zearalenol into porcine milk. Acta Vet. Acad. Sci. Hung. Tomus 28:217-222.

Patterson, D. S. P., Shreeve, B. J., Roberts, B. A., Berrett, S., Brush, P. J., and Glancy, E. M. 1981. Effect on calves of barley naturally contaminated with ochratoxin A and groundnut meal contaminated with low concentrations of aflatoxin B_1. Res. Vet. Sci. 31:213-218.

Pollmann, D. T., Koch, B. A., Seitz, L. M., Mohr, H. E., and Kennedy, G. A. 1985. Deoxynivalenol-contaminated wheat in swine diets. J. Anim. Sci. 60:239-247.

Prelusky, D. B., and Trenholm, H. L. 1991. Tissue distribution of deoxynivalenol in swine dosed intravenously. J. Agric. Food Chem. 39:748-751.

Prelusky, D. B., and Trenholm, H. L. 1992. Non-accumulation of residues in swine tissues following extended consumption of deoxynivalenol-contaminated diets. J. Food Sci. 57:801-802.

Prelusky, D. B., Trenholm, H. L., Lawrence, G. A., and Scott, P. M. 1984. Non-transmission of deoxynivalenol (vomitoxin) to milk following oral administration to dairy cows. J. Environ. Sci. Health Part B 19:593-609.

Prelusky, D. B., Hamilton, R. M. G., Trehnolm, H. L., and Miller, J. D. 1986. Tissue distribution and excretion of radioactivity following administration of ^{14}C-labeled deoxynivalenol to White Leghorn hens. Fundam. Appl. Toxicol. 7:635-645.

Prelusky, D. B., Trenholm, H. L., Hamilton, R. M. G., and Miller J. D. 1987a. Transmission of [^{14}C]deoxynivalenol to eggs following oral administration to laying hens. J. Agric. Food Chem. 35:182-186.

Prelusky, D. B., Veira, D. M., Trenholm, H. L., and Foster, B. C. 1987b. Metabolic fate and elimination in milk, urine and bile of deoxynivalenol following administration to lactating sheep. J. Environ. Sci. Health B22:125-148.

Prelusky, D. B., Hartin, K. E., Trenholm, H. L., and Miller, J. D. 1988. Pharmacokinetic fate of ^{14}C labeled deoxynivalenol in swine. Fundam. Appl. Toxicol. 10:276-286.

Prelusky, D. B., Hamilton, R. M. G., and Trenholm, H. L. 1989. Transmission of residues to eggs following long-term administration of ^{14}C-labelled deoxynivalenol to laying hens. Poult. Sci. 68:744-748.

Prelusky, D. B., Scott, P. M., Trenholm, H. L., and Lawrence, G. A. 1990. Minimal transmission of zearalenone to milk of dairy cows. J. Environ. Sci. Health B25:87-103.

Prelusky, D. B., Trenholm, H. L., and Savard, M. E. Pharmacokinetic fate of ^{14}C-labeled fumonisin B_1 in swine. Nat. Toxins (In press.)

Prior, M. G., and Sisodia, C. S. 1978. Ochratoxicosis in White Leghorn hens. Poult. Sci. 57:619-623.

Prior, M. G., O'Neil, J. B., and Sisodia, C. S. 1980. Effects of ochratoxin A on growth response and residues in broilers. Poult. Sci. 59:1254-1257.

Pritchard, D. G., Markson, L. M., Brush, P. J., Sawtell, J. A. A., and Bloxham, P. A. 1983. Haemorrhagic syndrome of cattle associated with the feeding of sweet vernal (*Anthoxanthum odoratum*) hay containing dicoumarol. Vet. Rec. 113:78-84.

Reichmann, K. G., Blaney, B. J., Connor, J. K., and Runge, B. M. 1982. The significance of aflatoxin and ochratoxin in the diet of Australian chickens. Aust. Vet. J. 58:211-

212.

Ribelin, W. E., Fukushima, K., and Still, P. E. 1978. The toxicity of ochratoxin to ruminants. Can. J. Comp. Med. 42:172-176.

Robison, T. S., Mirocha, C. J., Kurtz, H. J., Behrens, J. C., Chi, M. S., Weaver, G. A., and Nystrom, S. D. 1979a. Transmission of T-2 toxin in bovine and porcine milk. J. Dairy Sci. 62:637-641.

Robison, T. S., Mirocha, C. J., Kurtz, H. J., Weaver, J. C., and Chi, M. S. 1979b. Distribution of tritium-labeled T-2 toxin in swine. J. Agric. Food Chem. 27:1411-1413.

Ross, P. F., Rice, L. G., Plattner, R. D., Osweiler, G. D., Wilson, T. M., Owens, D. L., Nelson, H. A., and Richard, J. L. 1991. Concentrations of fumonisin B_1 in feeds associated with animal health problems. Mycopathologia 114:129-135.

Rousseau, D. M., and Van Peteghem, C. H. 1989. Spontaneous occurrence of ochratoxin A residues in porcine kidneys in Belgium. Bull. Environ. Contam. Toxicol. 42:181-186.

Rutqvist, L., Bjorklund, N. E., Hult, K., and Gatenbeck, S. 1977. Spontaneous occurrence of ochratoxin residues in kidneys of fattening pigs. Zentralbl. Veterinaermed. Reihe A 24:402-408.

Sandor, G., Glavits, R., Vajda, L., Vanyi, A., and Krogh, P. 1982. Epidemiologic study of ochratoxin A-associated porcine nephropathy in Hungary. Proc. IUPAC Int. Symp. Mycotoxins Phycotoxins 5:349-352.

Sharp, G. D., and Dyer, I. A. 1972. Zearalenol metabolism in steers. J. Anim. Sci. 34:176-179.

Shephard, G. S., Thiel, P. G., Sydenham, F. W., Alberts, J. F., and Gelderblom, W. C. A. 1992. Fate of a single dose of the ^{14}C-labelled mycotoxin, fumonisin B_1, in rats. Toxicon 30:768-770.

Shreeve, B. J., Patterson, D. S. P., and Roberts, B. A. 1979. The carry-over of aflatoxin, ochratoxin and zearalenone from naturally contaminated feed to tissues, urine and milk of dairy cows. Food Cosmet. Toxicol. 17:151-152.

Sundlof, S. F., and Strickland, C. 1986. Zearalenone and zeranol: Potential residue problems in livestock. Vet. Hum. Toxicol. 28:242-250.

Szebiotko, K., Chelkowski, J., Dopierala, B., Godlewska, B., and Radomyska, W. 1981. Mycotoxins in cereal grains. Part I. Ochratoxin, citrinin, sterigmatocystin, penicillic acid and toxigenic fungi in cereal grain. Nahrung 25:415-421.

Thiel, P. G., Marasas, W. F. O., Sydenham, E. W., Shephard, G., and Gelderblom, W. C. A. 1992. The implications of naturally occurring levels of fumonisins in corn for human and animal health. Proc. Symp. Fumonisins: A current perspective and a view to the future. Mycopathologia 117:3-9.

Vanyi, A., Batra, A., and Sandor, G. S. 1983. Metabolism of zearalenone in pregnant sows. Proc. Int. Symp. Mycotoxins pp. 311-315.

Visconti, A., and Mirocha, C. J. 1985. Identification of various T-2 toxin metabolites in chicken excreta and tissues. Appl. Environ. Microbiol. 49:1246-1250.

Vudathala, D. K., Prelusky, D. B., Ayroud, M., Trenholm, H. L., and Miller, J. D. Pharmacokinetic fate and pathological effects of ^{14}C-labeled fumonisin B_1 in laying hens. Nat. Toxins (In press.)

Whittemore, C. T., Macer, R. C. F., Miller, J. K., and Mantle, P. G. 1976. Some consequences of the ingestion by young and growing pigs of feed contaminated with ergot. Res. Vet. Sci. 20:61-69.

Yoshizawa, T., Swanson, S. P., and Mirocha, C. J. 1980. T-2 metabolites in the excreta of broiler chickens administered ^{3}H-labeled T-2 toxin. Appl. Environ. Microbiol.

39:1172-1177.

Yoshizawa, T., Mirocha, C. J., Behrens, J. C., and Swanson, S. P. 1981. Metabolic fate of T-2 toxin in a lactating cow. Food Cosmet. Toxicol. 19:31-39.

Young, J. C., and Marquardt, R. R. 1982. Effects of ergotamine tartrate on growing chickens. Can. J. Anim. Sci. 62:1181-1191.

Chapter 11

Decontamination of *Fusarium* Mycotoxins

L. L. Charmley and D. B. Prelusky

During certain environmental conditions of temperature and humidity, the contamination of various cereal grains with *Fusarium* fungi and mycotoxins can be an unavoidable problem for grain producers. Consequently, the search for a commercially applicable procedure for the decontamination of mycotoxin-contaminated grain has been a goal of food and agriculture industries throughout the world. While certain treatments have been found to reduce concentrations of specific mycotoxins, unfortunately, no single method has been developed that is equally effective against the wide variety of mycotoxins that may be present in a particular contaminated grain. Methods that decrease one or several mycotoxins may have little or no effect on other co-occurring mycotoxins.

For a decontamination procedure to be commercially viable, not only does it need to be effective against a variety of mycotoxins, it also needs to be simple and inexpensive, and, ideally, to use existing technology (Young 1985). In addition, it should work without producing new toxins or altering the nutritional and palatability properties of the grain or grain products. Although no single decontamination method has been developed to date, numerous methods have been tested, and several show potential for commercial application. For the purpose of this review decontamination methods have been divided into three categories: physical, chemical, and biological.

Physical Methods

At low levels of fungal and mycotoxin contamination (0.05 to 1.0 mg of toxin per kilogram of grain), mycotoxins such as deoxynivalenol (DON), are typically found predominantly near the exterior surface of the kernel (Young et al 1984, 1986; Tanaka et al 1986; Lee et al 1987; Trenholm et al 1991), whereas at high levels of contamination (more than 4 mg toxin/ kg grain) mycotoxins may be more evenly distributed throughout the kernel (Scott et al 1983, Hart and Braselton 1983). However, this is not invariably the case, and in some instances DON appears to be relatively evenly distributed throughout individual kernels regardless of the total grain DON concentration (Seitz et al 1985, 1986). DON exists only at the sites of fungal growth; there is little translocation of DON, once produced, to other sites in the kernel (Young et al 1984). Young et al (1984, 1986) therefore speculated that the distribution of DON would be determined by the distribution of *Fusarium* fungi within the kernel, which in turn may depend on the degree of fungal contamination.

Several simple, physical methods for decontaminating mycotoxin-contaminated grain have been tried. These include cleaning and washing, dehulling, milling, separation of contaminated from noncontaminated kernels, and various forms of heat treatment. The success of some of these methods depends, to a large extent, on the degree of contamination and the distribution of the mycotoxins throughout the grain.

Cleaning soft wheat containing 0.03 to 2.89 mg/kg of DON with a combination of screening and air flow, similar to that of most commercial cleaners, reduced the DON concentration by only 16% from that in uncleaned wheat (Seitz et al 1985). It was suggested that the apparently modest and inefficient removal of DON was due to variation in the degree of *Fusarium* infection in kernels and to the possibility that even kernels with nearly normal weight and size (which would not be selectively removed by the cleaner) could have significant fungal invasion and DON contamination. With hard red spring wheat containing 0.64 to 5.10 mg/kg of DON, several cleaning methods (including removal of screenings with or without increased air flow or an additional water washing) reduced DON concentrations by up to 40% (Seitz et al 1986). The efficacy of these cleaning methods appeared to depend on the proportion of total DON present in the shriveled, lighter weight, severely infected kernels compared with those of normal weight and size. Cleaning and removing the dockage reduced the DON concentration by 35% in hard red spring wheat contaminated with 7.1 mg/kg of DON (Scott et al 1983); scouring reduced the DON concentration by 22% in Canada western red spring wheat contaminated with 12.5 mg/kg of DON (Nowicki et al 1988); and cleaning and removing the dust and screenings reduced the DON concentration by 7 to 23% in soft white winter wheat contaminated with approximately 0.07 mg/kg of DON (Young et al 1984). Processing through a rotary grain cleaner equipped with 1.27- and 0.64-cm screens and removing the small kernels, broken kernels, and "red dog," reduced the DON concentration by 33% in moldy corn containing 18 mg/kg of DON, for a total loss of only 9.4% of the material (Patterson and Young 1992b). However, when the screened, moldy corn was fed to young pigs, it resulted in similar losses in body weight as when the unscreened, moldy corn was fed. In another study, removing the outer portion of the kernel in a dehuller resulted in a 40 to 100% reduction in DON and zearalenone (ZEN) concentration, with a concomitant loss of 13 to 19% of the grain material in barley, wheat, and rye contaminated with 5 to 23 mg/kg of DON and 0.5 to 1.2 mg/kg of ZEN (Trenholm et al 1991).

Simple washing procedures, using distilled water, resulted in 65 to 69% reductions in DON concentration and 2 to 61% reductions in ZEN concentration in barley and corn contaminated with 16 to 24 mg/kg of DON and 0.9 to 1.6 mg/kg of ZEN (Trenholm et al 1992). Substituting a $1M$ sodium carbonate solution for distilled water in the first washing step reduced the DON concentration by 72 to 74% and ZEN concentration by 80 to 87%. Washing procedures may be suitable as a cleaning step prior to wet milling operations and for grain to be used in ethanol fermentation, but they would not be suitable for commercial applications involving dry milling because of the prohibitive additional costs of drying the grain after decontamination.

The effectiveness of milling practices to reduce mycotoxin concentration in the flour fractions from that in the whole grain varies according to the relative distribution of mycotoxins throughout the kernels and, in some cases, with the degree of contamination (Young et al 1984, 1986).

In Korean wheat contaminated with 0.3, 0.9, and 2.1 mg/kg of nivalenol (NIV), DON, and ZEN, respectively, milling resulted in a 20 to 69% reduction in NIV, a 24 to 41% reduction in DON, and a 48 to 66% reduction in ZEN concentration in the flour fraction (Lee et al 1987). Similarly, milling wheat contaminated with 0.51, 0.17, and 0.08 mg/kg of NIV, DON, and ZEN was found to result in approximately 66, 33, and 100% respective reductions in toxin concentrations in the flour fractions (Tanaka et al 1986). In soft white winter wheat (Seitz et al 1985, Young et al 1984) and hard red spring wheat (Young et al 1984) contaminated with approximately 0.05 to 3 mg/kg of DON, milling reduced the DON concentration by 27 to 31% in the flour fraction. However, in a highly contaminated hard red spring wheat (8.7 mg/kg of DON) milling reduced the DON concentration in the flour fraction by only 15% (Young et al 1984). DON was found throughout all the milling fractions of hard red winter wheat contaminated with 0.64 to 5.10 mg/kg of DON (Seitz et al 1986), of hard red spring wheat contaminated with 1.5 to 7.5 mg/kg of DON (Scott et al 1983), and of wheat contaminated with 5 mg/kg of DON (Hart and Braselton 1983). However, there was a tendency towards slightly lower DON concentrations in the flour fractions and higher DON concentrations in the shorts, red dog, dockage, and bran fractions. Nowicki et al (1988) concluded that the distribution of DON throughout the various milling fractions of wheat depended, to a large extent, on the degree of fungal penetration of the endosperm. Thus, wheat with a higher overall DON concentration, but with the contamination located predominantly at the surface of the kernel, could produce a flour with a lower DON concentration than wheat having a lower overall DON concentration but with a greater degree of fungal and mycotoxin contamination of the endosperm. For example, these authors found a greater degree of penetration of *Fusarium graminearum* into the interior of the individual kernels of Canada western amber durum (CWAD) than in Canada western red spring (CWRS) wheat, although the DON concentration of CWAD was lower (9.6 mg/kg) than that of CWRS (12.5 mg/kg) wheat. Moreover, milling resulted in respective 50 and 70% reductions in DON concentration in the flour fractions of CWAD and CWRS wheat from that in the conditioned wheat (cleaned in a Carter dockage tester).

In some cases, mold-damaged, mycotoxin-contaminated kernels exhibit differing physical properties from nondamaged kernels and so may be separated by relatively simple procedures, such as density segregation in certain liquids, or fractionation by specific gravity table. In corn contaminated with 1.7 or 6.0 mg/kg of DON and in wheat contaminated with 0.6 or 2.4 mg/kg of DON or approximately 0.02 mg/kg of ZEN, density segregation and removal of kernels buoyant in water and 30% sucrose solutions led to reductions in DON concentrations of 53 to 77% in corn and 67 to 96% in wheat, and reductions in ZEN concentrations of 41 to 55% in wheat (Huff and Hagler 1985). The resulting loss of grain material was 11 to 13% for corn and 5 to 10% for wheat. Similar results were found in sorghum contaminated with DON and ZEN (respectively, 0.006 and 0.05 mg/kg in samples taken at harvest and 0.05 and 0.44 mg/kg in samples taken from storage bins). Density segregation and removal of kernels buoyant in water and saturated sodium chloride solutions led to a 100% removal of DON and ZEN from the more lightly contaminated samples and respective 64 and 84% reductions in DON and ZEN concentrations in the more heavily contaminated material (Babadoost et al 1987). The resulting losses in grain material were approximately 9.5 to 13.5%. Fractionation of wheat kernels naturally infected with *Fusarium* fungi and contaminated with 4 to 7 mg/kg of DON by specific gravity table and removal of the least dense fractions containing the tombstone kernels, reduced the degree of DON contamination by 68 to 85% (Tkachuk et al 1991). Furthermore, removal of the least dense fractions improved the visual grade and milling properties of the contaminated wheat.

Mycotoxins may be destroyed by certain heat treatments. For example, making cookies and doughnuts from flours contaminated with approximately 0.5 mg/kg of DON resulted in an approximately 35% decrease in DON concentration from that prior to baking (Young et al 1984). This decrease, however, only occurred in doughnuts containing no yeast. In yeast-containing doughnuts, a 118 to 189% increase in DON concentration occurred during the baking step, which was speculated to be due to the enzymatic conversion of an unidentified precursor into DON. A 20 to 40% reduction in DON concentration was observed when dough containing straight-grade flours contaminated with 0.2 to 0.9 mg/kg of DON was baked into bread (Seitz et al 1986). However, when dough containing flour contaminated with higher concentrations of DON (1 to 7 mg/kg) was baked in a similar fashion (Scott et al 1983, 1984; Seitz et al 1986), or when the flour was used to make cookies and doughnuts (Scott et al 1984), little or no reduction in DON concentration was observed. Similarly, no reduction in DON concentration occurred when Egyptian bread dough, made from flour contaminated with approximately 2 to 3.5 mg/kg of DON, was baked at 350°C for 2 min (El-Banna et al 1983). No appreciable reductions in NIV, DON, and ZEN concentrations were observed when flour contaminated with 0.19, 0.38, and 0.69 mg/kg, respectively, was used to make sponge cakes baked at 170°C for 30 min (Tanaka et al 1986). Autoclaving caused a 12% reduction in the DON concentration of corn contaminated with 540 mg/kg of DON (Young et al 1987). Microwave oven heating, at 100 to 230°C or higher, of corn contaminated with 20 to 50 mg/kg of DON resulted in a 50 to 100% reduction in DON concentration (Young 1986a, Stahr et al 1987). Convection oven heat appeared to have a greater effect than microwave oven heat, because the temperature could be maintained for longer periods (Young 1986a). It was found that T-2 toxin also could be destroyed by microwave oven heat, but relatively higher temperatures were required to achieve a similar degree of destruction as obtained for DON (Stahr et al 1987). Roasting corn contaminated with 20 mg/kg of T-2 toxin and wheat contaminated with 30 mg/kg of DON in a commercial gas-fired roaster for two cycles or one cycle, respectively, resulted in an approximate 50% reduction in toxin concentration and a small reduction in available protein (Stahr et al 1987).

Chemical Methods

Numerous chemicals have been tested for their ability to decontaminate *Fusarium* mycotoxin-contaminated grain or feed, and while some have proved successful, many have been found to have little or no effect on mycotoxin concentrations.

Calcium hydroxide monomethylamine effectively decontaminated feeds containing T-2 toxin, diacetoxyscirpenol (DAS), or ZEN at 10 to 20 mg/kg (Bauer et al 1987). The success of the procedure was dependent on the moisture content of the feed and the processing temperature. For example, with corn meal containing 20 mg/kg of T-2 toxin, the toxin concentration was reduced by 45 to 55% in 4 h at 10% moisture and a processing temperature of 20 or 30°C. When the moisture content was increased to

25%, T-2 toxin concentration was reduced by 95 to 99% in 4 h. At 60 and 100°C, T-2 toxin was reduced by 99% (below the detection limits of 0.15 mg/kg) in 15 min. However, the less toxic HT-2 toxin, formed during the detoxification process as a result of alkaline hydrolysis of T-2 toxin, remained at detectable levels. A moisture content of 25% and a processing temperature of 100°C had to be maintained for 1 h to destroy all the HT-2 toxin present. Under treatment conditions capable of reducing T-2 and HT-2 toxin concentrations below detection limits, all residual toxicity (measured using the chick embryo assay or skin irritation tests) was removed. Likewise, with corn meal contaminated with 10 mg/kg of DAS, a greater reduction in toxin concentration was observed at increased temperatures and moisture content. For example, at 10% moisture and 100°C, a 2-h processing time was necessary to reduce the toxin concentration below the level of detection (0.5 mg/kg). However, at 25% moisture, DAS was reduced below detection limits within 15 min at only 20°C. In this case, the chick embryo assay and skin irritation tests both indicated the presence of some residual toxicity, which was speculated to be due to the transformation of some DAS, first into 15-monoacetoxyscirpenol and then into scirpentriol, by alkaline hydrolysis. Calcium hydroxide monomethylamine was found to be capable of reducing ZEN concentrations in contaminated barley, the reaction being enhanced as the moisture content of the feed increased from 10 to 21%. In addition to reducing the ZEN concentration, this treatment also reduced the estrogenic effects of the contaminated feed on prepubertal gilts.

Sodium bisulfite solutions were capable of reducing DON concentrations in contaminated corn (Swanson et al 1984, Young 1986a, Young et al 1987) and wheat (Young et al 1986). In small quantities of corn (10 g) containing 540 mg/kg of DON, reductions in DON concentration of up to 95% were achieved when the corn was autoclaved at 121°C for 1 h in the presence of 8.33% aqueous sodium bisulfite (Young et al 1987). As autoclaving contributed only 12% of the reduction in DON concentration, most of the reduction was speculated to be due to the sodium bisulfite treatment. Sodium bisulfite reacts with DON to form nontoxic DON-sulfonate (DON-S) conjugate (Young 1986b). This treatment might not be suitable for direct application to human foods because it affects the rheological properties of flour (Young et al 1986) and because DON-S is unstable in the presence of alkali, being hydrolyzed to the parent DON under certain processing and baking conditions (Young et al 1987). Because DON-S appeared to be nontoxic to pigs, and sodium bisulfite treatment appeared to reduce the short-term toxic effects of DON-contaminated corn (7.2 mg/kg) on feed intake and body weight gain in pigs (Young et al 1987), this treatment might be useful for decontaminating DON-contaminated corn destined for use in pig feeds. To date, no reports have been published of attempts to scale up this procedure to decontaminate large quantities of grain on a commercial basis.

Other chemicals tested for their effectiveness in reducing DON concentrations in contaminated grains include hydrochloric acid, hydrogen peroxide, sodium hypochlorite, ascorbic acid, ammonium hydroxide (Young et al 1986), and ammonium carbonate (Young 1986a). None of these chemicals proved to be as effective as sodium bisulfite (Table 11.1), and some had little or no effect on DON concentration. The apparent increase in DON concentration with hypochlorite treatment was speculated to be due to conversion of a precursor into DON (Miller et al 1983).

Gaseous chemicals used for decontaminating DON-contaminated corn and wheat include sulfur dioxide (Young et al 1986), moist ozone (Young 1986a, Young et al 1986), chlorine, and ammonia (Young 1986a).

Dry gaseous sulfur dioxide, percolated for 3 h through 1 kg of wheat (contaminated with 1 mg/kg of DON), had no effect on DON concentration. Sulfur dioxide gas, moistened by bubbling through water prior to being passed through wheat, caused a 38% reduction in DON concentration (Young et al 1986). Moist and dry ozone (1:1 mol % in air) reduced the DON concentration in corn (contaminated with 100 mg/kg) by 90 and 70%, respectively, in 1 h (Young 1986a), whereas moist ozone had little effect on DON-contaminated wheat (Young et al 1986). This difference was thought to be due to the lower moisture content of wheat (12%, compared with approximately 50% in corn), and the larger sample size of the wheat (1 kg compared with 2 g).

Chlorine, at a gas concentration of 30% (v/v in nitrogen), completely removed the DON from corn contaminated with 1,000 mg/kg within 30 min (Young 1986a). It was suggested that the lack of a significant reduction in DON concentration in contaminated wheat under commercial milling and bleaching conditions (Young et al 1984) was likely due to use of low levels of chlorine (0.1 to 0.5%) and relatively short contact times (several minutes) (Young 1986a).

Ammonia, at a concentration of 100% at room temperature, also caused substantial reductions in DON concentrations in corn contaminated with

TABLE 11.1
Percentage of Deoxynivalenol (DON) Remaining in Wheat Contaminated
with 1 mg/kg of DON After Treatment with Chemical Reagents [a]

Reagent	Concentration	Reagent Volume (ml/kg)	Percent DON Remaining
Hydrogen peroxide	5–6%	33	92–102
Hydrogen peroxide	6%	750	34
Sodium hypochlorite	1%	33	124
Ascorbic acid	2%	600	53
Ammonium hydroxide	5%	600	65
Hydrochloric acid	$0.1 M$	600	65
Sodium bisulfite	10% SO_2	750	<2

[a] Adapted from Young et al (1986).

1,000 mg/kg of DON, but only after a lengthy exposure (Young 1986a). A 9% reduction in DON concentration was observed after a 1-h ammonia treatment, and an 85% reduction after 18 h.

Various chemicals have been tested for their effectiveness in reducing ZEN concentrations in samples of white corn grits spiked to contain 3 or 5 mg/kg of ZEN, or naturally contaminated yellow corn meal containing 33.5 mg/kg of ZEN (Bennett et al 1980). Most of the chemicals used had little or no effect on ZEN concentration (Table 11.2). However, formaldehyde (0.7 or 3.7%), in vapor form from paraformaldehyde crystals, or in aqueous solutions, was able to cause a significant reduction in the ZEN concentration in both the spiked corn grits and the naturally contaminated corn meal. In addition, concentrations as high as 10 mg/kg of ZEN in animal feed were reduced to less than 0.5 mg/kg by formaldehyde treatment. However, the samples treated with aqueous formaldehyde had to be heated to 50°C or above to cause effective reductions in ZEN concentration; corn grits treated with formaldehyde at room temperature for 16 h still contained 50% of the toxin concentration. Ammonium hydroxide (3%) also was found to be capable of reducing ZEN concentrations in naturally contaminated corn; a 64% reduction was achieved in 16 h (Table 11.2).

Physical and chemical treatments have been used in combination for the decontamination of grains contaminated with *Fusarium* mycotoxins. In some cases, combinations of treatments increased the effectiveness of the decontamination process from that observed with either type of treatment alone. Chemical decontamination procedures enhanced by the application of heat include formaldehyde treatment of corn grits or ground corn to

TABLE 11.2
Percent Reduction in Zearalenone (ZEN) Concentration in Corn
After Various Chemical Treatments[a]

Treatment	Conditions		Grain[b]	ZEN (mg/kg)	Reduction in ZEN (%)
	Time	°C			
3% Propionic acid	3 days	22	Corn grits	3 or 5	0
3% Acetic acid	3 days	22	Corn grits	3 or 5	0
1.85% Hydrochloric acid	3 days	22	Corn grits	3 or 5	0
10% Sodium bicarbonate	16 h	50	Corn grits	3 or 5	0
3% Hydrogen peroxide	3 days	22	Corn grits	3 or 5	0
3% Hydrogen peroxide	16 h	50	Corn grits	3 or 5	0
3.7% Formaldehyde soln.	16 h	50	Corn grits	3 or 5	100
Formaldehyde vapors	10 days	22	Corn grits	3 or 5	96
3% Ammonium hydroxide	16 h	50	Corn grits	3 or 5	80
3% Hydrogen peroxide	16 h	50	Ground corn	33.5	5
3% Ammonium hydroxide	16 h	50	Ground corn	33.5	64
3.7% Formaldehyde soln.	16 h	50	Ground corn	33.5	94
0.7% Formaldehyde soln.	16 h	50	Ground corn	33.5	84

[a] Adapted from Bennett et al (1980).
[b] Corn grits were spiked to contain 3 or 5 mg/kg of ZEN. Ground corn was naturally contaminated, containing 33.5 mg/kg of ZEN.

reduce the ZEN concentration (Bennett et al 1980); calcium hydroxymono-methylamine treatment of animal feed to reduce the T-2 toxin, DAS, and ZEN concentrations (Bauer et al 1987); and sodium bisulfite treatment of corn to reduce the DON concentration (Young et al 1987).

Abbas et al (1988) showed that a combination of heat and treatment with lime water (2% Ca[OH]$_2$), in the process of making tortillas, reduced ZEN by 59 to 100%, DON by 72 to 82%, and 15 acetyl-DON (15-ADON) concentrations by 100% in two corn samples contaminated, respectively, with 0.23, 3.28, and 1.49 mg/kg and 4.23, 12.26, and 9.83 mg/kg of ZEN, DON, and 15-ADON.

Ammonia treatment, combined with heat and pressure, was used to decontaminate corn contaminated with 86 mg/kg of fumonisin (Park et al 1992). In this study a 79% reduction in fumonisin concentration was observed when contaminated corn was subjected to two sequential ammonia treatments. The first treatment consisted of 2% ammonia, 15% moisture, 60 psi pressure, and 20°C temperature (referred to as the high-pressure, low-temperature treatment), and the second of 2% ammonia, 13% moisture, 17 psi pressure, and 125°C temperature (referred to as the low-pressure, high-temperature treatment). However, the same reduction in fumonisin concentration (80%) was observed when the second treatment was carried out alone.

Biological Methods

An alternative approach to the decontamination of mycotoxin-contaminated grain is to minimize the effects of the mycotoxins on the animal by modifying the contaminated diet. Such dietary manipulations have included diluting contaminated grain with noncontaminated grain; improving the nutritional content of the diet to compensate for mycotoxin-induced reductions in feed intake; addition of mold inhibitors or flavoring agents to the diet to improve feed intake; addition of potential mycotoxin-binding agents to the diet to inhibit absorption and stimulate metabolism of mycotoxins by the animal; and other, experimental, treatments.

Dilution of mycotoxin-contaminated grain with noncontaminated grain, because of its simplicity, has been one of the most widely used methods for minimizing the effects of contaminated grain on animals, particularly for improving feed intake and weight gain in pigs (Patterson and Young 1991). However, the success of this approach depends on the degree of contamination, the dilution achievable, and the ready availability of a source of suitable, noncontaminated grain.

When propionate, a mold inhibitor, was added to the diet of pigs, it improved their overall performance, but did not specifically reduce the adverse effects of a DON-contaminated corn-based diet (Foster et al 1987).

Similarly, the addition of mold inhibitors and flavoring agents did not increase the growth rate of pigs fed moldy corn diets (McNear et al 1981).

Increasing the energy, crude protein, and mineral and vitamin content of a DON-contaminated diet by 20% improved weight gain in pigs, but only if feed intake was reduced by 20% or less (Chavez and Rheaume 1986).

Substances investigated as potential mycotoxin-binding agents have included alfalfa (Smith 1980a,b; James and Smith 1982; Carson and Smith 1983b; Stangroom and Smith 1984), synthetic cation or anion exchange zeolite (Smith 1980a,b; Carson and Smith 1983a), bentonite (Carson and Smith 1983a, Smith 1984), spent canola oil bleaching clays (Smith 1984), Antitox Vana (polyvinylpyrrolidone) (Friend et al 1984), and a hydrated sodium calcium aluminosilicate (Wyatt 1987, Orr 1987, Patterson and Young 1992a).

At a dietary concentration of 15 to 25%, alfalfa overcame the adverse effects of a diet containing 250 mg/kg of ZEN on feed consumption and body weight gain in rats (Smith 1980a, James and Smith 1982, Stangroom and Smith 1984). This concentration of alfalfa had no effect on uterine enlargement in young gilts fed a diet containing 50 mg/kg of ZEN (Smith 1980a), although 15% alfalfa partially overcame this sign in those fed a diet containing 10 mg/kg of ZEN (James and Smith 1982). Similarly, at a dietary concentration of 20%, alfalfa overcame the adverse effects of a diet containing 3 mg/kg of T-2 toxin on feed consumption and body weight gain in rats (Carson and Smith 1983b).

Synthetic anion exchange zeolite was found to alleviate the adverse effects of ZEN in rats by reducing the intestinal absorption and the enterohepatic circulation of this mycotoxin (Smith 1980b).

Bentonite and spent canola oil bleaching clays appear to be effective in alleviating the toxicity of diets containing T-2 toxin. Bentonite and spent canola oil bleaching clays, at a dietary concentration of 10%, individually overcame the adverse effects of a diet containing T-2 toxin at 3 mg/kg of feed in rats (Carson and Smith 1983b, Smith 1984). Bentonite and spent canola oil bleaching clays appear to exert their beneficial effects by adsorbing the T-2 toxin present in the diet and inhibiting its absorption from the gastrointestinal tract.

Antitox Vana, the chemical binding agent polyvinylpyrrolidone, was found to have no effect on reductions in feed intake and body weight gain in pigs fed diets containing 14 mg/kg of DON (Friend et al 1984).

There is limited evidence that, in some cases, a hydrated sodium calcium aluminosilicate (HSCAS; Novasil, Englehard Corp., Cleveland, Ohio) can protect animals against the adverse effects of some *Fusarium* mycotoxins. For example, HSCAS at a dietary concentration of 1% caused improvements in body weight gain in young chicks fed moldy corn-based rations containing low concentrations (trace to 0.35 mg/kg) of T-2 toxin, DON, and ZEN

(Wyatt 1987). However, HSCAS, at a dietary concentration of 0.5%, did not diminish the adverse effects of a diet containing 8 mg/kg of T-2 toxin on male broiler chicks (Kubena et al 1990). Similarly, for one- to 14-day-old broiler chicks, HSCAS (at 0.5% of the diet) did not alleviate the adverse effects of a diet containing 300 mg/kg of fumonisin B_1 on body weight gain, liver weight relative to body weight, and histological changes in the liver (Brown et al 1992). HSCAS did not significantly improve average daily gain and feed efficiency in pigs fed diets containing 0.5 to 2.5 and 0.2 to 1.5 mg/kg of DON and ZEN, respectively, although there did appear to be a trend towards improved values in pigs fed 0.5% HSCAS (Orr 1987). Similarly, dietary concentrations of 0.5 and 1.0% HSCAS did not alleviate the adverse effects of a diet containing 4 to 7 mg/kg of DON on body weight gain in young pigs (Patterson and Young 1992a).

Treatment of moldy corn contaminated with approximately 5 mg/kg of DON with microbial inoculum from the digestive tract of poultry reduced the DON concentration by 54 to 56% (Ping et al 1992). This decontamination process also partially alleviated the toxic effects of this diet on feed intake and body weight gain in young pigs. It was concluded, therefore, that this treatment was effective in reducing the toxicity of DON and that the altered DON was nontoxic to pigs.

Administration of monoclonal antibodies specific for T-2 toxin can neutralize the in vitro inhibitory effects of the toxin on protein synthesis in human B-lymphoblastoid cultures and protect rats from lethal toxicosis (Feuerstein et al 1988). Moreover, monoclonal antibodies were shown to induce a net efflux of T-2 toxin from poisoned human B-lymphoblastoid cells in vitro and restore protein synthesis. Administration of monoclonal antibodies 30 min before infusion of a lethal dose (1 mg/kg) of T-2 toxin to rats caused sequestration of the toxin in the plasma (Hunter et al 1990). When administered 35 min after the toxin, monoclonal antibodies facilitated the migration of T-2 toxin back into the plasma from the tissues, reducing its toxic effects in the rats.

Summary

Many compounds and treatments have been tested for reducing the toxicity of *Fusarium* mycotoxin-contaminated grains and feeds. Contaminated grain may contain a wide variety of mycotoxins of differing chemical characteristics including heat stability, solubility, and adsorbent affinity (Foster et al 1986a,b). Consequently, it is difficult to develop a decontamination method that will be equally effective against each of the numerous mycotoxins that may be present in a naturally contaminated grain. Some of the treatments described in this chapter have promising prospects for commercial application. Moreover, some treatments, such as screening and air separation, dilution of contaminated grain with noncontaminated grain, and use of binding agents and roasting processes have had commercial applications already. However, until reliable, cost-effective, and commercially applicable methods of decontamination or detoxification (including minimizing the effects of mycotoxins on the animal) are more widely available, problems associated with mycotoxin contamination of grains and feedstuffs will continue to be seen in the food and agricultural industries.

Literature Cited

Abbas, H. K., Mirocha, C. J., Rosiles, R., and Carvajal, M. 1988. Decomposition of zearalenone and deoxynivalenol in the process of making tortillas from corn. Cereal Chem. 65:15-19.

Babadoost, M., Hagler, W. M., and Bowman, D. T. 1987. Field contamination of sorghum with zearalenone and deoxynivalenol in North Carolina: Density segregation to remove mycotoxins. Pages 99-109 in: Biodeterioration Research 1. G. C. Llewellyn and C. E. O'Rear, eds. Plenum Press, New York.

Bauer, J., Gareis, M., Detzler, W., Gedek, B., Heinritzi, K., and Kabilka, G. 1987. Detoxification of mycotoxins in animal feeds. Tierarztl. Umsch. 42:70-77.

Bennett, G. A., Shotwell, O. L., and Hesseltine, C. W. 1980. Destruction of zearalenone in contaminated corn. J. Am. Oil Chem. Soc. 57:245-247.

Brown, T. P., Rottinghaus, G. E., and Williams, M. E. 1992. Fumonisin mycotoxicosis in broilers: Performance and pathology. Avian Dis. 36:450-454.

Carson, M. S., and Smith, T. K. 1983a. Role of bentonite in prevention of T-2 toxicosis in rats. J. Anim. Sci. 57:1498-1509.

Carson, M. S., and Smith, T. K. 1983b. Effect of feeding alfalfa and refined plant fibers on the toxicity and metabolism of T-2 toxin in rats. J. Nutr. 113:304-313.

Chavez, R. R., and Rheaume, J. A. 1986. The significance of the reduced feed consumption observed in growing pigs fed vomitoxin-containing diets. Can. J. Anim. Sci. 66:277-287.

Collins, C. J., and Rosen, J. D. 1981. Distribution of T-2 toxin in wet-milled corn products. J. Food Sci. 46:877-879.

El-Banna, A. A., Lau, P.-Y., and Scott, P. M. 1983. Fate of mycotoxins during processing of foodstuffs. II. Deoxynivalenol (vomitoxin) during making of Egyptian bread. J. Food Prot. 46:484-488.

Feuerstein, G., Powell, J. A., Knower, A. T., and Hunter, K. W. 1988. Monoclonal antibodies to T-2 toxin. In vitro neutralization of protein synthesis inhibition and protection of rats against lethal toxemia. J. Clin. Invest. 76:2134-2138.

Foster, B. C., Neish, G. A., Lauren, D. R., Trenholm, H. L., Prelusky, D. B., and Hamilton, R. M. T. 1986a. Fungal and mycotoxin content of slashed corn. Microbiol. Aliment. Nutr. 4:199-203

Foster, B. C., Trenholm, H. L., Friend, D. W., Thompson, B. K., and Hartin, K. E. 1986b. Evaluation of different sources of deoxynivalenol (vomitoxin) fed to swine. Can. J. Anim. Sci. 66:1149-1154.

Foster, B. C., Trenholm, H. L., Friend, D. W., Thompson, B. K., and Hartin, K. E. 1987. The effect of a propionate feed preservative in deoxynivalenol (vomitoxin) containing corn diets fed to swine. Can. J. Anim. Sci. 67:1159-1163.

Friend, D. W., Trenholm, H. L., Young, J. C., Thompson, B. K., and Hartin, K. E. 1984. Effect of adding potential vomitoxin (deoxynivalenol) detoxicants or a *F. graminearum* inoculated corn supplement to wheat diets fed to pigs. Can. J. Anim. Sci. 64:733-741.

Hart, L. P., and Braselton, W. E. 1983. Distribution of vomitoxin in dry milled fractions of wheat infected with *Gibberella zeae.* J. Agric. Food Chem. 31:657-659.

Huff, W. E., and Hagler, W. M. 1985. Density segregation of corn and wheat naturally contaminated with aflatoxin, deoxynivalenol and zearalenone. J. Food Prot. 48,416-420.

Hunter, K. W., Brimfield, A. A., Knower, A. T., Powell, J. A., and Feuerstein, G. 1990. Reversal of intracellular toxicity of the trichothecene mycotoxin T-2 with monoclonal antibody. J. Pharmacol. Exp. Ther. 255:1183-1187.

James, L. J., and Smith, T. K. 1982. Effect of dietary alfalfa on zearalenone toxicity and metabolism in rats and swine. J. Anim. Sci. 55:110-118.

Kubena, L. F., Harvey, R. B., Huff, W. E., Corrier, D. E., Phillips, T. D., and Rottinghaus, G. E. 1990. Efficacy of a hydrated sodium calcium aluminosilicate to reduce the toxicity of aflatoxin and T-2 toxin. Poult. Sci. 69:1078-1086.

Lee, U. S., Jang, H. S., Tanaka, T., Oh, Y. J., Cho, C. M., and Ueno, Y. 1987. Effect of milling on decontamination of *Fusarium* mycotoxins nivalenol, deoxynivalenol and zearalenone in Korean wheat. J. Agric. Food Chem. 35:126-129.

McNear, R. M., Wilson, R. F., and Stitzlein, G. A. 1981. Short term swine appetite responses to dietary additives in moldy corn. Ohio Swine Res. Ind. Rep. Anim. Sci. Ser. 81-2:79-82.

Miller, J. D., Young, J. C., and Trenholm, H. L. 1983. Can. J. Bot. 61:3080.

Nowicki, T. W., Gaba, D. G., Dexter, J. E., Matsuo, R. R., and Clear, R. M. 1988. Retention of the *Fusarium* mycotoxin deoxynivalenol in wheat during processing and cooking of spaghetti and noodles. J. Cereal Chem. 8:189-202.

Orr, D. E. 1987. Field studies on swine with a selected aluminosilicate. Pages G1-G11 in: Proc. Recent Developments in the Study of Mycotoxins. Kaiser Aluminum and Chemical Corporation, Rosemont, IL.

Park, D. L., Rua, S. M., Mirocha, C. J., Abd-Alla, E.-S. A. M., and Weng, C. Y. 1992. Mutagenic potentials of fumonisin contaminated corn following ammonia decontamination procedure. Mycopathologia 117:105-108.

Patterson, R., and Young, L. G. 1991. Diluting mouldy corn still best bet. The Market Place, Fall 1991, pp. 36-39.

Patterson, R., and Young, L. G. 1992a. Using Novasil® to alleviate the effects of vomitoxin in moldy corn diets. Ontario Swine Research Review. Ontario Agricultural College Publ. 0292, pp. 18-19.

Patterson, R., and Young, L. G. 1992b. Screening moldy corn to alleviate the effects of vomitoxin in moldy corn diets. Ontario Swine Research Review. Ontario Agricultural College Publ. 0292, pp. 20-21.

Ping, H. E., Young, L. G., and Forsberg, C. 1992. Microbial detoxification of vomitoxin contaminated corn for pigs. Ontario Swine Research Review. Ontario Agricultural College Publ. 0292, pp. 21-22.

Scott, P. M., Kanhere, S. R., Lau, P.-Y., Dexter, J. E., and Greenhalgh, R. 1983. Effects of experimental flour milling and breadbaking on retention of deoxynivalenol (vomitoxin) in hard red spring wheat. Cereal Chem. 60:421-424.

Scott, P. M., Kanhere, S. R., Dexter, J. E., Brennan, P. W., and Trenholm, H. L. 1984. Distribution of the trichothecene mycotoxin deoxynivalenol (vomitoxin) during the milling of naturally contaminated hard red spring wheat and its fate in baked products. Food Addit. Contam. 1:313-323.

Seitz, L. M., Yamazaki, W. T., Clements, R. L., Mohr, H. E., and Andrews, L. 1985. Distribution of deoxynivalenol in soft wheat mill streams. Cereal Chem. 62:467-469.

Seitz, L. M., Eustace, W. D., Mohr, H. E., Shogren, M. D., and Yamazaki, W. T. 1986. Cleaning, milling, and baking tests with Hard Red Winter wheat containing deoxynivalenol. Cereal Chem. 63:146-150.

Smith, T. K. 1980a. Influence of dietary fiber, protein and zeolite on zearalenone toxicosis in rats and swine. J. Anim. Sci. 50:278-285.

Smith, T. K. 1980b. Effect of dietary protein, alfalfa and zeolite on excretory patterns of 5',5',7',7'-[3H] zearalenone in rats. Can. J. Physiol. Pharmacol. 58:1251-1255.

Smith, T. K. 1984. Spent canola oil bleaching clays: Potential for treatment of T-2 toxicosis in rats and short-term inclusion in diets for immature swine. Can. J. Anim. Sci. 64:725-732.

Stahr, H. M., Osweiler, G. D., Martin, P., Domoto, M., and Debey, B. 1987. Thermal detoxification of trichothecene contaminated commodities. Pages 231-238 in: Biodeterioration Research 1. G. C. Llewellyn and C. E. O'Rear, eds. Plenum Press, New York.

Stangroom, K. E., and Smith, T. K. 1984. Effect of whole and fractionated alfalfa meal on zearalenone toxicosis and metabolism in rats and swine. Can. J. Physiol. Pharmacol. 62:1219-1224.

Swanson, S. P., Hagler, W. M., and Rood, H. D. 1984. Destruction of deoxynivalenol (vomitoxin) with sodium bisulfite. (Abstr.) Ann. Meet. Am. Soc. Microbiol. 84:192.

Tanaka, T., Hasegawa, A., Yamamoto, Y. M., and Ueno, Y. 1986. Residues of *Fusarium* mycotoxins nivalenol, deoxynivalenol and zearalenone, in wheat and processed food after milling and baking. J. Food Hyg. Soc. Jpn. 27:653-655.

Tkachuk, R., Dexter, J. E., Tipples, K. H., and Nowicki, T. W. 1991. Removal by specific gravity table of tombstone kernels and associated trichothecenes from wheat infected with *Fusarium* head blight. Cereal Chem. 68:428-431.

Trenholm, H. L., Charmley, L. L., Prelusky, D. B., and Warner, R. M. 1991. Two physical methods for the decontamination of four cereals contaminated with deoxynivalenol and zearalenone. J. Agric. Food Chem. 39:356-360.

Trenholm, H. L., Charmley, L. L., Prelusky, D. B., and Warner, R. M. 1992. Washing procedures using water or sodium carbonate solutions for the decontamination of three cereals contaminated with deoxynivalenol and zearalenone. J. Agric. Food Chem. 40:2147-2151.

Wyatt, R. D. 1987. The relationship of *Fusarium* and other mold produced toxins and a selected aluminosilicate. Pages D1-D15 in: Proc. Recent Developments in the Study of Mycotoxins. Kaiser Aluminum and Chemical Corporation, Rosemont, IL.

Young, J. C. 1985. Pages 119-121 in: Decontamination in mycotoxins: A Canadian perspective. Natl. Res. Counc. Can. Assoc. Comm. Sci. Criter. Environ. Qual. Publ. 11848.

Young, J. C. 1986a. Reduction in levels of deoxynivalenol in contaminated corn by chemical and physical treatment. J. Agric. Food Chem. 34:465-467.

Young, J. C. 1986b. Formation of sodium bisulfite addition products with trichothecenes and alkaline hydrolysis of deoxynivalenol and its sulfonate. J. Agric. Food Chem. 34:919-923.

Young, J. C., Fulcher, R. G., Hayhoe, J. H., Scott, P. M., and Dexter, J. E. 1984. Effect of milling and baking on deoxynivalenol (vomitoxin) content of eastern Canadian wheats. J. Agric. Food Chem. 32:659-664.

Young, J. C., Subryan, L. M., Potts, D., McLaren, M. E., and Gobran, F. H. 1986.

Reduction in levels of deoxynivalenol in contaminated wheat by chemical and physical treatment. J. Agric. Food Chem. 34:461-465.

Young, J. C., Trenholm, H. L., Friend, D. W., and Prelusky, D. B. 1987. Detoxification of deoxynivalenol with sodium bisulfite and evaluation of the effects when pure mycotoxin or contaminated corn was treated and given to pigs. J. Agric. Food Chem. 35:259-261.

Part Five
Consequences for Humans

Chapter 12

Prevention of Human Mycotoxicoses Through Risk Assessment and Risk Management

Tine Kuiper-Goodman

Exposure to mycotoxins has been associated with the observation of adverse effects in humans and livestock. In Canada, no known human illnesses have been attributed to the dietary intake of mycotoxins, the major route of concern. This is mainly due to the low levels of mycotoxins generally present in food crops. As well, the Canadian marketing and distribution system—in which the produce of many farms is mixed or blended, distributed to many outlets, and consumed by many different people—further reduces the likelihood of exposure to high levels in food. Elsewhere, however, human mycotoxicoses caused by the ingestion of high levels of mycotoxins in contaminated food for a relatively short period of time (up to several months) are either well documented or are suspected from unconfirmed reports or inadequate epidemiological studies. One must also keep in mind that the actual incidence of illness could be more common than suspected, since slight signs or symptoms of acute mycotoxicoses could be overlooked or attributed to other causes. There are also generally problems in establishing the etiology of chronic diseases that may be caused by long-term ingestion of low levels of mycotoxins or other contaminants.

Health concerns related to dietary exposure to mycotoxins depend on the levels of mycotoxins in the food as consumed, on the amount of food consumed, on the body weight and physiological state of the individual, and on the bioavailability and toxicity of the compound to humans. Other dietary factors may increase or decrease the toxicity.

Human exposure to mycotoxins can also occur through the dermal route or through inhalation, as during grain handling on the farm, in grain elevators, and in food and feed processing. Little is known about the potential for human mycotoxicoses resulting from these occupational exposures, and there are no regulations in Canada regarding the acceptable levels of myco-toxins in airborne dust.

In this chapter it will be shown how through risk assessment one can determine whether mycotoxins present a health concern, and how, when deemed necessary, human mycotoxicoses can be prevented through risk management. The general aspects of risk assessment and risk management will be discussed next. These general discussions will be followed by dis-cussion of specific mycotoxins, chosen for their relevance to the Canadian situation.

General Aspects of Risk Assessment and Risk Management of Mycotoxins

Our risk assessments and regulatory efforts are directed towards ensuring that mycotoxins in foods are kept at levels well below those that may constitute potential acute and chronic health hazards. In relative terms, risk assessments are the expression of exposure, obtained through the exposure assessment, relative to the dose that is considered "safe," obtained through the hazard assessment.

EXPOSURE ASSESSMENT

When initially confronted with the task of estimating the level of myco-toxins in various foods, as consumed, one will have to err on the side of caution until more precise data are available. In Canada, this involves mostly the monitoring of residues of mycotoxins in food crops and food products at various stages in the manufacturing process, also known as environmental monitoring. There are other approaches for assessing human exposure, such as "measuring at the plate" and human tissue monitoring. These methods are especially important in epidemiology, which will be discussed in more detail later.

Environmental Monitoring

Direct Intake. When present in staple food crops, exposure to mycotoxins can be continuous over a long period of time. Mycotoxin levels can be reduced by specific (chemical) decontamination practices (Chapter 11), which would have to be reviewed for safety; at present none are being used for food in Canada. Normal processing of raw food either at the manufacturing level (drying, heating, sorting, milling) or at home may reduce the level of mycotoxins in the food, as consumed, depending on their heat stability and solubility (Scott 1984). Blending, as part of the food manu-facturing process, will reduce peak exposure but not average exposure.

Processing may also result in isomerization and breakdown products, the safety of which need to be considered.

Indirect Intake. To assess the indirect intake of mycotoxins through the consumption of food products derived from animals that were fed mycotoxin-contaminated feed, feed-food "carryover" studies are required. In such studies, the mycotoxin of concern is fed to different types of livestock at a specific level in the feed for a specific time period. The resulting level of that mycotoxin or its metabolites in the animal-derived food products (muscle, liver, milk, eggs) is measured (Kuiper-Goodman 1991b). Such studies have been conducted for aflatoxin B_1, ochratoxin A, trichothecenes, and zearalenone; studies for some of the fumonisins are ongoing. It is obvious that if the carryover is slight, analytical detection may be difficult and require the development of new methodology, such as the use of tracer techniques for staple foods. Another problem is that sometimes the identity and hazard of the metabolites are not known, adding to difficulties in the determination and interpretation of analysis results.

Estimates of Food Intake and Probable Daily Intake. Estimates for the intake of various food commodities differ from country to country. Within a country there may also be regional and ethnic differences in food preference. Where this is known to have an impact, these differences must also be considered. Several types of surveys are used to obtain estimates of food commodity intake.

In Canada, use is most often made of data on food commodity consumption patterns from the Nutrition Canada Survey (Health and Welfare Canada 1976). These data were obtained by dietary recall covering a 24-h or one-week period, for males and females, and for the different age groups. Both eaters (those eating the commodity of concern) and noneaters were accounted for. Most often the mean intake of eaters is used. Occasionally, particularly when considering acute health hazards, a higher estimate is examined, with 90% of the eaters consuming less than this figure. However, the 90th percentile eaters-only data from the Nutrition Canada Survey are based on inadequate sample size, and in many cases are unreliable. Use is also made of "Canada disappearance figures" (Statistics Canada), which provide a mean per capita intake but which do not take into account noneaters or nonfood uses of the commodity. Other North American surveys and specific surveys geared towards special segments of the population are also available, e.g., for infants or preschoolers (Yeung et al 1979, Leung et al 1984). None of these surveys are complete for each food item.

Combining the estimates of food intake with the estimated level of mycotoxins found or estimated to be present in food products gives an estimate of the probable daily intake (PDI) of a mycotoxin. The PDI is generally expressed on a body weight basis. It is often different for males and females and for different age groups. Normally the highest PDI, based on the group with highest consumption on a body weight basis, is used in the risk assessment, since this ensures that the subpopulation that is at highest risk is protected.

Human Tissue Monitoring

Human tissue monitoring involves the analysis of human body fluids such as blood, urine, or milk. If properly validated, human tissue data provide the ultimate evidence that exposure has taken place. These measurements also allow for quantification of exposure at an individual level, which is particularly important when conducting epidemiological studies. When one also has information on exposure estimates from environmental monitoring and/or intake estimates at the plate and time since last exposure, then it may be possible to estimate the amount absorbed, as well as the amount metabolized and eliminated by humans. For some mycotoxins, these data may also be used to estimate duration of exposure at an individual level.

With these data, it may be possible to determine which clinical signs and symptoms are associated with various human exposures, so that a dose-response relationship may be derived.

HEALTH HAZARD ASSESSMENT

Health hazard assessment generally involves the following: 1) an assessment of experimental toxicity studies, usually conducted in laboratory animals, as well as any relevant human data or livestock studies; 2) an estimation of the no observed (adverse) effect level(s), abbreviated NO(A)EL, or lowest observed (adverse) effect levels, LO(A)EL, where appropriate; and 3) an extrapolation from high to low doses and from test species to humans in order to establish an estimated "safe dose" for humans, such as a tolerable daily intake (Kuiper-Goodman 1990).

Infants and young children may be more susceptible to adverse effects because their basal metabolic rate is greater, and they may not possess the biochemical mechanisms to detoxify toxic substances. In children, organs such as the brain continue development for many years after birth, and this makes them more susceptible to substances affecting the central nervous system. Therefore, a separate risk assessment is sometimes made for children.

Toxicity Studies

In general, laboratory animals are exposed to mycotoxins to examine treatment-related adverse effects. For an ideal assessment, data are needed which address acute toxicity, 30–90 day toxicity, metabolic disposition, reproductive effects, teratogenicity, mutagenicity, and chronic toxicity or carcinogenicity. Special studies on target organ toxicity may also be desirable. Studies are usually conducted in a rodent (preferably rat) and a nonrodent species. Where possible, the species selected is the one that metabolizes the compound similarly to man. For these studies, oral administration of the mycotoxin is the preferred route. Adverse effects observed in natural outbreaks of disease in livestock and experimental studies in livestock are also taken into account in the assessment (Chapter 9).

Human Data

If epidemiological data indicate an association of disease with exposure to a mycotoxin (e.g., liver cancer with aflatoxin exposure), such data may also be used in the hazard assessment. In the interpretation of epidemiological studies, one must also consider the influence of other (risk) factors such as nutritional status, endemic infections, intake of other toxic substances, etc.

Extrapolation and Estimation of a Tolerable Daily Intake

For mycotoxins that are a health concern, a tolerable daily intake (TDI) should be estimated when the toxicological data base is adequate. Generally, this is accomplished by extrapolating the relevant toxicological endpoints, such as a NO(A)EL, LO(A)EL, or statistically derived no effect level (NEL), to humans using appropriate safety and/or uncertainty factors, generally ranging from 50 to 5,000, or other means of extrapolation and evaluation (Kuiper-Goodman 1990).

For nonthreshold and other carcinogenic mycotoxins, a virtual safe dose may be determined using a mathematical approach. A discussion of the various mathematical models used to estimate the risk at very low doses (nonobservable region of the dose-response curve) is outside the scope of this article. In previous risk assessments, we have used non-model-dependent robust linear extrapolation on the responses in the NO(A)EL region of the dose-response curve to obtain the lower confidence limits of the virtual safe dose at a risk level of $1:10^6$ or $1:10^5$ (Krewski et al 1986, Kuiper-Goodman 1990). The higher risk level of $1:10^5$ is considered as more appropriate for unavoidable natural contaminants.

Because there are many uncertainties in extrapolating from high to low doses and between species, the risk level of $1:10^5$ should not be thought of as giving the true incidence of anticipated adverse effects with exposure at the virtual safe dose, but rather as providing an "essentially negligible risk." This value is meant to be used for comparison with different mycotoxins and other food chemical substances and in priority setting; it has no implications with regard to a socially acceptable risk.

The above-mentioned approaches use points from the NO(A)EL region of the dose-response curve and ignore valuable data from the response region. Therefore, for carcinogens we have used, as an additional approach for estimating the TDI, the dose at which 50% of the animals would have developed tumors (the TD_{50}) divided by 50,000 (for rationale see Kuiper-Goodman 1990). The TD_{50} is a numerical estimate of carcinogenic potency and assumes that the test compound is administered for a "standard lifespan," with corrections made for intercurrent mortality and background incidence of tumors (Peto et al 1984, Ames et al 1987); for many studies the TD_{50} lies within the observed response region of the dose-response curve. By dividing the TD_{50} (risk of 0.5) by 50,000, one essentially makes a linear extrapolation, and the TDI so derived would have a risk level of 10^{-5}.

Thus a variety of approaches can be used to estimate the TDI. One can either choose the most appropriate estimate, depending on the underlying data, or consider the range of estimates. In establishing a TDI, information on species differences in metabolic disposition and biological aspects of target organ toxicity, such as possible mechanisms of action, is also taken into account when extrapolating the significance of animal data on tumor induction or other adverse effects to humans.

RISK ASSESSMENT

For risk assessment, the results of the exposure assessment are compared with the hazard assessment. A comparison of the estimated PDI (see Estimates of Food Intake and Probable Daily Intake) of a mycotoxin and the TDI (both expressed on a body weight basis) gives us an indication of the degree of concern.

There are many uncertainties in both the exposure assessment and the hazard assessment that can greatly influence the overall risk assessment (Kuiper-Goodman 1990). Of necessity, regulatory risk assessments tend to err on the side of safety until scientific data gaps have been resolved, and our understanding of the issues has improved. Thus the actual health risks, as a result of lifetime exposure to mycotoxins, are likely to be somewhat less than estimated. The risk assessment, therefore, represents a status document that needs to be updated as new data and new methods of extrapolating these data become available.

RISK MANAGEMENT AND REGULATION

When the estimated PDI of a mycotoxin exceeds its TDI, steps may be taken, under the authority of the Canadian Food and Drugs Act, to reduce the intake of that mycotoxin. These may range from completely removing the offending food from the marketplace to the establishment of tolerances (listed under schedule B-01-046 of the Food and Drugs Act and Regulations) or interim guideline levels (not listed in the Food and Drugs Act and Regulations). Such actions limit the amount or level of the mycotoxin that can be tolerated in the raw food crop or in a finished food. Analytical methods have to be available to ensure compliance with a tolerance or a guideline level.

In addition, the provision of information to consumers on the avoidance of visibly moldy foods has been used as an approach to reduce the intake of mycotoxins (Health and Welfare Canada 1990).

In Canada, government regulations or guidelines for mycotoxins exist only with respect to aflatoxins and deoxynivalenol (see below). In general, no tolerances or guideline levels have been adopted if the levels of mycotoxins in the food are consistently low (as measured through monitoring of food products) and have not been identified as a reason for health concern. International regulations with respect to mycotoxins were recently reviewed (van Egmond 1989).

Similarly, to limit the indirect intake of mycotoxins, tolerances may be set for the maximum levels of mycotoxins that can be allowed in feed, to ensure negligible residues of mycotoxins and their metabolites in animal-derived food products (Kuiper-Goodman 1991b). These tolerances may be restricted to certain types of livestock (e.g., dairy versus beef cattle, layer versus broiler chickens). In the United States, where aflatoxin-contaminated feed has been a problem, it was found difficult to ensure adherence to the restrictions for different livestock species.

Specific Mycotoxins

The individual mycotoxins that have been considered in Canada for possible health implications are aflatoxins, ochratoxin A, zearalenone, deoxynivalenol (DON, vomitoxin), fumonisins, patulin, and the ergot alkaloids. Detailed risk assessments, from a Canadian perspective, have been made for ochratoxin A (Kuiper-Goodman and Scott 1989) and zearalenone (Kuiper-Goodman et al 1987) and were compared with the aflatoxins (Kuiper-Goodman 1990, 1991a). The results of these risk assessments are summarized below. Although preliminary, risk assessments on deoxynivalenol and fumonisins are presented below as well. Risk assessments on patulin and the ergot alkaloids are in progress.

AFLATOXINS

Concern about the potential hazards posed by dietary aflatoxins started in the 1960s after some 100,000 turkey poults in Great Britain died as a result of aflatoxin exposure from their feed (for a review, see Busby and Wogan 1984). When it became evident that aflatoxin exposure caused cancer in many species, most countries, including Canada, established tolerances for aflatoxin levels in food or feed in order to limit exposure to this group of mycotoxins (van Egmond 1989).

Hazard Assessment

Toxicology. Of the aflatoxins present in food, aflatoxin B_1 and the metabolites aflatoxin M_1 and aflatoxicol represent possible health concerns. All three substances are carcinogenic, with aflatoxin B_1 the most potent. Very low dietary levels of aflatoxin B_1 (1 $\mu g/kg$ of diet) have caused liver cancer in the Fischer rat (Wogan et al 1974). Although a clear biological no-effect level NOAEL was not demonstrated in these studies, the NOAEL for aflatoxin B_1 was estimated as 0.75 $\mu g/kg$ body weight (bw) per day, using the Fisher exact (statistical) test (Kuiper-Goodman 1990, 1991a). Similarly, for aflatoxicol and aflatoxin M_1, the respective NOAELs were 1.25 and <2.5 $\mu g/kg$ bw per day. In these studies, the TD_{50}, as an indicator of carcinogenic potency, was estimated as 1.15, 3.10, and 10.38 $\mu g/kg$ bw per day for aflatoxin B_1, aflatoxicol, and aflatoxin M_1, respectively (Kuiper-Goodman 1990, 1991a). Although aflatoxin B_1 is carcinogenic in all species tested, there are significant differences with regard to species susceptibility. These differences have been explained in part by species differences in metabolic disposition. Nevertheless, aflatoxin B_1 has been found to be mutagenic and to induce cytogenetic changes in all in vitro and in vivo genotoxicity tests (for a review of the older literature, see Busby and Wogan 1984, IARC 1993). Furthermore, recent studies have shown that it induces a dose-related increase in unscheduled DNA synthesis in primary hepatocytes from rats, mice, monkeys, and humans, with human and rat hepatocytes giving the strongest response (Steinmetz et al 1988). Aflatoxin B_1 also induces DNA adduct formation, binding preferentially to guanine residues in guanine+cytosine-rich regions, and induces guanine to thymine substitutions almost exclusively (see references cited in Bressac et al 1991 and Hsu et al 1991). Of considerable interest, therefore, are recent findings by two independent research groups of a mutational "hot spot" in the DNA of a large proportion of hepatocellular carcinoma (HCC) patients from Quidong, China, and southern Africa, where both aflatoxin B_1 and hepatitis B virus (HBV) infection are important risk factors for HCC (Bressac et al 1991, Hsu et al 1991) (see Epidemiology). These researchers found selective guanine to thymine mutations in base 3 of the 249 codon of p53, a putative human tumor suppressor gene. Both groups considered it likely, based on the above cited similar effects of aflatoxin B_1 in mutation assays, that the mutations in human HCC were also caused by aflatoxin B_1, since HBV is unlikely to cause point mutations. Furthermore, studies by Challen et al (1992), Hayward et al (1991), and Hosono et al (1991), cited in IARC (1993) have shown that in areas where exposure to aflatoxin is low but exposure to HBV is relatively high, there were no point mutations at codon 249. The mutant p53 protein may be responsible for the selective clonal expansion of hepatocytes in the development of HCC.

Epidemiology—Acute and Subchronic Effects. Short-term exposure to high levels of aflatoxin B_1 in humans may lead to fatal aflatoxicosis, with jaundice, ascites, and portal hypertension (Krishnamachari et al 1975, Denning 1987). In addition, it has been suggested that aflatoxin B_1 is associated with the development of kwashiorkor (Hendrickse 1985) and Reye's syndrome (Ryan et al 1980). Acute outbreaks of disease can be prevented if tolerances to aflatoxin B_1 in foods are introduced and adhered to. In countries with low exposure to aflatoxin B_1, efforts are mainly directed towards preventing a chronic disease, such as HCC.

Epidemiology—Chronic Effects. Epidemiological studies have shown an association between dietary aflatoxin exposure and the high incidence of HCC in a number of African and Southeast Asian countries. The earlier studies were population-based (correlational, ecological) studies and only considered exposure to aflatoxin B_1. More recent studies have included an examination of the role of other factors, such as HBV. Overall, recent studies indicate that HCC has a multifactorial etiology, in which HBV, aflatoxin B_1, alcohol consumption, smoking, and other environmental risk factors can all play a role (Bosch and Munoz 1988). Seveal of these studies were recently reviewed (IARC 1993). Further analytical epidemiological studies (case-control and cohort) are ongoing, in which exposure to both aflatoxin B_1 (based on intake "at the plate" and an analysis of serum and urinary biomarkers) and HBV (based on HBsAg-carrier status) and disease outcome are determined at the level of the individual (Groopman 1988, Wild et al 1990, Kuiper-Goodman 1991a).

Extrapolation. On the basis of epidemiological data obtained in other countries, the estimated TDI for aflatoxin B_1, at a risk level of 10^{-5}, ranges from 0.11 to 0.19 ng/kg bw per day (Kuiper-Goodman 1991a). In the absence of endemic HBV infection, this figure is probably about one order of magnitude higher but cannot be estimated more precisely from the presently available epidemiological data, including a reassessment of a study in China by Yeh et al (1989). It has not yet been possible to include the contribution of other positive or negative risk factors.

Exposure Assessment

In Canada, human exposure to aflatoxin B_1 is mainly from nuts and nut products, and for children mainly from peanut butter. In the processing of peanut butter, major industries subject peanuts to electronic sorting with the aim of removing poor-quality nuts. This reduces aflatoxin levels in the final product, as monitored by in-house batch sampling. Roasting peanuts also reduces their aflatoxin level by 50% or more (see review, Scott 1984). Furthermore, the mixing and blending process involved in making peanut butter reduces high peak exposure, but not average exposure, to aflatoxin B_1. Based on monitoring data by the Health Protection Branch,

Health and Welfare Canada, about 60% of the total aflatoxins in peanuts is aflatoxin B_1.

The intake of aflatoxin B_1 is estimated to be between 1 and 2 ng/kg bw per day for one- to 11-year-old children eating peanuts or peanut butter. This estimate is based on the results from Health Protection Branch monitoring of residues in nuts and nut products (1985–1987) and the consumption patterns of these foods for both males and females and various age groups of Canadians (Kuiper-Goodman 1990). These exposure estimates, based on actual monitoring data, were considered more valid than estimates obtained by using the tolerance level for nuts and nut products of 15 ng/g.

Supplies of Brazil nuts and pistachio nuts have periodically been found to contain levels of aflatoxins as high as several micrograms per gram in individual nuts. When nuts are sold in the shell, it is not possible for the manufacturers to identify individual nut meats that are contaminated. In this instance, the responsibility for handsorting and removing nuts of poor quality is shifted to consumers. Canadians have been advised to inspect shelled nuts prior to consumption and to discard moldy nuts (Health and Welfare Canada 1990).

Canadian corn does not tend to be contaminated with aflatoxin B_1. Since corn from some U.S. states sometimes does contain higher levels of aflatoxin, a data-gathering project on imported corn food products has been proposed (Health Protection Branch, Field Operations Directorate, Food/Field Projects). Feed ingredients that are vulnerable to contamination by aflatoxin B_1, such as peanut meal and cottonseed, are generally not used in Canada, and an Agriculture Canada/Health Protection Branch survey of dairy feed (102 samples) conducted in 1981–1982 was negative (*unpublished*). Therefore, aflatoxin metabolites in dairy products are unlikely to represent a health concern in Canada (Kuiper-Goodman 1991b). Nevertheless, to ensure that this is indeed the case, a survey for the presence of aflatoxins in Canadian dairy products is currently underway.

Risk Assessment

Overall, the exposure to aflatoxins in Canada is one of the lowest in the world; it was estimated to be between 1 and 2 ng/kg bw per day for one- to 11-year-old children eating peanuts or peanut butter (Kuiper-Goodman 1990). Other age groups of Canadians have an even lower intake of aflatoxin B_1. The intake estimate is about 10-fold higher than the estimated TDI of 0.11 to 0.19 ng/kg bw per day at a carcinogenic risk level of 1:100,000, estimated from epidemiological studies conducted in certain Asian and African countries, where the consumers may in addition be exposed to HBV infection. Since HBV infection is not endemic in Canada, the estimated TDI may be one order of magnitude higher, and the estimated intake of aflatoxins is unlikely to pose a health hazard.

Tolerances in Effect in Canada

A tolerance of 15 ng/g (total aflatoxins) for nuts and nut products is in effect (Health and Welfare Canada 1981), but this is at present under review. The tolerance for aflatoxins was set with respect to analytical methods available in 1976 and to levels of detection that were technically achievable at that time. For the purpose of this regulation, the aflatoxin content of a nut or nut product shall be determined excluding the shell. Since compliance to this tolerance is enforced by the Health Protection Branch's Field Operations Directorate, there is little likelihood that the Canadian population is subjected to peak exposure levels from aflatoxin in peanut butter that significantly exceed the tolerance in those products. In fact, Health Protection Branch monitoring data (1985–1987) indicate that on average the level of aflatoxin B_1 in commercial peanut butter is 1.62 ng/g.

ZEARALENONE

Toxicology and Hazard Assessment

Studies in various species (rodents, rabbits, pigs, monkeys), including humans (Verdeal and Ryan 1979), have shown that zearalenone and related resorcylic acid lactones have estrogenic and anabolic activity. The major effects are on the reproductive system, leading to hyperestrogenism and affecting reproductive organ structure and function. Using the closely related α-zearalenol, the no-hormonal-effect level was estimated to be 50 μg/kg bw per day, based on a 90-day oral study in ovariectomized monkeys with vaginal cornification as the endpoint (Griffin et al 1984, Kuiper-Goodman et al 1987).

The results of carcinogenicity bioassays for zearalenone in rats and mice were considered by the U.S. Department of Health and Human Services National Toxicology Program (NTP) to demonstrate "positive evidence of carcinogenicity" (NTP 1982). In these studies, there was an increased incidence of pituitary adenomas in both male and female mice, with progression to malignancy, as indicated by the presence of pituitary carcinomas in some of the animals. Further studies will be required to confirm whether zearalenone should be considered a potential human carcinogen (Kuiper-Goodman et al 1987).

The estimated TDI based on the NOAEL in the carcinogenicity study (9,000 μg/kg bw per day in female mice) divided by a safety factor of 5,000 is 1,800 ng/kg bw per day; the corresponding estimates for the virtual safe dose based on a mathematical approach (robust linear extrapolation with a carcinogenic risk level of 1:100,000) is 500 ng/kg bw per day. The carcinogenic potency of zearalenone, as indicated by its TD_{50} value, was estimated as 20,000 μg/kg bw per day (female mice). Dividing this value by 50,000 (see Extrapolation and Estimation of a Tolerable Daily Intake) gives an estimated TDI of 400 ng/kg bw per day (Kuiper-Goodman et al 1987, Kuiper-Goodman 1990).

Because they were seen at much lower levels, the hormonal effects due to zearalenone are likely to be of greater importance than tumorigenic effects. Dividing the no-hormonal effect level by a safety factor of 500, which takes into account inter- and intraspecies differences and uncertainties in the experimental model, gives an estimated TDI of 100 ng/kg bw per day (Kuiper-Goodman et al 1987).

Overall, a TDI of 100 ng/kg bw per day is therefore suggested for zearalenone.

Exposure Assessment

We previously reviewed data on the occurrence of zearalenone in animal-derived food products (Kuiper-Goodman et al 1987). Wide variations in carryover data were reported for dairy cattle, even from the same laboratory. Further studies have indicated that generally transmission to the milk in cows does not pose a significant human health risk. Cows normally obtain no more than 50–165 mg of zearalenone daily from their protein rations. At this intake level, no residues of zearalenone (detection limit 0.3–0.5 ng/mL) or its metabolites (detection limit 1.5 ng/mL) were found (Prelusky et al 1990).

The human intake in Canada of zearalenone and other resorcylic lactones was recently estimated to be approximately 50–100 ng/kg bw per day (eaters only, highest relative exposure groups) and was considered to be more or less equally divided between plant-based (cereal crops) and animal-derived food products (Kuiper-Goodman et al 1987). The estimates for animal-derived food products were based on carryover studies, since no monitoring studies of zearalenone in animal-derived food products are available.

Risk Assessment

In Canada, the estimated intake of zearalenone is in the same range as the estimated TDI, indicating that this intake is unlikely to pose a health risk. Since we are dealing however with an estrogenic compound, its contribution to the total external estrogenic burden should also be considered.

Regulation

In Canada no regulations or guideline levels are in effect with regard to zearalenone.

OCHRATOXIN A

Ochratoxin A (OTA) is a mycotoxin which has been found to occur in foods of plant origin and in animal-derived food products, as well as in human blood sera and tissues. The ability of OTA to move up the food chain is aided by its long half-life in certain edible animal species. A detailed review and risk assessment of OTA by Kuiper-Goodman and Scott (1989) discusses its chemistry, mycology, and natural occurrence as well as its toxicity and metabolic disposition, and its role in porcine, avian, and human nephropathies. The WHO Joint Expert Committee on Food Additives also recently reviewed this mycotoxin with the aim of arriving at a TDI (WHO 1991) and prepared a monograph (Kuiper-Goodman and Grant 1991). OTA was also reviewed by IARC (1991, 1993) and was the subject of a recent symposium (Creppy et al 1993).

Hazard Assessment

Toxicology. The major target for OTA toxicity in all mammalian species tested is the kidney, and endemic nephropathies affecting livestock as well as humans have been attributed to OTA. OTA is also teratogenic, and in the fetus the major target is the developing central nervous system. Recent studies have provided "clear evidence" of the carcinogenicity of OTA in two rodent species. OTA was initially found to be nonmutagenic in various microbial and mammalian gene mutation assays, but weak genotoxic activity to mammalian cells was noted. Recently, a clear hepatocyte-mediated response was observed in a modified *Salmonella*-microsome assay, as well as an increased incidence of sister chromatid exchange in human lymphocytes (Hennig et al 1991). Furthermore, using a modified ^{32}P-post-labeling method, OTA-DNA adducts were found in kidney, liver, and spleen of mice orally exposed to OTA (Pfohl-Leskowicz et al 1991, 1993). In addition, OTA was found to suppress both humoral and cell-mediated immune function. Based on a carcinogenicity study of OTA in rats conducted by the NTP, the estimated TDI in humans is 1.5 ($TD_{50}/50,000$), 1.8 (virtual safe dose, risk level 10^{-5}), 4.2 (NOAEL/5,000), or 5.7 (NEL/5,000) ng/kg bw per day, depending on the method of extrapolation used (Kuiper-Goodman 1991c).

Epidemiology. Widespread human exposure to low levels of this mycotoxin is evident from surveys of human serum (<0.1–40 ng OTA per milliliter of serum) conducted in several European countries (Hald 1991), as well as in Canada (Frohlich et al 1991, Kuiper-Goodman 1991c, Kuiper-Goodman et al 1993). Recent studies have also indicated the presence of OTA in randomly collected human milk samples in Germany (range 0.017–0.03 ng/mL in four of 36 samples) (Gareis et al 1988) and Italy (range 1.7–6.6 ng/mL in nine of 50 samples) (Micco et al 1991). The widespread occurrence in human milk and serum is not surprising and can be attributed to the probable long serum half-life of OTA in humans (Hagelberg et al 1989). Estimates of human exposure based on serum or milk samples reflect both recent and chronic exposure at the level of the individual and are very useful in epidemiological studies. Comparisons are useful in identifying populations at higher risk due to higher exposure to OTA. By using a pharmacokinetic model, Breitholz et al (1991) estimated that the daily intake (expressed in nanograms per kilogram of body weight per day) was about 40% greater than the human plasma concentration (expressed in nanograms per milliliter). The human data implicating a role for OTA in Balkan nephropathy was the topic of a symposium (IARC 1991) and was recently reviewed again by the same organization (IARC 1993), but at the time of this writing, the evidence is insufficient for consideration under the health hazard assessment. Evidence suggesting that elevated exposure to OTA is associated with human nephropathies in Algeria and Tunisia was recently presented (Khalef et al 1993, Bacha et al 1993).

Exposure Assessment

For Canadians, the worst-case estimate of daily exposure to OTA from consumption of pork-derived food products and cereal foods is probably less than 3.5 ng/kg bw per day (mean of eaters) for young children, the group with highest consumption on a body weight basis (Kuiper-Goodman et al 1993). In a limited survey conducted in 1993 in the province of Manitoba, the same authors estimated a mean serum level of OTA of 1.29 ng/ml for 1991, with 18.3% of the samples indicating exposure that exceeded 4.2 ng/kg bw per day (Kuiper-Goodman et al 1993).

Risk Assessment and Regulations

As indicated above, the overall range of estimated TDIs for OTA is 1.5–5.7 ng/kg bw per day, based on the NTP carcinogenicity study. At present, the Health Protection Branch attempts to ensure that the PDI in Canada (obtained from ongoing surveys) does not exceed 4 ng/kg bw per day. This is slightly lower than the TDI of 5 ng/kg bw per day suggested for use in Nordic countries (Nordic Council of Ministers 1991), and about fourfold lower than the provisional weekly tolerance of 112 ng/kg bw suggested by the Joint Expert Committee on Food Additives (WHO 1991). In Canada, the worst-case estimate of exposure to OTA is probably less than 3.5 ng/kg bw per day, but further studies involving more food commodities are ongoing to better define the overall residue profile of OTA in cereal grains, animal feeds, animal food products, and human blood. Such data are required to better assess dietary exposure and to ascertain the need for regulatory controls and other control mechanisms.

DEOXYNIVALENOL

In Canada, concern with regard to deoxynivalenol started in 1980 when pink wheat kernels and elevated levels of deoxynivalenol were found in Ontario winter wheat (soft wheat) and Quebec spring wheat (hard wheat) (Scott et al 1981, Trenholm et al 1983). Deoxynivalenol has also been found in some corn and barley.

Hazard Assessment

Toxicology. Deoxynivalenol belongs to a group of more than 50 naturally occurring 12,13-epoxytrichothecenes (Ueno 1984). The trichothecene mycotoxins, as a whole, possess similar chemical and biological features, characterized by antibiotic, antifungal, and cytostatic activities. Many were isolated as causative agents of naturally occurring intoxications associated with moldy foods and feeds. The acute symptoms of trichothecene poisoning are characterized by skin irritation, feed refusal, vomiting, diarrhea, hemorrhages, neural disturbance, abortion, and death. Deoxynivalenol belongs to the B group of trichothecenes. Biochemically, trichothecenes, including deoxynivalenol, are inhibitors of protein and DNA synthesis. Results with the Ames test have been negative (Wehner et al 1978). Negative results were also obtained in the hepatocyte-mediated mutation assay with V-79 Chinese hamster lung cells at a range of concentrations well into the cytotoxic range (Rogers and Héroux-Metcalf 1983). Short-term and long-term effects of oral intake of deoxynivalenol by rodents (gavage, diet), as well as effects on reproduction, including teratogenic potential, were assessed at the Health Protection Branch (Khera et al 1982, 1984, 1986; Tryphonas et al 1984, 1986; Arnold et al 1986a,b) and elsewhere. These studies showed a preliminary NOAEL of 0.60 mg/kg bw. At higher dosage levels, reproductive, embryotoxic, and possible teratogenic effects were noted. Deoxynivalenol also affected food intake and body weight and suppressed humoral and cellular immune function (Tryphonas et al 1984, 1986; Pestka et al 1987; Robbana-Barnat et al 1988; Pestka and Bondy 1990; Pestka et al 1990a,b; Bondy and Pestka 1991). More than a 50% decrease in relative thymus weight and cardiac lesions was observed in mice given deoxynivalenol in the diet at 1.3 mg/kg bw per day for two to three weeks. Intraperitoneal administration of deoxynivalenol at the same dose level resulted in a 30% decrease in cardiac protein synthesis (Robbana-Barnat et al 1985). No carcinogenic effects were observed in a recently completed two-year carcinogenicity study in mice conducted at the Health Protection Branch in Canada (F. Iverson, *personal communication*). Feeding studies in livestock and poultry species are discussed elsewhere (Chapter 9).

Epidemiology. Outbreaks of red mold toxicoses that affect humans and involve *Fusarium graminearum* have been reported in Japan, China, and India. Luo (1988) reported 35 outbreaks, attributed to consumption of scabby wheat and moldy corn, that have occurred between 1961 and 1985 and have affected at least 7,818 victims. Typically, people became ill from 5 to 30 min after consumption, and the following symptoms were reported: nausea, vomiting, diarrhea, abdominal pain, headache, dizziness, and fever. No deaths were reported. In one of the outbreaks that occurred in 1984 in Xingtai county, 362 of 383 people consuming moldy corn became ill (attack rate 94.5%). Analysis of six samples associated with food poisoning in this outbreak indicated the presence of deoxynivalenol (3.75–92.8 $\mu g/g$) and zearalenone (0.128–0.587 $\mu g/g$) in five samples; one sample contained 0.34 $\mu g/g$ of deoxynivalenol and 0.004 $\mu g/g$ of zearalenone. No T-2 toxin or nivalenol were found. The same author reported deoxynivalenol levels of 1 to 40 $\mu g/g$ in scabby wheat collected from three different villages and showed that wheat samples collected during food poisoning had significantly higher levels of deoxynivalenol than samples not associated with food poisoning. No acute human illness was observed in Henan, China, in 1985 among 191 peasant families who consumed scabby wheat containing deoxynivalenol at a level of 0.016 to 3.33 $\mu g/g$ (Hong-Wei et al 1989). Similarly, no acute human illness was observed from the consumption of corn and corn meal containing deoxynivalenol at levels of 0.36 to 12.67 $\mu g/g$ (mean 5.38 $\mu g/g$) and nivalenol at levels of 0.054 to 2.76 $\mu g/g$ (mean 0.76 $\mu g/g$); these were consumed as the staple food by esophageal cancer patients in 1985–1986 in Linxian, China, a high-risk area for esophageal cancer (death rate 132 per 100,000 compared to <5 per 100,000 in the United States) (Hsia et al 1988). Subsequent studies, conducted in 1989, comparing randomly collected corn samples from families with esophageal cancer in Linxian with samples from nonesophageal cancer families in Shangqiu (esophageal cancer death rate 15.7 per 100,000) indicated that the mean level of deoxynivalenol in corn in Linxian was 0.57 $\mu g/g$ (range 0.017–3.51 $\mu g/g$) compared with 0.099 $\mu g/g$ (range 0.011–0.61 $\mu g/g$) in Shangqiu (Luo et al 1990). In most of these studies no information was provided on the amount of wheat or corn consumed in the areas being compared, so that no exposure estimates could be made.

In 1987, an outbreak of disease affecting about 50,000 people in the Kashmir Valley in India was attributed to the consumption of bread made from rain-damaged wheat. The wheat was reported to contain several trichothecenes, including deoxynivalenol (0.34–8.4 $\mu g/g$ in 11 of 24 samples), acetyldeoxynivalenol (0.6–2.4 $\mu g/g$ in four of 24 samples), nivalenol (0.03–0.1 $\mu g/g$ in two of 24 samples), and T-2 toxin (0.55-4 $\mu g/g$ in five of 24 samples) (Bhat et al 1989a,b). In retrospective interviews of about 150 affected families, it was found that only individuals who had consumed wheat products had become ill, exhibiting mainly mild gastrointestinal tract symptoms for about two days. The authors of the study presumed that consumption of any rain-damaged wheat was associated with symptoms, but this may not have been the case. Based on the lowest level of deoxynivalenol found in wheat (0.34 $\mu g/g$), an average intake of 67 g of wheat products, and a mean body weight of 52 kg, a NOEL of 0.438 $\mu g/$ kg bw per day was estimated by the authors of the study (Bhat et al 1987). There are, however, problems with this NOEL. Because samples were not collected until four months after the outbreak, a clear association of specific samples with specific cases of illness or lack of illness was not established. As well, many of the collected samples contained other, more toxic, trichothecenes, the contribution of which to the overall disease profile needs to be considered. Therefore, the actual NOEL for trichothecene-associated human illness, expressed as equivalents of deoxynivalenol, has not been established. Nevertheless, this mild outbreak, as well as the reports of human illness due to deoxynivalenol in China, indicates a need for caution, especially with regard to the potential health effects of chronic exposure.

It is evident that more detailed analytical epidemiological studies, such as case versus control or exposed versus nonexposed, need to be conducted in those areas of the world where the presence of scabby wheat or moldy corn is a cyclical endemic event, in order to establish a dose response relationship and a human NOAEL for exposure to deoxynivalenol and other trichothecenes and acute illness. It will be more difficult to establish a causal relationship for a chronic disease such as esophageal cancer, for which there are likely to be several factors, including possibly several mycotoxins, that interact.

Exposure Assessment

In Canada, deoxynivalenol has been found mainly in wheat grown in the eastern provinces and, in particular, in soft wheat, which is chiefly used in the production of nonstaple foods such as cakes, cookies, and biscuits. In addition, durum wheat, which is used in the manufacture of pasta, has been found to contain detectable levels of deoxynivalenol. Products made from kernel corn (e.g., corn breakfast cereal, popcorn) may also be of concern. The major Canadian wheat crop, western hard wheat, which is used for the most part in staple foods (bread), is generally not affected (Scott et al 1984, Scott 1990).

Studies by Agriculture Canada and private industry have shown that, in general, deoxynivalenol levels decrease by about 40% during industrial processing from uncleaned soft wheat to the flour portion of the finished food product; yeast doughnuts were an exception, appearing to contain increased amounts of deoxynivalenol (Young et al 1984). Analyses of bran showed no significant reduction in deoxynivalenol between uncleaned wheat and the finished product (Young et al 1984).

From time to time, estimates of deoxynivalenol intake have been made to most accurately reflect the human exposure in Canada from various food categories (bread, cakes, cookies, cereals, etc.). These estimates are based on year-to-year variations in deoxynivalenol levels in the various Canadian grains, as monitored by the Health Protection Branch, and take into account information on processing. In the fall of 1982, for instance, the total estimated direct mean and 90th percentile intakes of deoxynivalenol for 12- to 19-year-old males were 0.4 and 1.2 μg/kg bw per day, respectively.

Regulation

Based on the toxicity information on deoxynivalenol available early in 1983, a provisional TDI of 3.0 μg/kg bw for adults and 1.5 μg/kg bw for infants was established and is still in effect (Kuiper-Goodman 1985).

Current guideline levels for deoxynivalenol are 2.0 μg/g in uncleaned soft wheat used for nonstaple foods, including bran, except for soft wheat destined for infant food where the guideline level is 1.0 μg/g. As a result of processing, levels of 1.2 and 0.6 μg/g of deoxynivalenol would be expected in the flour portion of finished foods and infant foods, respectively.

For domestic products, the Health Protection Branch restricted its regulatory action to the final blend of uncleaned soft wheat prior to milling. Imported products, on the other hand, are subject to regulatory action based on the level of deoxynivalenol found in the flour portion of the finished product. If the level of deoxynivalenol in any sample of (hard) wheat grown in eastern Canada for use in staple foods exceeds 0.1 μg/g, an exposure assessment needs to be made. In case the estimated PDI exceeds the provisional TDI, options and guidelines may be developed to ensure that the overall PDI does not exceed the provisional TDI.

FUMONISINS AND OTHER *FUSARIUM MONILIFORME* TOXINS

The fumonisins (fumonisin B_1, B_2, B_3, B_4, A_1, A_2) are a group of mycotoxins produced by *Fusarium moniliforme* that have recently been isolated and chemically characterized. Research with this group of toxins and other *F. moniliforme* toxins was initiated in order to isolate the chemical substances responsible for the toxicity associated with the presence of *F. moniliforme* in corn-based feeds and foods.

Hazard Assessment of Fumonisins

Epizootics. The presence of *F. moniliforme* has been related to at least two diseases in livestock: equine leukoencephalomalacia (ELEM), which affects the brain and which is known to occur in South Africa, the United States, and Brazil (Ross et al 1991, Wilson et al 1990, Thiel et al 1992) and porcine pulmonary edema, known to occur in the United States (Colvin and Harrison 1992). As well, a possible association of fumonisins and mystery swine disease has been observed, characterized by adverse effects on reproduction as well as respiratory disease, and also occurring in the

United States (Bane et al 1992). Of the several mycotoxins isolated from
F. moniliforme, it has been possible to experimentally induce ELEM in
young horses by oral administration of pure fumonisin B_1 at 1–4 mg/kg
bw per day (Kellerman et al 1990) or by feeding naturally contaminated
feed equivalent to 0.18 mg/kg bw per day to ponies (Wilson et al 1992).
Since the effective oral doses of pure versus naturally occurring fumonisins
differ, additional factors may be involved. Haschek et al (1992) were able
to induce porcine pulmonary edema by oral administration of corn screenings naturally contaminated with fumonisins B_1 and B_2 to give a dose of
4.5–6.6 mg/kg bw per day (total fumonisins).

Epidemiology. In the Transkei, southern Africa, a high prevalence of
esophageal cancer has been noted, accounting for almost half of the reported
cancer cases (Jaskiewicz et al 1987a). In this area, liver cancer was the
second most prevalent type of cancer. In a district of the Transkei with
a high esophageal cancer rate, Marasas et al (1988) reported significantly
greater *F. moniliforme* contamination of home-grown corn in households
in which adults showed cytological evidence of premalignant esophageal
lesions compared with "unaffected" households from a district with a lower
esophageal cancer rate. In an ecological study in the Transkei, samples
from a district that had high rates of esophageal cancer also showed a
higher incidence of *F. moniliforme* contamination of healthy and moldy
corn than samples from a district with a lower rate (Sydenham et al 1990).
Significant differences in fumonisin B_1 content were also noted in the healthy
corn (1,600 versus 60 ng/g) and moldy corn (23,900 versus 6,500 ng/g)
samples from the two regions, suggesting an association between the
incidence of esophageal cancer and the incidence of *F. moniliforme* and
fumonisins in corn at the population level. Further epidemiological studies
at the level of the individual, such as case control or cohort studies, are
required to more precisely define the role of *F. moniliforme* and its
metabolites in the development of esophageal cancer in the Transkei and
other areas where the incidence of esophageal cancer is high.

Toxicology—Acute, Subacute, and Subchronic Toxicity. The fumonisins
are only moderately toxic in acute and subacute toxicity studies. Three
out of four young male BD IX rats died after receiving three daily doses
of 237 mg/kg bw of fumonisin B_1 by gavage. These rats developed toxic
hepatitis, with single-cell necrosis, as well as lesions in the heart (severe
disseminated acute myocardial necrosis in two rats), lungs (severe pulmonary
edema in two rats), kidney (proximal tubular necrosis), and gastrointestinal
tract (lymphoid necrosis in Peyer's patches and scattered epithelial necrosis)
(Gelderblom et al 1988). Rats that were orally administered by gavage about
one-fourth of this dose for 21 days or in the diet for 33 days had a significant
reduction in weight gain and developed hepatitis, which was more advanced
in the rats receiving fumonisin in the diet. Kidney lesions were similar
but less extensive than those seen in rats that died after three days. No

other organs were affected (Gelderblom et al 1988). The reduced weight gain and the pathological changes in the livers of rats caused by fumonisin B_1 were similar to those caused by *F. moniliforme* MRC 826 cultures.

Toxicology—Metabolic Disposition. In preliminary studies, the fate of ^{14}C-labeled or unlabeled fumonisin B_1 was determined in four BD IX rats given 7.5 mg/kg bw by gavage. Within 24 hr, more than 99% of the fumonisin B_1 was eliminated unmetabolized in the feces, with only trace amounts in urine, liver, kidney, and red blood cells (Shephard et al 1992a,b). These results suggest that fumonisin B_1 is either poorly absorbed or rapidly eliminated through biliary excretion.

Toxicology—Carcinogenicity. Fumonisin B_1 in the diet of BD IX rats at a level of 50 mg/kg bw per day was found to produce positive results in a short-term (four-week) liver cancer initiation and/or promotion bioassay. In the bioassay, fumonisin B_1 induced gamma-glutamyl-transpeptidase-positive (GGT+) foci both in diethylnitrosamine-initiated rats and control (DMSO treated) rats. This bioassay was actually used as a monitoring system to isolate cancer-promoting compounds from *F. moniliforme* culture material, and it formed the basis for the isolation of the fumonisins (Gelderblom et al 1988). The authors concluded that fumonisin B_1 was a promoter as well as possibly a "weak initiator." To further investigate the initiating potential of fumonisin B_1, a "resistant hepatocyte" model was used (Solt and Farber 1976), whereby the ability of cells to proliferate in an environment that inhibits the growth of normal hepatocytes is measured. In this assay, fumonisin B_1 was found to be a weak initiator of carcinogenesis only when incorporated in the diet for 26 days or longer (Gelderblom et al 1992b)

Further studies have shown that fumonisins B_2 and B_3 closely mimic the toxic and tumor-promoting effects seen with fumonisin B_1 in the short-term liver bioassay; like fumonisin B_1, they are able to induce resistant hepatocytes (Gelderblom et al 1992a). In contrast, no adverse effects could be detected for fumonisin A_1 under identical experimental conditions (Gelderblom et al 1992a).

A carcinogenicity study, considered preliminary, has been completed in two groups of 25 BD IX rats given fumonisin B_1 at 0.5 (controls) or 50 mg/kg in a semipurified corn-based diet, equivalent to 0.037 and 3.75 mg/kg bw per day, respectively. Five rats per group were killed at six, 12, 20, and 26 months (Gelderblom et al 1991, Thiel et al 1992). The major target of fumonisin B_1 was the liver. All fumonisin B_1-treated rats that died (five rats) or were killed (10 rats) from 18 months onward suffered from "chronic toxic hepatitis," which had progressed to cirrhosis. Ten out of 15 of these rats had developed primary hepatocellular carcinoma, with metastases to the heart, lungs, or kidneys present in four of the rats with hepatocellular carcinoma. Large areas of cholangiofibrosis were present at the hilus of the liver. All fumonisin-treated livers contained regenerative nodules as well as GGT+ foci. Relative liver weight was significantly increased in treated rats by almost fourfold at 26 weeks. No neoplastic changes were observed in control rats, and their survival rate was 96%. Serum levels of aspartate aminotransferase, gamma-glutamyl-transpeptidase, and bilirubin (total, conjugated, and unconjugated) in treated rats were significantly higher at 20 and 26 months. The kidneys of the fumonisin-treated rats were markedly affected at 26 months. The lesions included fibrosis, retention cysts, and proximal tubular necrosis. Contrary to observations in rats fed culture material of *F. moniliforme* MRC 826 (Jaskiewicz et al 1987b, Marasas et al 1984), there were no lesions in the esophagus, heart, or forestomach.

Further carcinogenicity studies are being conducted in South Africa (using lower dosage levels) and are planned to be conducted by the U.S. Department of Health and Human Services National Toxicology Program once sufficient amounts of fumonisins are available.

Toxicology—Genotoxicity. Fumonisin B_1 and B_2 were found to be negative in the Ames test (Gelderblom and Snyman 1991, Park et al 1992). Fumonisin B_1 (at concentrations of 0.04–250 μM) and fumonisin B_2 (at 0.04–40 μM) also lacked genotoxic effects in the in vitro DNA repair assays in primary rat hepatocytes (Gelderblom et al 1992b, Norred et al 1992a). Similarly, no DNA repair was induced in rat hepatocytes that were obtained 13–14 h after rats had been orally administered fumonisin B_1 (Gelderblom et al 1992b).

Toxicology—Cytotoxicity. In vitro studies, measuring release of lactate dehydrogenase by primary hepatocytes, showed that fumonisins B_1 and B_2 are cytotoxic only at high dose levels (175 and 87 μM, respectively) when compared with aflatoxin B_1 (<1 μM) (Gelderblom et al 1992b). In another more sensitive bioassay measuring cell viability in a hepatoma cell line (H4TG) and in the MDCK dog kidney epithelial cell line, fumonisins B_1 and B_2 (on a molar basis, after four days of culture) were about 1,000 to 3,000 times less toxic than T-2 toxin, with inhibitory concentrations (IC_{50}s) in MDCK cells of 3.5 μM, 2.8 μM, and 1 nM, respectively (Shier et al 1991).

Toxicology—Studies into the Mechanism of Action. The fumonisins are structurally similar to the long-chain base sphingosine, a component of the long-chain backbone of sphingolipids. It is therefore not surprising that fumonisins B_1 and B_2 are the first discovered naturally occurring specific inhibitors of de novo sphingolipid biosynthesis and sphingolipid turnover. The site of inhibition is the formation of ceramides catalyzed by sphingosine- and sphinganine-N-acetyltransferase (ceramide synthase). In primary rat hepatocytes this inhibition occurs at concentrations of fumonisin that are not toxic to the cells ($IC_{50} = 0.1$ μM)(Wang et al 1991, Norred et al 1992b). Similarly, fumonisins B_1 and B_2 inhibited de novo sphingosine biosynthesis ($IC_{50} = 10$–15 μM) in a proliferating cell line of pig kidney cells (LLC-

PK1) and caused a remarkable (128-fold) increase in cellular levels of sphinganine (Yoo et al 1992, Norred et al 1992b). In these cells inhibition of sphingolipid biosynthesis was an early event in the toxicity of fumonisins and preceded inhibition of cell proliferation and cytotoxicity. Inhibition of sphingolipid biosynthesis was also observed in yeast cells (Kaneshiro et al 1992).

In pigs dosed orally for five to 15 days with fumonisin-containing culture material (equivalent to 4.5–6.6 mg/kg bw per day of fumonisins), or intravenously for four to nine days with fumonisin B_1 (0.9–1.2 mg/kg bw per day), the primary target was the liver (Haschek et al 1992). On the basis of electron-microscopic observations, the authors hypothesized that altered sphingolipid metabolism resulted in hepatocellular damage, leading to the release of membranous material into circulation. It was further hypothesized that this material is engulfed by pulmonary macrophages, where it could trigger the release of mediators, leading to porcine pulmonary edema. Further studies in pigs fed diets containing corn screenings and formulated to contain 23 to 175 ppm total fumonisins (equivalent to about 1.15 to 8.75 mg/kg bw per day) for up to 14 days, showed that only the liver was affected at 1.15 mg/kg bw per day. At the highest dose level, both liver and kidney were affected (Riley et al 1993). The ratio of free sphinganine to free sphingosine in the sera and tissues (liver, lung, and kidney) of these pigs was increased in a dose-related manner at five and 14 days compared with baseline values for each pig. This ratio was a very sensitive biomarker for fumonisin exposure and was seen before detectable liver or lung lesions in pigs, or before other serum biochemistry changes. Similar observations have also been made in studies with ponies (Wang et al 1992).

Inhibition of sphingolipid biosynthesis may be the primary effect of fumonisin toxicity in ELEM and porcine pulmonary edema. It may also be responsible for the tumor-promoting ability of the fumonisins. In the latter, fumonisin-induced inhibition of sphingosine biosynthesis may lead to a deregulation of protein kinase C, which could lead to the proliferation of initiated cells (Norred et al 1992b).

Toxicology—Other Studies. No teratology, reproduction, or immuno-toxicity studies have been conducted or reported for the fumonisins.

Extrapolation. The available information on toxicity of fumonisins is inadequate. Therefore TDIs for this group of mycotoxins cannot be established.

Exposure Assessment of Fumonisins

Preliminary studies have shown that fumonisins do occur in Canadian corn (*unpublished*). At present, however, there is insufficient information on the levels of fumonisin in the overall Canadian food supply. Therefore an exposure assessment for Canada cannot be made. Studies by Agriculture Canada and Health and Welfare Canada are ongoing and will focus on corn-based products, including milled corn products imported from the United States. Studies on the transmission of fumonisin residues from feed to animal-derived food products are also ongoing at Agriculture Canada.

Regulations for Fumonisins

In Canada, no regulations or guideline levels are in effect with respect to fumonisins. In Arizona, a maximum residue level of 5 ppm has been set for fumonisin B_1 in horse feed (Frank Ross, *personal communication*).

Other *Fusarium moniliforme* Toxins

A number of other mycotoxins have been isolated from *F. moniliforme* cultures and been assessed for their possible role in *F. moniliforme* associated diseases. Fusarin C was found to be mutagenic in the Ames test (Gelderblom and Snyman 1991) but was not positive in a short-term initiation and promotion bioassay in rat liver (Gelderblom et al 1986). Rapid conjugation with glutathione and excretion are possibly related to its lack of carcinogenicity (Norred et al 1992a). Inconclusive results with fusarin C were obtained with a DNA repair assay in primary hepatocytes (Norred et al 1992a). Moniliformin and bikaverin were not genotoxic in a DNA repair assay with primary hepatocytes (Norred et al 1992a). Additional mycotoxins continue to be isolated from *F. moniliforme* cultures, and the possibility for interaction between the various *F. moniliforme* toxins in the genesis of both acute and chronic effects needs to be considered in the overall risk assessment and future regulation of this group of toxins.

Conclusion

To some extent, the presence of small amounts of mycotoxins or their metabolites in some plant-based food crops and animal-derived food products is unavoidable. By assessing the level in food at which these substances may pose a potential risk to human health, it is possible to devise appropriate risk management strategies. In some cases this requires the establishment of tolerances or guideline levels, which determine the maximum residue levels of individual mycotoxins in certain food commodities. These levels are based on the TDI derived from the hazard assessment and the likely exposure (PDI) to foods containing various levels of the mycotoxin. When the toxicological data base is inadequate, tolerances or guideline levels may for a limited time be based on technically achievable limits or on analytical limitations. A similar approach is sometimes used for carcinogenic mycotoxins.

Many other factors are considered in making a rational risk management decision. For example, if establishment of a guideline resulted in the complete elimination of an essential food from the marketplace, this could have nutritional repercussions of greater adversity to human health than the potential harm from the mycotoxin. Socioeconomic factors are also considered in the risk management phase. Because food intake patterns and levels of mycotoxins in food commodities differ between countries, as well as approaches to establish what risks are acceptable, risk management practices tend to vary from country to country (van Egmond 1989). The Canadian perspective was addressed in this chapter.

There are also guidelines in Canada for the use of feed containing deoxynivalenol. These relate to concentrations that may not be exceeded without risk of adverse economic effects, such as decreased performance and lowered fertility. For pigs, calves, lambs, and lactating cows and ewes the recommendation is 1 $\mu g/g$ in the diet. For adult cows, adult sheep, and poultry it is 5 $\mu g/g$ in the diet.

The United States Food and Drug Administration issued advisories concerning deoxynivalenol in 1982 and again in 1983. The 1993 recommendations for deoxynivalenol in finished wheat products intended for human consumption, such as flour, bran, and germ, is 1 $\mu g/g$. The recommended concentrations for animals are as follows: 1) 10 $\mu g/g$ in grains and by-products destined for chickens or ruminating beef and feedlot cattle older than four months provided these products do not exceed 50% of the diet; 2) 5 $\mu g/g$ in products that will make up no more than 20% of the diet for swine; and 3) 5 $\mu g/g$ for all other animals in products that will be no more than 40% of the diet.

Literature Cited

Ames, B. N., Magaw, R., and Gold, L. S. 1987. Ranking possible carcinogenic hazards. Science 236:271-280.

Arnold, D. L., Karpinski, K. F., Mcguire, P. F., Nera, E. A., Zawidzka, Z. Z., Lok, E., Campbell, J. S., Tryphonas, L., and Scott, P. M. 1986a. A short-term feeding study with deoxynivalenol (vomitoxin) using rats. Fundam. Appl. Toxicol. 6:691-696.

Arnold, D. L., Mcguire, P. F., Nera, E. A., Karpinski, K. F., Bickis, M. G., Zawidzka, Z. Z., Fernie, S., and Vesonder, R. F. 1986b. The toxicity of orally administered deoxynivalenol (vomitoxin) in rats and mice. Food Chem. Toxicol. 24:935-941.

Bacha, H., Maaroufi, K., Achour, A., Hamammi, M., Ellouz, F., and Creppy, E. E. 1993. Ochratoxines et ochratoxicoses humaines en Tunisie. Pages 111-121 in: Human Ochratoxicosis and Its Pathologics. E. E. Creppy, M. Castegnaro, and G. Dirheimer, eds. Colloque INSERM, John Libbey Eurotext, France.

Bane, D. P., Neumann, E. J., Hall, W. F., Harlin, K. S., and Slife, R. L. N. 1992. Relationship between fumonisin contamination of feed and mystery swine disease— A case-control study. Mycopathologia 117:121-124.

Bhat, R. V., Ramakrishna, Y., Sashidhar, R. B., and Nahdi, S. 1987. Trichothecene mycotoxicosis: Report on the investigations on the aetiology of a suspected disease outbreak due to products made from rain damaged wheat in the Kashmir Valley. Food Drug Toxicol. Res. Centre, Natl. Inst. Nutr. Indian Council Med. Res., Hyderabad, India.

Bhat, R. V., Beedu, S. R., Ramakrishna, Y., and Munshi, K. L. 1989a. Outbreak of trichothecene mycotoxicosis associated with consumption of mould-damaged wheat production in Kashmir Valley, India. Lancet, January 7:35-37.

Bhat, R. V., Ramakrishna, Y., and Sashidhar, R. B. 1989b. Outbreak of mycotoxicosis in Kashmir Valley, India. Nutr. News (Natl. Inst. Nutr. India) 10(1):1

Bondy, G. S., and Pestka, J. J. 1991. Dietary exposure to the trichothecene vomitoxin (deoxynivalenol) stimulates terminal differentiation of Peyer's patch B cells to IgA secreting plasma cells. Toxicol. Appl. Pharmacol. 108:520-530.

Bosch, F. X., and Munoz, N. 1988. Prospects for epidemiological studies on hepatocellular cancer as a model for assessing viral and chemical interactions. IARC (Int. Agency Res. Cancer) Sci. Publ. 89:427-438.

Breitholz, A., Olsen, M., Dahlb C., A., and Hult, K. 1991. Plasma ochratoxin A levels in three Swedish populations surveyed using an ion-pair HPLC technique. Food Addit. Contam. 8:183-192.

Bressac, B., Kew, M., Wands, J. and Ozturk, M. 1991. Selective G to T Mutations of p53 gene in hepatocellular carcinoma from southern Africa. Nature 350:429-431.

Busby, W. F., Jr., and Wogan, G. N. 1984. Studies of naturally occurring carcinogens in the diet 1. Aflatoxins. ACS (Am. Chem. Soc.) Monogr. 182:945-1136.

Colvin, B. M., and Harrison, L. R. 1992. Fumonisin-induced pulmonary edema and hydrothorax in swine. Mycopathologia 117:79-82.

Creppy, E. E., Castegnaro, M., and Dirheimer, G. 1993. Human Ochratoxicosis and its Pathologies. E. E. Creppy, M. Castegnaro, and G. Dirheimer, eds. Colloque INSERM, John Libbey Eurotext, France.

Denning, D. W. 1987. Aflatoxin and human disease. Adv. Drug React. Accid. Pois. Rev. 4:175-209.

Frohlich, A, A., Marquardt, R. R., and Ominski, K. H. 1991. Ochratoxin A as a contaminant in the human food chain: a Canadian perspective. Pages 139-143 in: Mycotoxins, Endemic Nephropathy and Urinary Tract Tumours. M. Castegnaro, R. Plestina, G. Dirheimer, I. Chernozemsky, and H. Bartsch, eds. IARC (Int. Agency Res. Cancer) Sci. Publ. 115.

Gareis, M., Martlbauer, E., Bauer, J., and Gedek, B. 1988. Bestimmung von ochratoxin A in muttermilch. (Determination of ochratoxin A in human milk). Z. Lebensm. Unters. Forsch. 186: 114-117.

Gelderblom, W. C. A., and Snyman, S. D. 1991. Mutagenicity of potentially carcinogenic mycotoxins produced by *Fusarium moniliforme*. Mycotoxin Res. 7:46-52.

Gelderblom, W. C. A., Thiel, P. G., Jaskiewicz, K., and Marasas, W. F. O. 1986. Investigations on the carcinogenicity of fusarin C—A mutagenic metabolite of *Fusarium moniliforme*. Carcinogenesis 7:1899-1901.

Gelderblom, W. C. A., Jaskiewicz, K., Marasas, W. F. O., Thiel, P. G., Horak, R. M., Vleggaar, R., and Kriek, N. P. J. 1988. Fumonisins: Novel mycotoxins with cancer-promoting activity produced by *Fusarium moniliforme*. Appl. Environ. Microbiol. 54:1806-1811.

Gelderblom, W. C. A., Kriek, N. P. J., Marasas, W. F. O., and Thiel, P. G. 1991. Toxicity and carcinogenicity of the *Fusarium moniliforme* metabolite, fumonisin B_1, in rats. Carcinogenesis 12:1247-1251.

Gelderblom, W. C. A., Marasas, W. F. O., Vleggaar, R., Thiel, P. G., and Cawood, M. E. 1992a. Fumonisins—Isolation, chemical characterization and biological effects.

Mycopathologia 117:11-16.

Gelderblom, W. C. A., Semple, E., Marasas, W. F. O., and Farber, E. 1992b. The cancer-initiating potential of the fumonisin B mycotoxins. Carcinogenesis 13:433-437.

Griffin, T. B., Singh, A. R., and Coulston, F. 1984. No hormonal effect in non-human primates of oral zeranol. Report to the International Minerals and Chemical Company, Terre Haute, IN.

Groopman, J. D. 1988. Do aflatoxin-DNA adduct measurements in humans provide accurate data for cancer risk assessment? Methods for detecting DNA damaging agents in humans: Applications in cancer epidemiology and prevention. IARC (Int. Agency Res. Cancer) Sci Publ. 89:55-62.

Hagelberg, S., Hult, K., and Fuchs, R. 1989. Toxicokinetics of ochratoxin A in several species and its plasma-binding properties. J. Appl. Toxicol. 9:91-96.

Hald, B. 1991. Ochratoxin A in human blood in European countries. Pages 159-164 in: Mycotoxins, Endemic Nephropathy and Urinary Tract Tumours. M. Castegnaro, R. Plestina, G. Dirheimer, I. Chernozemsky, and H. Bartsch, eds. IARC (Int. Agency Res. Cancer) Sci. Publ. 115.

Haschek, W. M., Motelin, G., Ness, D. K., Harlin, K. S., Hall, W. F., Vesonder, R. F., Peterson, R. E., and Beasley, V. R. 1992. Characterization of fumonisin toxicity in orally and intravenously dosed swine. Mycopathologia 117:83-96.

Health and Welfare Canada. 1976. Nutrition Canada Food Consumption Patterns Report. Department of Supply and Services, Ottawa.

Health and Welfare Canada. 1981. Departmental consolidation of the food and drug regulations, with amendments. Department of Supply and Services, Ottawa.

Health and Welfare Canada. 1990. Health Protection Branch Issues: Aflatoxins. Health and Welfare Canada, Ottawa.

Hendrickse, R. G. 1985. Kwashiorkor: 50 years of myth and mystery. Do aflatoxins provide a clue? Acta Leiden. 53:11-30.

Hennig, A., Fink-Gremmels, J., and Leistner, L. 1991. Mutagenicity and effects of ochratoxin A on the frequency of sister chromatid exchange after metabolic activation. Pages 255-260 in: Mycotoxins, Endemic Nephropathy and Urinary Tract Tumours. M. Castegnaro, R. Plestina, G. Dirheimer, I. Chernozemsky, and H. Bartsch, eds. IARC (Int. Agency Res. Cancer) Sci. Publ. 115.

Hong-Wei, G., Qi-Pei, L., Zuo-Han, H., Da-Dao, X., Tanaka, T., Ueno, Y., and Qi-Fu, W. 1989. The contamination of *Fusarium* toxins in wheat and the intake of these toxins by farmers. (In Chinese.) Chin. J. Food Hyg. 1:20-24.

Hsia, C. C., Wu, J. L., Lu, X. Q., and Li, Y. S. 1988. Natural occurrence and clastogenic effects of nivalenol, deoxynivalenol, 3-acetyl-deoxynivalenol, 15-acetyl-deoxynivalenol, and zearalenone in corn from a high-risk area of esophageal cancer. Cancer Detect. Prev. 13:79-86.

Hsu, I. C., Metcalf, R. A., Sun, T., Welsh, J. A., Wang, N. J. and Harris, C. C. 1991. Mutational hotspot in the p53 gene in human hepatocellular carcinomas. Nature 350:427-428.

IARC. 1987. Overall evaluation of carcinogenicity: An updating of IARC monographs volumes 1 to 42. IARC (Int. Agency Res. Cancer) Monogr. Eval. of Carcinogen. Risk Hum. Suppl. 7.

IARC. 1991. Pages 1-336 in: Mycotoxins, Endemic Nephropathy and Urinary Tract Tumours. M. Castegnaro, R. Plestina, G. Dirheimer, I. Chernozemsky, and H. Bartsch, eds. IARC (Int. Agency Res. Cancer) Sci. Publ. 115.

IARC. 1993. Some naturally occurring substances: Food items and constituents, heterocyclic aromatic amines and mycotoxins. World Health Organization. IARC (Int.

Agency Res. Cancer) Monogr. Eval. Carcinog. Risks Hum. 56.

Jaskiewicz, K., Marasas, W. F. O., and Van Der Walt, F. E. 1987a. Oesophageal and other main cancer patterns in four districts of Transkei, 1981-1984. S. Afr. Med. J. 72:27-30.

Jaskiewicz, K., Van Rensburg, S. J., Marasas, W. F. O., and Gelderblom, W. C. A. 1987b. Carcinogenicity of *Fusarium moniliforme* culture material in rats. J. Natl. Cancer Inst. 78:321-325.

Kaneshiro, T., Vesonder, R. F., and Peterson, R. E. 1992. Fumonisin-stimulated N-acetyldihydrosphingosine, n-acetylphytosphingosine, and phytosphingosine products of *Pichia-(hansenula)-ciferri,* NRRL Y-1031. Current Microbiol. 24:319-324.

Kellerman, T. S., Marasas, W. F. O., Thiel, P. G., Gelderblom, W. C. A., Cawood, M., and Coetzer, J. A. W. 1990. Leukoencephalomalacia in two horses induced by oral dosing of fumonisin B_1. Onderstepoort J. Vet. Res. 57:269-275.

Khalef, A., Zidane, C., Charef, A., Gharbi, A., Tadjerouna, M., Betbeder, A. M., and Creppy, E. E. 1993. Ochratoxicose humaine en Algerie. Pages 123-127 in: Human Ochratoxicosis and Its Pathologies. E. E. Creppy, M. Castegnaro, and G. Dirheimer, eds. Colloque INSERM, John Libbey Eurotext, France.

Khera, K. S., Whalen, C., Angers, G., Vesonder, R. F., and Kuiper-Goodman, T. 1982. Embryotoxicity of 4-deoxynivalenol (vomitoxin) in mice. Bull. Environ. Contam. Toxicol. 29:487-491.

Khera, K. S., Arnold, D. L., Angers, G., Scott, P. M., and Whalen, C. 1984. Vomitoxin (4-deoxynivalenol): Effects on reproduction of mice and rats. Toxicol. Appl. Pharmacol. 74:345-356.

Khera, K. S., Whalen, C., and Angers, G. 1986. A teratology study on vomitoxin (4-deoxynivalenol) in rabbits. Food Chem. Toxicol. 24:421-424.

Krewski, D., Murdoch, D., and Dewanji, A. 1986. Statistical modelling and extrapolation of carcinogenesis data. Pages 259-282 in: Modern Statistical Methods in Chronic Disease Epidemiology. S. H. Moolgavkaar and R. L. Prentice, eds. Wiley-Interscience, New York.

Krishnamachari, K. A. V. R., Bhat, R. V., Nagajan, V., and Tilak, T. B. G. 1975. Hepatitis due to aflatoxicosis. Lancet, May 10:1061-1063.

Kuiper-Goodman, T. 1985. Potential human health hazards and regulatory aspects. Pages 103-111 in: Mycotoxins, a Canadian perspective. P. M. Scott, H. L. Trenholm, and M. D. Sutton, eds. NRCC no. 22848. National Research Council, Ottawa.

Kuiper-Goodman, T. 1990. Uncertainties in the risk assessment of three mycotoxins: Aflatoxin, ochratoxin, and zearalenone. Can. J. Physiol. Pharmacol. 68:1017-1024.

Kuiper-Goodman, T. 1991a. Approaches to the risk assessment of mycotoxins: Aflatoxins. Pages 65-86 in: Pennington Center Nutrition Series. Vol. 1, Mycotoxins, Cancer, and Health. G. Bray and D. Ryan, eds. Louisiana State University, Baton Rouge.

Kuiper-Goodman, T. 1991b. Risk assessment to humans of mycotoxins in animal-derived food products. Vet. Hum. Toxicol. 33:325-333.

Kuiper-Goodman, T. 1991c. Risk assessment of ochratoxin A residues in food. Pages 307-320 in: Mycotoxins, Endemic Nephropathy and Urinary Tract Tumours. M. Castegnaro, R. Plestina, G. Dirheimer, I. Chernozemsky, and H. Bartsch, eds. IARC (Int. Agency Res. Cancer) Sci. Publ. 115.

Kuiper-Goodman, T., and Grant, D. L. 1991. Ochratoxin A. Pages 365-417 in: J. Herman, ed. Toxicological evaluation of certain food additives and contaminants. WHO Food Addit. Ser. 28.

Kuiper-Goodman, T., and Scott, P. M. 1989. Risk assessment of the mycotoxin ochratoxin A. Biomed. Environ. Sci. 2:179-248.

Kuiper-Goodman, T., Scott, P. M., and Watanabe, H. 1987 . Risk assessment of the mycotoxin zearalenone. Regul. Toxicol. Pharmacol. 7:253-306.

Kuiper-Goodman, T., Ominski, K., Marquardt, R. R., Malcolm, S., McMullen, E., Lombaert, G. A., and Morton T. 1993. Estimating human exposure to ochratoxin A in Canada. Pages 167-174 in: Human Ochratoxicosis and Its Pathologies. E. E. Creppy, M. Categnaro, and G. Dirheimer, eds. Colloque INSERM, John Libbey Eurotext, France.

Leung, M., Yeung, D. L., Pennell, M., and Hall, J. 1984. Dietary intake of preschoolers. J. Am. Diet Assoc. 84:551-554.

Luo, X. Y. 1988. Outbreaks of moldy cereal poisonings in China. Pages 56-63 in: Issues in Food Safety. Toxicology Forum, Washington, DC.

Luo, Y., Yoshizawa, T., and Katayama, T. 1990. Comparative study on the natural occurrence of *Fusarium* mycotoxins (trichothecenes and zearalenone) in corn and wheat from high- and low-risk areas for human esophageal cancer in China. Appl. Environ. Microbiol. 56:3723-3726.

Marasas, W. F. O., Kriek, N. P. J., Fincham, J. E., and Van Rensburg, S. J. 1984. Primary liver cancer and oesophageal basal cell hyperplasia in rats caused by *Fusarium moniliforme*. Int. J. Cancer 34:383-387.

Marasas, W. F. O., Jaskiewicz, K., Venter, F. S., and Van Schalkwyk, D. J. 1988. *Fusarium moniliforme* contamination of maize in oesophageal cancer areas in Transkei. S. Afr. Med. J. 74:110-114.

Micco, C, Ambruzzi, M. A., Miraglia, M., Brera, C., Onori, R., and Benelli, L. 1991. Contamination of human milk with ochratoxin A. Pages 105-108 in: Mycotoxins, Endemic Nephropathy and Urinary Tract Tumours. M. Castegnaro, R. Plestina, G. Dirheimer, I. Chernozemsky, and H. Bartsch, eds. IARC (Int. Agency Res. Cancer) Sci. Publ. 115.

Nordic Council of Ministers. 1991. Health evaluation of ochratoxin A in food products. Nord. Sem. Arbejdsrapporter 545:1-26.

Norred, W. P., Plattner, R. D., Vesonder, R. F., Bacon, C. W., and Voss, K. A. 1992a. Effects of selected secondary metabolites of *Fusarium moniliforme* on unscheduled synthesis of DNA by rat primary hepatocytes. Food Chem. Toxicol. 30:233-237.

Norred, W. P., Wang, E., Yoo, H., Riley, R. T., and Merrill, A. H. 1992b. In vitro toxicology of fumonisins and the mechanistic implications. Mycopathologia 117:73-78.

NTP (National Toxicology Program). 1982. Carcinogenesis bioassay of zearalenone in F 344/N rats and B6C3F1 mice. Natl. Toxicol. Program Tech. Rep. Ser. 235.

Park, D. L., Rua, S. M., Mirocha, C. J., Abdalla, E. S. A. M., and Cong, Y. W. 1992. Mutagenic potentials of fumonisin contaminated corn following ammonia decontamination procedure. Mycopathologia 117:105-108.

Peto, R., Pike, M. C., Bernstein, L., Gold, L. S., and Ames, B. W. 1984. The TD_{50}: A proposed general convention for the numerical description of the carcinogenic potency of chemicals in chronic-exposure animal experiments. Environ. Health Perspect. 58:1-8.

Pestka, J. J., and Bondy, G. S. 1990. Alteration of immune function following dietary mycotoxin exposure. Can. J. Physiol. Pharmacol. 68:1009-1016.

Pestka, J. J., Tai, J. H., Witt, M. F., Dixon, D. E., and Forsell, J. H. 1987. Suppression of immune response in the B6C3F1 mouse after dietary exposure to the *Fusarium* mycotoxins deoxynivalenol (vomitoxin) and zearalenone. Food Chem. Toxicol. 25:297-304.

Pestka, J. J., Dong, W., Warner, R. L., Rasooly, L., Bondy, G. S., and Brooks, K. H. 1990a. Elevated membrane IgA positive and CD4 positive (T helper) populations in murine Peyer's patch and splenic lymphocytes during dietary administration of the trichothecene vomitoxin (deoxynivalenol). Food Chem. Toxicol. 28:409-420.

Pestka, J. J., Dong, W., Warner, R. L., Rasooly, L., and Bondy, G. S. 1990b. Effect of dietary administration of the trichothecene vomitoxin (deoxynivalenol) on IgA and IgG secretion by Peyer's patch and splenic lymphocytes. Food Chem. Toxicol. 28:693-699.

Pfohl-Leskowicz, A., Chakor, K., Creppy, E. E., and Dirheimer, G. 1991. DNA adduct formation in mice treated with ochratoxin A. Pages 245-253 in: Mycotoxins, Endemic Nephropathy and Urinary Tract Tumours. M. Categnaro, R. Plestina, G. Dirheimer, I. Chernozemsky, and H. Bartsch, eds. IARC (Int. Agency Res. Cancer) Sci. Publ. 115.

Pfohl-Leskowicz, A., Grosse, Y., Obrecht, S., Kane, A., Castegnaro, M., Creppy, E. E., and Dirheimer, G. 1993. Preponderance of DNA-adducts in kidney after ochratoxin exposure. Pages 199-207 in: Human Ochratoxicosis and Its exposure. Pages 199-207 in: Human Ochratoxicosis and Its Pathologies. E. E. Creppy, M. Castegnaro, and G. Dirheimer, eds. Colloque INSERM, John Libbey Eurotext, France.

Prelusky, D. B., Scott, P. M., Trenholm, H. L., and Lawrence, G. A. 1990. Minimal transmission of zearalenone to milk of dairy cows. J. Environ. Sci. Health Part B Pestic. Food Contam. Agric. Wastes 25:87-104.

Riley, R. T., An, N.-H., Showker, J. L., Yoo, H.-S., Norred, W. P., Chamberlain, W. J., Wang, E., Merrill, A. H., Jr., Motelin, G., Beasley, V. R., and Haschek, W. M. 1993. Alteration of tissue and serum sphinganine to sphingosine ratio: An early biomarker of exposure to fumonisin-containing feeds in pigs. Toxicol. Appl. Pharmacol. 118:105-112.

Robbana-Barnat, S., Cohen, H., Loridon, B., Lafarge-Frayssinet, C., and Frayssinet, C. 1985. Heart injury induced by deoxynivalenol a mycotoxin which may contaminate cereals. Sci. Aliment. 5:33-40.

Robbana-Barnat, S., Lafarge-Frayssinet, C., Cohen, H., Neish, G. A., and Frayssinet, C. 1988. Immunosuppressive properties of deoxynivalenol. Toxicology 48:155-166.

Rogers, C. G., and Héroux-Metcalf, C. 1983. Cytotoxicity and absence of mutagenic activity of vomitoxin (4-deoxynivalenol) in a hepatocyte mediated mutation assay with V-79 Chinese hamster lung cells. Cancer Lett. 20:29-36.

Ross, P. F., Rice, L. G., Plattner, R. D., Osweiler, G. D., Wilson, T. M., Owens, D. L., Nelson, H. A., and Richard, J. L. 1991. Concentrations of fumonisin B_1 in feeds associated with animal health problems. Mycopathologia 114:129-135.

Ryan, N. J., Hogan, G. R., Hayes, A. W., Unger, P. D., and Siraji, M. Y. 1980. Role of aflatoxins in Reye's syndrome. Food Cosmet. Toxicol. 18:743-749.

Shier, W. T., Abbas, H. K., and Mirocha, C. J. 1991. Toxicity of mycotoxins fumonisins B_1 and B_2 and *Alternaria alternata* f. sp. *lycopersica* toxin (AAL) in cultured mammalian cells. Mycopathologia 116: 97-104.

Scott, P. M. 1984. Effects of food processing on mycotoxins. J. Food Prot. 47:489-499.

Scott, P. M. 1990. Trichothecenes in grains. Cereal Foods World 35:661-666.

Scott, P. M., Kanhere, S. R., Dexter, J. E., Brennan, P. W., and Trenholm, H. L. 1984. Distribution of the trichothecene deoxynivalenol (vomitoxin) during the milling of naturally contaminated hard red spring wheat and its fate in baked products. Food Addit. Contam. 1:313-323.

Scott, P. M., Lau, P.-Y., and Kanhere, S. R. 1981. Gas chromatography with electron capture and mass spectrometric detection of deoxynivalenol in wheat and other grains. J. Assoc. Offic. Anal. Chem. 64:1364-1371.

Shephard, G. S., Thiel, P. G., Sydenham, E. W., Alberts, J. F., and Gelderblom, W. C. A. 1992a. Fate of a single dose of the ^{14}C-labelled mycotoxin, fumonisin B$_1$, in rats. Toxicon 30:768-770.

Shephard, G. S., Thiel, P. G., and Sydenham, E. W. 1992b. Initial studies on the toxicokinetics of fumonisin B$_1$ in rats. Food Chem. Toxicol. 30:277-279.

Solt, D. B., and Farber, E. 1976. New principle for the analyses of chemical carcinogenesis. Nature 263:702-703.

Steinmetz, K. L., Green, C. E., Bakke, J. P., Spak, D. K., and Mirsalis, J. C. 1988. Induction of unscheduled DNA synthesis in primary cultures of rat, mouse, hamster, monkey, and human hepatocytes. Mutat. Res. 206:91-102.

Sydenham, E. W., Thiel, P. G., Marasas, W. F. O., Shephard, G. S., Van Schalkwyk, D. J., and Koch, K. R. 1990. Natural occurrence of some *Fusarium* mycotoxins in corn from low and high esophageal cancer prevalence areas of the Transkei, southern Africa. J. Agric. Food Chem. 38:1900-1903.

Thiel, P. G., Marasas, W. F. O., Sydenham, E. W., Shephard, G. S., and Gelderblom, W. C. A. 1992. The implications of naturally occurring levels of fumonisins in corn for human and animal health. Mycopathologia 117:3-9.

Trenholm, H. L., Cochrane, W. P., Cohen, H., Elliot, J. I., Farnworth, E. R., Friend, D. W., Hamilton, R. M. G., Standish, J. F., and Thompson, B. K. 1983. Survey of vomitoxin contamination of 1980 Ontario, white winter wheat crop: Results of survey and feeding trials. J. Assoc. Off. Anal. Chem. 66:92-97.

Tryphonas, H., O'Grady, L., Arnold, D. L., Mcguire, P. F., Karpinski, K., and Vesonder, R. F. 1984. Effect of deoxynivalenol (vomitoxin) on the humoral immunity of mice. Toxicol. Lett. 23:17-24.

Tryphonas, H., Iverson, F., So, Y., Nera, E. A., Mcguire, P. F., O'Grady, L., Clayson, D. B., and Scott, P. M. 1986. Effects of deoxynivalenol (vomitoxin) on the humoral and cellular immunity of mice. Toxicol. Lett. 30:137-150.

Ueno, Y. 1984. Toxicological features of T-2 toxin and related trichothecenes. Fundam. Appl. Toxicol. 4:S124- S132.

van Egmond, H. P. 1989. Current situation on regulations for mycotoxins. Overview of tolerances and status of standard methods of analysis. Food Addit. Contam. 6:139-188.

Verdeal, K., and Ryan, D. S. 1979. Naturally-occurring estrogens in plant foodstuffs— A review. J. Food Protect. 42:577-583.

Wang, E., Norred, W. P., Bacon, C. W., Riley, R. T., and Merrill, A. H. 1991. Inhibition of sphingolipid biosynthesis by fumonisins: Implications for diseases associated with *Fusarium moniliforme*. J. Biol. Chem. 26:14486-14490.

Wang, E., Ross, P. F., Wilson, T. M., Riley, R. T., and Merrill, A. H., Jr. 1992. Increases in serum sphingosine and sphinganine and decreases in complex sphingolipids in ponies given feed containing fumonisins, mycotoxins produced by *Fusarium moniliforme*. Am. Inst. Nutr. 122:1706-1716.

Wehner, F. C., Marasas, W. F. O., and Thiel, P. G. 1978. Lack of mutagenicity to *Salmonella typhimurium* of some *Fusarium* mycotoxins. Appl. Environ. Microbiol. 35:659-662.

WHO. 1991. Evaluation of certain food additives and contaminants. WHO Tech. Rep. Ser. 806:29-31.

Wild, C. P., Jiang, Y. Z., Sabbioni, G., Chapot, B., and Montesano, R. 1990. Evaluation of methods for quantitation of aflatoxin-albumin adducts and their application to human exposure assessment. Cancer Res. 50:245-51.

Wilson, T. M., Ross, P. F., Rice, L. G., Osweiler, G. D., Nelson, H. A., Owens, D. L.,

Plattner, R. D., Reggiardo, C., Noon, T. H., and Pickrell, J. W. 1990. Fumonisin B$_1$ levels associated with an epizootic of equine leukoencephalomalacia. J. Vet. Diagn. Invest. 2:213-216.

Wilson, T. M., Ross, P. F., Owens, D. L., Rice, L. G., Green, S. A., Jenkins, S. J., and Nelson, H. A. 1992. Experimental reproduction of ELEM—A study to determine the minimum toxic dose in ponies. Mycopathologia 117:115-120.

Wogan, G. N., Paglialunga, S., and Newberne, P. M. 1974. Carcinogenic effects of low dietary levels of aflatoxin B$_1$ in rats. Food Cosmet. Toxicol. 12:681-685.

Yeh, F. S., Yu, M. C., Mo, C. C., Luo, S., Tong, M. J., and Henderson, B. E. 1989. Hepatitis B virus, aflatoxins, and hepatocellular carcinoma in southern Guangxi, China. Cancer Res. 49:2506-2509.

Yeung, D. L., Leung, M., Hall, J., and Medina, D. 1979. Longitudinal survey of infant nutrition. J. Can. Diet. Assoc. 40:1-288.

Yoo, H. S., Norred, W. P., Wang, E., Merrill, A. H., and Riley, R. T. 1992. Fumonisin inhibition of de novo sphingolipid biosynthesis and cytotoxicity are correlated in LLC-PK1 cells. Toxicol. Appl. Pharmacol. 114:9-15.

Young, J. C., Fulcher, R. G., Hayhoe, J. H., Scott, P. M., and Dexter, J. E. 1984. Effect of milling and baking on deoxynivalenol (vomitoxin) content of eastern Canadian wheats. J. Agric. Food Chem. 32:659-664.

Chapter 13

Factors Responsible for Economic Losses due to *Fusarium* Mycotoxin Contamination of Grains, Foods, and Feedstuffs

L. L. Charmley, A. Rosenberg, and H. L. Trenholm

The Food and Agriculture Organization of the United Nations estimates that 25% of the world's food crops are affected by mycotoxins each year (Mannon and Johnson 1985). Although the mycotoxins produced by *Fusarium* species of fungi account for only a portion of this contamination, the economic impact of *Fusarium* fungal infestation can be substantial. The presence of mycotoxins in corn and small-grained cereals may have serious economic implications for crop, livestock, and poultry producers, grain handlers, and food and feed processors. *Fusarium* mycotoxins, although of widespread occurrence globally, tend to be more prevalent in the more temperate regions of the world. *Fusarium* mycotoxin contamination has been found to occur in wheat, corn, other cereals, and, in a few instances, in food products in certain regions of several countries including South Africa, Zambia, China, India, Poland, the Netherlands, northwest Spain, New Zealand, the United States, and Canada. Economically, in Canada and the northern United States, the most important mycotoxins are zearalenone (ZEN), and the trichothecene mycotoxins deoxynivalenol (DON), T-2 toxin, and HT-2 toxin. Other mycotoxins may be important also, but their impact on the agriculture and food industries is not known. In other countries, aflatoxins, ochratoxins, and other mycotoxins have more serious economic influences than the *Fusarium* mycotoxins.

Economic losses caused by *Fusarium* fungi and mycotoxin contamination originate from reduced crop and animal production, and from the cost of programs designed to monitor and regulate mycotoxin concentrations in crops to minimize health risks to humans and animals. Direct economic losses result from lower crop yields, poorer grain quality, reduced animal performance and reproductive capability, and increased disease incidence. Indirect economic losses, although more difficult to quantify, still represent major economic losses. Crop producers may receive lower revenues from reductions in the nutritive value and downgrading of grain crops, restricted markets, and, in extreme cases, the production of an unmarketable feed or food product. Added costs to feed and food processors derive from extra handling, distribution, and processing requirements, chemical analysis to monitor mycotoxin contamination, and in extreme cases, the disposal of contaminated grain. Indirect losses for livestock and poultry producers derive from increased production costs associated with searching for an uncontaminated grain supply, decontamination or dilution of contaminated grain with clean grain, and increased veterinary bills. During seasons of extensive contamination, prices may be higher for quality feeds, but net returns for livestock products are lower. The overall economic result for crop and animal producers is that output is reduced, production costs increase, and returns are reduced. Costs are passed on to consumers as higher commodity prices and, if regulations fail, health risks. Added to the costs for crop and livestock producers and feed and food processors are the social costs of mycotoxins to the national economy. These include the cost of research and development, monitoring and regulation, extension services, legal suits, loss of consumer confidence in the product, and health care and social welfare costs. Regulations set by different countries to limit mycotoxin concentrations of imported grain may serve as nontarrif barriers in the international markets for grain products. The following list categorizes the economic costs of *Fusarium* mycotoxins at the various levels of food and feed production, as well as costs to the consumer and the national economy.

Costs to grain producers: Reduced yields; reduced nutritive value; downgraded or nonmarketable product; restricted domestic and foreign markets; increased processing costs (drying, monitoring, detoxification); increased transportation costs; disposal of nonmarketable crops; possible health risks to grain harvesters and handlers.

Costs to animal producers (livestock and poultry): Higher feed costs; reduced animal growth and reproductive performance; increased disease incidences; unsafe mycotoxin residues in animal products; loss of markets; increased monitoring, testing, and detoxification costs; possible health risks to farm staff.

Costs to handlers and distributers: Increased processing costs (drying, detoxification); costs of mycotoxin monitoring; unused storage capacity; increased transportation costs; loss of markets; costs of law suits initiated by livestock producers receiving contaminated grain; health risks to handlers and distributors.

Costs to processors: Restricted product markets; product loss; monitoring and testing costs.

Costs to consumers: Higher product prices; loss of income and productivity if health problems occur due to consumption of contaminated food products; long-term chronic effects of continued low-level contamination.

Costs to society: Cost of regulatory procedures and quality control (establishment of maximum permissible and safe levels, monitoring, and enforcement); costs of research and extension; costs of information and training services (how to minimize extent of contamination, how to recognize the problem, detoxification methods); lower foreign exchange earnings due to loss of export markets; increased import costs.

Despite the obvious economic importance of mycotoxins, losses cannot be quantified adequately. However, in terms of reduced financial returns or increased costs for grain, livestock, and poultry producers and from the overall extent of *Fusarium* infestation and mycotoxin contamination, it was estimated that multimillion dollar losses due to mycotoxin contamination of grain and feedstuffs may occur annually in Canada (Trenholm et al 1985b). The purpose of this review is to identify and, whenever possible, quantify the economic losses due to mycotoxin contamination.

Effects of *Fusarium* Mycotoxins on Grain Crops

Fusarium fungal growth and mycotoxin production, to a large extent, are dependent on the prevailing weather conditions (James and Smith 1982, James 1987, Tubbs and Dekich 1989). Therefore, the disease incidence is sporadic and varies in severity from year to year (Nelson et al 1981, Sutton 1982) and from region to region (Andrews et al 1981, Jelinek et al 1989). Due to these variations, predicting financial costs to producers in any particular season is extremely difficult. Other factors also influence the degree of fungal and mycotoxin contamination of a crop. These include the susceptibility of a particular plant species to fungal invasion and mycotoxin contamination, which seems to have increased as a result of selectively breeding varieties for higher yield potential and disease resistance rather than for fungal and mycotoxin resistance (CAST 1989, Pomeranz et al 1990); the stages of growth at which the plant is exposed to fungi; crop management techniques (including harvesting techniques and storage and milling practices); and cultural practices (including rotation and tilling practices).

Fusarium infection of corn or grain results in reduced crop yields. For example, crop yields of wheat were reduced by an average of 30% in Germany in 1980 and 1981 due to reduced size and number of kernels caused by *Fusarium* (Meyer et al 1985). The presence of *Fusarium* fungi on wheat kernels reduces germination rates and seedling vigor to such a degree that it has been recommended that even lightly contaminated grain should not be used for seed stock (Bechtel et al 1985). *Fusarium* infestation also is associated with the production of inferior quality grain (Shotwell et al 1985, Bechtel et al 1985). It affects the physical qualities of grain and the chemical and technological qualities of grain and flour. The breadmaking potential (Berova and Mladenov 1974, Bockmann 1984) and the technological and baking qualities of wheat are substantially impaired if as little as 15% of the grain is contaminated with *Fusarium* (Berova and Mladenov 1974). Mycotoxin contamination may lead to downgrading of the grain and its diversion from use in quality foods to animal feed products. It may necessitate the use of additional cleaning and milling procedures to remove some of the contamination, finding an alternative use for the crop such as ethanol production, or it may lead to total rejection of the grain on delivery to a grain elevator or by the feed mill. Some kernels may contain very high concentrations of DON (2,000 to 3,000 mg/kg of grain). Therefore, less than 1% of these highly contaminated kernels would be necessary to render the entire crop's mycotoxin content above the recommended safe level. In instances of extremely high and extensive mold and mycotoxin contamination, it may be more economical for the crop producer to plough the crop back into the ground rather than to harvest it at all. Furthermore, improper drying and storage of lightly contaminated grain can lead to increased mycotoxin contamination. Depending on the nature and extent of contamination, a reduction in cash income with a concomitant increase in expenditure (to deal with contaminated grain) may be incurred by crop producers.

ECONOMIC EFFECTS OF MYCOTOXINS FOR GRAIN PRODUCERS

The impact of mycotoxins on the net revenues of grain producers can be predicted using the following equation:

$$\text{NCR} = [(100 - X)/100 \times \text{AP} \times \text{PP}] + (X/100 \times \text{AP} \times \text{PC}) \qquad (1)$$

where NCR is the net cash receipts; X is the percentage of the crop that is contaminated; AP is the annual grain production in metric tons per year; PP is the average price of premium grain in dollars per metric ton; and PC is the average price of contaminated grain in dollars per metric ton.

Net loss of cash income is estimated by calculating NCR in equation 1 when X is 0, and then subtracting the actual NCR.

Equation 1 does not account for losses caused by reduced yields in fungus-infected crops and production of unmarketable crops and so in some instances will underestimate the actual economic losses incurred.

CORN PRODUCTION

Many cultivars of corn worldwide are very sensitive to *Fusarium* infestation and mycotoxin contamination. In Canada, a large percentage of the corn production area in southern Ontario is adjacent to the Great Lakes and so experiences periods of high humidity and excessive rainfall at times conducive to mycotoxin contamination. Surveys of the corn crop grown in eastern and southern Ontario and western Quebec, in 1986, demonstrated that the corn hybrids used were susceptible to maize ear rot and mycotoxin contamination. In fact, throughout eastern Canada, serious outbreaks of *Fusarium graminearum* infestation have occurred in corn, and the lack of resistant, high-yielding varieties has had serious consequences on domestic corn markets and the revenues of corn producers. Although hot dry summer periods may encourage fumonisin contamination and the related problems to livestock, surveys of fumonisin contamination of grain have not been completed. Corn grown in southern Ontario is mainly used in animal feeds and is not of food grade.

The Food Protection and Inspection (FPI) Branch of Agriculture Canada conducted a survey, in the fiscal year 1989-1990, to monitor the presence of certain mycotoxins in corn and corn-containing feeds from farms across Canada. Their findings are shown in Table 13.1. Although concentrations of DON, ZEN, T-2 toxin, and HT-2 toxin were generally low, in some cases they would have been high enough to cause problems if such corn were fed to susceptible species, such as pigs, and particularly if combinations of toxins were present.

In 1990, an epidemic of Gibberella ear rot, caused by *Fusarium graminearum*, occurred in the corn crop grown in New York and in other locations in the northeastern and north central United States and eastern Canada. As a consequence, the corn from these regions was found to be extensively contaminated with low to moderate levels of DON and, in a few cases, also with low concentrations of ZEN (Bergstrom 1991). Because of this contamination, during the late winter of 1990 and spring and summer of

TABLE 13.1
Mycotoxin Concentration in Feed Samples
from Farms Across Canada in Fiscal 1989–1990[a]

Mycotoxin	Samples Positive (%)	Concentration (mg/kg)			Commodity Most Highly Contaminated
		Mean	Minimum	Maximum	
Deoxynivalenol	53.9	0.40	0.10	2.96	21% Corn gluten feed
Zearalenone	12.7	0.07	0.03	0.32	Whole corn
T-2	2.0	0.37	0.31	0.43	Whole corn
HT-2	4.9	0.36	0.20	0.64	Corn

[a] Data were obtained from random samples taken by the Food Protection and Inspection Branch of Agriculture Canada in 1990.

1991, several major grain buyers either stopped purchasing corn from certain regions of New York, purchased only certified DON-free corn (containing less than 0.5 mg/kg of DON), or purchased DON-contaminated (0.5 to 1.0 mg/kg of DON) corn at substantially discounted prices. Corn containing more than 1.0 mg/kg could not be marketed at all. Representatives of the New York Corn Growers Association estimated, conservatively, that approximately $12 million had been lost by grain sellers due to discounted prices, and nearly 40% of the 1990 crop in western and central New York had not been sold.

WHEAT PRODUCTION

In southern Ontario, winter wheat is another crop that has suffered from episodes of infestation with *Fusarium* fungi. Most of the wheat grown in this region is used in specialized food products such as cookies and biscuits rather than in staple food products, such as bread and cereals. Canadian food regulatory agencies ensure that levels of DON in winter wheat are maintained below a certain level (1 and 2 mg/kg in soft wheat for infant and adult foods, respectively). When these DON concentrations are exceeded, the crop cannot be used in food products for human consumption, and the wheat is downgraded to a lower value, feed grade, by millers. This feed, in turn, may be fed to livestock in accordance with feed regulatory guidelines depending on the safety and nutritional value of the grain and feedstuff. Crop insurance data suggest that 35% of winter wheat losses in Canada are due to contamination by mold and mycotoxins. In 1980, when Ontario winter wheat was extensively infested with *Fusarium*, producers lost domestic sales, shiploads of wheat were embargoed, and the wheat in storage had to be repositioned, all at additional expense. To maintain the international good reputation of Canadian grain, Health and Welfare regulatory officials stopped the exportation of winter wheat containing more than 2 mg/kg of DON and the use of such wheat in food products for human consumption. In addition, Agriculture Canada officials have advised that contaminated wheat should be blended 90% with uncontaminated grain for animal feeds. Many Canadian food industries subsequently have set very strict standards to ensure that mycotoxin-contaminated grain is not used in their food products.

Most of the spring wheat in Canada for human consumption is produced in the Prairie Provinces (Manitoba, Saskatchewan, and Alberta), where the environment is very dry and where there is little fungal infestation. The bulk of grain for export, which is under strict regulation regarding permitted mycotoxin concentrations, is produced in the Prairies. If this grain were to become contaminated with mycotoxins it could have a drastic impact on Canadian wheat export markets and concomitantly on the Canadian grain production industry.

During seasons of extensive mycotoxin contamination, grain shortages may ensue that, in turn, lead to elevated grain prices and consequentially to increased costs for livestock and poultry producers and for consumers of grain products. Additional expenses due to analytical monitoring to ensure mycotoxin concentrations do not exceed the maximum permissible levels are also added to the cost of the grain.

Effects of Mycotoxins on Animals and Humans

The consumption of mycotoxin-contaminated feeds can cause a wide variety of signs in farm animals depending on the nature and concentration of mycotoxin present, the duration of exposure to contaminated feed, and the animal species and its nutritional and health status prior to exposure. Signs attributable to *Fusarium* mycotoxins include reduced feed intake and feed conversion efficiency, vomiting, diarrhea, reduced weight gains or weight loss, reproductive problems, immunosuppression, organ damage, and death. Overt toxicosis, morbidity, and death occur in only a few cases of mycotoxicoses in animals and humans, and most economic losses result from lost production and increases in the incidence of other diseases (Thompson 1991). The fumonisins, a relatively recently discovered group of *Fusarium* mycotoxins, have been associated with several syndromes and mycotoxicoses in a variety of animal species. These include pulmonary edema syndrome in pigs (Ross et al 1990, 1991), leukoencephalomalacia in horses (Marasas et al 1988; Plattner et al 1990; Kellerman et al 1990; Ross et al 1990, 1991; Thiel et al 1991), and cancer in other species (Gelderblom et al 1988).

In humans, ingestion of mycotoxin-contaminated food may lead to increased incidence of infections and disease (due to immunosuppression), poor performance at work, and in extreme cases death. Prolonged exposure to mycotoxins may lead to cancer-related deaths. Farmers and grain handlers also are at risk from contamination if they do not protect themselves adequately from skin contact and inhalation of moldy grain and mycotoxin-contaminated dust (Trenholm et al 1988). Lung diseases, such as "farmers lung," may be due to prolonged inhalation of mycotoxin-contaminated dusts.

EFFECTS OF *FUSARIUM* MYCOTOXINS ON SWINE

Swine are extremely sensitive to the presence of *Fusarium* mycotoxins in their diet. The majority of documented cases of mycotoxicoses in swine have been due to the presence of DON or DON and ZEN in the feed. It is quite common for a feedstuff contaminated with DON to contain variable concentrations of ZEN and other mycotoxins because both DON and ZEN are produced by several common species of *Fusarium* fungi

(Marasas et al 1984, Bottalico et al 1988, Abdelhamid 1990, Sydenham et al 1991). DON causes reduced feed intake, feed refusal, and, at high concentrations in the diet, vomiting in swine (Pier 1981, Young et al 1983, Lun et al 1985, Trenholm et al 1988). All these effects lead to reduced body weight gains (or even body weight losses) and hence a delay in the time taken for pigs to reach market weight. Since the new Canadian grading system for hogs strongly discourages marketing animals with carcass weights of less than 70 kg, and increases focus on "core" weight areas of 75 to 85 kg (Pelletier 1992), a delay in time taken to reach this weight or marketing pigs at a lower body weight would have major economic consequences for producers. A more subtle effect of DON is immunosuppression, which in turn may lead to increased incidence of infections and diseases (Vanyi and Sandor 1988). There is some speculation that the "swine reproductive failure syndrome"—characterized by high neonatal mortality rates, with increased stillbirths and fetal mummification, and high preweaning mortalities—may be due to the presence of mycotoxins in the diet and, in particular, DON. DON has been associated with abortions, stillbirths, and weak piglets (Reotutar 1989).

ZEN affects reproduction in swine (James and Smith 1982, Friend and Trenholm 1988). Prepubertal gilts ingesting moldy corn containing ZEN exhibit signs of hyperestrogenism, including swollen vulva, mammary development, and, in severe cases, vaginal and rectal prolapse. Mature sows and gilts are less susceptible than prepubertal gilts to the adverse effects of ZEN. However, reproductive problems such as pseudopregnancy, anestrus, constant estrus, and production of weak or stillborn piglets have been associated with the ingestion of *Fusarium*-contaminated feedstuffs by mature sows and gilts (Mirocha et al 1977, Long et al 1983). Not only is the reproductive efficiency decreased due to failed pregnancy, but in most cases it will take several months for females to return to estrus and be able to be bred again (Cantley et al 1982). ZEN-contaminated diets may adversely affect prepubertal boars, possibly by causing decreases in the blood concentrations of testosterone (Berger et al 1981). Therefore, the effects of ZEN on the reproductive performance of swine would be expected to cause a disruption in the breeding program and so have a major influence on profit returns for the producer, particularly in farrow-to-finish operations. Pigs produced per sow per year and the number of times a sow is farrowed each year are the two most important financial aspects of a farrow-to-finish operation. Moreover, the number of pigs produced per sow per year is of utmost importance for determining the profit margins of farrow-to-finish and feeder pig enterprises (Bunn 1992; Table 13.2). Producers need to maintain a steady production cycle, because a lull in breeding disrupts the process, and the losses incurred are very difficult to recover. If gilts and sows are fed a ZEN-contaminated diet, they may show signs of pregnancy but ultimately not farrow. This may, in turn, result in the culling of normally reproductively efficient animals whose reproductive failure is entirely due to the presence of ZEN in their feed. The presence of ZEN in corn in Ontario resulted in the enforcement of new management practices in the hog industry. Producers now are reluctant to feed home-grown corn to breeding sows and gilts. This has led to a disruption of the total integration system and has forced the purchase of grain from unaffected areas such as Saskatchewan and Alberta, where there is no history of major mycotoxin contamination (Andrews et al 1981), or the importation of corn from the United States at great expense. This in itself has increased costs drastically for Ontario hog producers because there are, unfortunately, no viable crop alternatives grown in Ontario. Due to the extreme sensitivity of swine to both DON and ZEN and the impact that these mycotoxins can have on the profit returns of swine producers, it is recommended that each new batch of feed be tested on a small number of prepubertal gilts before being fed to the farrowing herds (Trenholm et al 1988). Moreover, it is recommended that such feed be diluted or treated before being used if mycotoxin contamination is suspected.

ECONOMIC IMPACT OF MYCOTOXINS FOR SWINE PRODUCERS

A model was developed to assess the theoretical economic impact of *Fusarium* mycotoxin-contaminated feed on a farrow-to-finish pig operation in Ontario (A. Rosenberg, *unpublished*; Table 13.3). The size of the operation and production parameters were defined following the guidelines of the National Tripartite Stabilization Program model for hogs. The control, as defined in this model, consists of a 130-sow farrow-to-finish operation marketing 2,251 hogs per year. The model calculates revenues and cash costs of production based on 1991 prices. Mycotoxins were assumed to reduce daily body weight gains and prolong the interval between farrowing. The model relates a worst-case scenario because both growth and

TABLE 13.2
Estimated Net Return to Labor
for Different Numbers of Pigs Marketed Per Sow[a]

| | Estimated Net Return to Labor[b] | |
Pigs Marketed per Sow per Year	Feeder Pig Enterprise	Farrow-to-Finish Enterprise
16	17,472	33,152
18	26,226	46,332
20	34,460	68,540
22	43,714	72,644
24	52,464	85,848

[a] Adapted from Bunn 1992.
[b] Figures were calculated (not including stabilization premiums or payouts) for a 100-sow enterprise using 1991 average input costs and returns in dollars.

reproduction are influenced simultaneously, implying that a mixture of toxins (probably DON and ZEN) are present. From the model it was calculated that a 10 or 20% reduction in farrowing rate, combined with a 10 or 20% reduction in growth, can cause an approximately 17 to 44% reduction in profit margins. The reduction in margins is explained by increased feeding and veterinary costs per head as well as a decrease in the number of hogs marketed.

EFFECTS OF *FUSARIUM* MYCOTOXINS ON POULTRY

Poultry are extremely tolerant of the presence of DON in their diet, which even at relatively high concentrations (18 mg/kg) causes few adverse effects in hens (Kubena et al 1987). In addition, there appears to be little transmission of DON into eggs (Prelusky et al 1987, Moran et al 1987). Poultry are somewhat more sensitive to the more toxic T-2 toxin, which at a level of 8 mg/kg of diet causes reductions in feed consumption, egg production, shell thickness, and hatchability in laying hens (James 1987). Moreover, consumption of feed containing a combination of toxins as occurs in naturally contaminated feeds (Mannion and Blaney 1988) or that has been experimentally controlled (Kubena et al 1989) has greater adverse

TABLE 13.3
Theoretical Economic Impact of *Fusarium* Mycotoxins
on a Farrow-to-Finish Operation, Using 1991 Prices[a]

Revenue and Cost Structure ($/year)	Control	Farrowing Rate −10%		Farrowing Rate −20%	
		Daily Gain −10%	Daily Gain −20%	Daily Gain −10%	Daily Gain −20%
Revenue					
Sales	269,156	242,232	228,134	215,427	215,427
Tripartite payment	19,995	17,987	16,802	15,989	15,989
Cost structure					
Feed	134,816	123,009	119,887	117,304	120,321
Tripartite premium	8,301	7,467	6,975	6,638	6,638
Repairs and maintenance	10,636	10,636	10,636	10,636	10,636
Utilities	11,306	11,306	11,306	11,306	11,306
Property taxes	940	940	940	940	940
Insurance	2,954	2,954	2,954	2,954	2,954
Replacements	2,150	2,150	2,150	2,150	2,150
Veterinary and health	10,782	10,670	11,038	9,915	10,346
Labor	6,989	6,989	6,989	6,989	6,989
Marketing	2,579	2,320	2,179	2,062	2,062
Interest	5,504	5,130	5,033	4,913	5,012
Cost total	196,957	183,571	180,087	175,807	179,354
Cash margin	92,194	76,648	64,849	55,609	52,062
Difference from control, %	...	−16.9	−29.7	−39.7	−43.5

[a] Data from A. Rosenberg, *unpublished*.

effects on the growth rate of poultry than the effects observed for each of the toxins in isolation.

On some New York poultry farms, decreased egg production, poor shell quality, lesions in the mouth, leg problems, prolapse, cage layer fatigue, inferior albumin quality, variations (increases and decreases) in feed consumption, inferior growth, and higher mortality were reported in poultry consuming *Fusarium*-contaminated feeds (Keshavarz 1991). However, these field observations did not establish a positive relationship between the concentration of DON in the finished feed (between 0.5 and 10 mg/kg) and the severity of the production losses. As research results support the claim that poultry are relatively tolerant to the presence of DON in their diet, it may be assumed that the adverse effects of the *Fusarium*-contaminated feed were due to the presence of other unidentified mycotoxins or due to a synergistic effect between DON and unidentified mycotoxins.

EFFECTS OF *FUSARIUM* MYCOTOXINS ON CATTLE

Cattle also are quite tolerant of the presence of *Fusarium* mycotoxins in their diet (Trenholm et al 1985a, Cote et al 1986). However, clinical signs, including feed refusal, diarrhea, decreased growth and milk production, lowered immune response, and abnormal estrous cycles have been reported in cattle consuming *Fusarium*-contaminated diets. In addition, cattle were found to exhibit slight reductions in feed consumption when fed a wheat-oat diet containing 6 mg/kg of DON at a level of 1% of body weight (Trenholm et al 1985b). Cows receiving diets containing 200 mg/kg of DON exhibited a reduction in milk production of 0.45 kg per day and also an eight-day delay in breeding (Behlow 1986). Similarly, in a survey of commercial dairy farms in North Carolina, milk production was found to decline in herds as the concentration of DON in the diet increased (Whitlow et al 1986, Whitlow and Hagler 1987). At a dietary level of 0.8 mg/kg of DON dry matter, milk production was reduced by as much as 2 kg per day.

As the effects of *Fusarium* mycotoxins on poultry and cattle, in general, are minor, no extensive economic assessments have been made.

EFFECTS OF *FUSARIUM* MYCOTOXINS ON HUMANS

The presence of numerous mycotoxins from different species of *Fusarium* in certain types of grain is a matter of some concern from a public health standpoint (Jelinek et al 1989). DON concentrations are under strict regulatory control in food in Canada and the United States. The Canadian government has set guidelines of 1 and 2 mg/kg of DON in uncleaned soft wheat destined for use in nonstaple foods for infants and adults, respectively (Kuiper-Goodman 1985). Outbreaks of *Fusarium* head blight in China and India caused severe human health problems (Bhat et al 1989, Luo 1988). Epidemiological studies of esophageal cancer in Transkei and parts of China revealed a link between this disease and exposure to *Fusarium moniliforme*. This fungus produces a variety of toxins including fusarin C and fumonisins, which are mutagens and neural toxins (Gelderblom et al 1988, Marasas et al 1988). Ingestion of food contaminated with *Fusarium* mycotoxins also has been associated with other diseases such as alimentary toxic aleukia, a fatal disease of animals and humans, and akakabi (red mold disease), characterized by nausea, vomiting, diarrhea, feed refusal, congestion, hemorrhages, and destruction of bone marrow.

If humans are exposed to *Fusarium* mycotoxins—either in their food supply or by working with moldy substances without adequate protection—and suffer health problems, the economic impact of this is felt by the country as a monetary drain on the medical and welfare systems as well as by the individual as a loss of income.

Summary

Fusarium mycotoxin contamination of grain and cereal crops can cause major economic losses at all levels of food and feed production (including crop production, distribution, and processing, and animal production) and on the national economy as a whole. Losses could amount to millions of dollars each year in a grain- and livestock-producing country. The economic costs of mycotoxin contamination of crops are not limited to crop and livestock production losses, but also include effects on other sectors of society. The national economy is affected by losses incurred by grain and livestock producers and the multiplier effect this has on other industries as a result of the lower spending power of the producers. This multiplier effect may be several orders of magnitude greater than the value of the direct losses incurred by crop and livestock producers. The costs of chemical analyses and quality control progams, research and development, extension services, law suits and out-of-court settlements, regulatory expenses, and the costs of human illness and medical treatments must all be borne by the national economy. Of course the actual value of the losses encountered each year depends on several factors, such as grain, animal, and animal product prices; interest rates; the degree and extent of contamination, which would depend on the prevailing weather conditions; and many other economic variables. In Canada, grain and livestock production represent a $10 billion industry, with food industries representing additional revenues. Even during the more favorable seasons, it seems likely that millions of dollars are lost from the Canadian economy as a result of regions of mycotoxin contamination of grain and corn crops, and Canada is not unique. Worldwide, *Fusarium* mycotoxins cause major economic losses in agricultural and food industries and have serious effects on human health and the quality of life of populations that have no choice but to eat mycotoxin-contaminated grain and food.

Literature Cited

Abdelhamid, A. M. 1990. Occurrence of some mycotoxins (aflatoxin, ochratoxin A, citrinin, zearalenone and vomitoxin) in various Egyptian feeds. Arch. Anim. Nutr. 40:647-664.

Andrews, R. I., Thompson, B. K., and Trenholm, H. L. 1981. A national survey of mycotoxins in Canada. J. Am. Oil Chem. Soc. 989A-991A.

Bechtel, D. B., Kaleikau, L. A., Gaines, R. L., and Seitz, L. M. 1985. The effects of *Fusarium graminearum* infection on wheat kernels. Cereal Chem. 62:191-197.

Behlow, R. F. 1986. Deoxynivalenol said costing millions. Poult. Dig. 45(538):512-513.

Berger, T., Esbenshade, K. L., Diekmann, M. A., Hoagland, T., and Tuite, J. 1981. Influence of prepubertal consumption of zearalenone on sexual development of boars. J. Anim. Sci. 53:1559.

Bergstrom, G. C. 1991. Situation in New York State. Pages 1-2 in: Plant Pathology Extender. *Fusarium* Molds and Mycotoxins Associated with Corn. G. C. Bergstrom and L. J. Thompson, eds. Plant Pathol. Ext. Rep. 91-1. Cornell Cooperative Extension, Ithaca, NY.

Berova, S., and Mladenov, M. 1974. Influence of wheat ear and grain fusariosis (*Fusarium graminearum* (Schwabe)) on the chemical, technological and baking qualities. Rastenievud. Nauki 11:125-133.

Bhat, R. V., Beedu, S. R., Ramakrishna, Y., and Munshi, K. L. 1989. Outbreak of trichothecene toxicosis associated with consumption of mould-damaged wheat products in Kashmir Valley, India. Lancet 8628:35-37.

Bockmann, H. 1984. Qualitat und backfahigkeit von weizen bei befall mit septoria nodorum berk und *Fusarium culmorum*. Nachrichtenbl. Dtsch. Pflanzenschutzdienstes (Braunschweig) 16:5-10.

Bottalico, A., Logrieco, A., and Visconti, A. 1988. Mycotoxins produced by *Fusarium crookwellense* from plant products. Proc. Jpn. Assoc. Mycotoxins Suppl. 1:227-228.

Bunn, A. J. 1992. Pork News and Views, July/August, Ontario Ministry of Agriculture and Food, Animal Industry Branch.

Cantley, T. C., Redmer, D. A., Osweiler, G. D., and Day, B. N. 1982. Effect of zearalenone mycotoxin on luteal function in gilts. J. Anim. Sci. 55(Suppl. 1):104.

CAST. 1989. Mycotoxins: Economic and Health Risks. Rep. 116. Council for Agricultural Science and Technology, Ames, IA.

Cote, L. M., Dahlem, A. M., Yoshizawa, T., Swanson, S. P., and Buck, W. B. 1986. Excretion of deoxynivalenol and its metabolite in milk, urine and feces of lactating dairy cows. J. Dairy Sci. 69:2416-2423.

Friend, D. W., and Trenholm, H. L. 1988. Mycotoxins in pig nutrition. Pig News Inf. CAB Int. 9:395-401.

Gelderblom, W. C. A., Jaskiewicz, K., Marasas, W. F. O., Thiel, P. G., Horak, R. M., Vleggaar, R., and Kriek, N. P. J. 1988. Fumonisins—Novel mycotoxins with cancer-promoting activity produced by *Fusarium moniliforme*. Appl. Environ. Microbiol. 54:1806-1811.

James, L. 1987. Mycotoxins and poultry performance. Chic Chat 3:1-2.

James, L. J., and Smith, T. K. 1982. Effect of dietary alfalfa on zearalenone toxicity and metabolism in rats and swine. J. Anim. Sci. 55:110-118.

Jelinek, C. F., Pohland, A. E., and Wood, G. E. 1989. World wide occurrence of mycotoxins in foods and feeds—An update. J. Assoc. Off. Anal. Chem. 72:223-230.

Kellerman, T. S., Marasas, W. F. O., Thiel, P. G., Gelderblom, W. C. A., Cawood, M., and Coetzer, J. A. W. 1990. Leukoencephalomalacia in two horses induced by oral dosing of fumonisin B_1. Onderstepoort J. Vet. Res. 57:269-275.

Keshavarz, K. 1991. Mycotoxicosis in poultry. Pages 3-4 in: Plant Pathology Extender. *Fusarium* Molds and Mycotoxins Associated with Corn. G. C. Bergstrom and L. J. Thompson, eds. Plant Pathol. Ext. Rep. 91-1. Cornell Cooperative Extension, Ithaca, NY.

Kubena, L. F., Harvey, R. B., Corrier, D. E., Huff, W. E., and Phillips, T. D. 1987. Effects of feeding deoxynivalenol (DON, vomitoxin)-contaminated wheat to female white leghorn chickens from day old through egg production. Poult. Sci. 66:1512-1518.

Kubena, L. F., Huff, W. E., Harvey, R. B., Phillips, T. D., and Rottinghaus, G. E. 1989. Individual and combined toxicity of deoxynivalenol and T-2 toxin in broiler chicks. Poult. Sci. 68:622-626.

Kuiper-Goodman, T. 1985. Potential human health hazards and regulatory aspects. Pages 103-111 in: Mycotoxins: A Canadian Perspective. P. M. Scott, H. L. Trenholm, M. D. Sutton, eds. Publ. 22848. National Research Council of Canada, Ottawa, ON.

Long, G. G., Diekmann, M. A., Tuite, J. F., Shannon, G. M., and Vesonder, R. F. 1983. Effect of *Fusarium roseum* (*Gibberella zeae*) on pregnancy and the estrous cycle in gilts fed molded corn on days 7–17 postestrus. Vet. Res. Comm. 6:199.

Lun, A. K., Young, L. G., and Lumsden, J. H. 1985. The effects of vomitoxin and feed intake on the performance and blood characteristics of young pigs. J. Anim. Sci. 61:1178.

Luo, Y. 1988. Fusarium toxins contaminations of cereals in China. Proc. Jpn. Assoc. Mycotoxicol. Suppl. 1:97-98.

Mannion, P. F., and Blaney, B. J. 1988. Responses of meat chickens offered 4-deoxynivalenol- and zearalenone-containing wheat, naturally infected with *Fusarium graminearum*. Aust. J. Agric. Res. 39:533-540.

Mannon, J., and Johnson, E. 1985. Fungi down on the farm. New Sci. 105(1446):12-16.

Marasas, W. F. O., Nelson, P. E., and Toussoun, T. A. 1984. Toxigenic *Fusarium* Species: Identity and Mycotoxicology. Pensylvania State University Press, University Park, PA.

Marasas, W. F. O., Kellerman, T. S., Gelderblom, W. C. A., Coetzer, J. A. W., Thiel, P. G., and Vanderlugt, J. J. 1988. Leukoencephalomalacia in a horse induced by Fumonisin-B_1 isolated from *Fusarium moniliforme*. Onderstepoort J. Vet. Res. 55:197-203.

Meyer, D., Weipert, D., and Mielke, H. 1985. Beeinflussung der qualitat von weizen durch den befall mit *Fusarium*. Pages 73-78 in: Quality of *Fusarium*-Attacked Wheat. Proc. Getreide-Tagung Arbeitsgemein. Getreideforsch. 12th. Granum Verlag, Detmold, Germany.

Mirocha, C. J., Pathre, S. V., and Christensen, C. M. 1977. Zearalenone. Pages 345-364 in: Mycotoxins in Human and Animal Health. J. V. Rodricks, C. W. Hesseltine, and M. A. Mehlam, eds. Pathox Publications, Forest Park South, IL.

Moran, E. T., Ferket, P. R., and Lun, A. K. 1987. Impact of high dietary vomitoxin on yolk yield and embryonic mortality. Poult. Sci. 66:977-982.

Nelson, P. E., Toussoun, T. A., and Cook, R. J. 1981. Page 457 in: *Fusarium* Diseases, Biology and Taxonomy. Pensylvania State University Press, University Park, PA.

Pelletier, M. 1992. Meat Probe 8(4).

Pier, A. C. 1981. Mycotoxins and animal health. Adv. Vet. Sci. Comp. Med. 25:185.

Plattner, R. D., Norred, W. P., Bacon, C. W., Voss, K. A., Paterson, R., Shackelford, D. D., and Weisleder, D. 1990. A method of detection of fumonisins in corn samples associated with field cases of equine leukoencephalomalacia. Mycologia 82:698-702.

Pomerantz, Y., Bechtel, D. B., Sauer, D. B., and Seitz, L. M. 1990. Fusarium head blight (scab) in cereal grains. Adv. Cereal Sci. Technol. 10:373-433.

Prelusky, D. B., Trenholm, H. L., Hamilton, R. M. G., and Miller, J. D. 1987. Transmission of [14][C]deoxynivalenol to eggs following oral administration to laying hens. J. Agric. Food Chem. 35:182-186.

Reotutar, R. 1989. Swine reproductive failure syndrome mystifies scientists. J. Am. Vet. Med. Assoc. 195:425-428.

Ross, P. F., Nelson, P. E., Richard, J. L., Osweiler, G. D., Rice, L. G., Plattner, R. D., and Wilson, T. M. 1990. Production of fumonisins by *Fusarium moniliforme* and *Fusarium proliferatum* isolates associated with equine leukoencephalomalacia and a pulmonary edema syndrome in swine. Appl. Environ. Microbiol. 56:3225-3226.

Ross, P. F., Rice, L. G., Plattner, R. D., Osweiler, G. D., Wilson, T. M., Owens, D. L., Nelson, H. A., and Richard, J. L. 1991. Concentrations of fumonisin B_1 in feeds associated with animal health problems. Mycopathologia 114:129-135.

Shotwell, O. L., Bennett, G. A., Stubblefield, R. D., Shannon, G. M., Kwolek, W. F., and Plattner, R. D. 1985. Deoxynivalenol in hard red winter wheat: Relationship between toxin levels and factors that could be used in grading. J. Assoc. Off. Anal. Chem. 68:954-957.

Sutton, J. C. 1982. Epidemiology of wheat head blight and maize ear rot caused by *Fusarium graminearum*. Can. J. Plant Pathol. 4:195-209.

Sydenham, E. W., Marasas, W. F. O., Thiel, P. G., Shephard, G. S., and Nieuwenhuis, J. J. 1991. Production of mycotoxins by selected *Fusarium graminearum* and *Fusarium crookwellense* isolates. Food Addit. Contam. 8:31-41.

Thiel, P. G., Shephard, G. S., Sydenham, E. W., Marasas, W. F. O., Nelson, P. E., and Wilson, T. M. 1991. Levels of fumonisins B_1 and B_2 in feeds associated with confirmed cases of equine leukoencephalomalacia. J. Agric. Food Chem. 39:109-111.

Thompson, L. J. 1991. *Fusarium* mycotoxins and animal effects. Pages 2-3 in: Plant Pathology Extender. *Fusarium* Molds and Mycotoxins Associated with Corn. G. C. Bergstrom and L. J. Thompson, eds. Plant Pathol. Ext. Rep. 91-1. Cornell Cooperative Extension, Ithaca, NY.

Trenholm, H. L., Thompson, B. K., Hartin, K. E., Greenhalgh, R., and McAllister, A. J. 1985a. Ingestion of vomitoxin (deoxynivalenol)-contaminated wheat by nonlactating dairy cows. J. Dairy Sci. 68:1000-1005.

Trenholm, H. L., Thompson, B. K., Standish, J. F., and Seamen, W. L. 1985b. Mycotoxins in feeds and feedstuffs. Pages 43-49 in: Mycotoxins: A Canadian Perspective. P. M. Scott, H. L. Trenholm, and M. D. Sutton, eds. Publ. 22848. National Research Council of Canada, Ottawa, ON.

Trenholm, H. L., Prelusky, D. B., Young, J. C., and Miller, J. D. 1988. Reducing mycotoxins in animal feeds. Agriculture Canada Publ. 1827E.

Tubbs, R. C., and Dekich, M. A. 1989. Handling mycotoxin problems in swine herds. Vet. Med. 84:925-926, 928-930.

Vanyi, A., and Sandor, G. 1988. Deoxynivalenol (DON) toxicosis II Effect of deoxynivalenol fusariotoxin (vomitoxin) on pig fattening. Magy. Allatorv. Lapja 43:503-507 (Chem. Abstr. 110:74001n).

Whitlow, L. M., and Hagler, W. M. 1987. The association of productivity losses in dairy cows with deoxynivalenol. Page E1 in: Recent Developments in the Study of Mycotoxins. Kaiser Chemicals, Rosemont, IL.

Whitlow, L. M., Nebel, R. L., Behlow, R. F., Hagler, W. M., and Brownie, C. F.-G. 1986. Mycotoxins in North Carolina dairy feeds—A survey of 100 dairy farms. J. Dairy Sci. Suppl. 1:223.

Young, L. G., McGuirr, L., Vallee, V. E., Lumsden, J. H., and Lun, A. 1983. Vomitoxin in corn fed to young pigs. J. Anim. Sci. 57:655-664.

Chapter 14

Diseases in Humans with Mycotoxins as Possible Causes

J. M. Beardall and J. D. Miller

Ergot and Ergotism

The oldest and best known human mycotoxicosis is ergotism. The first clear mention of ergot was made by Adam Lonicer of Frankfurt in 1582 although older references to ergot may still be uncovered (Bove 1970). An epidemic occurred in 430 B.C. among the Spartans that could conceivably have been due to ergot. Outbreaks of a disease resembling ergotism have been reported in Central Europe since 857 and reached epidemic proportions in the Middle Ages (Van Rensburg and Altenkirk 1974). The disease became known as Saint Anthony's fire because of the miraculous cures reported by afflicted pilgrims paying homage to St. Anthony's shrine in Dauphiné, France, during the Middle Ages (Bove 1970). Relief from the symptoms was probably due to a difference in diet consumed at the shrine.

It was not until 1673 that Denis Dodart linked the etiology of Saint Anthony's fire to ergoty rye (Bove 1970). He described rye grains that were hard and longer than normal rye kernels and black outside while rather white inside. Ergot bodies are the sclerotia of the fungi involved and can measure up to several centimeters in length. Up to seven or eight of these sclerotia could be found on a single spike of rye. In addition, Dodart observed that ergot was more plentiful in warm summers preceded by rainy springs. Etienne François Geoffroy in 1711 remarked that ergot resembled a fungus, but it was not until 1853 that Louis René Tulasne established the three stages of its life cycle (Bove 1970). About 50 species of the ergot fungus *Claviceps* have been found to infect many different grasses.

Two types of ergotism have been described, convulsive ergotism and gangrenous ergotism. Initially, during the Middle Ages, accounts of outbreaks were of the necrotic, gangrenous type (Bove 1970). Victims experienced a prickling sensation in the limbs followed by severe muscular pain (Van Rensburg and Altenkirk 1974). Swollen limbs and an icy cold sensation alternating with a severe burning sensation ensued, hence the word *fire* in Saint Anthony's fire. Fingers and toes became necrotic as gangrene set in, and in the most severe cases, feet and entire limbs shriveled up and dropped off (Van Rensburg and Altenkirk 1974).

In 1085, an epidemic involving both gangrene and convulsions was reported (Bove 1970). Convulsive ergotism is characterized by a tingling sensation under the skin, itching, numbness of the hands and feet, twitching, and muscle cramps (Van Rensburg and Altenkirk 1974). Extreme pain, sustained spasms, and convulsions were experienced by those affected (Bove 1970). Hallucinations are also characteristic of convulsive ergotism (Marth 1990). Postmortem findings include bleeding and softening of the brain and lesions in the posterior horns of the spinal cord. Full mental recovery from a nonfatal case is rare (Van Rensburg and Altenkirk 1974).

The mortality rate is reported to vary between 11 and 60% (Van Rensburg and Altenkirk 1974). Bove (1970) cites a report of an epidemic in the Middle Ages in which 40,000 people are estimated to have died. In some severe cases, patients were reported to have died within hours of the onset of the disease. In other cases sufferers recover, although not always completely, after experiencing symptoms for up to two months (Van Rensburg and Altenkirk 1974). Previous victims of ergotism are reported to be more susceptible to recurrences.

Ergotism is not transmitted to breast-fed infants; however, children are generally more susceptible, particularly to convulsive ergotism. There is no preponderance among either sex (Van Rensburg and Altenkirk 1974).

In this century, outbreaks of ergotism have occurred in Russia in 1926, Ireland in 1929, France in 1953 (Patel et al 1958), and Ethiopia in 1978 (King 1979). Over 140 people were affected by gangrenous ergotism in the Ethiopia outbreak in which ergoty wild oats were implicated (King 1979).

Poisonings attributed to the consumption of infected pearl millet (*Pennisetum typhoideum*), a staple in India, have been reported by Patel et al (1958) and Krishnamachari and Bhat (1976). Clinical symptoms were different from those of classical gangrenous or convulsive ergotism and included nausea, vomiting, diarrhea, giddiness, and somnolence (Krishnamachari and Bhat 1976). Tulpule and Bhat (1978) described this type of erotism as enteroergotism. Symptoms occurred 1–4 h after ingestion of the millet and lasted up to 48 h (Krishnamachari and Bhat 1976). Only those who consumed the incriminated millet experienced symptoms, and no particular age group or sex was more susceptible than another (Patel et al 1958, Krishnamachari and Bhat 1976). In the outbreak reported by Krishnamachari and Bhat, 78 people in 14 households were affected late in the year 1975. It was not possible to ascertain the exact number of people affected in the series of outbreaks from October 1956 to February 1957 since ergotism was not a notifiable disease (Patel et al 1958).

Millet samples were collected from households where poisonings occurred in both the 1956–1957 and 1975 outbreaks. Ergot contamination was clearly visible in both, and in the case of the 1975 outbreak, the percentage of contaminated ergoty grain varied from 1.5 to 17.4% by weight, and concentrations of total alkaloids ranged from 15 to 199 mg/kg of millet. The extent of visible contamination of grains did not consistently correspond to the amount of total alkaloids present. Concentrations of total alkaloids in samples collected from households not affected by the disease ranged from 0.2 to 26 mg/kg of millet. The inconsistent manifestation of disease symptoms between total alkaloid concentrations of 15 and 26 mg/kg of millet could result from differences among households in cleaning of grain before use (Krishnamachari and Bhat 1976). In this process, the black sclerotia are removed from the grain by hand.

Upon examination of the alkaloids associated with pearl millet implicated in the 1975 outbreak, only alkaloids of the clavine group could be identified (Krishnamachari and Bhat 1976). Agroclavine, elymoclavine, chanoclavine, penniclavine, and setoclavine were isolated from the contaminated pearl millet. Bhat et al (1976) found that the ergot alkaloids of wheat and rye belong mainly to the ergotamine group, whereas the alkaloids of ergoty pearl millet belong mainly to the clavine group. Ergotamine and related alkaloids are derivatives of lysergic acid and are associated with peptides. Clavine alkaloids are not lysergic acid derivatives and are not associated with peptides. Upon hydrolysis, ergot alkaloids of wheat and rye yielded amino acids, whereas the alkaloids from ergoty pearl millet did not (Bhat et al 1976).

No circulatory effects were reported in either of these accounts (Patel et al 1958, Krishnamachari and Bhat 1976). Certain ergot alkaloids are vasoconstrictors (Van Rensburg and Altenkirk 1974).

Bhat et al (1976) conducted studies on roosters and found no effects when alkaloids from ergoty pearl millet were injected into the comb. When alkaloids of ergoty rye were injected, the combs and wattles started to turn pale and cyanotic within 30 min, indicating the presence of vaso-constrictive ergotamine and related compounds (Bhat et al 1974, 1976). According to Bhat et al (1974), this confirmed a difference in the nature of ergot alkaloids in rye compared with those in pearl millet. Bhat et al (1976) speculated that the fungus *Claviceps purpurea*, known to be associated with ergoty rye (Bove 1970), produced mainly ergotoxine-ergotamine-ergo-metrine group alkaloids, whereas *Claviceps fusiformis*, which is associated with ergoty pearl millet (Bhat 1977), produces mainly clavine alkaloids (Bhat et al 1976). Bhat et al (1976) attributed the findings of ergotamine alkaloids in ergoty pearl millet by Patel et al (1958) to the methodology used, which was unable to distinguish between alkaloid groups.

Rhesus monkeys were fed an alkaloid extract from ergoty pearl millet at a dose of 10 mg/kg of body weight (bw). They were more resistant to the toxic effects of the alkaloids than man and exhibited different symptoms (Bhat and Roy 1976). The animals experienced clinical symptoms of hyperexcitation but no vomiting. When injected intraperitoneally with a mixture of clavine alkaloids, the monkeys exhibited hyperexcitation and uncoordinated muscle activity. The animals recovered in approximately 60 min.

When dietary intakes of alkaloids associated with ergoty pearl millet implicated in the 1975 outbreak were taken into account, the quantity of alkaloid that could be tolerated in humans without toxic effects was about 28 μg/kg bw (Krishnamachari and Bhat 1976). This is significantly lower than the tolerance of rhesus monkeys, in which an oral dose of 10 mg/kg bw was necessary to evoke mild clinical symptoms (Bhat and Roy 1976). Based on this difference in sensitivity, as well as on the difference in the nature of clinical symptoms manifested in man and monkeys, Bhat and Roy (1976) suggested that safe limits for ergot-contaminated grain be fixed according to epidemiological studies rather than toxicity studies in animals.

An inconsistent relationship between the degree of visual contamination and the level of total alkaloids dictated that safety limits should be set based on alkaloid content rather than percentage by weight of ergoty grains (Krishnamachari and Bhat 1976). Differences in the nature and toxicity of ergot alkaloids associated with wheat and rye compared with those found in ergoty pearl millet also require consideration.

Alimentary Toxic Aleukia

Alimentary toxic aleukia (ATA) has been reported in the literature under many different names, including septic angina, alimentary panhematopathy, alimentary toxicosis, alimentary hemorrhagic aleukia, alitoxicosis, alimentary agranulocytosis, and endemic panmyelotoxicosis, to name a few (Mayer 1953a). Alimentary toxic aleukia is now the official name used for the disease in Soviet government documents.

ATA has been widely reported in the former U.S.S.R. since 1913 (Joffe 1978), particularly in the Orenburg district, and has been attributed to the consumption of grain contaminated with *Fusarium* fungi (Joffe 1971). The most severe outbreak occurred in the spring of 1944 in the Orenburg district, in which 10% of the population were affected and mortality rates were as high as 60% in some counties (Joffe 1978). Outbreaks of ATA were again reported in the U.S.S.R. in the years 1952, 1953, and 1955 (Joffe 1986).

The clinical features of ATA include leukopenia; agranulocytosis; bleeding from the nose, throat, and gums; necrotic angina; a hemorrhagic rash; sepsis; exhaustion of the bone marrow; and fever. The first stage of the disease occurs shortly after the consumption of moldy grain and may last from three to nine days. The patient may experience vomiting, diarrhea, abdominal pain, and a burning sensation in the mouth, esophagus, and stomach. The second stage of the disease is characterized by hematopoietic disturbances including leukopenia, agranulopenia, and relative lymphocytosis, which may decrease the patient's resistance to bacterial infection. These symptoms may extend over a period from two to eight weeks.

Petechial hemorrhages develop on the skin up to a few centimeters in diameter and may be accompanied by hemorrhages in the oral cavity (mouth, tongue, tonsils) in the third stage. Necrotic lesions in the oral cavity follow, frequently accompanied by enlargement of the submandibular and cervical lymph nodes such that the patient may experience difficulty in opening his mouth. Death by suffocation can occur in about 30% of the patients as a result of edema and eventual blockage of the upper air passage.

Blood abnormalities intensify such that leukocyte, thrombocyte, and erythrocyte counts may decrease to as low as 100, 5,000, and 1,000,000 per cubic millimeter, respectively. The blood sedimentation rate also increases. If the victim survives, recovery of the blood-forming capacity of the bone marrow to normal usually takes at least two months (Joffe 1971).

In investigating the epidemiology of ATA, it was found that the disease was not contagious, that it had a seasonal nature, and that it occurred in people who had consumed overwintered grain (Joffe 1986). In the war years from 1942 to 1944 when food was often scarce, people were sometimes forced to gather grain that had been left unharvested in the field throughout the winter. This grain included prosomillet, wheat, barley, rye, oats, and buckwheat, of which prosomillet and wheat were the most toxic when overwintered in the field (Joffe 1978).

Grain that was harvested in the autumn was nontoxic, whereas grain that was harvested after a winter of heavy snowfall and frequent freezing contained toxins (Mayer 1953b). Joffe (1963) reported the greatest number of toxic samples in the spring. The toxicity of the overwintered grain may vary greatly even within samples from the same field (Joffe 1978). In addition, the toxicity of overwintered cereals was higher in low-altitude areas (70–200 m above sea level) than in high altitude areas (350–400 m above sea level). This may have been due to lower average temperatures in higher altitude areas (Joffe 1971). The toxicity of grain samples was generally evaluated by the application of a small amount of grain or culture extract onto the skin of rabbits (see following).

People consuming at least 2 kg of toxic grain usually developed ATA two to three weeks after ingestion, and death occurred six to eight weeks after the consumption of at least 6 kg (Joffe 1978). The nutritional status of the victim may play a part in the etiology of ATA, since those subsisting on a staple diet of overwintered grain were more severely affected than those eating balanced diets (Joffe 1978). It was once hypothesized that deficiencies in vitamins B-1, C, and riboflavin were of etiological significance (cited in Joffe 1986). Both sexes and all age groups were afflicted by the disease although some researchers found the disease more prevalent in middle-aged women (cited in Joffe 1978). It seemed that the toxic principle was not secreted in breast milk, since breast-fed babies less than one year old were not affected (Joffe 1978).

Toxigenic strains of *Fusarium, Aspergillus, Phoma, Hymenopsis, Mucor, Piptocephalis, Mortierella, Gliocladium,* and *Trichoderma* have been isolated from grains in various regions of the former U.S.S.R. (cited in Joffe 1978). *Alternaria* has also been isolated from toxic wheat in Siberia (cited in Joffe 1978). *Penicillium, Alternaria, Fusarium, Cladosporium,* and *Mucor* were the most prevalent genera of fungi isolated from overwintered cereal in the Orenburg district from 1943 to 1949 (Joffe 1971). Of these, *Fusarium* constituted the greatest percentage of highly toxigenic isolates followed by *Cladosporium. Fusarium poae* and *F. sporotrichioides* were commonly isolated (Joffe 1971).

A variety of animals were used for non-toxin-specific skin bioassays during the 1940s to evaluate relative toxicity of various ether or ethanol extracts from overwintered grains or culture material. Nonpigmented rabbits were frequently used. At an interval of 24 h, a small amount of extract was twice applied to the rabbit on shaved patches of skin measuring 9–20 cm^2 (Joffe 1960). Positive and negative control patches were also included (Joffe 1960). An extract was considered toxic if edema, hemorrhage, and necrosis resulted (Joffe 1978).

The early work of Soviet scientists in determining toxic fractions from fungal cultures and overwintered grain was reviewed by Joffe (1986). Much of the research done before the 1950s yielded conflicting theories on the chemical composition of *Fusarium* toxins. Olifson purified and isolated toxins from the lipid fraction of inoculated cultures and overwintered grain samples that caused a strong inflammatory reaction in the rabbit skin test and death in a number of small mammals fed a diet containing this lipid fraction. The steroid glycosides sporofusarin and poaefusarin were extracted from cultures of *Fusarium sporotrichioides* and *F. poae,* respectively.

When tested for toxicity in cats, these two derivatives produced a disease characterized by leukopenia, similar to ATA in humans. Olifson et al (cited in Szathmary et al 1976) suggested that the toxins responsible for ATA were sporofusarin and poaefusarin and their aglycones, sporofusariogenin and poaefusariogenin. However, Mirocha and Pathre (1973) found no trace of steroids in an authentic sample of "poaefusarin" but rather detected sufficient amounts of T-2 toxin to explain the toxicity found in the rat and rabbit skin tests. Szathmary et al (1976) also attributed the toxicity of extracts of *Fusarium poae* and *F. sporotrichioides* obtained from the U.S.S.R. to T-2 toxin and suggested that the biological activity observed by Olifson was due to T-2 toxin rather than steroids.

Six steroidal metabolites from a strain of *F. sporotrichioides (F.*

sporotrichioides 921) implicated in fatal outbreaks of ATA were cultured on millet and found to be neither toxic nor irritating to skin (Yagen et al 1977). However, large quantities of T-2 toxin were isolated from this culture. Furthermore, 95% of 131 isolates of *F. sporotrichioides* and *F. poae* implicated in ATA outbreaks were able to produce T-2 toxin in various amounts. T-2 toxin was proposed as the principal cause of rabbit skin irritation (Joffe and Yagen 1977, Yagen and Joffe 1976).

Cats dosed with either pure T-2 toxin or crude extracts containing T-2 toxin displayed changes in blood-producing tissue similar to those seen in ATA. Clinical symptoms in these cats included vomiting, leukopenia, and hemorrhagic diathesis and postmortems revealed enlarged and hemorrhagic lymph nodes (Yagen et al 1977). In a similar experiment on cats, clinical features were similar to the disease syndrome of ATA in man (Lutsky and Mor 1981a,b). Cats dosed orally with T-2 exhibited leukopenia, hemorrhagic diathesis, agranulocytosis, bone marrow aplasia, sepsis, and generalized intoxication (Lutsky and Mor 1981b).

Schoental and Joffe (1974) noted local cytotoxic effects and accompanying basal cell hyperplasia of the squamous epithelium when crude extracts of *F. sporotrichioides* and *F. poae* where applied to the skin, esophagus, and stomach of rodents. Depletion of lymphoid tissues followed by widespread infections was also observed. T-2 toxin is reported to be genotoxic to V79 cells (Chinese hamster lung fibroblasts) (Zhu et al 1987). Wright et al (1987) found a 50% reduction of DNA in cultured chondrocytes, Chang liver cells, and bovine pulmonary endothelial CPAE cells. In addition, low doses of T-2 toxin were found to induce single-strand DNA breaks in cultured splenic and thymic cells and to cause DNA breaks in splenic and thymic lymphocytes in vivo in mice administered T-2 toxin (Lafarge-Frayssinet et al 1981). Tumors in the brain and digestive tract were observed in rats dosed intragastrically with T-2 toxin (Schoental et al 1979).

According to Joffe (1986), T-2 toxin from overwintered cereals caused fatal outbreaks of ATA in the U.S.S.R. Marasas et al (1984a) noted that the detection of T-2 toxin in samples of overwintered grain actually associated with ATA outbreaks would strengthen the implication of this toxic metabolite as the causal agent of ATA.

Patients can recover from ATA if it is detected in the earlier stages of the disease (Joffe 1986). Blood transfusion, administration of nucleic acid, and calcium preparations, antibiotics, vitamins C and K, and a healthy diet are among the treatments employed for ATA (cited in Joffe 1986). It is important to avoid consumption of overwintered grain.

Kashin-Beck Disease

The name Kashin-Beck disease (KBD) derives from the two Russian scientists, Kashin and Beck, who studied the disease extensively from the

1860s until after the turn of the century. It is also known as Urov disease because its occurrence was first described along the Urov River in Russia in 1861 (Joffe 1986). More recently KBD has been found to exist in Taiwan, northern China, Korea, Japan, Sweden, and Holland (Joffe 1986). Although many of the Russian reports are not available, Joffe (1986) summarized some of the earlier research.

The clinical course of this acquired disease is described by Sokoloff (1985). Symptoms of the onset of the disease are seen in preadolescent and adolescent children. They include symmetric stiffness, pain, and swelling, most commonly in the fingers but also in the wrist joints. The hips are rarely affected (Marasas and Van Rensburg 1986). Inflammation is not involved (Sokoloff 1985). At this stage the disease may be reversible if the patient leaves the endemic area (Sokoloff 1988). The disease progresses until symptoms of generalized osteoarthritis manifest themselves, typically in the elbows, knees, and ankles, which may become locked in a later phase of the disease (Sokoloff 1985). Deformation of the joints occurs and flattening of the vertebral bodies is common (Sokoloff 1987a). Asymmetric shortening of the extremities and malformation of the skeleton may lead to disproportionate dwarfism (Sokoloff 1985). Mobility of the patient becomes limited such that work in the field becomes impaired.

The characteristic pathological changes observed in patients afflicted with KBD include degeneration and necrosis of chondrocytes in the growth plates and articular cartilages as well as proliferation or repair of cartilage following chondronecrosis (Zhang and Liu 1989). A rise in the activity of lactic dehydrogenase, aspartate transaminase, alanine transaminase, and alkaline phosphatase in serum accompanied by an increase in creatine and a decrease in creatinine in the urine are biochemical changes typical of the disease (Zhang and Liu 1989).

This incapacitating disease is neither heritable nor congenital in nature (Sokoloff 1987b, 1988). Although it is often not reported in the initial stages, two million people are estimated to be affected in China alone (Levander 1987a). KBD is reported predominantly among peasants in rural areas (Allander 1987); the number of affected individuals reaches up to 90% in some communities (Sokoloff 1988). The disease afflicts mainly individuals consuming maize or wheat, whereas those consuming rice, even in endemic areas, are generally spared (Levander 1987b).

Many hypotheses as to the etiology of the disease have been mooted in the over 100 years since its discovery. Much controversy has ensued throughout but no cause has been conclusively established. Included in the proposed theories were vitamin and mineral deficiencies, trace metal toxicity, disturbances in mineral metabolism, hereditary factors, and mycotoxins to name a few (Joffe 1986). Most of these hypotheses have been sharply criticized; however, the three independent views suggesting the involvement of selenium deficiency (Liang et al 1987), trace metal toxicity, and mycotoxins associated with *Fusarium*-contaminated grain are currently being investigated (Sokoloff 1990). *F. oxysporum* is the fungal species of contemporary etiological interest, although *F. poae* was thought responsible a short time ago (Sokoloff 1985).

In the U.S.S.R., the suggestion that the disease had an epidemiology of geographical and environmental basis, such that toxin formation in cereal grains was favored, was put forth by Sergiyevski in the mid 20th century (cited in Joffe 1986). Nesterov (1964), in accordance with this, pointed out that similar natural conditions exist in Northern China and Korea where the disease is known to occur. Endemic regions are hilly, soil composition is poor, and runoff water is contaminated with organic matter (Sokoloff 1985). Temperate climates with long, severe winters and hot, humid summers are characteristic of endemic areas (Sokoloff 1988).

Of the 425 strains of *Fusarium* isolated by Rubinstein during 1948-1951 from cereal grains in an endemic region of the U.S.S.R., strains of *F. sporotrichiella* var. *poae* (syn. *F. poae* (Peck) Wollenweb.) proved to be of interest when assayed by the rabbit skin test (Joffe 1986, see section on alimentary toxic aleukia). The search for animal models of KBD that paralleled clinical features in humans was initiated. Bone growth abnormalities including delayed lengthwise bone growth and deformation of the joints were demonstrated in pups and rats fed wheat contaminated with *F. poae* (Joffe 1986). The toxic principle responsible for the observed toxic response in pups and rats was not isolated, and it was argued that this animal model is not a suitable animal analogue of human KBD since no chondronecrosis or articular deformity was reported (Sokoloff 1985).

Zhang and Liu (1989) conducted studies on macaque monkeys in an effort to determine the possible role of water and grain from endemic foci. Monkeys on a diet of either grain (86% maize powder) or water or both from endemic areas exhibited chondronecrosis in the articular cartilage and zonal necrosis in the growth plate similar to those found in human KBD. In addition, monkeys fed either grain or both grain and water from endemic areas showed distinct biochemical changes closely resembling those typical of human KBD concurrently with chondronecrosis. Zhang and Liu (1989) suggested that pathogenetic factors in KBD relate both to grain and to water in the diet in endemic areas.

Although field outbreaks of KBD in domestic animals have not been reported even in high-incidence areas of human KBD, several animal diseases share pathological similarities to KBD in man. In particular, tibial dyschondroplasia in avian species and "osteochondrosis" in swine and other mammals are acquired disorders that affect animals in the growth stage (Sokoloff 1990). In the case of osteochondrosis in some domestic animals, osteochondritis dissecans and osteoarthritis result from damaged or necrosed chondrocytes. However, necrosis of the chondrocytes is not so marked as in the human disorder and has a focal rather than zonal character (Sokoloff

1985). Tibial dyschondroplasia in birds differs from human KBD in the lack of involvement of chondronecrosis and the articular cartilage. In this case, failure of growth plate cartilage to undergo calcification is the principal lesion (Sokoloff 1988). Thus both of these animal disorders have fundamental pathological differences from human KBD that render them medically distinct.

Avian tibial dyschondroplasia has been shown to be mycotoxin associated. Lee et al (1985) isolated a water-soluble fraction, referred to as TDP-1, from *F. graminearum* that caused tibial dyschondroplasia in chickens. The toxic compound was later identified as fusarochromanone (Pathre et al 1986). In cell culture experiments, fusarochromanone had no selective toxicity for cultured chondrocytes over Chang liver cells. These findings lend no support to the possible involvement of fusarochromanone in the etiology of human KBD (Wright et al 1987).

Intraperitoneal injections of T-2 into pregnant mice were shown to induce skeletal malformations in the fetuses (cited in Joffe 1986). T-2 toxicity has also been demonstrated in cultured human juvenile and rabbit chondrocytes (Wright et al 1987).

The occurrence of KBD is reported to be on the decline in some areas (Sokoloff 1989). This may be the result of improved hygiene and the importation of grain from nonendemic areas (Nesterov 1964). Despite the efforts of scientists located worldwide over a long period, efforts continue to conclusively determine the causal agent of this widespread, crippling disease. Whether the disease has a nutritional basis associated to selenium deficiency or whether the disease stems from a toxic principle such as a mycotoxin or trace metal has not been reliably established. If the disease is indeed mycotoxin related, further study is required to determine the causative fungus and the associated toxin(s). Investigations into the possible role of multiple etiological factors would be of value.

Mseleni Joint Disease

Mseleni joint disease (MJD) is a progressive, crippling osteoarthropathic disorder prevalent in blacks in the Mseleni area in Kwazulu, southern Africa. Three thousand people from the Zulu and Tonga tribes are estimated to be affected (Sokoloff 1985). Several reports on the clinical and epidemiological features of MJD appeared in the literature in the 1970s, but controversy exists as to when the disease first appeared (Fellingham et al 1973, Lockitch et al 1973, Nurse and Jenkins 1974).

The initial manifestations of MJD are characterized by pain primarily in the hip, but also in the knee and ankle joints (Lockitch et al 1973). Wrists, shoulders and elbows are less commonly affected. Radiological examinations revealed characteristic abnormalities in different age groups of affected individuals, suggesting a progressive symptomatology. Symmetrical, bilateral irregularities of the surface and density of the epiphyses of the hips, knees, hands, wrists, and sometimes ankles, shoulders, and elbows were characteristic abnormalities seen in younger patients. In older patients, symmetrical, bilateral osteoarthrosis in multiple joints was typical. In intermediate age groups, both joint structure irregularities and osteoarthrosis were prominent. A common lesion was protrusio acetabula. Anomalies in the radius and ulna have been reported, as have stubby digits (Lockitch et al 1973). Mild stunting was frequent in affected individuals; however, true achondroplasia was not evident (Fincham et al 1981).

The onset of MJD may occur at any age; however, it was particularly prevalent in women of older age groups. Although the incidence of MJD among children aged from newborn to 10 years was approximately evenly distributed between males and females, 86.9% of women in the endemic region aged 61–70 years were affected whereas only 31.7% of males in the same age group were affected. The overall occurrence of the disease in all age groups in the endemic region is 38.9% in women and 11.1% in men. The rate of progression of the disease and the stage at which it stabilizes may vary from patient to patient (Fellingham et al 1973). Consequently the mobility of affected individuals is limited to various degrees. While some victims are able to walk with the aid of a stick, others are reduced to propelling themselves by the arms. The disease generally progresses further in women than in men (Fellingham et al 1973), but life expectancy is not affected by MJD (Lockitch et al 1973).

Unlike the environmental conditions in the endemic areas of Kashin-Beck Disease, the MJD endemic area is characterized by a coastal plain of extremely sandy, infertile soil (Anonymous 1985). The climate is hot and humid in spite of the low annual rainfall.

The diet in the endemic region consists of maize, cowpeas, groundnuts, sweet potatoes, melons, pumpkins, and wild fruits and vegetables, while the consumption of animal protein is limited (Marasas and Van Rensburg 1986). Nutritional deficiencies have been investigated as possible causal agents in the etiology of the disease. Calcium, magnesium, and manganese deficiencies have been suggested (Fincham et al 1981, 1986).

Manganese deficiency is known to induce bone growth abnormalities in fast-growing domestic animals (Fincham et al 1981). No manganese was detected in water samples from Mseleni; however, because there is a lack of sufficient control data it is impossible to assess the significance of this finding. Manganese has not been analyzed for in food and soil from the Mseleni area. Serum calcium and magnesium levels in women affected by MJD were found to be low compared with levels found in healthy white and black women (Fincham et al 1986); however, control data for women from Mseleni not manifesting symptoms of MJD are not available. In addition, MJD does not manifest itself in other black populations known to be at risk of calcium deficiency (Fincham et al 1985).

Changlong et al (1987) addressed the question of whether the deficiency of an essential agent or the presence of a toxic or infective principle in the blood of Mseleni patients was responsible for the defective growth of chondrocytes. They conducted studies on cultures of rabbit and juvenile human chondrocytes using fetal bovine serum, serum from unaffected individuals, and serum from MJD affected patients as undefined media supplements. No difference in DNA or sulfated proteoglycan synthesis by chondrocytes was detected between the experimental and control serum (Changlong et al 1987).

Toxicology studies were conducted on rats fed one of three different diets: 90% maize and 9.7% groundnuts, both from the high-incidence MJD area; 90% commercial maize and 9.7% commercial groundnuts; or commercial nutritionally balanced rat ration (Fincham et al 1985). Rats fed maize-based diets invariably developed fibrous osteodystrophy and nutritional hyperparathyroidism thought to be due to a calcium deficiency. Other foodstuffs eaten by the people of Mseleni were not represented in the diets.

In a similar experiment involving rats fed maize and groundnuts from the Mseleni area, Marasas and Van Rensburg (1986) found that the histopathological and chemical changes observed in experimental rats were not characteristic features of MJD. They could not confirm the presence of a toxic principle in the diet responsible for causing MJD. It is to be noted that the maize and groundnut samples used to prepare the diets were collected during a period of abnormally low rainfall and that the possible toxic principle(s), particularly if it were associated with a fungus, may have been absent from the samples.

Genetic factors were suspected to be etiologically significant since MJD has a tendency for familial aggregation (Fellingham et al 1973). Serogenetic investigations revealed that the Mseleni population is not genetically isolated. Moreover, the uneven distribution of MJD between males and females is not compatible with genetic heritability theory, thus rendering the hypothesis of genetic etiology remote (Nurse and Jenkins 1974).

Environmental toxins were also put forth as possible etiological agents. Mycotoxicological investigations were conducted by Marasas and Van Rensburg (1986) on a small number of samples of two staple foods, maize and groundnut, collected from affected and nonaffected households in Mseleni. Unfortunately, these samples were collected during a period of severe drought, and therefore the results may not be representative of the case in a period of average or above average rainfall. The predominant fungus associated with maize from the endemic region was *Fusarium moniliforme*. It was detected in 96.3% of samples.

Other species of fungi isolated in significant numbers from maize in the Mseleni area included *Acremonium* spp., *Lasiodiplodia theobromae*, *Penicillium* spp., and *Nigrospora* spp. From groundnut in the Mseleni region, *Penicillium* spp., *L. theobromae*, *Macrophomina phaseolina*, and *F. oxy-*

sporum were frequently isolated. *F. compactum, F. oxysporum, L. theo-bromae,* and *M. phaseolina* were more prevalent in maize and groundnut from MJD-affected households than from nonaffected households. Several species of *Fusarium* that have been suggested to be implicated in the etiology of bone growth diseases of animals and man were isolated, including *F. equiseti, F. moniliforme, F. oxysporum,* and *F. poae* (Marasas and Van Rensburg 1986).

Toxicity studies were conducted in ducklings fed cultures of fungi isolated from the same maize and groundnut samples as above (Marasas and Van Rensburg 1986). Characteristic symptoms of MJD were not experimentally reproduced in the ducklings; however, 120 of 322 cultures isolated caused the death of at least three out of four ducklings within 14 days. The most toxic cultures to duckling were those of *F. compactum, F. moniliforme, F. oxysporum,* and *Phomopsis* spp.

The genetic etiological hypothesis for MJD is unconvincing. The possible involvement of nutritional deficiencies in the etiology of the disease requires further investigation. Toxicology studies conducted with foodstuffs collected from the Mseleni district during a period of climatic conditions more representative of the norm may help to clarify the role of mycotoxins as possible causal determinants in the etiology of the disease.

Endemic Familiar Arthritis of Malnad

Endemic familial arthritis of Malnad (EFAM) is yet another syndrome of abnormal bone growth in which environmental factors are suspected to be etiological agents. It was reported from the years 1965–1975 in several villages in the Malnad district of Karnataka State in southern India (Krishnamachari and Bhat 1978). The climatic conditions of this area are characterized by heavy rainfall (Bhat and Krishnamachari 1977).

Epidemiological studies have indicated that individuals of both sexes and assorted ages from 140 families were afflicted by the disease (Bhat and Krishnamachari 1977). Children often experienced more severe symptoms than adults, but EFAM has never been observed in children younger than five years old. The disease is noncongenital. Only the most impoverished castes are affected (Krishnamachari and Bhat 1978). These people subsist on a diet of rice and various fauna, in particular, fish and crabs that are abundant in the rice paddies (Bhat and Krishnamachari 1977).

Like Mseleni joint disease, bilateral, symmetrical lesions develop primarily in the hip where pain is experienced to variable degrees (Krishnamachari and Bhat 1978). The disease may progress to the knees, and the mobility of the patient may become severely impaired, but other joints are rarely involved (Bhat and Krishnamachari 1977). Radiological features include osteoarthritis in the hip joint, pelvis, and vertebrae. Destruction of the femoral neck and articular surfaces of the femur and tibia are characteristic. Also prominent are the cystic appearances of the femoral head with areas of sclerosis and rarefaction about the hip joint (Krishnamachari and Bhat 1978).

As the name implies, the disease tends to strike a number of people in the same family, although it is noninfectious. In some households, the onset of the disease appeared almost concurrently in individuals of three generations. Pedigree studies showed that all those affected in the studied group were interrelated. Furthermore, there was a high incidence of dwarfism in the studied districts of the endemic region, and all of the dwarfs were somehow related. Although only a small proportion of the dwarfs themselves were affected by the disease, all were related to affected individuals. EFAM does not appear to be genetically determined since the disease was not in existence before the mid-1960s (Krishnamachari and Bhat 1978).

Genetic susceptibility to a recently introduced environmental factor that precipitates the disease may be involved (Krishnamachari and Bhat 1978). During the endemic period, the use of potent pesticides and fertilizers increased significantly, killing fish and crab in the rice paddy immediately following spraying. It has been speculated that these potentially toxic agents could have entered the food chain and been concentrated in the various fauna consumed, thus precipitating EFAM in genetically susceptible individuals (Bhat and Krishnamachari 1977).

No incidents of EFAM manifesting in domestic animals have been reported (Bhat and Krishnamachari 1977).

Although no mycological surveys or toxicity studies have been conducted on foodstuffs consumed by affected individuals, EFAM bears certain epidemiological and pathological similarities to the other bone growth disorders described in which mycotoxins have been investigated as possible etiological agents. Like Kashin-Beck disease, osteoarthritic changes are characteristic pathological features, and as in Mseleni joint disease, the hip is the predominant joint affected. Both in Mseleni joint disease and EFAM a pattern of familial distribution is evident, and the occurrence of mild stunting and dwarfism in the respective endemic areas may have some significance. All three diseases are acquired conditions whose occurrence is restricted to geographically isolated areas. Further epidemiological, mycological, and toxicological investigations are necessary to identify the etiological determinant(s) of these three diseases.

Balkan Endemic Nephropathy

Balkan endemic nephropathy (BEN) is a chronic kidney disease of unknown etiology that was first described in the 1950s in the Balkan area (Vratza District) of Bulgaria. The disease is now prevalent in many rural areas of Yugoslavia, Bulgaria, and Romania that lie within the Danube Basin. It is estimated that some 20,000 people are affected by the disease (Barnes et al 1977).

BEN involves progressive renal failure and atrophia of the kidney. Anemia, polyuria, proteinuria, lassitude, yellowness of skin, headaches, anorexia, and uremia are common clinical features of BEN, although they may not all be experienced and are often features of other kidney diseases (Austwick 1981). Biochemical detection of increases in B_2-microglobulin secretions may facilitate the diagnosis of BEN, since this is a feature of the early stages of the disease. Marked reduction in kidney size and changes in the renal cortex such as interstitial fibrosis, hyalinization of glomeruli, and degeneration of the tubular epithelium as well as loss of the brush border in the tubule are pathological findings in BEN victims (Austwick 1981).

Up to 12% of the population in an endemic area may be affected by BEN (Austwick 1975). Endemic areas tend to be rural, and most sufferers are from an agricultural background. Socioeconomic status, nutrition, and hygiene are not thought to play a causal role in the etiology of BEN (Anonymous 1977). The disease is fatal; approximately half of the sufferers die within two years, although some have been known to survive 10 years (Austwick 1975). Approximately twice as many women as men are affected in the younger years, but this ratio equalizes in the later years (Anonymous 1977). BEN is encountered almost exclusively in individuals between the ages of 35 and 55 (Austwick 1975). This may be the consequence of a requirement for prolonged exposure to a causal agent in the high-incidence area.

BEN has a familial distribution, but a spouse entering a family with a history of BEN may develop the disease after at least 15 years of living with the affected family (Austwick 1975). This fact does not support a genetic etiological hypothesis, although some still argue that the disease is primarily a genetic disorder (Mihailov cited in Radovanovic 1989). Individuals emigrating from an endemic area have been known to be stricken with the disease long after having left the endemic area (Austwick 1981). Furthermore, immigrants entering an endemic area have been known to develop BEN after having lived in the endemic area for a sufficiently long time (usually over 20 years; Radovanovic 1989).

Radovanovic (1989) reviews the many possible causal factors that have been investigated in an attempt to establish the etiology of BEN. Heavy metals, trace elements, minerals, radioactive substances, viral infections, toxic plants, fertilizers, pesticides, even cosmic influences have been considered, but none has yielded a conclusive causal relationship. A significant correlation between mortality from BEN and higher than average rainfall in late summer and autumn in endemic areas has been reported (Austwick 1975). Austwick (1975), in accordance with the mycotoxin theory first put forth by Dimitrov in 1960 (cited in Barnes et al 1977), speculated that a fungal toxin associated with high autumn humidity may be responsible for the elevated number of ensuing deaths.

The hypothesis that fungal metabolites play a causal role in BEN is attractive in light of the discovery that the mycotoxin ochratoxin A (OTA) is a major disease determinant in the etiology of mycotoxic porcine nephropathy in Scandinavia (Krogh 1987). Ochratoxins are fungal metabolites produced by *Penicillium verrucosum* and members of the *Aspergillus ochraceus* group (Frisvad and Filtenborg 1989, Krogh 1987). OTA is the most common of the ochratoxins and consists of a dihydrocoumarin moiety linked to an L-β-phenylalanine functional group (Kuiper-Goodman and Scott 1989). Field outbreaks of mycotoxic porcine nephropathy are characterized by degeneration of the proximal tubules, followed by atrophy of the tubular epithelium, interstitial fibrosis in the renal cortex, and hyalinization of some glomeruli (Krogh 1987).

These morphological changes were experimentally reproduced in the kidneys of pigs fed pure OTA at a level of 1 mg/kg of feed for three months. Some similar renal lesions have also been observed in poultry ingesting feed contaminated with OTA. Residues of OTA have been found in the blood and tissues of pigs and poultry fed OTA-contaminated feed. Studies done in Danish slaughter houses revealed that 25–39% of the kidneys from pigs affected with porcine nephropathy contained levels of OTA in the range of 2–100 μg/kg (Krogh 1987). Mycotoxic porcine nephropathy does not occur in endemic areas of the human disease BEN (Radovanovic 1989).

OTA is cleaved by microorganisms living in the gut of ruminants leaving the nontoxic cleavage product ochratoxin-α. OTA residues are, therefore, not generally found in the tissues and blood of ruminants (Krogh 1987). The persistence of OTA and/or other mycotoxins through the food chain, either by direct consumption of contaminated grain or by the consumption of nonruminant meat contaminated with mycotoxins, however, represents a health risk to humans.

Mycological studies have been conducted in an effort to define nephrotoxin-producing fungi from BEN endemic areas. Barnes et al (1977) examined various moldy foodstuffs originating from BEN endemic areas in Yugoslavia, Bulgaria and Romania. *Penicillium verrucosum* var. *cyclopium* (now known as *P. aurantiogriseum*) was the most frequently isolated species. Other fungi isolated included *Phoma* sp., *Fusarium oxysporum*, *Penicillium expansum*, and another unidentified *Penicillium* sp. When rats were fed culture homogenates of these fungi, only one of the isolates of *P. aurantiogriseum* yielded a response at necropsy after approximately 30 days (Barnes et al 1977).

Renal lesions were observed in the lower reaches of the proximal convoluted tubules, and abnormalities of the tubular epithelium were prominent. The normal brush-border of the tubular epithelium was lacking, but other segments of the renal tubules and the glomeruli were normal. Barnes et al (1977) suggested that the proximal tubule lesions closely resembled the tubular changes in BEN sufferers. It was later argued that damage in the lower part of the proximal convoluted tubule has never been observed in human BEN, but that during short-term exposure the proximal tubule is the renal target for OTA in rats and pigs (Elling and Krogh 1977). However, recent studies have affirmed the potential importance of *P. aurantiogriseum* (see following).

In a five-year study, the mycoflora of stored cereals (maize, wheat, and beans) from affected and unaffected households in a BEN endemic region in Yugoslavia were studied (Pepeljnjak and Cvetnic 1984). *Penicillium* and *Aspergillus* spp. dominated in cereals and beans from both affected and unaffected households. Other fungal genera isolated included *Alternaria, Fusarium, Absidia, Rhizopus, Botrytis, Cladosporium, Nigrospora, Trichoderma, Scopulariopsis,* and several unidentified others. *A. niger* and *A. flavus* were the most frequently isolated species of *Aspergillus,* while *P. aurantiogriseum* and *P. expansum* were the predominant *Penicillium* species isolated. It was concluded that no significant differences in the distribution of fungi in cereal and bean stores from these affected and unaffected families were evident.

In another study by Pepeljnjak and Cvetnic (1985), contamination by toxigenic species occurred on maize and wheat in vegetation approximately equally in endemic and nonendemic areas in Yugoslavia. A slightly higher degree of *Aspergillus, Alternaria, Rhizopus,* and *Trichoderma* contamination was noted in stored maize and wheat in the endemic area compared with the nonendemic area. Species of *Penicillium* and *Fusarium* were prevalent in stored grain from both endemic and nonendemic areas (Pepeljnjak and Cvetnic 1985).

Studies were undertaken to determine differences in the toxigenicity of various strains of *Aspergillus ochraceus* isolated from cereals and smoke-dried meat products from a BEN endemic and nonendemic area in Yugoslavia (Cvetnic and Pepeljnjak 1990). One strain isolated from the endemic area was a strong OTA producer (240 mg/L on liquid yeast extract and sucrose [YES] media), whereas no strong OTA producers were isolated from the nonendemic area. Moderate producers of OTA (10–50 mg/L on YES media) were isolated from both the endemic and nonendemic areas. It was not possible to confirm any significant differences in the overall toxigenicity of *A. ochraceus* strains isolated from the endemic and nonendemic areas studied (Cvetnic and Pepeljnjak 1990).

Differences in distribution of toxigenic fungi and their associated toxigenicities between endemic and nonendemic areas may provide clues as to causal relationships in the etiology of diseases of possible mycotoxic origin. However, these investigations realistically only represent potentials for causal associations. Comparative studies of actual toxin levels in foodstuffs consumed by affected and unaffected individuals represent a more concrete data source from which to base hypotheses. A number of studies of this type have been conducted.

Pavlovic et al (1979) analyzed maize, wheat, barley, and bread from a nephropathic area in Yugoslavia for OTA. They found levels up to 140 μg/kg in some samples. The frequencies of contamination with OTA were higher than those reported previously (Krogh 1978); however, concurrent control data for a nearby nonendemic area were not included in this study. Petkova-Bocharova and Castegnaro (1985) detected no significant difference between the mean values of OTA in contaminated samples of beans and maize from endemic and control areas in Bulgaria. They did, however, note that a much greater proportion of the samples from the endemic area were contaminated with OTA. In a study by Pepeljnjak and Cvetnic (1985) average concentrations of OTA on stored maize and wheat were higher in areas of endemic nephropathy than in nonendemic areas in Yugoslavia. Concentrations of OTA over 2 mg/kg were detected in 45% of the positive samples from endemic areas but only in 15% of the positive samples from nonendemic areas. In addition, OTA residues were detected in the kidneys, liver, and blood of pigs from an endemic area in Yugoslavia but not from a nonendemic area (Pepeljnjak and Cvetnic 1985).

Levels of OTA in the blood of subjects living in endemic areas of BEN have been investigated as a measure of previous foodborne exposure to this mycotoxin. OTA was found more frequently and at higher concentrations in the blood of inhabitants of endemic villages than inhabitants of a nonendemic village in Yugoslavia (Plestina et al 1990). Levels of citrinin, another mycotoxin that may play a role in the etiology of porcine nephropathy (Krogh 1987), were also compared in staple foods from households of families affected and not affected by BEN in Bulgaria. Again a greater proportion and a greater degree of citrinin contamination was found in samples from affected families (Castegnaro et al 1990). In a study by Petkarova-Bocharova et al (1988), patients in Bulgaria with BEN and/ or urinary tract tumors had higher levels of OTA in the blood than healthy individuals from endemic regions.

Yeulet et al (1988), reinvestigated the nephrotoxicity of the toxin producing isolate of *Penicillium aurantiogriseum* isolated from maize in Bulgaria (Barnes et al 1977) as well as four other isolates of *P. aurantiogriseum* from Yugoslavia. Rats fed a mixture of rat ration and cultured mycelia developed histopathological renal lesions similar to those described by Barnes et al (1977). Both abnormally large and abnormally small cells, pyknotic nuclei, and mitotic figures were observed in epithelial cells of proximal tubules in the pars recta region. Nephrotoxicity was observed in rats fed fractionated extracts of fungal mycelium, and DNA synthesis inhibition was noted in kidney cells cultured with fungal fractions (Yeulet et al 1988). OTA has not been found as a metabolite of the Bulgarian isolate (Yeulet et al 1988).

Yeulet et al (1988) proposed that a novel *Penicillium* mycotoxin, possibly a glycopeptide fractionating in the molecular weight range of approximately 1,500, may have been responsible for the nephrotoxicity. Nephrotoxic strains of *P. aurantiogriseum* producing tubular damage at the cortico-medullary junction of rat kidneys were also isolated from air, cereals, and foodstuffs in Yugoslavia (MacGeorge and Mantle 1991).

Isolates of *P. commune* were frequently isolated but were either non-nephrotoxic or only mildly so (MacGeorge and Mantle 1991). Many other species of *Penicillium* and *Aspergillus* were also isolated (MacGeorge and Mantle 1991). No species commonly producing OTA were found (Mac-George and Mantle 1991).

Zhen et al (1991) reported the mutagenicity of extracts of *P. aurantiogriseum*; strains were isolated from samples of cereals in Linxian, China, and cultured. Unscheduled DNA synthesis, DNA synthesis inhibition, muta-genesis in the *Salmonella typhimurium* assay and *Escherichia coli* reversion assay, and increases in the frequency of sister chromatid exchanges were among the effects induced by culture extracts in various mutagenicity assay systems. The potential for contamination by penicillia producing a potent, as yet uncharacterized, nephrotoxin is an important question in the BEN situation (MacGeorge and Mantle 1991).

A strikingly high incidence of urinary tract tumors has been reported in the BEN endemic regions, and it is speculated that the two diseases may have a common causal relationship (Castegnaro et al 1990). Papillomas and/or carcinomas of the renal pelvis, ureter, or bladder have been reported in approximately a third of the patients dying from BEN (Castegnaro and Chernozemsky 1987 cited in Kuiper-Goodman and Scott 1989).

Kuiper-Goodman and Scott (1989) reviewed reports of the carcinogenicity, genotoxicity, immunotoxicity, and teratogenicity of OTA with the aim of assessing the risk of OTA exposure to human health. Renal tumors and hepatic cell tumors occurred in male mice fed a diet containing 40 ppm of OTA for 44 weeks. Nephropathy involving cystic dilation of renal tubules and sometimes hyperplasia of the lining epithelium was also observed in male rats fed 40 ppm of OTA in their diets. The TD_{50} (dose at which 50% of the animals would be expected to bear a specific tumor type) for mice was found to be 4,050 $\mu g/kg$ body weight (bw) of OTA per day. In rats fed OTA in their diets, the incidence of renal tumors was very common and was accompanied by a high frequency of metastases, particularly in the lungs and lymph nodes. An increased incidence in multiplicity of fibroadenomas in the mammary glands of female rats was noted. Rats were found to be much more sensitive to OTA than mice and had a TD_{50} of only 74 $\mu g/kg$ bw of OTA per day (Kuiper-Goodman and Scott 1989).

OTA is not considered to be a mutagen; however, weak positive responses for unscheduled DNA synthesis in cultured rat and mouse primary hepatocytes indicated mild genotoxicity (Kuiper-Goodman and Scott 1989).

OTA is a potent teratogen. Gross malformations, fetal mortality, and occasionally brain necrosis have been observed in mice and rats dosed with OTA by various means and at different times of gestation. Teratogenic effects of OTA have also been observed in hamsters and chickens but not in pigs (Kuiper-Goodman and Scott 1989).

A wide variety of immunotoxic effects have been induced in animals administered OTA. Bone marrow depression, decrease in lymphoid cell population, reduction in size of the thymus, decreased relative concentration of lymphocytes, and reduction of globulins and immunoglobulins in the blood plasma are among the many responses observed in various animal species dosed with OTA (Kuiper-Goodman and Scott 1989).

Kuiper-Goodman and Scott (1989) speculated that competitive inhibition by OTA of protein synthesis at the postranscriptional level may be the underlying mechanism of action responsible for many of these toxic effects.

With the intent of estimating intake levels for OTA in humans at which toxic effects are likely to be negligible, Kuiper-Goodman and Scott (1989) extrapolated animal toxicity data and applied different safety factors, which take into account the seriousness of the effect, uncertainty in the data, and inter- and intraspecies extrapolation factors. The estimated no observed effect level for nephrotoxicity to humans of 0.016 μg/kg bw of OTA per day was based on chronic toxicity in pigs. It was not possible to arrive at a tolerable daily intake (TDI) level that would account for possible subtle functional or other postnatal effects in the teratogenicity of OTA. Estimates of TDI to account for the carcinogenic effects of OTA were calculated in two ways, one using a safety factor of 5,000 and the other using a 95% confidence level. The resulting TDIs were 4.2 and 0.2 ng/kg bw per day (Kuiper-Goodman and Scott 1989).

The etiology of BEN is still not known. No specific correlation between environmental factors, chemical or infective agents, and the disease has been established. Climatic conditions that favor the growth and toxigenicity of fungi were suggested to be associated with the occurrence of BEN. Animal data show conclusively that the mycotoxin OTA has a number of toxic effects, including nephrotoxicity, and almost certainly also represents a serious health risk to humans. It is logical to pursue OTA as having a possible etiological relationship with the human nephropathic disease BEN in view of the causal association between OTA and porcine nephropathy.

Although striking pathological similarities between OTA-induced porcine nephropathy and BEN have been observed, no positive evidence to link the two diseases has been presented. Indeed, some still argue that no close correlation between the incidence of OTA and human nephropathy has yet been established (Yeulet et al 1988). According to Kuiper-Goodman and Scott (1989), studies in BEN endemic areas on the presence of OTA in the blood of BEN patients and foodstuffs do not provide sufficient data for estimating the quantitative relationship between the extent of OTA exposure and the severity of human nephropathy. However, animal toxicity data have been extrapolated to estimate tolerable daily intakes of OTA for humans.

Further research into the role of other known and unknown nephrotoxic mycotoxins in the etiology of BEN is warranted. Epidemiological characteristics suggest a causal association between urothelial cancers and BEN; however, further investigations are required to substantiate this hypothesis. The International Agency for Research on Cancer recently reviewed the link between OTA and urothelial cancers in BEN endemic areas. OTA was rated as a possible human carcinogen (2B) as opposed to a "higher" classification, primarily because the epidemiological data were not conclusive (IARC 1993).

Red Mold Disease in Japan

Interest in the *Fusarium* toxin deoxynivalenol first arose as a result of reports of human and animal toxicoses associated with the ingestion of grain contaminated with *Fusarium graminearum* in Japan (Yoshizawa 1983). *Fusarium* contamination of grain in Japan was referred to as *red mold* disease because of the reddish discoloration of the seeds infected by the fungus. *F. graminearum* has since been identified as the predominant fungus in scabby cereals in Japan, and the degree of visual contamination of grain was found to correlate with the detection rates of *Fusarium* species (Yoshizawa 1983). Japanese reports of field outbreaks of suspected mycotoxicoses in animals and humans and the early mycotoxicological research leading to the identification and characterization of toxic fungal metabolites has been reviewed by Yoshizawa (1983).

Epidemics of red mold disease in wheat and barley were reported as early as 1914. In 1933, the suggestion was made that grain contaminated with red mold may be the causative agent responsible for certain clinical symptoms in animals and humans (cited in Yoshizawa 1983). During the years from 1946 to 1963, sporadic outbreaks of a disease characterized by nausea, vomiting, diarrhea, abdominal pain, fever, and throat irritation were reported in humans in Japan and Korea. In two of these outbreaks symptoms occurred 5 to 30 min after the consumption of wheat noodles made from grain contaminated with red mold. Wheat, rice, and barley were implicated in the other outbreaks, in which symptoms generally appeared 1 to 2 h after ingestion of the grain products. In total, over 500 people were affected by the disease, although there were no fatalities (Yoshizawa 1983).

Mycological surveys of grain suspected to be implicated in two human outbreaks of toxicoses revealed heavy infection by *Fusarium graminearum* and *F. nivale*. Species of *Fusarium* found to be predominant in third-class Japanese rice and wheat included *F. graminearum, F. moniliforme, F. nivale,* and *F. kuhunii* in rice and *F. graminearum, F. nivale, F. poae,* and *F. oxysporum* in wheat (cited in Yoshizawa 1983).

In animals, it was suspected that damaged feedstuffs were responsible for field outbreaks in horses of a disease characterized by diarrhea, loss of appetite, increased temperature and heart rate, changes in the color and viscosity of the blood, lameness, and in severe cases, death. *F. graminearum* contamination was present in virtually all toxic moldy feed samples examined. Also predominant were *Alternaria alternata* and *Cladosporium herbarum* (cited in Yoshizawa 1983).

Toxicology studies in animals fed grain artificially contaminated with *Fusarium* and *Gibberella* (the sexual stage of *Fusarium*) soon substantiated the hypothesis of the involvement of these fungi in disease outbreaks. Symptoms of feed refusal, loss of body weight, hemorrhages in various organs, and death were observed in guinea pigs fed *Gibberella saubinetii* cultures of oats and wheat. *Gibberella zeae*, the sexual stage of *F. graminearum*, was isolated from samples of wheat associated with human toxicosis in 1946 and was found to elicit a toxic response in mice (cited in Yoshizawa 1983).

The next step in solving the red mold disease enigma was to isolate the toxic component in contaminated grain responsible for the observed toxicity so that toxicology studies using pure substances could be conducted and chemical structures of pure toxic metabolites be determined. Toxigenic strains of *F. nivale* and *F. graminearum* were isolated from the severe 1963 red mold infestation of wheat and barley in southern Japan. One of the strains of *F. nivale* has since been shown to be a toxigenic strain of *F. sporotrichioides* (Marasas et al 1984a).

Fractions of crude extracts from cultures of these fungi were tested for toxicity by various animal toxicity experiments and cell culture bioassays until eventually the toxic principles were isolated. In this fashion, nivalenol, fusarenon-X, and 4,15-diacetylnivalenol were isolated and structurally determined in the late 1960s. Similarly, toxic metabolites were isolated from rice cultures inoculated with *Fusarium* spp. isolated from the 1970 red mold grain infestation in Japan. The structures were determined to be deoxynivalenol, 3-acetyldeoxynivalenol, and 3,15-diacetyldeoxynivalenol. These mycotoxins are trichothecenes.

The mycotoxins deoxynivalenol, nivalenol, and zearalenone are known to occur naturally in grain, flour, and commercial foods in Japan (Yoshizawa 1983, Yoshizawa and Hosokawa 1983, Tanaka et al 1985) and may represent a health risk to humans. The toxicity of various pure trichothecene metabolites has been investigated in animals (chapter 9).

Determining the exact toxicity of the *F. graminearum*-contaminated grain is complex because of the possible synergistic nature of the metabolites. For example, the monogastric animal, swine, is sensitive to deoxynivalenol and displayed a 20% reduction in feed consumption when fed a diet to which 3.6 ppm of deoxynivalenol had been added and a 90% reduction at 40 ppm. In addition, naturally contaminated maize was found to have a greater feed refusal effect on the pigs than feed to which the same concentration of deoxynivalenol had been added. This suggests the possibility of another factor acting in synergy with deoxynivalenol (Forsyth et al 1977; chapter 9).

Fusariotoxicoses in China

Frequent outbreaks of toxicosis associated with the consumption of moldy wheat and maize in humans have been reported in China since 1961 (Luo 1988a). Clinical symptoms include nausea, abdominal pain, diarrhea, dizziness, and headaches. Fever and disturbances of the nervous system have also been reported. At least 7,818 people are reported to have been affected with positive correlations between those consuming the implicated grain and those manifesting symptoms ranging from 35 to 100% (Luo 1988a). In only two of the outbreaks reported by Luo (1988b) is an asymptomatic interval mentioned. Victims experienced symptoms between 5 and 30 min after ingesting incriminated grain. Of the 35 outbreaks reported, 26 occurred between the months of March and July, the remaining nine being unspecified (Luo 1988b).

In the May 1973 outbreak in Anhui province involving 51 people, men and women were equally affected. There was no mention of predisposition for age or sex in any of the other outbreaks in the report by Luo (1988b).

Strains of *F. graminearum* were isolated from moldy wheat and maize associated with food poisoning and tested for deoxynivalenol toxigenicity in rice culture. Of the 139 isolates tested, 76 were able to produce deoxynivalenol, and 12 of these were able to produce 500–2,500 ppm of deoxynivalenol in rice culture (Luo 1988b).

Moldy maize samples from Hebei Province were analyzed for the *Fusarium* toxins deoxynivalenol, nivalenol, T-2 toxin, and zearalenone (Luo 1988b). No nivalenol or T-2 were detected; however, in one of the five samples deoxynivalenol was detected at a level of 92.8 ppm. The average concentration of deoxynivalenol in these five samples of maize was 22.9 ppm. Zearalenone concentrations in three of the five samples were determined and ranged from 0.004 to 0.587 ppm (Luo 1988b).

Wheat samples collected during food poisoning outbreaks had significantly higher deoxynivalenol concentrations than those of the nontoxic wheat samples. Deoxynivalenol concentrations in 19 samples of moldy wheat collected from three different villages in Henan Province ranged from 1.0 to 40.0 ppm with an average of 17.4 ppm. Zearalenone was detected in two samples at levels of 0.25 and 0.5 ppm (Luo 1988b). The reports by Luo (1988a,b) provide evidence for the involvement of deoxynivalenol-contaminated grain in the etiology of human food poisoning.

Luo (1988a) suggested a tolerance level of 1.0 ppm for deoxynivalenol in wheat in China. This recommendation is based on tolerance levels set in other countries as well as occurrence data for mycotoxins in China.

Deoxynivalenol Toxicosis in India

In the months of July through September of 1987, human food poisonings in the Kashmir Valley in northwestern India were reported (Bhat et al 1989a,b). An epidemiological study in the affected area was conducted by the National Institute of Nutrition in Hyderabad, India.

The disease was found to be noncommunicable (Bhat 1989), and 50,000 people were estimated to have been affected irrespective of age or sex (Bhat et al 1989a). Of the 150 families surveyed, 97 members from 39 households were afflicted and experienced symptoms 15 min to 1 h after the consumption of breads prepared from certain consignments of wheat (Bhat et al 1989b). Only those who had consumed wheat were affected, and symptoms usually persisted for only as long as the implicated wheat product was consumed. One to six of the breads containing 10 to 30 g of wheat were generally consumed. All of those affected experienced abdominal pain or a feeling of fullness in the abdomen, 63% experienced throat irritation, and 39% experienced diarrhea. Vomiting, blood in the stool, and, in a small percentage, allergic reactions were also reported (Bhat et al 1989b). In addition, the epidemiological survey revealed a dose dependency, whereby the more incriminated wheat product consumed the more symptomatic the victim (Bhat, *personal communication*). No deaths were attributed to the disease.

The sequence of events leading to the occurrence of mold-damaged wheat in the local marketplace is noteworthy and emphasizes the economic element in such outbreaks, particularly in developing countries. During the May wheat harvest season in 1987, unseasonal rains occurred in the area from which the incriminated wheat originated (Bhat et al 1989b). The wheat crop was left standing in the field before being harvested and became moldy. This rain-damaged wheat was sold at a much reduced price to the local millers, who mixed it with good wheat to the extent of 50% (Bhat 1989). The resulting wheat flour was bought by the local bakers, who in turn sold it to consumers either in the form of flour or bread. Consumers noted a distinct difference in the consistency of bread baked from the affected wheat and described it as "chewing gum" wheat flour (Bhat et al 1989a).

Samples of wheat grain and wheat flour actually implicated in the outbreak were collected from households and traders along with samples that had been seized from dealers by government authorities under the Prevention of Food Adulteration Act (PFA). In all, 24 samples were obtained. Microscopic examination of these samples revealed that they conformed to the regulations set out by the PFA in terms of moisture content, total ash, ash insoluble in dilute HCl, alcoholic acidity, white worms, black beetles, webs, rodent hairs, excreta, and gluten (Bhat et al 1989b).

Mold infestation in these samples was assessed by mycological examination as well as by analysis for ergosterol, a chemical constituent of fungal cell walls that is indicative of the degree of mold contamination (Martin et al 1990). Upon mycological examination *Fusarium* species were isolated from all samples of wheat grain and flour and *Aspergillus flavus* from eight of the 24 samples. Other fungi isolated included *A. niger, A. candidus, A. terreus,* and *Penicillium* sp., with a total seed infection rate ranging from 31 to 36%. Ergosterol was found to be present in the 24 samples at concentrations ranging from 1.8 to 5.4 ppm (Bhat et al 1989b).

In the 24 samples of wheat grain and wheat flour, deoxynivalenol was detected in 11 samples at levels in the range of 0.346–8.38 ppm, nivalenol in two samples in the range of 0.03–0.1 ppm, acetyldeoxynivalenol in four samples in the range of 0.6–2.4 ppm, and T-2 toxin in five samples in the range of 0.55–4 ppm. Several trichothecenes could not be identified. Tests for pesticide residues were negative, and neither aflatoxin nor ergot alkaloids could be detected (Bhat et al 1989b).

Some 30 samples of wheat flour and wheat grain, which served as controls, were obtained from a local market in Hyderabad in south central India. Although both *Fusarium* and *A. flavus* could be isolated from these samples, no ergosterol was detected (Bhat et al 1989b).

Ramakrishna et al (1989) examined the deoxynivalenol producing potential of 37 strains of *Fusarium* spp. isolated from the wheat grain and wheat flour samples from Kashmir. Only two strains of *F. graminearum* and four strains of *F. culmorum* were able to produce deoxynivalenol in liquid medium, and concentrations were reported in the range of 5–186.0 and 0.37–5.0 μg/g in liquid medium, respectively, for the two species.

Toxicology studies were conducted on two-month-old pups fed bread containing the toxins extracted from the contaminated wheat samples at a dose of 0.2 mg/kg bw. While the control pups displayed no toxic signs, all those that had consumed the bread containing toxins vomited within 1 to 3 h of ingestion (Bhat et al 1989b). Deoxynivalenol levels of 0.3 mg/kg bw were necessary to evoke an emetic response in monkeys (Bhat 1989). Pigs dosed with deoxynivalenol exhibited an emetic response at an oral dose of 0.1 mg/kg bw and an intraperitoneal dose of 0.05 mg/kg bw (Forsyth et al 1977).

This example of possible human mycotoxicosis resulting from the ingestion of contaminated grain that conformed to existing regulations reflects the need for legislation of tolerance levels of toxic fungal metabolites in foods and feeds. Tolerance levels have been set or recommended for deoxynivalenol in Canada, Romania, the former U.S.S.R., and the United States (Van Egmond 1989).

Bhat et al (1989b) estimated that mild symptoms were caused by the consumption of 30–150 g of wheat flour containing a minimum of 0.34 $\mu g/g$ of deoxynivalenol. Based on a 10-fold safety factor, average body weight, average consumption of wheat products, and the minimum concentration of deoxynivalenol in naturally contaminated wheat products needed to evoke disease symptoms in humans, Bhat et al (1989a) suggested a tolerance level of 34 ppb for deoxynivalenol in wheat and wheat products.

Some of the evidence implicating deoxynivalenol and related compounds as etiological principles associated with the human mycotoxicoses in Japan, China, and India is equivocal. However, epidemiological data in all three cases point towards deoxynivalenol-contaminated grain products as the causative agent.

Esophageal Cancer

In parts of southern Africa, China, and northeastern Iran, the incidences of esophageal cancer (EC) are the highest in the world. This type of cancer is relatively rare in other parts of the world. Within the high-incidence areas, substantial variations in EC rates may be separated by only short geographical distances. Over the last few decades, escalations in EC rates have occurred in certain of these high-incidence areas. Epidemiological characteristics have prompted researchers to investigate environmental contaminants in the etiology of the affliction.

Epidemiological studies of EC were initiated in China in 1959. Cancer of the esophagus is particularly prevalent in certain provinces that border the Taihang mountains in northern China, notably, Henan, Hebei, and Shanxi provinces (Yang 1980). In Linxian county, Henan Province, the average age-adjusted mortality rates per 100,000 were reported as 161.33 for males and 102.88 for females. These were the highest rates reported in an epidemiological study done in China between 1973 and 1975 (Li et al 1980). The geographic distribution pattern is a series of irregular concentric circles decreasing in mortality with distance from Linxian county (B. Q. Liu and Li 1984). Other regions of high EC incidence in China can be found in the provinces of Hubei, Anhui, Jiangsu, Sichuan, Fujian, Guangdong, Xinjiang, Heulongjian, Gansu, Qinghai, Neimenggu, and Xinzang (Yang 1980).

Similarly, certain areas of northeastern Iran that border the Caspian Sea are noted for their extremely high prevalence of EC. The district of Gonbad has age-adjusted incidence rates per 100,000 of 109 for males and 174 for females (Kmet and Mahboubi 1972).

The esophagus is the most common or second most common site of tumor development in many parts of eastern and southern Africa, whereas in parts of west Africa it is almost unknown (Cook 1971). The highest EC rates in the Transkei, southern Africa, are reported in the district of Kentani in which the age-standardized incidence rates per 100,000 are 45.0 for males and 29.3 for females (Jaskiewicz et al 1987a).

These areas in China, Iran, and Africa are extremely high risk areas for EC compared with low-risk populations in which incidence rates for males are generally less than 2/100,000 (Van Rensburg 1985).

Some of these high-incidence areas share various common epidemiological features. In northern China, the male-to-female ratio of patients with EC ranges from 1.44:1 to 2.63:1 (Yang 1980). Cook (1971) reported EC male-to-female ratios varying from 1.5:1 in the Transkei to 12:1 in western Kenya. In Iran, however, EC was more common in women than in men in the highest incidence area, Gonbad district, with a male-to-female ratio of 0.6:1 (Kmet and Mahboubi 1972). In the lower incidence areas of Iran, the situation is reversed such that EC is 1.2 to 3.1 times more common in men than in women.

Kmet and Mahboubi (1972) reported distinct agricultural practices that correlated with the distribution of EC. In very high incidence areas in Iran, wheat and barley were cultivated, whereas in low-incidence areas rice paddy, fruit, and tea were the main crops. In Africa, high-risk areas are always associated with the consumption of maize or wheat as the dietary staple (Van Rensburg 1985). The consumption of wheat is increasing in Linxian, China, replacing maize bread and millet as the staples (Li and Cheng 1984).

Low annual rainfall and saline soils are environmental factors common to the high-incidence areas in Iran and Africa (Van Rensburg 1985). Yang (1980) claimed that there was no correlation between EC mortality and rainfall, altitude, soil type, or temperature in China. However, Van Rensburg (1985) suggested that the sharp EC incidence gradient around the Taihang Mountains, China, may be related to the hydromorphic properties of the soils.

Trace elements have been extensively studied in China for their possible etiological role in the development of EC. Levels of molybdenum, manganese, zinc, magnesium, silicon, nickel, iron, bromium, iodine, chlorine, potassium, sodium, phosphorus, and carbonic acid were investigated in drinking water and were found to have an inverse correlation with EC mortality (Yang 1980). Van Rensburg (1985) reported the tendency of water in the Transkei to accumulate certain heavy metals that may act antagonistically to the utilization of essential minerals. The deficiency of some beneficial minerals in the water of high risk regions of the Transkei was also pointed out (Van Rensburg 1985).

Analysis of zinc and copper in serum of EC and cardial cancer patients in China revealed a lower average level of zinc but a higher level of copper in cancer patients than in normal individuals (Yang 1980). Zinc was found in higher concentrations in the urine of cancer patients than in normal individuals. The reverse was true for magnesium (Yang 1980). Correlations between EC mortality rates and levels of certain trace elements in the hair of individuals from high- and low-incidence areas have also been asserted (Yang 1980).

The consumption of hot tea, a thermal irritant, was not found to be a principal carcinogenic factor in China, since tea drinking is not as common in some high-incidence areas as it is in some low-incidence areas (Yang 1980). Inhabitants of some high-incidence areas habitually consume their food at temperatures of 60–70° C. Hyperplasia and mild dysplasia were observed in mice force-fed hot water at 80° C but not with water at 60–70° C (Yang 1980).

Mechanical irritants, such as thorns, spicules, sharp particles from the milling process, and siliceous bodies formed in some crop plants have been suggested as causal agents in EC, since piercing particles such as these have been found in grains and milled products from high-incidence areas. Oral health, the consumption of tannins in herbal medicines, and exposure to polycyclic hydrocarbons from poorly ventilated coal heating systems may also play contributory causal roles (Yang 1980).

The advancement of screening techniques has allowed for the detection of cytological abnormalities in the epithelium of the esophagus that are believed to be associated with the development of esophageal cancer. Epithelial dysplasia and esophagitis are believed to precede carcinoma of the esophagus (Crespi et al 1979, Yang 1980). Chronic esophagitis was found in 80% of the individuals examined from an area of high EC prevalence in Iran (Crespi et al 1979). Fungal infection of esophageal tissue was also found to be very common in hyperplastic epithelium near early carcinoma lesions (Li and Cheng 1984). Fungal infection, especially with *Candida albicans*, may be associated with precancerous lesions of the esophagus (Li and Cheng 1984).

Whether genetic susceptibility plays an etiological role in EC has been investigated. Although 25–60% of EC patients surveyed in northern China had a family history of the disease, most of these individuals had lived for more than 20 years with other patients in their families (Li and Cheng 1984). Li and Cheng (1984) suggested that common exposure to a carcinogen is the more likely explanation for the familial occurrence but recommended further research to determine the possible role of genetic susceptibility.

The Chinese have investigated nitrosamines as causal agents of EC. *N*-nitroso compounds are known chemical carcinogens, and some nitrosamines are known to selectively induce esophageal and forestomach cancer in rats (Li and Cheng 1984). Nitrates, nitrites, and secondary and tertiary amines are precursors to nitrosamines. Nitrate and nitrite were detected in most drinking water sources tested from a high-incidence area in Linxian county,

China. In addition, higher levels of salivary nitrites were detected in individuals suffering from epithelial dysplasia or EC than in the normal controls. Higher average levels of secondary amines were detected in foods from Linxian than from Fanxian, a low-incidence area. Secondary amines have been demonstrated to form nitrosamines when incubated with human gastric juice and sodium nitrite under specific pH conditions (Yang 1980). When food sources were analyzed for nitrosamines, food from high-incidence areas contained nitrosamines much more frequently than did foods from low-incidence areas (Yang 1980).

The association of EC with consumption of alcohol and tobacco smoking is unclear. Neither of these factors is likely to play a causal role in EC in Iran or China (Kmet and Mahboubi 1972). In a study done in the Transkei, smoking habits were found to correlate with EC distribution. Some 56% of males in low-incidence areas were nonsmokers, whereas only 14% of males were nonsmokers in high-incidence areas. However, Van Rensburg (1985) reported more recent epidemiological data that contradicted the theory of tobacco smoking as a major causal determinant in EC.

Cook (1971) considered the consumption of beer made from maize in the etiology of EC in Africa. She claimed that the geographic distribution of EC and the use of maize to brew beer show similar distribution patterns. This may be of particular interest in view of the observation that moldy maize is selectively used in beer brewing (Marasas et al 1981). Furthermore, the introduction of maize to Africa and its use as a constituent of beer corresponded with the temporal increase in incidence of EC (Cook 1971). The consumption of home-distilled spirits has not been associated with EC in Africa (Cook 1971).

Poor nutritional status is suspected to be a contributing factor to the incidence of EC. Insufficient intakes of animal protein, vitamin A, riboflavin, and vitamin C were reported in the high-incidence area of Gonbad in Iran (Kmet and Mahboubi 1972). In Africa, nicotinic acid, magnesium, vitamin C, carotene, riboflavin, calcium, and zinc intakes were described as being inadequate but not necessarily deficient (Van Rensburg 1985). Potential deficiencies in vitamins A and C may occur in the winter in China when the availability of fresh fruit and vegetables is limited (Yang 1980). Insufficient intakes of riboflavin and calcium are also a possibility in China.

Van Rensburg (1985) suggested that adverse climatic conditions and poor soil properties limit the variety and quality of foods that can be produced in high-incidence areas. This has a direct effect on the nutritional status of communities living by subsistence farming. In addition, food crops may have an increased susceptibility to invasion by opportunistic fungi (Van Rensburg 1985), further taxing their nutritional quality and presenting the possibility of introducing toxic fungal metabolites into the food chain.

It is likely that several contributory factors act together or in succession in the progress of EC. Several of the possible contributory factors mentioned above point towards the involvement of fungi. Studies in the high-risk areas of Iran, China, and Africa have been conducted to determine the possible role of several genera of fungi. *Fusarium, Alternaria,* and *Penicillium* are of particular interest.

Mold contamination may occur in foodstuffs as a result of poor storage. The consumption of pickled vegetables and other moldy foods is common in some areas of high EC incidence in China (Yang 1980). Cereals from high-incidence areas in China were reported to be more frequently contaminated by *Fusarium moniliforme, Penicillium aurantiogriseum, Aspergillus flavus, A. nidulans, A. fumigatus, Alternaria alternata, Mucor* spp., and *Rhizopus* spp. than cereals from low-incidence areas (Li and Cheng 1984, G. Liu et al 1991). Lacey (1988) investigated the mycoflora infecting wheat and barley in northeastern Iran and found that *Alternaria* spp. were common contaminants of freshly harvested grain, as were *Penicillium* spp. *Aspergillus restrictus* was predominant in stored grain. Van Rensburg et al (1982) found *Alternaria* spp. to be the dominant fungi infecting wheat collected from the homes of EC patients in the Gonbad district of Iran.

In the Transkei, a significantly higher number of maize kernels were infected by *F. verticillioides* (Sacc.) Nirenberg (syn. *F. moniliforme* J. Sheld.) in high-incidence areas than in low-incidence areas (Marasas et al 1981). Furthermore, a positive correlation was found between the logs of the mean incidence of *F. moniliforme* and the rates of EC in the respective areas (Marasas 1982). *F. sacchari* var. *subglutinans* (syn. *F. subglutinans*) was usually also isolated more frequently from maize kernels in high-incidence areas than in low-incidence areas (Marasas et al 1981).

Further to this study, Marasas et al (1988a) compared the degree of *F. moniliforme* contamination in maize from "affected" and "nonaffected" households in high- and low-prevalence areas of the Transkei. Affected households were classified as such if one or more of the occupants showed esophageal cytological abnormalities (mild to advanced cellular changes) as determined by brush biopsy. Mild cellular changes included folic acid deficiency, atypia, and mild dysplasia, whereas advanced cellular changes included moderate to severe dysplasia and cancer. *F. moniliforme* was significantly more prevalent in maize from cytologically affected households in the high-incidence area than cytologically nonaffected households in the low-incidence area. Thus, a positive correlation was found between *F. moniliforme* contamination of maize and cytological abnormalities in the esophagus of affected individuals (Marasas et al 1988a).

In order to assess the possible association between specific fungal metabolites and the incidence of EC, levels of known fungal toxins were compared in areas of high and low EC prevalence (Marasas 1982). Higher levels of both the *Fusarium* mycotoxins deoxynivalenol and zearalenone were detected in visibly moldy maize from high-incidence areas of the Transkei. However, toxicity studies on ducklings using culture material from isolates of three species of *Fusarium* (*F. moniliforme, F. subglutinans,* and *F. graminearum*) revealed no difference in acute toxigenicity of extracts of fungi isolated from maize in high-incidence areas and the corresponding species isolated from maize in the low-incidence areas in the Transkei. Approximately equal percentages of isolates of *F. subglutinans* from both the high- and low-incidence areas produced moniliformin. Highly toxigenic isolates of *F. graminearum* that produced zearalenone were obtained from both areas as well; however, zearalenone is not acutely toxic. The nature of the mycotoxin(s) responsible for the observed toxic effect of these *F. graminearum* isolates in ducklings was not reported.

Levels of *F. graminearum* toxins in wheat and maize in Linxian were compared with those in Shangqui, a low-risk area 250 km from Linxian (Luo et al 1990). The incidence and mean levels of deoxynivalenol and 15-acetyldeoxynivalenol were higher in maize samples from Linxian than Shangqui. No T-2 toxin was detected in maize from either area. Deoxynivalenol levels in Linxian wheat were also higher than in wheat from Shangqui. Hsia et al (1988) analyzed samples of maize from households of patients suffering from EC in Linxian and found that most samples contained very high levels of deoxynivalenol and nivalenol.

Purified deoxynivalenol and nivalenol extracts from these samples induced chromosome aberrations in V-79 cells (Chinese hamster lung fibroblasts) at very low levels (in nanograms per milliliter of medium). The pure *Fusarium* toxins nivalenol, deoxynivalenol, 3-acetyldeoxynivalenol, and T-2 toxin also induced chromosome aberrations in V-79 cells (Hsia et al 1986). Hsia et al (1988) referred to unpublished data in which T-2 toxin was identified in maize from Linxian. T-2 toxin is known to induce cardiovascular lesions and pancreatic, gastrointestinal, and brain tumors in rats (Schoental et al 1979).

The consistent feature of maize in high EC areas in Africa and China is contamination by *F. moniliforme.* A strain of *F. moniliforme* isolated from the high-prevalence area in the Transkei was grown on maize and then fed to rats (Marasas et al 1984b). Hepatic lesions including cirrhosis, nodular hyperplasia, and bile-duct proliferation developed in rats fed 8% culture material. The mortality rates of these rats was 100%; mean times to death was 75 or 57 days, respectively, for rats fed oven-dried or freeze-dried culture material. In addition, hyperplasia of the esophageal basal cells was observed particularly in the rats fed the freeze-dried culture material. Hepatocellular carcinomas and ductular carcinomas of the liver were observed in 80 and 63%, respectively, of rats fed a lower percentage of culture material in their diets.

Fusarin C, a *Fusarium* metabolite, has been shown to occur naturally on maize from Linxian (Cheng et al 1985) and the Transkei (Gelderblom et al 1984a). It is also produced by *Fusarium* spp. from North America (Farber and Sanders 1986, Savard and Miller 1992), Taiwan (Tseng et al

1990), and Europe (Thrane 1988). Although this compound is known to be a potent mutagen (Gelderblom et al 1983, 1984b), it is not believed to be associated with hepatocarcinogenicity in rats (Jaskiewicz et al 1987b, Marasas 1993).

Fumonisin B_1, a recently described mycotoxin, was isolated from a strain of *F. moniliforme* isolated from maize in a high-incidence area of the Transkei and fed to rats that had previously been exposed to the initiating carcinogen diethylnitrosamine (DEN) and to rats that had not been exposed to DEN (Gelderblom et al 1988). The cancer-promoting potential of fumonisin B_1 in DEN-initiated and non-DEN-initiated rats was then assessed as measured by the induction of gamma-glutamyl-transpeptidase-positive (GGT+) liver cell foci. Induction of GGT+ foci was observed in DEN-initiated and to a lesser degree in non-DEN-initiated rats fed dietary levels of 0.1% fumonisin B_1 (Gelderblom et al 1988).

In view of the induction of GGT+ foci in both initiated and noninitiated rats, Gelderblom et al (1988) suggested that fumonisin B_1 might be a complete carcinogen. Pathological changes including proliferation of bile ducts and hyperplastic nodules in the liver, lesions in the kidney, and necrosis in the epithelial mucosa of the stomach and myocardia were also observed in rats fed fumonisin B_1. No comparable liver or kidney lesions or induction of GTT+ foci were observed in control rats not fed dietary fumonisin B_1. Further experiments have shown this toxin to be a hepatocarcinogen (Gelderblom et al 1991, 1992).

Rats treated with fumonisin B_1 showed a significant reduction in weight gain compared with control rats, indicating a toxic effect of this mycotoxin (Gelderblom et al 1988). Equine leukoencephalomalacia is a naturally occurring mycotoxicosis (Marasas et al 1984b) associated with *F. moniliforme*-contaminated feed. It has been experimentally induced in horses dosed intravenously and orally with fumonisin B_1 (Marasas et al 1988b, Kellerman et al 1990). Fumonisin B_1 production by isolates of other *Fusarium* species, including *F. proliferatum* and *F. nygami,* has been demonstrated (Thiel et al 1991).

Sydenham et al (1990) reported comparative levels of fumonisin B_1 and fumonisin B_2 in healthy maize from areas of high and low EC incidence in the Transkei. Fumonisins B_1 and B_2 were detected more frequently and at higher levels in maize from high-incidence areas than maize from low-incidence areas. More extensive surveys found very high concentrations of fumonisins in corn from high EC areas in Africa (Thiel et al 1992).

The relationship between *Alternaria* toxins and EC has been extensively studied in China. G. Liu et al (1988, 1991) and Zhen et al (1991) reviewed the Chinese research on *Alternaria* and cite studies in which extracts of *A. alternata* cultures showed mutagenic activity.

Extracts sometimes required metabolic activation by liver microsomal fraction, S-9 mixture. *A. alternata* extracts induced chromosomal aberrations and sister chromatid exchanges in human peripheral blood cells. Claims have also been made that *Alternaria* culture extracts were able to induce 8-azaguanine-resistant mutations in V-79 cells, unscheduled DNA synthesis in human amnion epithelial cells, and transformations in NIH-3T3 cells (mouse fibroblasts) (Dong et al 1987, G. Liu et al 1991). Extracts of *Alternaria* culture material have been reported to be mutagenic to *Salmonella typhimurium* in the Ames test (Van Rensburg et al 1982). Forestomach papillomas and esophageal papillomas have been reported in rats fed maize inoculated with *A. alternata* (G. Liu et al 1988). Van Rensburg et al (1982) reported a significant enhancement in the number of esophageal tumors in nitrosamine-induced rats fed 2.5% *Alternaria* culture material over the control rats.

Scott and Stoltz (1980) reported alternariol methyl ether, an *Alternaria* metabolite, to be weakly mutagenic without metabolic activation in the *Salmonella typhimurium* assay. Alternariol monomethyl ether was found to be mutagenic to V-79 cells and *Escherichia coli* ND160 cells. Alternariol monomethyl ether is also reported to have transformed NIH-3T3 cells such that they could grow into colonies in soft agar and into tumors after implantation into the subcutaneous tissues of nude mice. Cultures of human fetal esophageal epithelial tissues were found to undergo hyperplasia and papillary growth after exposure to alternariol monomethyl ether (G. Liu et al 1991). Zhang et al (1991) verified the binding of alternariol monomethyl ether to the DNA of human fetal esophageal cells and suggested that this binding action may play a role in the mutagenicity of alternariol monomethyl ether.

Alternariol, another *Alternaria* metabolite, was also found to be mutagenic in several cell bioassays tested (Zhen et al 1991). Chinese scientists reported the activation of oncogenes in human fetal esophageal epithelium treated with alternariol (Zhang and Dong 1991). NIH-3T3 cells were transformed by transfection of DNA from the human fetal esophageal epithelium (Zhang and Dong 1991).

The *A. alternata* metabolites altertoxins I, II, and III have also been shown to be mutagenic in the Ames *Salmonella typhimurium* assay both with and without microsomal activation (Stack and Prival 1986).

As noted, some other fungi have been found in grains from high EC areas. Zhen et al 1991 reported the mutagenicity of extracts of *Penicillium aurantiogriseum*. Strains of *P. aurantiogriseum* were isolated from samples of cereals in Linxian and cultured. Unscheduled DNA synthesis, DNA synthesis inhibition, mutagenesis in the *S. typhimurium* assay and *E. coli* reversion assay, and increases in the frequency of sister chromatid exchanges were among the effects induced by culture extracts in various mutagenicity assay systems.

Another fungus, *Geotrichium candidum,* was present in 83% of pickled food samples in northern China (Li 1982), and a positive correlation between

EC mortality rate and the frequency of consumption of pickled vegetables was found in some but not all of the communes studied (Yang 1980). Adenocarcinoma of the glandular stomach, angioendothelioma of the thoracic wall, fibrosarcoma of the liver, and epithelial dysplasia lesions in the esophagus and forestomach were induced in rats fed extracts and concentrated liquid from pickled vegetables (Yang 1980). Roussin red methyl ester ($Fe_2[SMe]_2[NO]_4$) was found to occur in pickled vegetables from Linxian and was found to promote skin tumors in DEN-initiated rats (J.-G. Liu and Li 1989).

The presence of nitrosamines and their precursors in foods is suspected to be attributable to soil composition and mold contamination (Yang 1980). Concentrations of nitrates, nitrites, secondary amines, and the promotion of nitrosamines from their precursors have been reported to increase with mold contamination by various fungi on foodstuffs (Yang 1980). Li and Cheng (1984) presented results from earlier studies in which a 17-fold increase in secondary amines occurred on maize bread inoculated with *F. moniliforme*. The formation of the carcinogenic nitrosamines, dimethylnitrosamine, diethylnitrosamine, methylbenzylnitrosamine, and N-1-methyl-acetonyl-N-3-methylbutylnitrosamine (MAMBNA), have been demonstrated on *F. moniliforme* inoculated maize bread supplemented with $NaNO_2$ (Li and Cheng 1984).

Feeding experiments were conducted on rats fed moldy maize bread that was either supplemented or not supplemented with $NaNO_2$ (Li and Cheng 1984). Rats fed maize bread inoculated with *F. moniliforme* developed papillomas and early carcinomas in the forestomach after 554–701 days of treatment. Other lesions observed included epithelial hyperplasia in the esophagus, forestomach, and glandular stomach as well as fibroadenomas and adenomas of the mammary glands. More papillomas and carcinomas in the forestomach were observed when the rats were fed $NaNO_2$ supplemented, *F. moniliforme*-inoculated maize bread. MAMBNA was formed in the maize bread supplemented with $NaNO_2$. In retrospect, it is probable that the *F. moniliforme*-inoculated maize bread contained fumonisins and fusarins.

Xia (1984) reported a significant enhancement of benzylmethylnitrosamine formation when *Candida albicans,* the pathogenic fungus frequently associated with esophageal infections, was cultured in the presence of benzylmethylamine and $NaNO_2$. Van Rensburg (1985) questioned the involvement of foodborne nitrosamines in the etiology of EC and referred to studies in Iran and the Transkei in which it was shown that exposure was minimal and did not differ between regions. In addition, he pointed out that exposure to nitrosamines through tobacco smoke is much more significant than from foodborne exposure (Van Rensburg 1985).

No single factor or combination of factors has been identified as a causal determinant in human esophageal cancer. High-incidence areas in a number of countries have been identified and are characterized by sharp gradients of incidence. Differences in agricultural practices, climatic factors, soil characteristics, cultural practices, levels of micronutrients in the food and water, and nutritional status of inhabitants, as well as potential carcinogenic compounds, mycoflora, and mycotoxins in foods have been recognized between these high- and low-prevalence areas.

Cultures of fungi isolated from areas of high EC prevalence have been shown to produce mycotoxins, especially fumonisins and fusarins. The toxicity, mutagenicity, and cancer-promoting potential of these fungal metabolites have been established in cell and tissue culture systems and animal models. The International Agency for Research on Cancer recently reviewed the carcinogenicity of cultures and grain contaminated by *F. moniliforme* containing fumonisins and fusarins. This material was considered a possible human carcinogen (IARC 1993). Animal toxicoses of fumonisins, especially in equine species, have long been documented testifying to the importance of these compounds in nature (Marasas 1993).

It is the general belief of researchers that the etiology of human esophageal cancer probably involves several factors. Nutritional factors may play a role before or during the initiation process and natural protective mechanisms may not function when certain nutrients are absent or deficient. Xia (1984) suggested that vitamin A might protect DNA from the action of chemical carcinogens and vitamin C might block the synthesis of nitrosamines. Certainly the possibility of multiple etiological factors complicates the elucidation of the mechanism of carcinogenesis. A better understanding of the individual parts played by the various possible etiological factors mentioned here, and perhaps others, is necessary before their combined or successive roles are to be determined. However, it is clear that exposure to fumonisins in the high EC areas of Africa and China is a significant factor (Marasas 1993).

Onyalai

Environmental causes have been investigated in the etiology of a disease known as *onyalai* (Rabie et al 1975). Many reports of this disease have appeared in the literature since its first description in 1904 in Angola, several of which are cited in Lewis and Lurie (1953) and Wicks (1972). Onyalai occurs almost exclusively in African races, where it is widespread south of the Sahara (Lurie et al 1969). Several cases have also been reported in Europeans and Chinese (Wicks 1972).

The typical lesions in patients with onyalai are distinctive hemorrhagic bullae that appear suddenly on the mucous membranes of the oral cavity and sometimes on the skin (Hesseling 1985). Hematuria and profuse bleeding from the nose, mouth, and conjunctiva follow (Lewis and Lurie 1953, Hesseling 1985). Hemorrhagic shock and cerebral hemorrhages occur in severe cases and may result in death within a few days of the onset of the disease (Hesseling 1985). Shortened platelet life span and thrombocytopenia are characteristic in onyalai patients (Rabie et al 1975). A significant proportion of first time sufferers recover spontaneously within two months of the onset, and their platelet levels often return to normal; others experience recurrent exacerbations (Lewis and Lurie 1953). The severity and mortality of onyalai vary depending on the area of its occurrence. Mortality in one case study was reported to be 14% (Wicks 1972).

The disease may occur in individuals of all ages and in both sexes (Hesseling 1985), although some authors report a predominance in males (Wicks 1972). It is neither heritable nor infectious (Rabie et al 1975). The incidence of onyalai is not limited to any particular season, but it is reported to occur most frequently in the summer (Lewis and Lurie 1953). In one year in the early 1980s the incidence of the disease in northeastern Namibia was reported to be 151.4 per 100,000 (cited in Hesseling 1985).

Onyalai has been compared to a disease in Caucasians known as idiopathic thrombocytopenic purpura, in which decreased platelet life span is also a feature along with an immunological pathogenesis (Rabie et al 1975); however, the two are now considered to be separate disease entities (Hesseling 1985). Brink et al (1981) studied the difference in the immune responses of onyalai and immune thrombocytopenic purpura, a condition in which antibodies against platelets are present in the serum. They found that while patients with immune thrombocytopenic purpura had immunoglobulin G (IgG) platelet antibodies in their serum, most patients with onyalai had both IgG and IgM antibodies in their serum. According to Brink et al (1981) the presence of IgM antibodies in the serum of onyalai patients suggests the possibility of a toxin, perhaps acting as a hapten, in the etiology of onyalai. Furthermore, they proposed a difference in etiologies for the two different immune responses, thus providing further indication for the distinctness of the two diseases. Onyalai patients have been reported to respond well to steroid administration, platelet enriched blood, and hospital diets, but no treatment of specific effect has been identified (Wicks 1972).

The cause of onyalai is not known. Vitamin deficiencies, local tribal medicines, poisons from trees of the Euphorbiaceae family, allergic reactions, hookworm, syphilis, malaria, and bilharzia are among the proposed causative factors in the disease (cited in Rabie et al 1975, Lewis and Lurie 1953). It has been hypothesized that a mycotoxin in the staple food of onyalai patients may be responsible for the disease (cited in Lewis and Lurie 1953).

It was noted that in Ovamboland, northern Namibia, the disease occurs only among eaters of millet (*Pennisetum typhoides*) (cited in Rabie et al 1975). A mycological study of fungi occurring in millet and sorghum (*Sorghum vulgare*) from Ovamboland was undertaken. Thirty-one different species of fungi were isolated from three samples of millet obtained from the house of an onyalai patient. *Phoma sorghina* and *Fusarium fusarioides* (syn. *F. chlamydosporum*) were the most prevalent (Rabie et al 1975). These two fungi along with another unidentified species of *Fusarium* and *Phoma jolyana* were highly toxic to day-old chicks fed a mash consisting of 50% of the respective fungal culture material grown on maize (Rabie et al 1975). *F. chlamydosporum* has since been shown to produce a number of slightly toxic compounds (Abbas et al 1992).

Aflatoxin, sterigmatocystin, and ochratoxin were analyzed for but not detected in the original millet samples. A toxigenic strain of *P. sorghina* was also predominant in a sample of sorghum obtained from the household of another victim of onyalai. *P. sorghina* was present in almost all field and stored grain samples of millet and sorghum collected from Ovamboland during the 1973 harvest (Rabie et al 1975).

Day-old chicks fed *P. sorghina* contaminated material developed hemorrhages in the gastrointestinal and respiratory tract commencing on the third day, and all died within seven days with trace blood around the beak and cloaca (Rabie et al 1975). Eight of 11 isolates of *P. sorghina* in millet from Ovamboland proved to be toxic when tested in this fashion, while the other three as well as an isolate from India proved to be nontoxic. It was later found that not more than 50% of the chickens used in the experiments were susceptible to the toxin; thus, the nontoxic results may be false negatives (Rabie et al 1975).

Anorexia; thrombocytopenia; hematuria; epistaxis; blood in the feces, intestine, stomach, and mouth; and death were observed in rats fed a diet containing 10–50% *P. sorghina* culture material mixed with rat ration. Although no hemorrhagic bullae were observed in the oral cavity, damage to the vascular system and small hemorrhages in a wide variety of tissues were prominent (Rabie et al 1975).

With the aim of characterizing the toxic principle, a methanol fraction that was toxic to day-old chicks was isolated from *P. sorghina*-inoculated maize culture. From a further extraction of this toxic methanol fraction, calcium-, magnesium-, and sodium-tenuazonate were isolated. Acid treatment of the tenuazonate powder yielded tenuazonic acid (Steyn and Rabie 1976), a fungal metabolite also known to be produced by certain species of *Alternaria* (Meronuck et al 1972).

Of 85 isolates of *Alternaria alternata,* 57 were toxic to rats fed maize-rice cultures of the fungus. Hemorrhages in the gastrointestinal lumen were among the lesions observed in the post mortem examination of the rats. Most of the isolates of toxigenic *Alternaria* investigated produced tenuazonic acid (Meronuck et al 1972). Sauer et al (1978) conducted toxicity studies using *A. alternata* cultures that either produced alternariol monomethyl ether, alternariol, and altenuene, or these three toxins in addition to tenuazonic acid and altertoxin I. Necrosis and hemorrhages in the gizzard and isthmus were observed in chicks fed diets consisting of 50% *Alternaria* culture material that produced all five toxins. No deleterious effects or lesions in these organs were observed in chicks fed diets consisting of 50% *Alternaria* culture material that did not produce altertoxin I and tenuazonic acid. Although altertoxin I may have contributed to the toxicity of the culture material, Sauer et al (1978) concluded that "tenuazonic acid is responsible for most of the lethal effect observed in laboratory feeding tests of *Alternaria* isolates grown on grain substrates."

Emetic responses and hemorrhages in vital organs and the gastrointestinal tract were observed in mice, rats, and dogs dosed orally and intravenously with sodium tenuazonate (Smith et al 1968). Two primary toxic actions for tenuazonic acid have been suggested, namely, an emetic action and a cardiovascular action (Smith et al 1968). Steyn and Rabie (1976) proposed that the toxicity of tenuazonic acid may be in part related to its ability to selectively complex with trace metals in vivo but suggest further investigation in the determination of its role in onyalai.

Onyalai is becoming less common (Goldin and Gelfand 1980), probably because there is better nutritional status in the endemic regions.

Conclusive evidence for the mycotoxin theory in the etiology of onyalai is lacking. The fact that onyalai presents in some individuals of a given family but not in others eating the same staple food from the same source is not consistent with the mycotoxin hypothesis. The argument given by Rabie et al (1975) to explain this apparent paradox is that of "considerable individual variation of tolerance." They suggest that a systematic histopathological study of the vascular system in human onyalai would be in order in view of the observed vascular damage in experimental rats.

Although certain disease symptoms and pathological lesions characteristic of onyalai have been experimentally reproduced in various animals fed *P. sorghina* and *Alternaria* culture material as well as in animals dosed with tenuazonate, in no case have the hemorrhagic bullae in the oral cavity been observed in these experimental animals. Furthermore, the evidence presented does not exclude the possibility that these lesions are being produced by environmental factors or even other mycotoxins.

Pellagra

Pellagra is a deficiency disease due to insufficient intake or failure of the body to absorb the B complex vitamin niacin (nicotinic acid) or its amide (niacinamide, nicotinamide) (Thomas 1977). Pellagra has traditionally been associated with the consumption of maize as a staple, since much of the niacin in maize is in an unavailable form, and the concentration of certain niacin precursors is low in maize (Srikantia 1978).

Endemic pellagra is known among people in certain communities in India whose staple diet is sorghum (Srikantia 1978). Sorghum is not deficient in either available niacin or its precursors (Srikantia 1978). Jaya Rao (1983) suggested that some other factor may be involved in the precipitation of pellagra among sorghum eaters since apparent inconsistencies were detected between the incidence of pellagra and dietary intakes of niacin. The diets in the sorghum-eating area of Andhra Pradesh, India, were not highly deficient in niacin even though pellagra was endemic. Niacin intake was even lower in rice-eating communities in which pellagra was unknown. Jaya Rao (1983) considered the possibility that pellagra is associated with mycotoxins and recommended a critical examination of the theory.

Schoental (1980) cited many outbreaks of pellagra that have occurred in the past as a result of people consuming food or drink made from moldy or "unwholesome" maize and put forth the theory of an etiology of mycotoxic basis. The possibility of trichothecene mycotoxins and perhaps, more specifically, T-2 toxin playing a causal role in pellagra was suggested. The hypothesis was made that historical accounts of variation in the severity of disease symptom manifestations could be attributed to variable susceptibility to mycotoxins depending on nutritional status. Schoental (1980) argued that deficiency of vitamin-dependent coenzymes would prevent the occurrence of detoxification processes, thus, nutritionally taxed individuals may suffer more acute effects of mycotoxin exposure.

Schoental (1980) accounts not only for the historical incidence of pellagra but also for its lack of occurrence among populations in which one would expect at least occasional mold contamination of grain. He proposed that pellagra was not seen in the maize-consuming civilizations of South and Central America because of the manner in which the maize bread (tortilla) was prepared. Traditional pretreatment of ground maize with slaked lime during tortilla making hydrolyzes the ester groups in some trichothecene mycotoxins. This would leave the less toxic hydroxy derivatives that do not induce pellagra (Schoental 1980). Others, also claiming that lime treatment was fundamental in preventing pellagra in South and Central American Indians, asserted that the lime released bound niacin from the maize substrate making it available for uptake and thus preventing niacin deficiency in tortilla consumers (cited in Schoental 1980). Schoental (1980) contested this explanation, arguing that soaking and washing of the maize paste following alkali treatment would probably remove the niacin.

Evidence presented by Schoental (1980) and Jaya Rao (1983) in favor of mycotoxins as the primary causal agents in the etiology of pellagra is entirely circumstantial. Future concrete investigations into the fungi contaminating staples consumed by pellagra sufferers may reveal mycotoxins as relevant considerations in the disease etiology. Animal studies in which clinical features of pellagra are replicated would do much to clarify opposing theories on the etiology of pellagra.

Sago Hemolysis

Several outbreaks of a disease known as sago hemolysis, which affects both males and females, have been reported in Papua New Guinea since

1974 (Taufa 1974; Donovan et al 1976, 1977). Clinical features include severe anemia, sudden onset of jaundice, and dark red urine. The color of the urine and clinical picture are suggestive of hemolysis. In some cases patients experience fever and vomiting and, in those worst affected, mental confusion and loss of consciousness.

In four outbreaks reported by Donovan et al (1977) and Taufa (1974), 21 people from five different families were affected, and four died. This represents a mortality rate of almost 20%. Those most severely affected were given blood transfusions and either died or recovered in a few days.

Several members of the same family may be affected by the disease. This suggests a common genetic defect, common exposure to some environmental agent, or both (Donovan et al 1977). It is interesting that in all reported cases of the disease, the affected people have consumed "stale" sago the day before the onset of symptoms. Donovan et al (1977) reported the appearance of the dry sago implicated in one of the outbreaks as greenish gray and mostly powdery. When examined microscopically, some bacteria and fungal cells were observed but no fungal hyphae were noted. When the sago was plated on potato dextrose agar the only fungi isolated were one colony of each of *Paecilomyces lilacinus* and *Tilletiopsis minor* (Donovan et al 1977). Two species of yeast as well as the bacteria *Bacillus megaterium, B. subtilis,* and *B. cereus* were also isolated.

The sample was tested for aflatoxin, ochratoxin, byssochlamic acid, zearalenone, sterigmatocystin, and citrinin with no positive results (Donovan et al 1977). An orange fluorescent compound with a high R_f value, believed to be a degradation product of aflatoxin, was found. However, it was noted that it would be surprising that this degradation product be present in the absence of the parent compound, aflatoxin (Donovan et al 1977).

It is known that the red cells of individuals with glucose-6-phosphate dehydrogenase (G-6-PD) deficiency are more susceptible to hemolysis than normal red cells. It is unlikely, but not impossible, that all four affected females reported in the Donovan et al (1977) study were G-6-PD deficient, since the defect for this deficiency is transmitted as an X-linked recessive (Donovan et al 1977). Donovan et al (1977) speculated that G-6-PD deficient individuals may be more susceptible to the toxins present in stale sago than normal individuals.

The possible relationship between contributory environmental factors and the cause of sago hemolysis is unknown; however, some investigators consider the evidence for the involvement of stale sago to be fairly strong. As yet, no pathogenic organism or associated toxin has been identified.

Frontoethmoidal Encephalomeningocele

Teratogens in grain staples contaminated by fungi are suspected in the etiology of frontoethmoidal encephalomeningocele (FEEM) in Myanmar (formerly Burma). FEEM is the result of a neural tube defect and is characterized by a tumor protrusion at the base of the nose or between the eyes (Thu and Kyu 1984). The size of the protrusion may vary from approximately 1.5 cm to approximately 8 cm in diameter. Hypertelorism is common and vision and smell can be affected. FEEM is not a seriously debilitating or disabling disease. Motor function was not reported to be hindered in a survey of 35 individuals suffering from the disease (Thu and Kyu 1984).

FEEM is known to occur in England, the United States, Germany, Australia, South Africa, Nigeria, Morocco, and India (Thu and Kyu 1984). In these countries its occurrence is relatively rare compared to that in Russia, Thailand, and Myanmar.

Extensive epidemiological studies were conducted in Myanmar showed the incidence of FEEM was high in certain parts of the country and rare in others (Thu and Kyu 1984). The disease did not show patterns of familial aggregation or genetic inheritance. Karyotypes in all seven individuals analyzed were normal. In one study, 64% of the patients surveyed were under two years of age; however, the disease has been reported to affect all age groups, even one individual 90 years of age. FEEM had only a slightly higher incidence in males than in females, and the disease occurred in members from all major ethnic groups.

In a study of 114 FEEM patients who had attended the neurosurgical outpatient department of a large referral hospital in Myanmar, a high proportion of patients were children of professionals or managerial or clerical workers. It was suggested that this may have been a consequence of these groups making greater use of available health services. In a more extensive epidemiological study involving 389 cases of FEEM, a definite preponderance among peasants and agricultural laborers was evident, particularly in rice paddy workers and those who consumed rice as a staple (Thu and Kyu 1984). Vitamin deficiencies are not believed to play a role in the etiology of FEEM.

Unlike some other diseases associated with neural tube closure defects, there does not appear to be any association with paternal age or birth rank in affected individuals. Mothers of patients had experienced a longer interval between birth of the affected child and the birth of the preceding child (Thu and Kyu 1984). It was noted that when compared to a representative portion of the general population, a disproportionate number of sufferers were conceived during the rainy season months from June to September (Thu and Kyu 1984). During the rainy season flooding and humid conditions lead to the molding of grain consumed as staples in the villages.

An etiological hypothesis was formulated on the basis of epidemiological and environmental indications. Thu and Kyu (1984) proposed that a teratogen present in fungus-infected rice is consumed by expectant mothers during a critical period of pregnancy. The teratogen interferes with the development of the embryo such that interposing bone is not formed, and intracranial substance herniates through the osseus.

Leck (1974) reviewed the possible causal determinants that have been investigated in various neural tube closure defects and suggested that environmental factors must, in part, play a role. The only mention of studies involving the possible role of fungi as causal agents was one of consumption of blight affected potatoes by pregnant women (Renwick 1972). No studies have shown a positive correlation (Clarke et al 1973, Roberts et al 1973).

The hypothesis of a mycotoxic origin is based on epidemiological studies and has no mycotoxicological justification as yet. The notion that seasonally determined environmental factors may act as causal agents of FEEM is relatively recent, thus comprehensive studies that consider other aspects of the environment are still required.

Arthrinium Sugarcane Poisoning

Deteriorated sugarcane poisoning (DSP) is a disease of the nervous system that has been reported in China in which fungal metabolites have been investigated as causal agents. The disease results in torsion spasms and may leave the victim permanently disabled. Of the 847 cases reported from 13 provinces in China between 1972 and 1988, 84 have died. The disease is most prevalent during the period from February to April (Liu et al 1989).

Fusarium poae, F. moniliforme, Penicillium aurantiogriseum, and *Cladosporium* spp. have been suggested as possible etiological agents (cited in Liu et al 1989, Zhen et al 1988); however, Liu et al (1989) claimed that certain *Arthrinium* species (*A. sacchari, A. saccharicola,* and *A. phaeospermum*) are the etiological fungi.

Sugarcane samples, collected from the homes of intoxicated patients and from storehouses, were incubated on plates containing potato-2% glucose agar. Of the 582 strains of fungi and bacteria isolated from the sugarcane samples implicated in food poisoning, *Arthrinium* species were predominant, representing 44% of the fungal isolates. Only 58 strains of fungi were isolated from the normal sugarcane sample and none were *Arthrinium* sp. (Liu et al 1989).

Bacterial and fungal isolates were individually cultured in sugarcane juice medium and tested for toxicity on weaning mice by gastric intubation of 0.5 ml of culture supernatant. Of the 81 fungal isolates tested, 17 strains were toxic to mice, of which 15 were *Arthrinium,* one was *Alternaria,* and one was *Candida.* Symptoms seen in the mice fed *Arthrinium* culture material included moving in circles and paralysis of limbs 1–2 h after administration, followed by death of all the mice within 3 h. Pathological examination revealed encephaledema. The brain was the only affected organ; no changes were found in the liver, kidney, heart, lung, stomach, or intestine. The supernatant from the *Candida* culture was reported to cause symptoms similar to alcoholic poisoning but the toxic effects of the *Alternaria* culture were not described. None of the bacterial cultures were toxic to mice (Liu et al 1989).

A toxic fraction from a culture of *Arthrinium* isolated from sugarcane implicated in human food poisoning was isolated by TLC and identified as 3-nitropropionic acid (3-NPA) (Liu et al 1989). The pure toxin was detected from juices of poisonous sugarcane at concentrations of up to 1,600 ppm. Liu et al (1989) claimed this level is sufficient to cause human food poisoning outbreaks. 3-NPA from both *Arthrinium* cultures and toxic sugarcane juice caused encephaledema in four of nine mice. The LD_{50} for oral dosing of mice with pure 3-NPA was found to be 110 mg/kg bw for males and 68.1 mg/kg bw for females (Liu et al 1989).

Zhen et al (1988) found *Fusarium moniliforme* and *Penicillium aurantiogriseum* to be the predominant fungi associated with mildewed sugarcane. The observed pathological changes in mice dosed intraperitoneally with ethyl acetate extracts of *F. moniliforme* cultures included encephalomalacia, edema, and hyperemia of the brain and spinal cord; vacuolar degeneration of the brain stem and spinal cord; granular degeneration of renal tubes; liver swelling; and pulmonary edema (Zhen et al 1988). Toxicity studies conducted on mice and ducklings with extracts from either mildewed sugarcane or *F. moniliforme* and *P. aurantiogriseum* cultures resulted in nervous system disorders and death. Authentic isolates of *F. moniliforme* produce fumonisin B_1 and of *P. aurantiogriseum* produce various toxic factors (see sections on Esophageal Cancer and Balkan Endemic Nephropathy).

The causal agent(s) of DSP remains uncertain. Several toxigenic fungi have been isolated from deteriorated sugarcane, and a different pathological picture presents in mice when dosed with ethyl acetate extracts of *Arthrinium* sp. and *F. moniliforme*. An animal model in which the specific clinical and pathological changes associated with the disease in humans can be reproduced would greatly aid in the determination of the etiology of DSP.

Glossary

Definitions are from from Thomas (1977) unless otherwise indicated.

Acetabulum—the rounded cavity on the external surface of the innominate bone that receives the head of the femur

Achondroplasia—a defect in the formation of cartilage at the epiphyses of long bones, producing a form of dwarfism

Adenocarcinoma—a malignant adenoma arising from the epithelium of a glandular organ

Adenoma—a neoplasm of glandular epithelium

Agranulocytosis—an acute disease in which the white blood cell count drops to extremely low levels and neutropenia becomes pronounced

Angina—any disease characterized by attacks of choking or suffocation

Bulla—a large blister or skin vesicle filled with fluid

Carcinoma—a new growth or malignant tumor enclosing epithelial cells in connective tissue and tending to infiltrate and give rise to metastases

Chondrocyte—a cartilage cell

Diathesis—constitutional predisposition to a certain disease, condition, or group of diseases

Dysplasia—abnormal development of tissue

Edema—a local or generalized condition in which the body tissues contain an excessive amount of tissue fluid

Encephalomeningocele—protrusion of membranes and brain substance

Ethmoid bone—the sievelike spongy bone that forms a roof for the nasal fossae and part of the floor of the anterior fossa of the skull

Ethmoid sinus—the air cells or spaces inside the ethmoid bone, opening into the nasal cavity through the cranium

Ethmoidal—pertaining to the ethmoid bone or sinuses

Epiphyseal plate—the thin layer of cartilage between the epiphysis and the shaft of a long bone (Lapedes 1978)

Epiphysis—the end portion of a long bone in vertebrates (Lapedes 1978)

Epistaxis—hemorrhage from the nose

Fibroadenoma—adenoma with fibrous tissue forming a dense stroma

Hapten—a simple substance that reacts like an antigen in vitro by combining with antibody (Lapedes 1978)

Hemapoiesis—blood formation

Hematuria—a pathological condition in which the urine contains blood (Lapedes 1978)

Hyperplasia—excessive proliferation of normal cells in the normal tissue arrangement of an organ

Hypertelorism ocular—abnormal width between the eyes

Idiopathic—pertaining to conditions without clear pathogenesis or disease without recognizable cause, as of spontaneous origin

Leukopenia—abnormal decrease of white blood corpuscles, usually below 5,000 mm^3

Lymphocytosis—excess of lymph cells

Osteoarthritis—a chronic joint disease characterized pathologically by degeneration of articular cartilage and hypertrophy of bone

Osteochondritis dissecans—a condition affecting a joint in which a fragment of cartilage and its underlying bone become detached from the articular surface

Osteochondrosis—a disease characterized by avascular necrosis of ossification centers followed by regeneration (Lapedes 1978)

Osteodystrophy—defective bone development

Papilloma—any benign epithelial tumor

Petechiae—small, purplish, hemorrhagic spots on the skin that appear in certain severe fevers and are indicative of great prostration, as in typhus; may be caused by an abnormality of the blood-clotting mechanism

Purpura—an affectation with various manifestations and diverse causes; characterized by hemorrhages into the skin, mucous membranes, internal organs, and other tissues

Literature Cited

Abbas, H. K., Mirocha, C. J., and Shier, T. 1992. Isolation and identification and biological activity of chlamodosporol from *Fusarium cumorum* HM-8. Mycopathologia 118:115-123.

Allander, E. 1987. The epidemiology of Kashin Beck disease. Some reflections on possibilities for international cooperation. Pages 72-75 in: Nutrition '87, Am. Inst. Nutr. Symp. Proc.

Anonymous. 1977. Balkan nephropathy (Editorial). Lancet 1(8013):683-684.

Anonymous. 1985. Mseleni joint disease (Editorial). Lancet 2(8453):483-484.

Austwick, P. K. 1975. Balkan nephropathy. Proc. R. Soc. Med. 68(4):219-221.

Austwick, P. K. 1981. Balkan nephropathy. Practitioner 225:1031-1038.

Barnes, J. M., Carter, R. L., Peristianis, G. C., Austwick, P. K. C., Flynn, F. V., and Aldridge, W. N. 1977. Balkan (endemic) nephropathy and a toxin-producing strain of *Penecillium verrucosum* var. *cyclopium*: An experimental model in rats. Lancet 1(8013):671-675.

Bhat, R. V. 1977. Identity of the pathogen causing ergot of pearl millet in India. Curr. Sci. 46(6):184-185.

Bhat, R. V. 1989. Recent Indian studies on food toxins. Bull. Nutr. Found. India 10(3):5-7.

Bhat, R. V., and Krishnamachari, K. A. V. R. 1977. Endemic familial arthritis of Malnad—An epidemiological study. Indian J. Med. Res. 66(5):777-786.

Bhat, R. V., and Roy, D. N. 1976. Toxicity study of ergoty bajra (pearl millet) in Rhesus monkeys. Indian J. Med. Res. 64(11):1629-1633.

Bhat, R. V., Roy, D. N., and Tulpule, P. G. 1974. Ergot contamination of bajra. Pages 7-11 in: Proc. Nutr. Soc. India 17th. B. Belavady and G. M. Desai eds. Nutrition Society of India, Hyderabad.

Bhat, R. V., Roy, D. N., and Tulpule, P. G. 1976. The nature of alkaloids of ergoty pearl millet and its comparison with alkaloids of ergoty rye and ergoty wheat. Toxicol. Appl. Pharmacol. 36(1):11-17.

Bhat, R. V., Ramakrishna, Y., and Sashidhar, R. B. 1989a. Outbreak of mycotoxicosis in Kashmir Valley. Nutr. News 10(1).

Bhat, R. V., Ramakrishna, Y., Beedu, S. R., and Munshi, K. L. 1989b. Outbreak of trichothecene mycotoxicosis associated with consumption of mold-damaged wheat products in Kashmir valley India. Lancet 1(8628):35-37.

Bove, F. J. 1970. The Story of Ergot. S. Karger, New York.

Brink, S., Hessling, P. B., Amadhila, S., and Visser, H. S. 1981. Platelet antibodies in immune thrombocyopenic purpura and onyalai. S. Afr. Med. J. 59(24):855-858.

Castegnaro, M., and Chernozemsky, I. N. 1987. Endemic nephropathy and urinary tract tumours in the Balkans. Cancer Res. 47:3608-3609.

Castegnaro, M., Chernozemsky, I. N., Hietanen, E., and Bartsch, H. 1990. Are mycotoxins risk factors for endemic nephropathy and associated urothelial cancers? Arch. Geschwulstforsch. 60(4):295-303.

Changlong, Y., Fincham, J. E., Wright, G. C., Taljaard, J. J. F., and Sokoloff, L. 1987. Mseleni disease serum is not harmful to cultured chondrocytes. Arthritis Rheum. 30(3):348-352.

Cheng, S., Jiang, Y. Z., and Lo, H. Z. 1985. A mutagenic metabolite produced by *Fusarium moniliforme* isolated from Linxian county, China. Carcinogenesis 6(6):903-905.

Clarke, C. A., McKendrick, O. M., and Sheppard, P. M. 1973. Spina bifida and potatoes. Br. Med. J. 3:251-254.

Cook, P. 1971. Cancer of the oesophagus in Africa. A summary and evaluation of the evidence for the frequency of occurrence, and a preliminary indication of the possible association with the consumption of alcoholic drinks made from maize. Br. J. Cancer 25:853-880.

Crespi, M., Grassi, A., Amiri, G., Munoz, N., Aramesh, B., Mojtabai, A., and Casale, V. 1979. Oesophageal lesions in northern Iran: A premalignant condition? Lancet 2(8136):217-220.

Cvetnic, Z., and Pepeljnjak, S. 1990. Ochratoxigenicity of *Aspergillus ochraceus* strains from nephropathic and nonnephropathic areas in Yugoslavia. Mycopathologia 110(2):93-99.

Dong, Z., Liu, G., Dong, Z., Qian, Y., An, Y., Miao, J., and Zhen, Y. 1987. Induction of mutagenesis and transformation by the extract of *Alternaria alternata* isolated from grains in Linxian, China. Carcinogenesis 8(7):989-991.

Donovan, K. O., Shaw, D. E., and Amato, D. 1976. Sago and haemolysis. Papua New Guinea Med. J. 19(3):183-184.

Donovan, K. O., Shaw, D. E., and Amato, D. 1977. Sago haemolysis: Clinical features and microbiological studies. Papua New Guinea Med. J. 20(4):167-174.

Elling, F., and Krogh, P. 1977. Fungal toxins and Balkan (endemic) nephropathy. Lancet 1(8023):1213.

Farber, J. M., and Sanders, G. W. 1986. Fusarin C production by North American isolates of *Fusarium moniliforme*. Appl. Environ. Microbiol. 51(2):381-384.

Fellingham, S. A., Elphinstone, C. D., and Wittmann, W. 1973. Mseleni joint disease: Background and prevalence. S. Afr. Med. J. 47:2173-2180.

Fincham, J. E., Van Rensburg, S. J., and Marasas, W. F. O. 1981. Mseleni joint disease a manganese deficiency? S. Afr. Med. J. 60:445-447.

Fincham, J. E., Hough, F. S., Taljaard, J. J. F., and Capatos, D. 1985. Mseleni joint disease. Part I. An animal model? S. Afr. Med. J. 67:51-57.

Fincham, J. E., Hough, F. S., Taljaard, J. J. F., Weidemann, A., and Schutte, C. H. J. 1986. Mseleni joint disease. Part II. Low serum calcium and magnesium levels in women. S. Afr. Med. J. 70:740-742.

Forsyth, D. M., Yoshizawa, T., Morooka, N., and Tuite, J. 1977. Emetic response in of deoxynivalenol to swine. Appl. Environ. Microbiol. 34:547-552.

Frisvad, J. C., and Filtenborg, O. 1989. Tetraverticillate penicillia: Chemotaxonomy and mycotoxin production. Mycologia 81:837-861.

Gelderblom, W. C. A., Thiel, P. G., Van der Merwe, K. J., Marasas, W. F. O., and Spies, H. S. C. 1983. A mutagen produced by *Fusarium moniliforme*. Toxicon 21(4):467-473.

Gelderblom, W. C. A., Marasas, W. F. O., Steyn, P. S., Thiel, P. G., Van der Merwe, K. J., Van Rooyen, P. H., Vleggaar, R., and Wessels, P. L. 1984a. Structure elucidation of fusarin C, a mutagen produced by *Fusarium moniliforme*. J. Chem. Soc. Chem. Commun. 122-124.

Gelderblom, W. C. A., Thiel, P. G., Marasas, W. F. O., and Van der Merwe, K. J. 1984b. Natural occurrence of fusarin C, a mutagen produced by *Fusarium moniliforme* in corn. J. Agric. Food Chem. 32:1064-1067.

Gelderblom, W. C. A., Jaskiewicz, K., Marasas, W. F. O., Thiel, P. G., Horak, R. M., Vleggaar, R., and Kriek, N. P. J. 1988. Fumonisins—Novel mycotoxins with cancer-promoting activity produced by *Fusarium moniliforme*. Appl. Environ. Microbiol. 54(7):1806-1811.

Gelderblom, W. C. A., Kriek, N. J. P., Marasas, W. F. O. and Thiel, P. G. 1991. Toxicity and carcinogenicity of the *Fusarium moniliforme* metabolite, fumonisin B_1, in rats. Carcinogenesis 12:1247-1251.

Gelderblom, W. C. A., Semple, E., Marasas, W. F. O., and Farber, E. 1992. The cancer-initiating potential of the fumonisin B mycotoxins. Carcinogenesis 13:433-437.

Goldin, R., and Gelfand, M. 1980. Idiopathic thrombocytopaenic purpura and onyalai in Zimbabwean Africans. Cent. Afr. J. Med. 26(11):236-239.

Hesseling, P. B. 1985. Onyalai in pregnancy—Effects on the mother and the newborn. S. Afr. Med. J. 67(7):252-253.

Hsia, C.-C., Gao, Y., Wu, J.-L., and Tzian, B.-L. 1986. Induction of chromosome aberrations by *Fusarium* T-2 toxin in cultured human peripheral blood lymphocytes and Chinese hamster fibroblasts. J. Cell. Physiol. 4(Suppl.):65-72.

Hsia, C.-C., Wu, J.-L., Lu, X. Q., and Li, Y. S. 1988. Natural occurrence and clastogenic effects of nivalenol, deoxynivalenol, 3-acetyl-deoxynivalenol, 15-acetyl-deoxynivalenol, and zearalenone in corn from a high-risk area of esophageal cancer. Cancer Detect. Prevent. 13:79-86.

IARC. 1993. Some naturally occurring substances: Some food items and constituents, heterocyclic aromatic amines and mycotoxins. IARC (Int. Agency Res. Cancer) Monogr. Eval. Carcinog. Risk Chem. Hum. 56.

Jaskiewicz, K., Marasas, W. F. O., and Van der Walt, F. E. 1987a. Oesophageal and other main cancer patterns in four districts of Transkei, 1981-1984. S. Afr. Med. J. 72:27-30.

Jaskiewicz, K., Van Rensburg, S. J., Marasas, W. F. O., and Gelderblom, W. C. A. 1987b. Carcinogenicity of *Fusarium moniliforme* culture material in rats. JNCI (J. Natl. Cancer Inst.) 78(2):321-325.

Jaya Rao, K. S. 1983. Pellagra in sorghum eaters: A mycotoxicosis? Ecol. Food Nutr. 13(1):59-62.

Joffe, A. Z. 1960. The mycoflora of overwintered cereals and its toxicity. Bull. Res.

Counc. Israel Sect. D Botany 9:101-126.

Joffe, A. Z. 1963. Toxicity of overwintered cereals. Plant Soil 18(1):31-44.

Joffe, A. Z. 1971. Alimentary toxic aleukia. Pages 139-190 in: Microbial Toxins—Algal and Fungal Toxins (Microbial Toxins 7). S. Kadis, A. Ciegler, and S. J. Ajl, eds., New York, Academic Press.

Joffe, A. Z. 1978. *Fusarium poae* and *F. sporotrichioides* as principal causal agents of alimentary toxic aleukia. Pages 21-86 in: Mycotoxic Fungi, Mycotoxins, Myco-toxicoses—An Encylopedic Handbook. vol. 3. Marcel Dekker, New York.

Joffe, A. Z. 1986. *Fusarium* Species—Their Biology and Toxicology. John Wiley and Sons, New York.

Joffe, A. Z., and Yagen, B. 1977. Comparative study of the yield of T-2 toxic produced by *Fusarium poae, Fusarium sporotrichioides* and *Fusarium sporotrichioides* var. *tricinctum* strains from different sources. Mycopathologia 60(2):93-97.

Kellerman, T. S., Marasas, W. F. O., Thiel, P. G., Gelderblom, W. C. A., Cawood, M., and Coetzer, J. A. W. 1990. Leukoencephalomalacia in two horses induced by oral dosing of fumonisin B$_1$. Onderstepoort J. Vet. Res. 57:269-275.

King, B. 1979. Outbreak of ergotism in Wollo, Ethiopia. Lancet 1(8131):1411.

Kmet, J., and Mahboubi, E. 1972. Esophageal cancer in the Caspian littoral of Iran: Initial studies. Science 175:846-853.

Krishnamachari, K. A. V. R., and Bhat, R. V. 1976. Poisoning by ergoty bajra (pearl millet) in man. Indian J. Med. Res. 64(11):1624-1628.

Krishnamachari, K. A. V. R., and Bhat, R. V. 1978. Endemic familial arthrities of malnad an outbreak in southern India. Trop. Geogr. Med. 30:33-37.

Krogh, P. 1978. Causal associations of mycotoxic nephropathy. Acta Pathol. Microbiol. Scand. Sect. A (Suppl. 269):28.

Krogh, P. 1987. Ochratoxins in food. Pages 97-121 in: Mycotoxins in Food. P. Krogh, ed. London, Academic Press.

Kuiper-Goodman, T., and Scott, P. M. 1989. Risk assessment of the mycotoxin ochratoxin A. Biomed. Environ. Sci. 2(3):179-248.

Lacey, J. 1988. The microbiology of cereal grains from areas of Iran with a high incidence of oesophageal cancer. J. Stored Prod. Res. 24(1):39-50.

Lafarge-Frayssinet, C., Decloitre, F., Mousset, S., Martin, M., and Frayssinet, C. 1981. Induction of DNA single strand-breaks by T-2 toxin, a trichothecene metabolite of *Fusarium*. Effect on lymphoid organs and liver. Mutat. Res. 88(2):115-123.

Lapedes, D. N., ed. 1978. Dictionary of Scientific and Technical Terms. 2nd ed. New York, McGraw-Hill.

Leck, I. 1974. Causation of neural tube defects. Br. Med. Bull. 30(2):158-163.

Lee, Y. W., Mirocha, C. J., Shroeder, D. J., and Walser, M. M. 1985. TDP-1, a toxic component causing tibial dyschondroplasia in broiler chickens, and trichothecenes from *Fusarium roseum* 'Graminearum'. Appl. Environ. Microbiol. 50:102-107.

Levander, O. A. 1987a. Kashin-Beck disease: Introduction. American Inst. Page 60 in: Nutrition '87, Am. Inst. Nutr. Symp. Proc.

Levander, O. A. 1987b. Etiological hypotheses concerning Kashin-Beck disease. Pages 67-71 in: Nutrition '87, Am. Inst. Nutr. Symp. Proc.

Lewis, S. M., and Lurie, A. 1953. Onyalai. J. Trop. Med. Hyg. 56:281-289.

Li, J.-Y. 1982. Epidemiology of esophageal cancer in China. Natl. Cancer Inst. Monogr. 62:113-120.

Li, M.-H., and Cheng, S. 1984. Etiology of carcinoma of the esophagus. Pages 26-52 in: Carcinoma of the Esophagus and Gastric Cardia. G. J. Huang and Y. K. Wu, eds. Springer-Verlag, New York.

Li, M.-H., Lu, S.-H., Ji, C., Wang, Y., and Wang, M. 1980. Experimental studies on the carcinogenicity of fungus-contaminated food from Linxian county. Pages 139-148.in: Genetic and Environmental Factors in Experimental and Human Cancer. H. V. Gelboin, ed. Japan Scientific Societies Press, Tokyo.

Liang, S.-T., Zhang, J.-C., Shang, X., Mu, S.-Z., and Zhang, F.-J. 1987. Effects of selenium supplementaion in prevention and treatment of Kashin-Beck disease. Pages 938-946 in: Selenium in Biology and Medicine, Proc. Int. Symp. 3. Part B. G. F. Combs, J. E. Spallholz, O. A. Levander, and J. E. Oldfield, eds. New York, Van Nostrand.

Liu, B. Q., and Li, B. 1984. Epidemiology of carcinoma of the esophagus in China. Pages 2-24 in: Carcinoma of the Esophagus and Gastric Cardia. G. J. Huang and Y. K. Wu, eds. Springer-Verlag, New York.

Liu, G., Miao, J., Zhen, Y., and Xu, Y. 1988. The progress of research on the carcinogenicity of fungi in the esophagus in Henan Province, China. J. Henan Med. Univ. 2:4-11.

Liu, G., Qian, Y., Zhang, P., Dong, Z., Shi, Z. Y., Zhen, Y., Miao, J., and Xu, Y. 1991. Relationships between *Alternaria alternata* and oesophageal cancer. Pages 258-262 in: Relevance to Human Cancer of N-Nitroso Compounds, Tobacco and Mycotoxins. I. K. O'Neill, J. Chen, and H. Bartsch, eds. IARC (Int. Agency Res. Cancer) Sci. Publ. 105.

Liu, J.-G., and Li, M.-H. 1989. Roussin red methyl ester, a tumor promoter isolated from pickled vegetables. Carcinogenesis 10(3):617-620.

Liu, X. J., Luo, X. Y., and Hu, W. J. 1989. *Arthrinium* spp. and the etiology of deteriorated sugarcane poisoning. Pages 109-119 in: Proc. IUPAC Int. Symp. Mycotoxins Phyco-toxins, 7th. (Bioactive Molecules 10). S. Natori, K. Hashimoto, and Y. Ueno, eds. Elsevier Science Publishers, Amsterdam.

Lockitch, G., Fellingham, S. A., Wittmann, W., de Wet, I. S., and du Toit, G. T. 1973. Mseleni joint disease: The pilot clinical survey. S. Afr. Med. J. 47:2283-2293.

Luo, X. Y. 1988a. *Fusarium* toxins contamination of cereals in China. Proc. Jpn. Assoc. Mycotoxicol. Suppl. 1:97-98.

Luo, X. Y. 1988b. Outbreaks of moldy cereals poisoning in China. Pages 56-63 in: Issues in Food Safety. Toxicology Forum, Washington D.C.

Luo, Y., Yoshizawa, T., and Katayama, T. 1990. Comparative study on the natural occurrence of *Fusarium* mycotoxins (trichothecenes and zearalenone) in corn and wheat from high- and low-risk areas for human esopageal cancer in China. Appl. Environ. Microbiol. 56(12):3723-3726.

Lurie, A., Katz, J., Ludwin, S. K., Seftel, H. C., and Metz, J. 1969. Platelet life-span and sites of platelet sequestration in Onyalai. Br. Med. J. 4:146-148.

Lutsky, I. I., and Mor, N. 1981a. Alimentary toxic aleukia (septic angina, endemic panmyelotoxicosis, alimentary hemorrhagic aleukia): T-2 toxin-induced intoxication of cats. Am. J. Pathol. 104(2):189-191.

Lutsky, I. I., and Mor, N. 1981b. Experimental alimentary aleukia in cats. Lab. Anim. Sci. 31(1):43-47.

Macgeorge, K. M., and Mantle, P. G. 1991. Nephrotoxic fungi in Yugoslavia community in which Balkan nephropathy is endemic. Mycol. Res. 95(6):660-664.

Marasas, W. F. O. 1982. Mycotoxicological investigations on corn produced in esophageal cancer areas in Transkei. Pages 29-40 in: Cancer of the Esophagus. vol 1. C. J. Pfeiffer, ed. CRC Press, Boca Raton, FL.

Marasas, W. F. O. 1993. *Fusarium*. Chapter 17 in: Foodborne Disease Handbook. Vol. 2. Y. M. Hui, J. R. Gorham, K. D. Murrell, and D. O. Cline. Marcel Dekker, New

York. (In press.).

Marasas, W. F. O., and Van Rensburg, S. J. 1986. Mycotoxicological investigations on maize and groundnuts from the endemic area of mseleni joint disease in Kwazulu. S. Afr. Med. J. 69, 369-374.

Marasas, W. F. O., Wehner, F. C., Van Rensburg, S. J., and Van Schalkwyk, D. J. 1981. Mycoflora of corn produced in human esophageal cancer areas in Transkei, southern Africa. Phytopathology 71:792-796.

Marasas, W. F. O., Nelson, P. E., and Toussoun, T. A. 1984a. Toxigenic Fungi: Their Toxins and Health Hazards. Pennsylvania State University Press, University Park.

Marasas, W. F. O., Kriek, N. P. J., Fincham, J. E., and Van Rensburg, S. J. 1984b. Primary liver cancer and oesophageal basal cell hyperplasia in rats caused by *Fusarium moniliforme*. Int. J. Cancer 34:383-387.

Marasas, W. F. O., Jaskiewicz, K., Venter, F. S., and Van Schalkwyk, D. J. 1988a. *Fusarium moniliforme* contamination of maize in oesophageal cancer areas in Transkei. S. Afr. Med. J. 74:110-114.

Marasas, W. F. O., Kellerman, T. S., Gelderblom, W. C. A., Coetzer, J. A. W., Thiel, P. G., and Van der Lugt, J. J. 1988b. Leukoencephalomalacia in a horse induced by fumonisin B$_1$ isolated from *Fusarium moniliforme*. Onderstepoort J. Vet. Res. 55(4):197-203.

Marth, E. H. 1990. Mycotoxins. Pages 137-157 in: Foodborne Diseases. D. D. Cliver, ed. Academic Press, New York.

Martin, F., Delaruelle, C., and Hilbert, J.-L. 1990. An improved ergosterol assay to estimate fungal biomas in ectomycorrhizas. Mycol. Res. 94(8):1059-1064.

Mayer, C. F. 1953a. Endemic panmyelotoxicosis in the Russian grain belt. Part one: the clinical aspects of alimentary toxic aleukia (ATA): a comprehensive review. Military Surgery 113:173-189.

Mayer, C. F. 1953b. Endemic panmyelotoxicosis in the Russian grain belt. Part two: The botany, phytopathology and toxicology of Russian cereal food. Mil. Surg. 113:295-315.

Meronuck, R. A., Steele, J. A., Mirocha, C. J., and Christensen, C. M. 1972. Tenuazonic acid, a toxin produced by Alternaria alternata. Applied Microbiology 23(3):613-617.

Mirocha, C. J., and Pathre, S. V. 1973. Identification of the toxic principle in a sample of poaefusarin. Appl. Microbiol. 26(5):719-724.

Nesterov, A. I. 1964. The clinical course of Kashin Beck disease. Arthritis Rheum. 7(1):29-40.

Nurse, G. T., and Jenkins, T. 1974. Mseleni joint disease: Population genetic studies. S. Afr. J. Sci. 70, 360-365.

Patel, T. B., Bowman, T. J., and Dallal, U. C. 1958. An epidemic of ergot poisoning through ingestion of infected bajra (*Pennisetum typhoideum*) in southern parts of Bombay state. Indian J. Med. Sci. 12:257-261.

Pathre, S. V., Gleason, W. B., Lee, Y. W., and Mirocha, C. J. 1986. The structure of fusarochromanone: New mycotoxin from *Fusarium roseum* "Graminearum". Can. J. Chem. 64:1308-1311.

Pavlovic, M., Plestina, R., and Krogh, P. 1979. Ochratoxin A contamination of foodstuffs in an area with Balkan (endemic) nephropathy. Acta Pathol. Microbiol. Scand. Sect. B Micorobiol. 87(4):243-246.

Pepeljnjak, S., and Cvetnic, Z. 1984. Distribution of moulds on stored grains in households in an area affected by endemic nephropathy in Yugoslavia. Mycopathologia 86:83-87.

Pepeljnjak, S., and Cvetnic, Z. 1985. The mycotoxicological chain and contamination of food by ochratoxin A in the nephrotic and non-nephrotic areas in Yugoslavia. Mycopathologia 90(3):147-153.

Petkova-Bocharova, T., and Castegnaro, M. 1985. Ochratoxin A contamination of cereals in an area of high incidence of Balkan endemic nephropathy in Bulgaria. Food Addit. Contam. 2(4):267-270.

Petkova-Bocharova, T., Chernozemsky, I. N., and Castegnaro, M. 1988. Ochratoxin A in human blood in relation to Balkan endemic nephropathy and urinary system tumours in Bulgaria. Food Addit. Contam. 5(3):299-302.

Plestina, R., Ceovic, S., Gatenbeck, S., Habazin-Novak, V., Hult, K., Hokby, E., Krogh, P., and Radic, B. 1990. Human exposure to ochratoxin A in areas of Yugoslavia with endemic nephropathy. J. Environ. Pathol. Toxicol. Oncol. 10(3):145-148.

Rabie, C. J., Van Rensburg, S. J., Van der Watt, J. J., and Lubben, A. 1975. Onyalai— The possible involvement of a mycotoxin produced by *Phoma sorghina* in the aetiology. S. Afr. Med. J. 49:1647-1650.

Radovanovic, Z. 1989. Aetiology of Balkan nephropathy: A reappraisal after 30 years. Eur. J. Epidemiol. 5(3):372-377.

Ramakrishna, Y., Bhat, R. V., and Ravindranath, V. 1989. Production of deoxynivalenol by *Fusarium* isolates from samples of wheat associated with a human mycotoxicosis outbreak and from sorghum cultivars. Appl. Environ. Microbiol. 55(10):2619-2620.

Renwick, J. H. 1972. Hypothesis—Anencephaly and spina bifida are usually preventable by avoidance of a specific but unidentified substance present in certain potato tubers. Br. J. Prev. Soc. Med. 26:67-88.

Roberts, C. J., Revington, C. J., and Lloyd, S. 1973. Potato cultivation and storage in South Wales and its relation to neural tube malformation prevalence. Br. J. Prev. Soc. Med. 27:214-216.

Sauer, D. B., Seitz, L. M., Burroughs, R., Mohr, H. E., West, J. L., Milleret, R. J., and Anthony, H. D. 1978. Toxicity of *Alternaria* metabolites found in weathered sorghum grain at harvest. J. Agric. Food Chem. 26(6):1380-1382.

Savard, M. E., and Miller, J. D. 1992. Characterization of fusarin F, a new fusarin from *Fusarium moniliforme*. J. Nat. Prod. 55:64-70.

Schoental, R. 1980. Mouldy grain and the aetiology of pellagra: The role of toxic metabolites of *Fusarium*. Biochem. Soc. Trans. 8:147-150.

Schoental, R., and Joffe, A. Z. 1974. Lesions induced in rodents by extracts from cultures of *Fusarium poae* and *Fusarium sporotrichioides*. J. Pathol. 112:37-42.

Schoental, R., Joffe, A. Z., and Yagen, B. 1979. Cardiovascular lesions and various tumors found in rats given T-2 toxin, a trichothecene metabolite of *Fusarium*. Cancer Res. 39, 2179-2189.

Scott, P. M., and Stoltz, D. R. 1980. Mutagens produced by *Alternaria alternata*. Mutat. Res. 78:33-40.

Smith, E. R., Fredrickson, T. N., and Hadidian, Z. 1968. Toxic effects of the sodium and the N,N'-dibenzylethylenediamine salts of tenuazonic acid (NSC-525816 and NSC-82260). Cancer Chemother. Rep. 52(5):579-585.

Sokoloff, L. 1985. Endemic forms of osteoarthritis. Clin. Rheum. Dis. 11(2):187-202.

Sokoloff, L. 1987a. Kashin-Beck disease. Rheum. Dis. Childhood 13(1):101-104.

Sokoloff, L. 1987b. Kashin-Beck disease: Historical and pathological perspectives. Pages 61-63 in: Nutrition '87. Am. Inst. Nutr. Symp. Proc.

Sokoloff, L. 1988. Kashin Beck disease: Current status. Nutr. Rev. 46(3):113-119.

Sokoloff, L. 1989. The history of Kashin Beck disease. N.Y. State J. Med. 89(6):343-351.

Sokoloff, L. 1990. Acquired chondronecrosis. Ann. Rheum. Dis. 49(4):262-264.

Srikantia, S. G. 1978. Endemic pellagra among jowar eaters. Indian J. Med. Res. 68(Suppl.):38-47.

Stack, M. E., and Prival, M. J. 1986. Mutagenicity of the *Alternaria* metabolites altertoxin I, II and III. Appl. Environ. Microbiol. 52(4):718-722.

Steyn, P. S., and Rabie, C. J. 1976. Characterization of magnesium and calcium tenuazonate from *Phoma sorghina.* Phytochemistry 15:1977-1979.

Sydenham, E. W., Thiel, P. G., Marasas, W. F. O., Shephard, G. S., Van Schalkwyk, D. J., and Koch, K. R. 1990. Natural occurrence of some *Fusarium* mycotoxins in corn from low and high esophageal cancer prevalence areas of the Transkei, southern Africa. J. Agric. Food Chem. 38(10):1900-1903.

Szathmary, C. I., Mirocha, C. J., Palyusik, M., and Pathre, S. V. 1976. Identification of mycotoxins produced by species of *Fusarium* and *Stachybotrys* obtained from eastern Europe. Appl. Environ. Microbiol. 32(4):579-584.

Tanaka, T., Hasegawa, A., Matsuki, Y., and Ueno, Y. 1985. A survey of the occurrence of nivalenol, deoxynivalenol and zearalenone in foodstuffs and health foods in Japan. Food Addit. Contam. 2(4):259-265.

Taufa, T. 1974. Sago haemolytic disease. Papua New Guinea Med. J. 17:227-228.

Thiel, P. G., Marasas, W. F. O., Sydenham, E. W., Shephard, G. S., Gelderblom, W. C. A., and Nieuwenhuis, J. J. 1991. Survey of fumonisin production by *Fusarium* species. Appl. Environ. Microbiol. 57(4):1089-1093.

Thiel, P. G., Marasas, W. F. O., Sydenham, E. W., Shephard, G. S., and Gelderblom, W. C. A. 1992. The implications of naturally-occurring levels of fumonisins in corn for human and animal health. Mycopathologia 117:3-9.

Thomas, C. L., ed. 1977. Taber's Cyclopedic Medical Dictionary. 13th ed. F. A. Davis Company, Philadelphia.

Thrane, U. 1988. Screeening for fusarin C production by European isolates of *Fusarium* species. Mycotoxin Res. 4:2-10.

Thu, A., and Kyu, H. 1984. Epidemiology of frontoethmoidal encephalomeningocoele in Burma. J. Epidemiol. Commun. Health 38:89-98.

Tseng, T.-C., Chung, C.-S., and Li, I. 1990. Production of fusarin C mycotoxin by *Fusarium moniliforme* isolates of Taiwan. Bot. Bull. Acad. Sin. (Taipei) 31:169-174.

Tulpule, P. G., and Bhat, R. V. 1978. Food toxins and their implication in human health. Indian J. Med. Res. 68(Suppl.):99-108.

Van Egmond, H. P. 1989. Current situation on regulations for mycotoxins. Overview of tolerances and status of standard methods of sampling and analysis. Food Addit. Contam. 6(2):139-188.

Van Rensburg, S. J. 1985. Recent studies on the etiology of oesophageal cancer. S. Afr. Cancer Bull. 29:22-31.

Van Rensburg, S. J., and Altenkirk, B. 1974. *Claviceps purpurea*—Ergotism. Pages 69-96 in: Mycotoxins. I. F. H. Purchase, ed. Amsterdam, Elsevier.

Van Rensburg, S. J., Marasas, W. F. O., Gelderblom, W. C. A., Thiel, P. G., and Rabie, C. J. 1982. Mycotoxins and oesophageal cancer. Pages 265-268 in: Proc. IUPAC Int. Symp. Mycotoxins and Phycotoxins, 5th.

Van Rensburg, S. J., Benade, A.·S., Rose, E. F., and du Plessis, J. P. 1983. Nutritional status of African populations predisposed to esophageal cancer. Nutr. Cancer 4(3):206-216.

Wicks, A. C. B. 1972. Onyalai—A disappearing disease entity. Cent. Afr. J. Med. 18(5):93-97.

Wright, G. C., Marasas, W. F. O., and Sokoloff, L. 1987. Effect of fusarochromanone and T-2 toxin on articular chondrocytes in monolayer culture. Fundam. Appl. Toxicol. 9(3):595-597.

Xia, Q. J. 1984. Carcinogenesis in the esophagus. Pages 53-76 in: Carcinoma of the Esophagus and Gastric Cardia. G. J. Huang and Y. K. Wu, eds. Springer-Verlag,

New York.

Yagen, B., and Joffe, A. Z. 1976. Screening of toxic isolates of *Fusarium poae* and *Fusarium sporotrichioides* involved in causing alimentary toxic aleukia. Appl. Environ. Microbiol. 32(3):423-427.

Yagen, B., Joffe, A. Z., Horn, P., and Lutsky, I. I. 1977. Toxins from a strain involved in ATA. Pages 329-336 in: J. V. Rodricks, C. W. Hesseltine, and M. A. Mehlman, eds. Mycotoxins in Human and Animal Health. Pathotoxx Publ., Park Forest South, IL.

Yang, C. S. 1980. Research on esophageal cancer in China: A review. Cancer Res. 40, 2633-2644.

Yeulet, S. E., Mantle, P. G., Rudge, M. S., and Greig, J. B. 1988. Nephrotoxicity of *Penicillium aurantiogriseum,* a possible factor in the aetiology of Balkan endemic nephropathy. Mycopathologia 102(1):21-30.

Yoshizawa, T. 1983. Red-mold diseases and natural occurrence in Japan. Pages 195-209 in: Dev. Food Sci. 4. Trichothecenes—Chemical, Biological and Toxicological Aspects. Y. Ueno, ed. Kodansha Ltd., Tokyo, Japan.

Yoshizawa, T., and Hosokawa, H. 1983. Natural cooccurrence of deoxynivalenol and nivalenol, trichothecene mycotoxins, in commercial foods. J. Food Hyg. Soc. Jpn. 24(4):413-415.

Zhang, G., and Liu, J. 1989. An experimental animal model of Kashin-Beck disease. Ann. Rheum. Dis. 48:149-152.

Zhang, P., and Dong, Z. 1991. Studies on the activation of oncogenes by alternariol in human fetal esophageal epithelium. Chin. J. Pathophysiol. 7(3):233-236.

Zhang, P., Dong, Z., Liu, G., Hao, H., Miao, J., and Zhen, Y. 1991. Studies on the binding action between alternariol monomethyl ether and DNA of human fetal esophagus and the identification of the type of binding. J. Henan Med. Univ. 16(2):115-119.

Zhen, Y., Xing, Y. D., Xu, M., Han, S.-Y., Liu, G., Liu, Z. P., and Chen, Y. F. 1988. Study of toxigenic fungi and their mycotoxins in mildewed sugarcane. Pages 37-38 in: in: Proc. IUPAC Int. Symp. Mycotoxins Phycotoxins, 7th. K. Aibara, S. Kumagai, K. Ohtsubo, and T. Yoshizawa, eds. Proc. Jpn. Assoc. Mycotoxicol. Suppl. 1.

Zhen, Y., Xu, Y., Liu, G., Miao, J., Xing, Y. D., Zheng, Q. L., Ma, Y. F., Su, T., Wang, X.-L., Ruan, L. R., Tian, J. F., Zhou, G., and Yang, S. L. 1991. Mutagenicity of *Alternaria alternata* and *Penicillium cyclopium* isolated from grains in an area of high incidence of oesophageal cancer—Linxian, China. Pages 253-257 in: Relevance to Human Cancer of N-Nitroso Compounds, Tobacco and Mycotoxins. I. K. O'Neill, J. Chen, and H. Bartsch, eds. IARC (Int. Agency Res. Cancer) Sci. Publ. 105.

Zhu, G.-F., Cheng, S., and Li, M.-H. 1987. The genotoxic effects of T-2 toxin, a trichothecene produced by *Fusarium* fungi. Acta Biol. Exp. Sin. 20(2):129-134.

Index

Italic type designates figures.